VOLUME NINETY ONE

Current Topics in DEVELOPMENTAL BIOLOGY

Plant Development

VOLUME NINETY ONE

Current Topics in DEVELOPMENTAL BIOLOGY

Plant Development

Edited by

MARJA C. P. TIMMERMANS
Cold Spring Harbor Laboratory
Cold Spring Harbor
New York, USA

ELSEVIER

AMSTERDAM • BOSTON • HEIDELBERG • LONDON
NEW YORK • OXFORD • PARIS • SAN DIEGO
SAN FRANCISCO • SINGAPORE • SYDNEY • TOKYO
Academic Press is an imprint of Elsevier

Academic Press is an imprint of Elsevier
525 B Street, Suite 1900, San Diego, CA 92101-4495, USA
30 Corporate Drive, Suite 400, Burlington, MA 01803, USA
32, Jamestown Road, London NW1 7BY, UK
Linacre House, Jordan Hill, Oxford OX2 8DP, UK

First edition 2010

ISBN: 978-0-12-380910-0
ISSN: 0070-2153

For information on all Academic Press publications
visit our website at elsevierdirect.com

Printed and bound by CPI Group (UK) Ltd, Croydon, CR0 4YY

Transferred to Digital Print 2011

Contents

CONTRIBUTORS

Maria C. Albani
Department of Plant Developmental Biology, Max Planck Institute for Plant Breeding Research, Cologne, Germany

Rachappa Balkunde
Botanical Institute III, University of Cologne, Köln, Germany

Tom Bennett
Department of Molecular Genetics, University of Utrecht, Utrecht, The Netherlands

Dominique C. Bergmann
Department of Biology, Stanford University, Stanford, California, USA

Holger Breuninger
Department of Cell & Developmental Biology, John Innes Centre, Norwich, UK

George Coupland
Department of Plant Developmental Biology, Max Planck Institute for Plant Breeding Research, Cologne, Germany

Juan Dong
Department of Biology, Stanford University, Stanford, California, USA

Christian Fankhauser
Center for Integrative Genomics, University of Lausanne, Lausanne, Switzerland

Jennifer C. Fletcher
Plant Gene Expression Center, USDA/UC Berkeley, Albany, California, USA and Department of Plant and Microbial Biology, University of California, Berkeley, California, USA

Chan Man Ha
Plant Gene Expression Center, USDA/UC Berkeley, Albany, California, USA and Department of Plant and Microbial Biology, University of California, Berkeley, California, USA

Ykä Helariutta
Department of Biological and Environmental Sciences, Institute of Biotechnology, University of Helsinki, Helsinki, Finland

Patricia Hornitschek
Center for Integrative Genomics, University of Lausanne, Lausanne, Switzerland

Martin Hülskamp
Botanical Institute III, University of Cologne, Köln, Germany

Thomas Jack
Department of Biological Sciences, Dartmouth College, Hanover, New Hampshire, USA

Ji Hyung Jun
Plant Gene Expression Center, USDA/UC Berkeley, Albany, California, USA and Department of Plant and Microbial Biology, University of California, Berkeley, California, USA

Chitose Kami
Center for Integrative Genomics, University of Lausanne, Lausanne, Switzerland

Catherine A. Kidner
Royal Botanic Garden Edinburgh, Edinburgh, UK, and Institute of Molecular Plant Sciences, University of Edinburgh, Edinburgh, UK

Daniel Koenig
Department of Plant Biology, University of California, Davis, California, USA

Michael Lenhard
Department of Cell & Developmental Biology, John Innes Centre, Norwich, UK

Cristina I. Llavata Peris
Laboratory of Biochemistry, Wageningen University, Wageningen, The Netherlands

Séverine Lorrain
Center for Integrative Genomics, University of Lausanne, Lausanne, Switzerland

Hong Ma
State Key Laboratory of Genetic Engineering and School of Life Sciences, Institute of Plant Biology, Fudan University, Shanghai, China, and Department of Biology, Pennsylvania State University, University Park, Pennsylvania, USA

Anwesha Nag
Department of Biological Sciences, Dartmouth College, Hanover, New Hampshire, USA

Martina Pesch
Botanical Institute III, University of Cologne, Köln, Germany

Eike H. Rademacher
Laboratory of Biochemistry, Wageningen University, Wageningen, The Netherlands

Enrico Scarpella
Department of Biological Sciences, University of Alberta, Edmonton, Canada

Ben Scheres
Department of Molecular Genetics, University of Utrecht, Utrecht, The Netherlands

Neelima Sinha
Department of Plant Biology, University of California, Davis, California, USA

Venkatesan Sundaresan
Departments of Plant Biology and Plant Sciences, University of California, Davis, California, USA

Marja C. P. Timmermans
Cold Spring Harbor Laboratory, Cold Spring Harbor, New York, USA

Dolf Weijers
Laboratory of Biochemistry, Wageningen University, Wageningen, The Netherlands

Preface

This book captures many of the major advances made in the field of plant developmental biology. Over the past 20 years, genetic screens in Arabidopsis and other species have yielded a wide array of developmental mutants whose analyses have greatly enhanced our understanding of plant morphogenesis. Perhaps because of an initial focus on mutants with clearly defined, non-pleiotropic phenotypes, transcription factors were prominently represented among the first genes isolated from these screens. The field has matured to a point, however, that complex networks comprising signal transduction pathways, transcription factors, and downstream target genes are becoming clear for an ever-increasing number of developmental processes. Significant progress has been made on analysis of the mobile positional signals critical to plant development, which has long remained a main challenge in contemporary plant development. Moreover, by utilizing real-time analyses and computer modeling developmental principles are being elucidated in the context of the whole organ or organism.

Given the wealth of tools and resources available in Arabidopsis, the analysis of development has progressed most substantially in this model species. Comparative studies at the molecular and genetic level are beginning to illuminate the history of morphogenesis in plants, revealing evolutionary changes in key molecular players that form the basis for the enormous morphological diversity present within the plant kingdom. Moreover, with the tremendous advances in genomics, more complex systems are becoming accessible. This will allow new insights to be tested in the context of incisive classical experiments that have been the cornerstone of developmental biology, and can provide new interpretations of such formative experiments at the molecular and cellular level. Several of the reviews provide an insight into this new trend.

The chapters presented in this book describe our current knowledge, prospective insights, and key outstanding questions of multiple fundamental facets of plant development. Chapter 1 outlines the formative events that sculpt the organization of the plant body during embryogenesis and discusses the molecular mechanisms that regulate these patterning processes. It emerges that auxin has a profound coordinating role in embryogenesis. One example of this is in the formation of the embryonic root (Chapter 3), where an auxin reflux loop specifies the stem cell niche. This chapter also presents the progress in our understanding of root meristem function, outlining the distinctive properties of stem cells within the root meristem and the network that controls root growth through the coordinate regulation of cell division and elongation. Germination as well as numerous other developmental transitions and adaptive growth responses are regulated through the perception of light. How plants sense light and coordinate the molecular events underlying photomorphogenesis is presented in Chapter 2. The signaling networks that maintain the position and activity of stem cells within the shoot apical meristem and that organize the initiation of lateral organs on the meristem periphery are discussed in Chapter 4. The complexity of these networks is remarkable, but with the development of live imaging approaches and computational tools, our understanding of the central principles of shoot apical meristem function has rapidly advanced.

The next three chapters focus on patterning, growth, and the morphological diversification of lateral organs in plants. Chapter 5 describes the evolutionary distinct contributions of three highly conserved interlocking pathways to leaf polarity and outlines the unique contributions of mobile small RNAs as positional signals that refine the boundary between the upper and lower sides of the leaf. Chapter 6 compares the mechanisms that regulate leaf shape in simple and compound leafed model species. Auxin is recognized as a key player also in this aspect of development. The emergent picture is one in which developmental pathways that modulate determinacy within the leaf can vary auxin-signaling outputs to generate diverse compound and simple leaf forms. Recent insights into how plant organs reach their characteristic final shape and size are presented in Chapter 7. In addition to the genetic networks that affect organ size at the cellular level by regulating cell proliferation and expansion, insights into the systems that coordinate such cellular processes across the entire organ or modulate them in response to environmental stimuli are discussed.

New insights into the patterning processes that give rise to the precise arrangements of vasculature, stomatal complexes, and trichomes are outlined in Chapters 8 through 10. Interestingly, these tissue or cell patterns are generated through entirely distinct mechanisms. Chapter 8 describes the role of polarized auxin transport in generating the distinctive arrangements of vascular strands within leaves, stems, and roots. In contrast, stereotyped asymmetric cell divisions are at the center of stomatal patterning and, as outlined in Chapter 9, the spacing of such instructive divisions is regulated by signaling pathways involving secreted peptide ligands and receptor-like kinases. Finally, Chapter 10 discusses current models for trichome patterning, presenting evidence for the involvement of two parallel spacing mechanisms; a lateral inhibition mechanism that relies in part on plasmodesmatal movement of trichome inhibitor proteins and an activator-depletion mechanism in which inducers of trichome differentiation are depleted around incipient trichome cells.

Chapter 11 reviews our understanding of the networks that regulate the switch from vegetative growth to flowering in response to intrinsic signals and environmental stimuli in *Arabidopsis thaliana*, and then broadens the discussion to the control of flowering in perennial species. The contribution of miRNAs in regulating the transition to flowering is described in Chapter 12. This chapter also outlines the varied contributions of small regulatory RNAs to the specification of floral organ identity and the control of floral organ shape and size. The final chapter of the book focuses on the progress in understanding the development of the male and the female gametophytes, and their interactions during the process of double fertilization. Signaling molecules and downstream genetic pathways that direct the differentiation of distinct cell types within the developing gametophytes, pollen tube attraction, or fertilization are beginning to emerge. An exciting example is the recent finding that auxin acts as a morphogenic determinant in the patterning of the female gametophyte.

Besides providing timely reviews of new advances in many fundamental aspects of plant development, the chapters presented in this book illustrate clearly the diversity of patterning mechanisms that drive plant morphogenesis. Mobile transcription factors, secreted peptides, hormones, and small regulatory RNAs function as instructive signals feeding into a wide array of signaling pathways and gene regulatory networks. This wealth of new data also allows us to draw parallels with animal development. It is now probably safe to state, rather than speculate, that many of the individual molecular components are either different or wired differently, but that the logic of developmental circuitry is conceptually quite similar between plants and animals. This book is testimony to the progress in our understanding of plant development that has been made in a surprisingly short time.

CHAPTER ONE

Green Beginnings – Pattern Formation in the Early Plant Embryo

Cristina I. Llavata Peris,* Eike H. Rademacher,* *and* Dolf Weijers

Contents

Abstract

Embryogenesis in plants transforms the zygote into a relatively simple structure, the seedling, which contains all tissues and organs that later form the mature plant body. Despite a profound diversity in cell division patterns among plant species, embryogenesis yields remarkably homologous seedling architectures. In this review, we describe the formative events during plant embryogenesis and discuss the molecular mechanisms that regulate these processes, focusing on *Arabidopsis*. Even though only a relatively small number of factors are known that regulate each patterning step, a picture emerges where locally acting transcription factors and intercellular signaling contribute to the specification and spatio-temporal coordination of the various cell types in the embryo.

Laboratory of Biochemistry, Wageningen University, Wageningen, The Netherlands
* These authors have contributed equally to this manuscript.

Current Topics in Developmental Biology, Volume 91
ISSN 0070-2153, DOI 10.1016/S0070-2153(10)91001-6

Notably, several patterning processes are controlled by the plant hormone auxin. Most regulators that were identified in *Arabidopsis* have orthologs in other sequenced plant genomes, and several of these are expressed in similar patterns. Therefore, it appears that robust conserved mechanisms may underlie pattern formation in plant embryos.

1. Introduction–Seeds and Embryos

During evolution, plants as largely immobile organisms have developed sophisticated mechanisms to disperse their offspring and explore new habitats. One way to expand the populated area is to grow into the surrounding by means of vegetative propagation via specialized structures such as rhizomes or stolons. However, these mechanisms do not reach over great distances and fail to overcome relatively small obstacles.

The innovation that surpassed such restrictions and allowed plants to use wind, water, animals, gravity, and even ballistic mechanisms as means of dispersal is the seed. Furthermore, seeds permit the enclosed embryo to remain in a dormant state and survive long periods of harsh environmental conditions, thereby mastering not only spatial but also temporal restrictions to its habitat. Finally, when conditions are beneficial for germination, the seed provides the embryo with the nutrients that support the first steps of post-embryonic development. To fulfill all of these different requirements, seeds develop in very different shapes and sizes. Adaptations in seed morphology have enabled plants to occupy new ecological niches with the result that plants now dominate habitats on all different kinds of land, in both hot and cold climates (Kesseler and Stuppy, 2006).

Seed-bearing plants (spermatophytes) can be subdivided into two major groups: gymnosperms and angiosperms. The name-giving difference between these lies in the organization of the reproductive organs. Gymnosperms bear ovules on scales, which are usually in cone-like structures as, for example, in pine. Angiosperms instead have their reproductive organs arranged in flowers. Another important difference is that flowering plants feature a double fertilization step to produce the (typically) diploid embryo and the triploid extra-embryonic endosperm (Lersten, 2004). Despite these differences in seed anatomy and function, the development and basic body organization of the embryo are very similar among most higher plants (Cairney and Pullman, 2007; Johri *et al.*, 1992). Accordingly, embryogenesis in both groups undergoes three common phases. Initially the fertilized egg cell (zygote) undergoes elongation and after a few cell division rounds develops an apico-basal and radial axis. During the next phase, further cell divisions occur and the primordia of the fundamental organs for post-embryonic growth are established. A final phase of desiccation prepares the mature embryo for dormancy and equips the seed with storage products that aid the seedling in the process of germination (Bewley and Black, 1994).

Typically, seeds carry a single embryo that emerges as the product of fertilization of the egg cell (Fig. 1.1, box 1). However, the formation of poly-embryonic seeds has been observed in many taxa (e.g., *Rutacea* (Citrus) or *Pinacea* (Pinus); Batygina and Vinogradova, 2007; Lakshmanan and Ambegaokar, 1984). Interestingly, the additional embryos within seeds do not always originate from cells that are derived from the zygote. More importantly, this is not even obligatory in mono-embryonic seeds. Hence, also unfertilized cells possess the potential to form embryos within a seed. Such asexual reproductive mechanisms of embryo formation have been summarized under the term *apomixis* (Fig. 1.1, box 2). We refer the reader to another review (Koltunow and Grossniklaus, 2003) that covers the various developmental origins of apomictic embryos. In addition to the naturally occurring origins of embryogenesis, male gametes (pollen grains) can give rise to viable but haploid embryos when cultured under certain

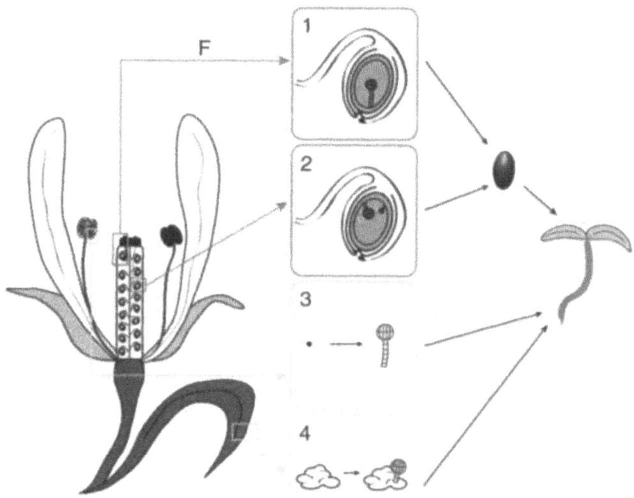

Figure 1.1 Origins of plant embryos. Embryogenesis normally occurs in the reproductive tissues within the flower. Shown here is a flower with subtending leaf. Zygotic embryogenesis (box 1) occurs when a pollen grain lands on the stigma of the carpel and fertilizes the egg cell in an ovule contained within the gynoecium. This ovule develops into a seed (brown oval) which, upon germination, generates the seedling (here dicotyledonous). Alternative modes of embryogenesis include the initiation of apomictic embryos within the ovule (box 2), microspore-derived embryos developing from pollen grains (box 3), and somatic embryos that are initiated from callus tissue derived from somatic cells (box 4). While only zygotic and apomictic embryogenesis pass through a seed stage, all these different modes of embryogenesis finally give rise to seedlings with the same body plan.

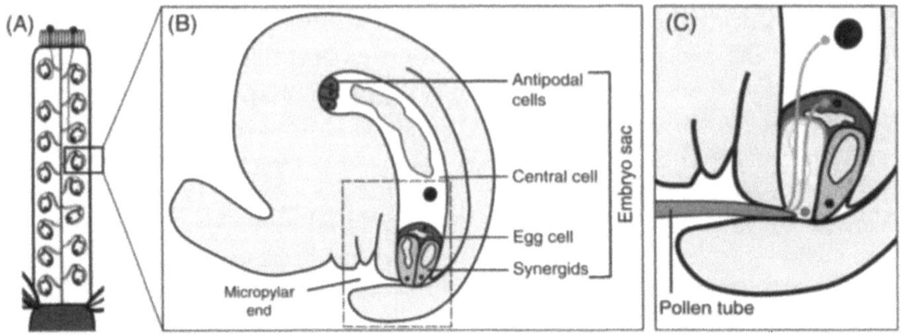

Figure 1.2 Fertilization in *Arabidopsis*. Fertilization occurs within the gynoecium (A), where ovules (B) are attached to the placental tissues. The *Arabidopsis* ovule consists of a gametophyte (embryo sac) covered with somatic integument cell layers. The egg cell, surrounded by two synergid cells, is localized at the micropylar end. Additionally, the embryo sac houses a central cell nucleus and three antipodal cells. (C) Upon germination and growth, the pollen tube is attracted to the ovule by the synergid cells. The pollen tube then releases its generative nuclei in the degenerating synergid, upon which the nuclei fuse with the egg cell and central cell nuclei. (B, C) Drawn after Sundaresan and Alandete-Saez (2010).

conditions (Seguí-Simarro and Nuez, 2008; Fig. 1.1, box 3). Finally, somatic cells can be forced into an embryogenic pathway by treatment of cultured explants with auxin (2,4-D) (Toonen and de Vries, 1996; Fig. 1.1, box 4).

As transpires from the above, a wide variety of cells can generate embryos either naturally or after experimental treatment. However, despite the distinct origins of embryos and the often different patterns of cell divisions within the embryo, the outcome is always a seedling that consists of an apico-basal axis with apical shoot and root meristems, a radial axis that defines the respective tissues, and one or two cotyledons positioned in the immediate surrounding of the shoot apical meristem (SAM). It is therefore likely that a robust set of conserved genetically encoded instructions underlies pattern formation during embryogenesis in all species.

Molecular processes underlying pattern formation and development are best described in *Arabidopsis thaliana*. Due to the regular pattern of cell divisions during embryogenesis (Jürgens and Mayer, 1994), this species has also become the model for studying zygotic embryogenesis. In this review we will discuss embryogenesis in *Arabidopsis* to exemplify basic mechanisms that sculpt the organization of the plant body during embryogenesis. In the following sections, we will first describe the developmental landmarks during embryogenesis, and then discuss the molecular mechanisms that control pattern formation, followed by a perspective on the evolutionary conservation of the regulatory mechanisms that have been found in *Arabidopsis*.

2. Landmarks of Embryo Pattern Formation

In angiosperms such as *Arabidopsis*, the female gamete (egg cell) is positioned within the embryo sac, which in turn is embedded in the protective maternal tissue of the ovule inside the carpel (Fig. 1.2A, B). After deposition of a male microgametophyte (pollen grain) on the stigma of the carpel, a pollen tube extends toward the ovule. This tube eventually enters the ovule via its micropylar end and facilitates the delivery of two haploid sperm nuclei. One of these nuclei fertilizes the egg cell while the other fuses with the two nuclei of the central cell to give rise to the triploid endosperm (Lersten, 2004; Fig. 1.2C). Embryo and endosperm develop simultaneously, and several mutant studies suggest that their growth patterns are interdependent (reviewed in Berger *et al.*, 2006). So far, however, there is no conclusive evidence that embryo pattern formation depends on the presence of the endosperm. Hence, consistent with the flexible origin of embryos, pattern formation in the embryo is driven by intrinsic factors rather than by environmental cues.

Embryo patterning follows a series of "landmarks" that each establishes part of the final organization. In the following, we briefly describe each such landmark.

2.1. Formation of the apico-basal axis

Embryo development occurs within the highly polarized environment of the ovule. Both the ovule and the embryo sac have a distinct polarized axis, and also the egg cell is intrinsically polar as judged from the localization of its organelles on the basal (e.g. vacuole) or apical end (e.g. nucleus) (Fig. 1.2C; Lersten, 2004). Fertilization induces stretching of the zygote, which is followed by an asymmetric division that gives rise to two daughter cells with different composition, shape, and developmental fate. The apical cell is small with a dense cytoplasm, while the basal cell is large and vacuolated. With three rounds of cell divisions, the apical cell generates a spherical proembryo consisting of a total of eight cells, while the basal cell only divides transversally and gives rise to a transient filamentous structure called the suspensor. This extra-embryonic suspensor connects the proembryo to maternal tissue and pushes it into the lumen of the ovule. Later during embryogenesis, only the uppermost suspensor cell, the hypophysis, becomes incorporated in the embryonic root meristem, as precursor of the quiescent center (QC) and central root cap cells (see below) (Fig. 1.3A).

At the eight-cell (octant) stage, four different domains can be distinguished along the apico-basal axis of the embryo; the proembryo consists of an upper and a lower tier of four cells each and is positioned on top of hypophysis and suspensor (Fig. 1.3A, B).

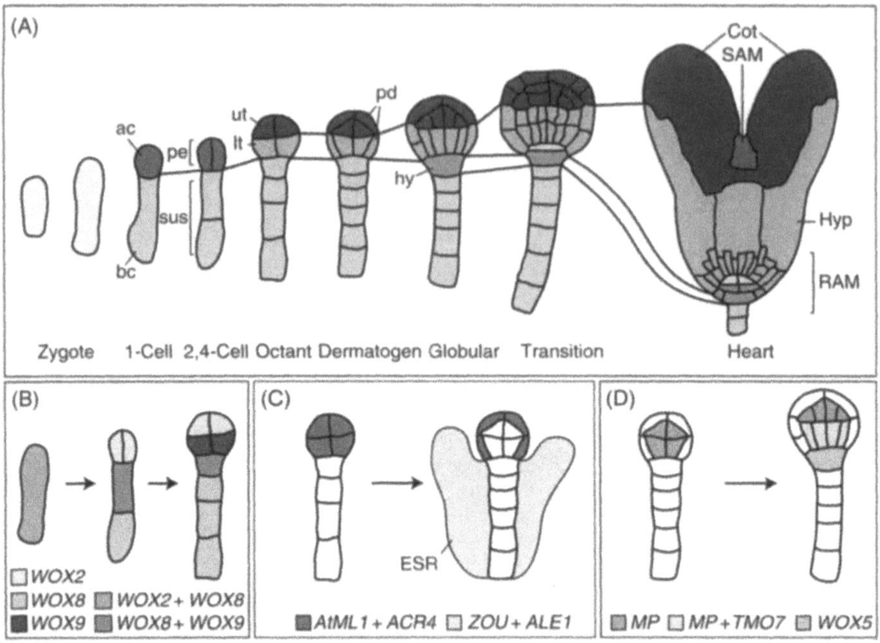

Figure 1.3 Embryo pattern formation and cell specification. (A) Stages of embryogenesis and cell lineages. The lines between embryos highlight the lineage relationships between different stages of embryogenesis. Stage nomenclature is given below each embryo. Colors represent clonally related regions. After fertilization the zygote elongates and undergoes an asymmetrical division generating a small apical cell (ac) and a larger basal cell (bc). From this point, two regions with different developing programs will be established, the embryonic proembryo (pe) and the extra-embryonic suspensor (sus). All cells in the four-celled proembryo divide to generate an upper tier (ut) and a lower tier (lt). Subsequently, a radial pattern is generated in the eight-cell proembryo when a layer of protoderm (pd) is established. At the globular stage, the uppermost cell of the suspensor is specified to become the hypophysis (hy). Through an asymmetrical division, the hypophysis gives rise to a lens-shaped cell from which the quiescent center will be generated and a basal cell from which stem cells and columella cells of the root tip will derive. Further divisions generate primordia for all seedling structures at the heart stage, including cotyledons (cot), shoot apical meristem (SAM), root apical meristem (RAM), and the hypocotyl (Hyp). (B–D) Molecular markers of cell fate decisions during pattern landmarks. Colors indicate the expression of genes as described in the box legends. (B) While *WOX2* and *WOX8* are coexpressed in the zygote, their mRNAs are found in different cells after two rounds of division. Combined with *WOX9* expression, different combinations of *WOX* transcripts mark each cell. (C) In the octant stage, *ATML1* and *ACR4* are both expressed in the entire proembryo. Upon division, their transcripts are only found in the protoderm. *ZOU* and *ALE1* transcripts are found in the embryo-surrounding endosperm region (ESR) and contribute to protoderm differentiation. (D) At the dermatogen stage, *MP* is expressed in all inner cells and remains so during two division rounds later. At the globular stage, the MP target *TMO7* is expressed only in lower tier inner cells, while the *WOX5* transcript is found in the hypophysis. (See Color Insert.)

2.2. Outside versus inside—radial pattern establishment

While the apico-basal axis of the embryo is preceded by polarity of the egg cell, the radial axis is established *de novo* after several rounds of cell division of the embryo.

All cells in the octant proembryo divide along a tangential plane, aligned along the apico-basal axis. This divides the proembryo in two different regions with different identities: an outer layer of eight cells, the protoderm, which is the precursor of the epidermis, and eight cells in the center of the proembryo, the inner cells, which are the precursors of ground and vascular tissues (Fig. 1.3A, C).

2.3. Stem cell niches—sources of all other cells

After several more rounds of cell division, the cells in the upper tier generate the SAM and the cotyledons (see below). Even though the shoot meristem can not be recognized by stereotypic cell divisions, gene expression markers for stem cells and for the organizing center (OC) become active in the area around the mid-globular stage (see below). The initiation of the root meristem, by contrast, is marked by a stereotypic asymmetric division of the hypophysis during the mid-globular stage. This initially extra-embryonic cell divides asymmetrically along the horizontal plane and generates two different cells: a small apical lens-shaped cell and a larger basal cell. Two vertical divisions of the apical cell form the QC. These cells control the undifferentiated state of the neighboring stem cells in the root apical meristem (RAM) (Jiang and Feldman, 2005). The descendants of the basal cell form the stem cells and the outer cell layers of the central root cap (columella) (Scheres *et al.*, 1994; Fig. 1.3A, D).

With the specification of the SAM and RAM at this stage of embryogenesis, the sources for all cells of the post-embryonic body are established.

2.4. Initiation of cotyledons

The radial symmetry that characterizes the embryo until the late globular stage is broken with the initiation of the cotyledons. In the dicotyledonous *Arabidopsis* embryo, cotyledons are specified from two lateral zones at the apical domain of the proembryo. The function of these structures differs between plants that develop a single cotyledon and those that develop two. In the former case, the embryonic leaf remains under the soil and serves as a storage organ. In the latter case, the cotyledons serve first as storage organ and subsequently as the first photosynthetic organs.

The sum of the patterning landmarks as described above is a mature embryo with concentric tissue types arranged along an apico-basal axis that carries the meristems at either end. Most of the cell identity specification events can be readily observed by unique cell division planes or expansion (shape) changes. In each case, however, including the anatomically indistinguishable cell types, the cell fate is marked by a unique set of transcripts (Fig. 1.3B–D). Some of these merely mark a cell fate, while others encode proteins that control cell fate, division, or differentiation. In the following section, we will discuss the mechanisms that control pattern formation in *Arabidopsis* embryos.

3. Regulation of Embryo Pattern Formation

Patterning of the embryo requires both specification of individual cell identities and the coordination of cell specification between cells in space and time. Among the regulators of patterning, one therefore expects both intrinsic cell fate determinants, such as transcriptional regulators, and molecules that act as intercellular signals to coordinate the specification of cells on both long and short ranges. In the following, we will describe, separately for each developmental landmark, the key regulators that have emerged mostly from genetic studies. However, as the plant hormone auxin appeared as an important regulator in most patterning steps, we will first briefly describe the mechanism of action of this signal molecule.

3.1. Auxin biosynthesis, transport, and response

Auxins are indole-derived molecules that can control the division and elongation rate, as well as the identity or differentiation state of cells. Accumulation of and response to auxin have been shown to regulate many growth and developmental processes throughout embryonic and post-embryonic development. These include tropic growth responses, vascular development, leaf and flower initiation, root growth, and lateral root formation (Taiz and Zeiger, 2002). Auxin is synthesized from indole directly or from tryptophan through several intermediates that are not all known. Two types of biosynthetic enzymes have been shown to produce auxin pools that are required for normal development (Woodward and Bartel, 2005). These are the TAA1/TAR tryptophan aminotransferases that catalyze the conversion of tryptophan to indole-3 pyruvic acid (Stepanova *et al.*, 2008; Tao *et al.*, 2008), and the YUCCA monooxygenases that convert tryptamine into *N*-hydroxyl tryptamine (Zhao *et al.*, 2001; Fig. 1.4). Mutations in either of these pathways interfere with most if not all auxin-dependent processes, including those in the embryo. Auxin biosynthesis is thought to be localized in specific areas while auxin-dependent

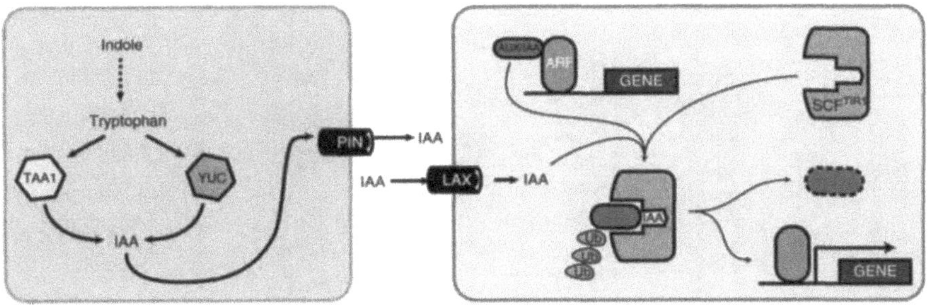

Figure 1.4 Auxin biosynthesis, transport, and response. Simplified schemes for the biosynthesis (left cell), transport (interface between cells), and response (right cell) of auxin. Note that many components have been omitted for clarity. Auxin is synthesized from indole either through the intermediate tryptophan (shown here) or independent of tryptophan (not shown). The TAA1 and YUCCA enzymes act in different branches to convert tryptophan into indole acetic acid (IAA). Auxin cannot exit cells freely, but its efflux is mediated by the rate-limiting PIN proteins, whose polar localization determines the direction of the flux. Auxin can enter cells by diffusion, yet uptake can be facilitated by the LAX influx carriers. Auxin directly binds to the SCF(TIR1) ubiquitin ligase and increases its affinity for the Aux/IAA proteins, targeting these for ubiquitin-mediated degradation. Under low auxin conditions, Aux/IAA proteins inhibit AUXIN RESPONSE FACTORS (ARFs) and DNA-binding transcription factors. Aux/IAA degradation releases the ARFs from inhibition and allows these to control gene expression.

growth is observed also in tissues that do not appear to produce auxin (reviewed in Woodward and Bartel, 2005). Directional transport of auxin provides auxin also to these areas of the plant. To achieve this, cell membranes carry several auxin influx and efflux carriers. While influx carriers of the LAX family are not generally thought of as providing direction to transport (Petrasek and Friml, 2009), this is certainly the case for their counterparts, proteins of the PIN family of efflux carriers (Friml, 2003; Fig. 1.4). The polar subcellular location of these auxin transporters determines the direction of auxin flux. Therefore, the polar targeting of PIN proteins is an important control element in the regulation of auxin accumulation (Wisniewska *et al.*, 2006).

Once accumulated, auxin is perceived by its receptor, the SCF-TIR1/AFB ubiquitin ligase. By binding to this enzyme complex, auxin facilitates the recognition of its substrates, transcriptional repressors of the Aux/IAA family, and their subsequent degradation by the proteasome (Chapman and Estelle, 2009; Fig. 1.4). Thus, auxin promotes degradation of transcriptional repressors. Without auxin, these Aux/IAA proteins bind to and inhibit another family of transcription factors, the AUXIN RESPONSE FACTORS (ARFs). Upon degradation of the Aux/IAA proteins, ARFs are released from inhibition and alter expression of their target genes (Fig. 1.4). Most of auxin's activity in controlling plant development can

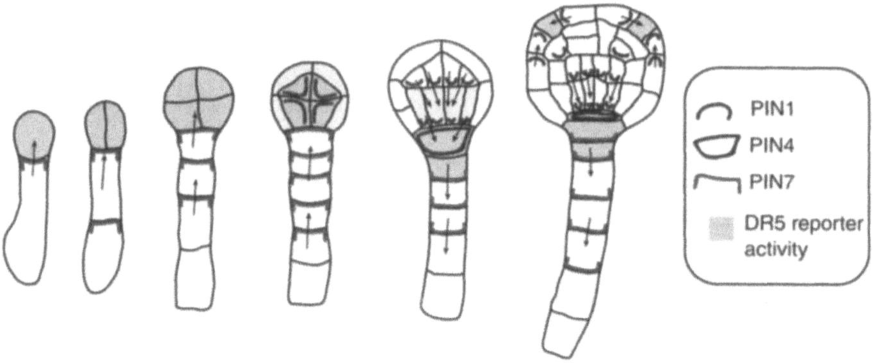

Figure 1.5 Auxin transport and response in *Arabidopsis* embryogenesis. Activity of the *DR5–GFP* reporter (green) and localization of PIN1 (blue lines), PIN4 (green lines), and PIN7 (red lines) during early embryogenesis. PIN7 is polar localized in the basal cell after zygote division and remains so until the globular stage (for stages, see Fig. 1.3). This is associated with *DR5–GFP* activity in the proembryo. PIN1 becomes polarly localized toward the basal end of the inner cells in the globular stage. This promotes auxin transport to the hypophysis (note *DR5–GFP* activity). Simultaneously, PIN7 polarity is switched to the basal end in suspensor cells and PIN4 is activated in the hypophysis. This constellation of PIN proteins and *DR5–GFP* activity is maintained at later stages. During the globular stage, PIN1 localization becomes polarized toward the flanks of the proembryo apex, eliciting new *DR5–GFP* activity maxima. Arrows indicate auxin flux direction as deduced from PIN polarity. (See Color Insert.)

be accounted for by this short signaling pathway, as mutations in the auxin receptors TIR1/AFB cause defects indistinguishable from those of the auxin biosynthesis mutants (Cheng *et al.*, 2007; Dharmasiri *et al.*, 2005; Stepanova *et al.*, 2008). The activity status of auxin signaling can be indirectly visualized by the expression of an ARF-dependent gene expression reporter, *DR5* (Ulmasov *et al.*, 1997). When coupled to GFP, green fluorescence marks cells that show auxin-dependent gene expression (Ottenschlager *et al.*, 2003; Friml *et al.*, 2003; Fig. 1.5). As discussed below, the landscape of auxin activity suggests that all patterning landmarks are closely mirrored by local auxin activity.

3.2. Specification of the apico-basal axis

With the first division of the zygote, an apical and a basal cell fate are differentiated and the basic dualism of proembryo and suspensor development is precast. Genetic studies have not identified many factors that control this first event of embryogenesis, but recently, several components have been found to regulate aspects of zygote division as well as correct establishment and maintenance of the newly generated cell identities.

3.2.1. The first asymmetrical division–SSP/YDA pathway

Loss-of-function mutations in *SHORT SUSPENSOR* (*SSP*), *YODA* (*YDA*), *MITOGEN ACTIVATED PROTEIN KINASES3* and *6* (*MPK3/6*), or *GROUNDED* (*GRD*) impair elongation of the zygote (Bayer *et al.*, 2009; Lukowitz *et al.*, 2004; Wang *et al.*, 2007). As a consequence, the basal cell is relatively small upon division of the zygote. Subsequently, the apical cell undergoes a normal pattern of divisions while aberrant division planes in the basal cell lineage produce a shorter and broadened cell file. This leads to a delayed establishment of the apico-basal axis perhaps due to the absence of signals from the malformed suspensor cells.

SSP encodes an interleukin-1 receptor-associated kinase (IRAK)/Pelle-like kinase (Bayer *et al.*, 2009), while *YDA* encodes a MAPKK Kinase (Lukowitz *et al.*, 2004). Genetic studies are consistent with these proteins acting in a linear kinase pathway that also includes MAP Kinases 3 and 6 (Wang *et al.*, 2007). Remarkably, *SSP* transcripts are delivered to the zygote and central cell by the pollen (Bayer *et al.*, 2009). Hence, SSP accumulates only transiently in the zygote and provides a temporal cue for zygote development.

The output of the proposed SSP–YDA–MPK3/6 pathway in the zygote has not been identified so far. However, potential analogies to YDA activity in the zygote can be drawn from stomatal development. Here YDA and MPK3/6 negatively regulate the asymmetric division of meristemoid mother cells that give rise to a larger cell and the smaller meristemoid that becomes the guard mother cell (see chapter 9; Bergmann *et al.*, 2004; Wang *et al.*, 2007). This signaling cascade is opposed by a set of bHLH transcription factors (SPEECHLESS (SPCH), INDUCER OF CBF EXPRESSION1 (ICE1), and SCREAM2 (SCRM2)) that promote initiation of asymmetric divisions to form meristemoids (Nadeau, 2009). The meristemoid-inhibiting YDA–MPK3/6 activity converges on the meristemoid-promoting ICE1/SCRM2/SPCH activities, as phosphorylation of SPCH by the YDA–MPK3/6 kinase cascade inhibits its activity (Lampard *et al.*, 2008). Hence, signaling via YDA and MPK3/6 is translated into transcriptional activity by the phosphorylation status of a downstream transcription factor that is expressed in a cell-specific manner. A main open question is the identity and function of this direct phosphorylation target of the SSP–YDA–MPK3/6 cascade in the zygote.

3.2.2. The WOX2/8/9 pathway

Zygote division gives rise to two cells with dramatically different cell fates. How these cell fates are established and maintained is currently not well understood, nor is it known which intrinsic determinants, if any, segregate during zygote division. The best-studied candidates for regulators of apical and basal cell fate specification are transcription factors of the

WUSCHEL-RELATED HOMEOBOX (WOX) family. *WOX2* and *WOX8/STIMPY-LIKE* (*STPL*) mRNAs are coexpressed in the zygote, but upon division *WOX2* transcripts are specific to the apical cell, while *WOX8* and its close relative *WOX9/STIMPY* (*STIP*) are found only in the basal cell. Until the octant stage, activity of these three genes is further refined to form four domains along the apico-basal axis. Top-down these domains are characterized by *WOX2* expression (apical tier), *WOX9* expression (basal tier), coexpression of *WOX8* and *WOX9* (uppermost suspensor cell), and *WOX8* expression (suspensor) (Fig. 1.3B). Whether *WOX* mRNAs are indeed segregating during zygote division remains to be demonstrated. Their function in specifying apical and basal cell fates, however, has been established (Breuninger *et al.*, 2008; Haecker *et al.*, 2004).

wox8 wox9 mutant embryos display normal zygote division but fail to correctly specify the apical and basal cell lineages. Subsequently, finger-like structures are formed by enlarged embryonic cells and, to a lesser extent, aberrant divisions in the suspensor cell file. Eventually, this leads to arrest of growth at around the heart stage phase of normal development. Analysis of several markers in *wox8 wox9* embryos demonstrates that the basal lineage is not properly established. In addition to this, properties of the apical domain are also not fully developed, suggesting a non-cell-autonomous control of apical development by *WOX8* and *WOX9*. This nonautonomy is partially mediated by promoting *WOX2* expression in apical cells. *wox2* mutant embryos display aberrant division planes during protoderm formation at the eight-cell stage of embryogenesis. However, this phenotype is enhanced to a higher frequency (30% vs. 80%), when closely related members of the WOX family (WOX1, WOX3) involved in further apical development are mutated as well. The lack of such phenotypes in *wox1*, *wox3*, and *wox1 wox3* embryos suggests that WOX2 is the main regulator of embryonic shoot development. Furthermore, expression of WOX2 in *wox8 wox9* embryos under control of the WOX9 promoter results in *yda*-like zygotic and embryonic phenotypes indicating that WOX2 expression is sufficient to establish various aspects of apical cell fates.

Activity of the SSP–YDA pathway as well as of WOX8 and WOX9 is crucial for the initial steps of embryo development. While the SSP–YDA pathway controls division of the zygote, the main outcome of WOX8 and WOX9 activity appears to be the establishment of basal cell fate. Consequently, zygotes in *yda wox8 wox9* triple mutants divide in a *yda*-like manner producing two small daughter cells before they arrest (Breuninger *et al.*, 2008). The absence of such an early developmental arrest in *yda* (Lukowitz *et al.*, 2004) or *wox8 wox9* (Breuninger *et al.*, 2008) mutants suggests that these pathways act in a nonlinear manner and are necessary for the correct specification of apical and basal cell fates.

3.2.3. Auxin control of polar axis establishment

Several lines of evidence suggest an important role for auxin in establishing apical cell fate after zygote division. Transcriptional activation by auxin, as reported by the ARF-dependent *DR5-GFP* gene, is localized to the apical cell upon zygote division and remains active in the proembryo during the divisions that follow (Friml *et al.*, 2003). This localized activity is likely the consequence of directional transport through PIN7, which is polarly localized in the basal cell toward the apical cell (Friml *et al.*, 2003; Fig. 1.5). Pharmacological interference with PIN7 localization or a mutation in *PIN7* impedes the apical auxin response and also interferes with normal subsequent divisions of the apical cell. Interestingly, mutations in *ARF5/MONOPTEROS* (*MP*) and *IAA12/BODENLOS* (*BDL*) show the same division defect, albeit at low frequencies (Berleth and Jürgens, 1993; Hamann *et al.*, 1999). Since MP and BDL are expressed in the apical lineage presumably already in the apical cell, this suggests that auxin is transported from the basal to the apical cell through PIN7 and elicits apical cell-specific characteristics through MP and BDL. Whether a similar auxin response is also involved in basal cell specification is not known, nor is it known what genes and processes are controlled by auxin in the apical cell or how auxin activity converges with SSP–YDA and/or WOX activities. Since *PIN1* expression and *DR5–GFP* activity are disrupted in later stages of *wox8 wox9* embryos (Breuninger *et al.*, 2008), a plausible scenario would be that the SSP–YDA and WOX modules act upstream of the auxin transport and response pathway, for example, by controlling the expression of *PIN1, PIN7, MP*, or *BDL*.

3.3. Establishment of the radial pattern

3.3.1. Delimiting territories—the epidermis cell fate

In the adult plant the epidermis will form the outermost boundary of the plant body to the surrounding environment and hence perform vital functions including the regulation of water status and gas exchange or the uptake of nutrients from the soil. Furthermore, the epidermis provides a means of mechanical restrictions that allow for directional growth. Hence, much attention has been given to epidermal structures of the adult plant such as stomata, trichomes, root hairs, the overlying cuticle, or the microtubule cytoskeleton of epidermal cells.

The fundamental question of how embryonic cells sense their outside position and are specified to become the founder cells of the epidermis remains mostly unanswered. The reason for this lies in the severe consequences of mutations in genes necessary for proper epidermal development (e.g., Johnson *et al.*, 2005). A failure to specify the protoderm most likely leads to an early arrest of embryogenesis, a phenotype that is also caused by mutations in basic cellular/metabolic pathways. On the other hand, milder defects that allow for embryogenesis and germination of the seedling might disturb the establishment of the epidermis as barrier for water retention. Hence, the growing seedling will dehydrate quickly and subsequently might be overlooked in genetic screens (Tanaka *et al.*, 2001 for examples).

Several genes have been identified that are involved in the formation of the embryonic cuticle and the physical separation of the embryo from the surrounding endosperm. Two of these genes, *ZHOUPI* (*ZOU*) and *ABNORMAL LEAF SHAPE1* (*ALE1*), are predominantly expressed in the embryo-surrounding region (ESR) (Tanaka *et al.*, 2001; Yang *et al.*, 2008; Fig. 1.3C). Genetic evidence suggests that the bHLH transcription factor ZOU controls expression of the secreted subtilisin-like serine protease ALE1. Morphological defects in *ale1* seedlings are enhanced by mutations in either one of the two receptor-like kinases ABNORMAL LEAF SHAPE2 (ALE2) and ARABIDOPSIS CRINKLY4 (ACR4). *ale2 acr4* double mutant seedlings not only show a lack of cuticle formation but also display a partial loss of epidermal cell fate specification (Tanaka *et al.*, 2007). This implies that ALE1 might be involved in the generation of endosperm-derived signals that trigger ALE2- and ACR4-dependent signaling in the epidermis to maintain epidermal cell fate (Tanaka *et al.*, 2007). Nonetheless, Brassica microspore embryos grown in culture develop a normal epidermal layer (Custers *et al.*, 1997), suggesting that, while endosperm-derived signals may support protoderm specification or differentiation in zygotic embryos, this signaling is not required *per se* to form the epidermis.

Mutations in the above genes, however, appear not to affect the initial establishment of the protoderm but rather to interfere with the maintenance of epidermal cell fate or with differentiation at later stages of development. Protoderm formation is therefore likely controlled by factors intrinsic to the developing embryo. The homeodomain-GLABRA2 (HD-GL2) transcription factor ARABIDOPSIS THALIANA MERISTEM LAYER1 (AtML1) and its closest homolog PROTODERMAL FACTOR2 (PDF2) have been found to play a role in the maintenance of shoot epidermal cells. A lack of both gene functions results in a failure of cell differentiation in the apical domain of the proembryo. Subsequently, embryos form a normal root and hypocotyl but develop a dome-shaped structure instead of proper cotyledons. Expression of both factors is found in all cells of the octant proembryo but is restricted to protodermal cells at later stages (Abe *et al.*, 2003; Fig. 1.3C). Thorough inspection by deletion analysis revealed complex regulation of the *AtML1* promoter through a number of different motifs (Takada and Jürgens, 2007). Interestingly, one of these motifs constitutes an L1 box which is also present in the *PDF2* promoter and can be bound by both factors. Hence, AtML1 and PDF2 might positively regulate their own expression. Furthermore, a model emerges in which this positive feedback loop is confined to the outer cells by factors in the central domain of the proembryo that effectively repress the expression of *AtML1* and *PDF2* (Abe *et al.*, 2003).

A factor that has been shown to be involved in the activation of epidermis-specific genes such as *AtML1* and *ACR4* (Fig. 1.3C) is the calpain-like cysteine protease *Arabidopsis thaliana* DEFECTIVE KERNEL1 (AtDEK1). *atdek1* embryos lack *ACR4* and *AtML1* expression and fail to develop an epidermis (Johnson *et al.*, 2005). However, the initial periclinal divisions that form the protoderm at the octant stage can be observed in *atdek1* embryos, and *AtDEK1* is expressed ubiquitously until globular stage. This suggests that also AtDEK1 is a factor involved in epidermis maintenance rather than its specification (Johnson *et al.*, 2005).

In conclusion, several signaling components have been identified that mediate the specification of outer and inner cells. Importantly, however, the nature of the information that is interpreted by the proembryo to define outer and inner is not at all known. This "signal" could be of chemical or physical nature, but in any event the normal epidermis specification in embryos grown in isolation suggests that specific external signals are probably not involved.

3.3.2. Definition of inner cell types

With the formation of the outer protodermal cell layer, an inner central domain is also set aside. These inner cells will give rise to the initials for ground tissue and provasculature, while protodermal cells are committed to only generate epidermis cells.

On one hand the inner domain is characterized by the exclusion of epidermis specific factors such as AtML1 and PDF2 (see above) (Fig. 1.3C), while on the other hand these cells are also marked by the presence of specific transcripts, such as *MP* (Hardtke and Berleth, 1998; Fig. 1.3D) or *PIN1* (Friml *et al.*, 2003). No mutants have been described that impair specifically the establishment of inner cells. One reason could be that the upper and lower tier inner cells are specified through independent mechanisms. This is conceivable because markers for the shoot and root apical meristems are activated in the upper tier or lower tier inner cells around this stage (see below; WUS and TMO5/7; Mayer *et al.*, 1998; Schlereth *et al.*, 2010). The only information on the mechanisms that might separate outer and inner cell fates comes from more pleiotropic mutants, such as mutations in RECEPTOR-LIKE PROTEIN KINASE 1 (RPK1) and TOADSTOOL2 (TOAD2) (Nodine *et al.*, 2007). The basal proembryo cell tier of *rpk1 toad2* embryos contains no cells that differentiate into protoderm or ground tissue initials but consists entirely of cells that express the provascular marker *SHORT-ROOT* (*SHR*). Consistent with this phenotype, *AtML1* is not expressed (Nodine *et al.*, 2007). Unfortunately, no ligands or kinase substrates are known for these two receptor-like kinases.

3.4. Initiation of meristems

Stem cells are located at the ends of the apico-basal axis of the plant in the SAM and the RAM, respectively. Although the meristems differ in their anatomical characteristics, and the organs or cells that are produced, the basic organization is similar. In both, an organizing center (OC in the shoot meristem, QC in the root meristem) is surrounded by stem cells for various tissues. Strikingly, the factors that are involved in the maintenance of shoot and root meristem also show a degree of similarity (Lenhard and Laux, 2003; Sarkar *et al.*, 2007; Stahl *et al.*, 2009). The initiation of the two meristems in the embryo, however, differs significantly.

3.4.1. The root meristem

The root meristem of *Arabidopsis* forms at the boundary of the apical and basal lineages. The proembryo contributes the stem cells for vascular, ground, and epidermal tissues, as well as the lateral root cap, while the QC and the columella root cap are derived from a former suspensor cell (Scheres *et al.*, 1994). Cell identity specification therefore needs to be tightly coordinated between the neighboring proembryo and suspensor cells in order to generate the QC and its adjacent stem cells. From genetic studies, several factors have emerged that play a critical role in the specification of the uppermost suspensor cell as hypophysis, precursor of the QC and columella. Their activities mostly converge upon auxin transport or response. Indeed, pharmacological inhibition of auxin transport (Friml *et al.*, 2003; Hadfi *et al.*, 1998), or genetic interference with auxin transport (*pin1,3,4,7* mutant; Friml *et al.*, 2003), synthesis (*yucca1,4,10,11*; Cheng *et al.*, 2007; or *taa1 tar1,2*; Stepanova *et al.*, 2008), or perception (*tir1 afb1, 2, 3*; Dharmasiri *et al.*, 2005) prevents root formation.

As described above, the output of auxin activity is a change of gene expression through the ARF transcription factors. One of the 23 ARFs in *Arabidopsis*, ARF5/MP, is critically required for root formation. Mutants do not make a root and show abnormal hypophysis specification (Berleth and Jürgens, 1993) and cell division. Interestingly, MP is not expressed in the hypophysis itself, but in the adjacent proembryo cells, which suggests non-cell-autonomous control of hypophysis specification through secondary signals (Weijers *et al.*, 2006). While PIN1 is present on all inner membranes of the proembryo until the 16-cell stage, immediately prior to hypophysis specification, PIN1 protein becomes polarly localized toward the future hypophysis (Fig. 1.5; Friml *et al.*, 2003). This suggests that auxin is transported from the proembryo to the uppermost suspensor cell. Indeed, the *DR5–GFP* reporter is active in this cell, and this activation depends on MP activity. Furthermore, PIN1 expression is strongly reduced in *mp* mutants. Since treatment with external auxin does not restore root formation in the *mp* mutant, other signal(s) were proposed to act in parallel (Weijers *et al.*,

2006). Recently, through the identification of MP target genes, a second MP-dependent cell–cell signal was identified. This is represented by the small bHLH transcription factor TARGET OF MONOPTEROS7 (TMO7). In addition, MP was shown to activate several other transcription factors, including the bHLH factor TMO5, which acts in the proembryo (Schlereth *et al.*, 2010). In summary, MP coordinates root meristem formation by activating both cell-autonomous factors (e.g. TMO5) and mobile signals (auxin and TMO7).

Upon the initiation of the root meristem, the identity of this area is specified through a family of AP2-type transcription factors encoded by the *PLETHORA* (*PLT*) genes (Aida *et al.*, 2004). *PLT* gene expression depends on MP, but probably not through direct binding (Aida *et al.*, 2004; Schlereth *et al.*, 2010). Mutations in *PLT* genes interfere with divisions of the hypophysis derivates, resulting in an absence of the QC at early heart stage (Galinha *et al.*, 2007). Strikingly, when ectopically expressed, PLT proteins can convert shoot cells to root identity, supporting their role as root identity specifiers (Aida *et al.*, 2004).

Regulators for later steps in root patterning have been identified (e.g., *WOX5* and *SCR*; see section 4 below). For reasons of brevity, we will not discuss these here, but refer to excellent reviews on this subject (see chapter 3; Petricka and Benfey, 2008; Ten Hove and Heidstra, 2008).

3.4.2. The shoot apical meristem

The SAM is initiated within the apical half of the proembryo and as such does not involve cell–cell signaling across the apico-basal boundary. Many factors are known to control aspects of shoot meristem function, but the vast majority of these acts in the homeostatic control of meristem size rather than the initiation of the meristem. The OC cells and the overlying stem cells mutually control the size of the other population, leading to a stable meristem size. The installation of the OC and stem cell area occurs during the mid-globular stage, while both are maintained throughout plant life. In this review, we will consider the mechanisms that are involved in positioning the OC and stem cells, and refer to chapter 4 and another review for discussions on meristem function (Tucker and Laux, 2007).

The OC is marked by the WOX family member WUSCHEL, whose activity is critical for SAM formation (Mayer *et al.*, 1998). WUS activity in the OC is later required to promote expression of the gene encoding the secreted peptide CLV3 in the stem cells (Schoof *et al.*, 2000). In turn, CLV3 and the membrane receptor-like kinases CLV1 and CLV2 suppress *WUS* expression in the OC. WUS activity is supported by the vascular ARGONOUTE10/ZWILLE (ZLL) protein, which functions in a small RNA pathway. While *ZLL, WUS*, and *CLV3* expression does not overlap in the mature SAM, *WUS* and *ZLL* are coexpressed in the upper tier inner cells, and *CLV3* is not activated until the heart stage of embryogenesis (Tucker *et al.*, 2008). How the coexpression state progresses into the state where *ZLL* and *WUS* are separated is not understood, and no factors that activate *WUS* locally have been identified, leaving the first step in SAM initiation elusive.

Several other factors act early in the embryo to specify the SAM. Among these are members of the class III HD-Zip transcription factors. These are expressed in the vascular tissues and extend to the future SAM area at the globular stage (Prigge *et al.*, 2005). Simultaneous loss of multiple members prevents SAM formation, although it has not been studied exactly at what stage the first defects are manifested (Emery *et al.*, 2003; Prigge *et al.*, 2005). Finally, the HD transcription factor SHOOT MERISTEMLESS (STM) accumulates in the entire SAM area at the late globular stage of embryogenesis (Long *et al.*, 1996). STM is required for SAM formation, presumably by preventing differentiation (Barton and Poethig, 1993). The expression domains of *STM* and *WUS* are established independently (Endrizzi *et al.*, 1996), but combined overexpression leads to functional ectopic meristem formation (Gallois *et al.*, 2002), suggesting that both are required to generate a proper meristem.

In summary, several genes are required for meristem initiation and further elaboration toward a functional meristem, but the mechanisms that position the OC and stem cells have not been identified to date.

3.5. Cotyledon initiation

At the globular stage, a few cells at the flanks of the proembryo apex are selected to become cotyledons, after which cells start to proliferate at these sites (Fig. 1.2A). The correct establishment of cotyledons requires that, from a zone of competence, only these two sites are selected and, subsequently, that these cells acquire cotyledon identity. Like root initiation, also this process appears to be under heavy auxin control. Most mutations or treatments that affect auxin transport, biosynthesis, perception, or response cause changes in the pattern of cotyledon initiation (see Möller and Weijers, 2009 for summary). Strikingly, while auxin transport during root initiation is mediated by the redundant activity of multiple PIN proteins, PIN1 is the dominant family member in cotyledon initiation (Benkova *et al.*, 2003; Liu *et al.*, 1993). Therefore, localization of PIN1 is particularly informative to define auxin streams that are required for cotyledon initiation. Ubiquitous presence of auxin in the proembryo apex causes the establishment of cotyledon identity around the entire circumference (Friml *et al.*, 2003; Weijers *et al.*, 2005); hence, auxin transport acts to direct the accumulation of the positively acting signal auxin only to those sites where cotyledons need to be. Indeed, activity of the *DR5–GFP* reporter shows peaks of auxin response in the cotyledon initiation sites. PIN1 is expressed in the protoderm, its polarity facing the cotyledon initiation sites, and in the vasculature, where polarity is facing downwards (Benkova *et al.*, 2003). The cotyledon initiation sites are marked by convergence of PIN1 in the adjacent membranes of neighboring cells (Fig. 1.5). How this convergence is regulated remains to be determined; at least this constellation explains the auxin accumulation points, as it does in mathematical simulations of the post-embryonic shoot apex (Jonsson *et al.*, 2006). Consistent with the central role of PIN1 in defining auxin accumulation sites, mutations that interfere with correct PIN1 expression and polarity lead to altered cotyledon numbers or positioning (Friml *et al.*, 2004; Treml *et al.*, 2005). The auxin accumulation is translated into cotyledon developmental programs through the ARFs MP and ARF7/ non-phototropic hypocotyl4 (NPH4). Information on target genes of these transcription factors is scarce, but includes the AP2 transcription factor DRN (Cole *et al.*, 2009), whose activity is required for proper cotyledon establishment (Chandler *et al.*, 2007). Importantly, auxin does not only act through activating cotyledon initiation genes, but also negatively regulates the expression of *CUP-SHAPED COTYLEDON* (*CUC*) genes in the cotyledon positions. In *mp* and *pid pin1* double mutants, these genes, which are normally expressed in the boundaries between SAM and cotyledons, expand into the cotyledon position, which correlates with reduced cotyledon initiation (Aida *et al.*, 2002). Conversely, *cuc* mutants develop cotyledon fate around the entire apical circumference, presumably due to the absence of specified boundaries (Aida *et al.*, 1997).

After selection of cells to become cotyledons, the developmental fate of these cells needs to be specified. Although several genes have been shown to be restricted to cotyledons (Fiers *et al.*, 2004; Long and Barton, 1998), very little is actually known about the identity of the initial specifying factors. This is in stark contrast to the number of genes that are known to act in the upper–lower axis definition in cotyledons (reviewed in Husbands *et al.*, 2009). The only reported case of alterations in the identity of the cells in the position of cotyledons is when either an apolar PIN1 protein or a dominant-negative Rab GTPase was expressed ubiquitously in embryos. In these genotypes, root-like structures occasionally appear in the position of cotyledons (Dhonukshe *et al.*, 2008). Intriguingly, this is accompanied by the misexpression of the "root specifier gene" *PLT1*, whose ectopic expression has been shown to convert cells to root identity (Aida *et al.*, 2004). *PLT1* belongs to the same transcription factor family as *ANT*, which is normally expressed in the cotyledon primordia (Long and Barton, 1998). Therefore, although purely hypothetical at present, it is conceivable that the different PLT/AINTEGUMENTA-LIKE members define various organ identities after their initial specification.

4. Evolutionary Aspects of Plant Embryogenesis

During evolution, plant genes have diverged dramatically, and so have the patterns of cell division in embryos. However, despite often very different cell divisions, the overall body plan of plants has been conserved. The identification of molecular mechanisms in *Arabidopsis* now allows addressing whether conserved cellular and molecular mechanisms underlie pattern formation in phylogenetically unrelated plant species.

When considering the regulatory factors in pattern formation, it appears that most factors that are instrumental in *Arabidopsis* embryogenesis have putative orthologs in other genomes (reviewed in Nardmann and Werr, 2007). Obviously, there are only a handful of examples where the expression pattern of these orthologs has been studied, and the actual function has not been addressed for any except a few. However, the limited expression data available suggests that, to some extent, the position-specific expression of orthologs is conserved.

For example, orthologs to *Arabidopsis WUS*, *WOX2*, and *WOX5* have been identified in maize (*ZmWUS1, ZMWOX2, ZmWOX5*; Nardmann *et al.*, 2007) and rice (*OsQHB*; Kamiya *et al.*, 2003b). Even though rice and maize embryos do not have the same shape or stereotypic cell division pattern as the *Arabidopsis* embryo (Fig. 1.6), in both cases, defined *WUS* and *WOX5/QHB* expression sites are found on either side of the vascular axis

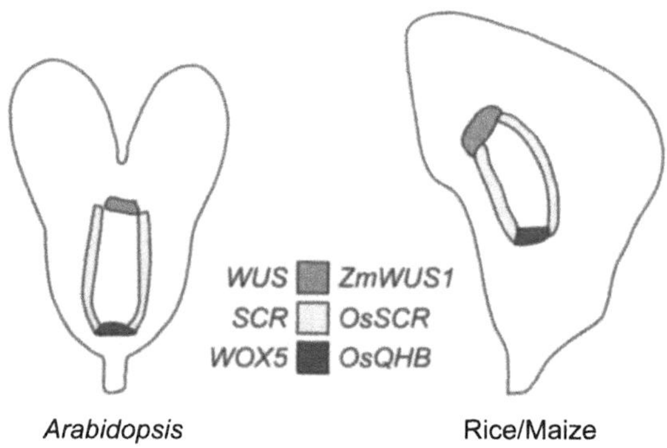

Figure 1.6 Conservation of plant embryo patterning. Comparison of the mRNA expression patterns of important *Arabidopsis* patterning regulators in monocotyledonous species. Embryos at approximately the same developmental stage are depicted. The rice/maize embryo does not have a "cotyledon primordium" that is as easily recognizable as the *Arabidopsis* cotyledon primordia but will initiate at the top left side. While the shape of *Arabidopsis* and maize/rice embryo is rather different, the relative position of *WUS* (or *ZmWUS1*), *SCR* (or *OsSCR*), and *WOX5* (or *OsQHB*) is conserved and marks the shoot (*WUS*) to root (*WOX5*) axis.

(Fig. 1.6). Also, in both monocot and dicot embryos, the endodermal layer surrounding the vascular axis expresses *SCR* or its rice ortholog *OsSCR* (Kamiya *et al.*, 2003a). Therefore, the relative position of the *WUS*-marked SAM OC, the vascular axis, and the *WOX5/QHB*-marked root QC area is analogous in distant species, which suggests common underlying mechanisms. An open question is whether the establishment of this rather advanced constellation of expression domains (Fig. 1.6) follows steps similar to the ones found in *Arabidopsis*. Evidence suggests that this may at least in part be the case since the maize *WOX2* ortholog *ZmWOX2* is expressed specifically in the proembryo-like structure similar to its *Arabidopsis* counterpart (Nardmann *et al.*, 2007). There are large collections of maize (Scanlon *et al.*, 1994) and rice (Hong *et al.*, 1995) embryo-defective mutants. Identifying causal genes and determining the expression of the conserved cell identity markers will undoubtedly shed light on the conservation of molecular mechanisms underlying plant embryo patterning.

At the very least, it appears that the profound role of auxin in embryo patterning in *Arabidopsis* is matched by similar roles in other species including Brassica (Hadfi *et al.*, 1998), wheat (Fischer *et al.*, 1997), maize, and the gymnosperm *Picea abies* (Hakman *et al.*, 2009). The genomes of each of these species, and those of all other embryophytes analyzed so far, contain genes for auxin biosynthesis, transport, perception, and response (reviewed in Lau *et al.*, 2008). Auxin has indeed also been detected in all land plants that were investigated (Cooke *et al.*, 2002). Interference with polar PIN protein localization through treatment with auxin transport inhibitors, or with auxin, causes strong embryo patterning defects in all the above species (Fischer *et al.*, 1997; Hadfi *et al.*, 1998; Hakman *et al.*, 2009). Even though the exact defects depend on the species analyzed, this suggests that one common theme in the regulation of embryo patterning is the profound role of auxin.

The fact that very different cell division patterns can still give rise to similarly patterned overall plant structures suggests that cell division planes are not causally connected to patterning. This makes extrapolation of the mechanisms in *Arabidopsis* to species that have embryos with many more cells somewhat problematic. However, two fundamental issues emerge from the study of *Arabidopsis* embryogenesis. First, there is an early distinction between apical (embryonic) and basal (extra-embryonic) lineages. Second, the root initiates at the boundary between these two lineages. Observation of embryos from a wide array of plant species (Johri *et al.*, 1992) shows that, even though the embryonic and extra-embryonic lineages are not always separated at the first division, the fundamental dualism is present in many plant species. Also, the site of root initiation relative to the embryo shape can vary substantially (see, e.g., Fig. 1.6). While the suspensor is easily recognizable in *Arabidopsis*, the shape and size are extremely variable in other species (Yeung and Meinke, 1993), which makes it difficult to define the suspensor without molecular markers.

Nonetheless, in the rice embryo, in which the suspensor measures several cell diameters and is more continuous in shape with the proembryo (Nardmann *et al.*, 2007), cells in the basal domain are more vacuolated similar to *Arabidopsis* suspensor cells. The *QHB* and *ZmWOX5* expression regions are positioned in close association with these vacuolated cells (Kamiya *et al.*, 2003b; Nardmann *et al.*, 2007), suggesting that a mechanism similar to the one in *Arabidopsis* could be involved in initiating the root meristem.

5. Concluding Remarks

Pattern formation in plant embryos involves several morphogenetic steps, during which cell type specification, asymmetric cell division, and cell–cell communication play critical roles. In this chapter, we have highlighted those patterning steps that we consider landmarks in embryo development and we have described the regulatory factors that are known to act in these processes. While critical regulators for each of these landmarks have been isolated mostly through genetics, it should be emphasized that with only a handful of factors for each patterning step, most of the factors involved remain to be identified. Yet, it emerges that one common theme in embryo patterning is the profound coordinating role of auxin. With its brief signal transduction pathway and a dedicated directional transport machinery, auxin is perfectly suited to coordinate cell fates both at long and at short ranges. A key open question is how specificity in its action is achieved such that the uppermost suspensor cell will be specified as hypophysis and the cells in the proembryo apex as cotyledon precursors. A further unifying theme in embryo patterning is the intersection between hormonal regulation and regional transcription factor activities. As an example, WOX transcription factors are required for proper auxin transport and response, while auxin activity in turn is required to specify the domain of WOX5 expression. The identification of the direct targets of the regionally acting transcription factors will soon provide insight into the wiring of the regulatory networks in both space and time.

Furthermore, phylogenetic and expression analysis of *Arabidopsis* regulators demonstrates that, despite profound differences in embryo anatomy, similar principles and mechanisms may underlie pattern formation across plant species. An important challenge will be the functional analysis of the role of these orthologs in other species, for example, through reverse genetics in model organisms.

ACKNOWLEDGMENTS

We thank our lab members Annemarie Lokerse and Barbara Möller for critical reading and helpful comments on the manuscript. Work in our laboratory is supported by grants from the Netherlands Organization for Scientific Research (NWO; ALW-VIDI 864-06.012) and the European Commission 7th Framework Programme (Initial Training Network "SIREN," Contract no. 214788).

REFERENCES

Abe, M., Katsumata, H., Komeda, Y., and Takahashi, T. (2003). Regulation of shoot epidermal cell differentiation by a pair of homeodomain proteins in Arabidopsis. *Development* **130**, 635–643.

Aida, M., Beis, D., Heidstra, R., Willemsen, V., Blilou, I., Galinha, C., Nussaume, L., Noh, Y. S., Amasino, R., and Scheres, B. (2004). The PLETHORA genes mediate patterning of the Arabidopsis root stem cell niche. *Cell* **119**, 109–120.

Aida, M., Ishida, T., Fukaki, H., Fujisawa, H., and Tasaka, M. (1997). Genes involved in organ separation in Arabidopsis: an analysis of the cup-shaped cotyledon mutant. *Plant Cell* **9**, 841–857.

Aida, M., Vernoux, T., Furutani, M., Traas, J., and Tasaka, M. (2002). Roles of PIN-FORMED1 and MONOPTEROS in pattern formation of the apical region of the Arabidopsis embryo. *Development* **129**, 3965–3974.

Barton, M. K., and Poethig, R. S. (1993). Formation of the shoot apical meristem in Arabidopsis thaliana: an analysis of development in the wild type and in the shoot meristemless mutant. *Development* **119**, 823–831.

Batygina, T. B., and Vinogradova, G. (2007). Phenomenon of polyembryony. Genetic heterogeneity of seeds. *Russ. J. Dev. Biol.* **38**, 126–151.

Bayer, M., Nawy, T., Giglione, C., Galli, M., Meinnel, T., and Lukowitz, W. (2009). Paternal control of embryonic patterning in Arabidopsis thaliana. *Science* **323**, 1485–1488.

Benkova, E., Michniewicz, M., Sauer, M., Teichmann, T., Seifertova, D., Jürgens, G., and Friml, J. (2003). Local, efflux-dependent auxin gradients as a common module for plant organ formation. *Cell* **115**, 591–602.

Berger, F., Grini, P. E., and Schnittger, A. (2006). Endosperm: an integrator of seed growth and development. *Curr. Opin. Plant Biol.* **9**, 664–670.

Bergmann, D. C., Lukowitz, W., and Somerville, C. R. (2004). Stomatal development and pattern controlled by a MAPKK kinase. *Science* **304**, 1494–1497.

Berleth, T., and Jürgens, G. (1993). The role of the monopteros gene in organising the basic body region of the Arabidopsis embryo. *Development* **118**, 575–587.

Bewley, J. D., and Black, M. (1994). *Seeds: Physiology of Development and Germination.* Plenum, New York.

Breuninger, H., Rikirsch, E., Hermann, M., Ueda, M., and Laux, T. (2008). Differential expression of WOX genes mediates apical-basal axis formation in the Arabidopsis embryo. *Dev. Cell* **14**, 867–876.

Cairney, J., and Pullman, G. S. (2007). The cellular and molecular biology of conifer embryogenesis. *New Phytol.* **176**, 511–536.

Chandler, J. W., Cole, M., Flier, A., Grewe, B., and Werr, W. (2007). The AP2 transcription factors DORNROSCHEN and DORNROSCHEN-LIKE redundantly control Arabidopsis embryo patterning via interaction with PHAVOLUTA. *Development* **134**, 1653–1662.

Chapman, E. J., and Estelle, M. (2009). Mechanism of auxin-regulated gene expression in plants. *Annu. Rev. Genet.* **43**, 265–285.

Cheng, Y., Dai, X., and Zhao, Y. (2007). Auxin synthesized by the YUCCA flavin monooxygenases is essential for embryogenesis and leaf formation in Arabidopsis. *Plant Cell* **19**, 2430–2439.

Cole, M., Chandler, J., Weijers, D., Jacobs, B., Comelli, P., and Werr, W. (2009). DORNROSCHEN is a direct target of the auxin response factor MONOPTEROS in the Arabidopsis embryo. *Development* **136**, 1643–1651.

Cooke, T. J., Poli, D., Sztein, A. E., and Cohen, J. D. (2002). Evolutionary patterns in auxin action. *Plant Mol. Biol.* **49**, 319–338.

Custers, J. B., Oldenhof, M. T., Schrauwen, J. A., Cordewener, J. H., Wullems, G. J., and van Lookeren Campagne, M. M. (1997). Analysis of microspore-specific promoters in transgenic tobacco. *Plant Mol. Biol.* **35**, 689–699.

Dharmasiri, N., Dharmasiri, S., Weijers, D., Lechner, E., Yamada, M., Hobbie, L., Ehrismann, J. S., Jürgens, G., and Estelle, M. (2005). Plant development is regulated by a family of auxin receptor F-box proteins. *Dev. Cell* **9**, 109–119.

Dhonukshe, P., Tanaka, H., Goh, T., Ebine, K., Mahonen, A. P., Prasad, K., Blilou, I., Geldner, N., Xu, J., Uemura, T., Chory, J., Ueda, T., *et al.* (2008). Generation of cell polarity in plants links endocytosis, auxin distribution and cell fate decisions. *Nature* **456**, 962–966.

Emery, J. F., Floyd, S. K., Alvarez, J., Eshed, Y., Hawker, N. P., Izhaki, A., Baum, S. F., and Bowman, J. L. (2003). Radial patterning of Arabidopsis shoots by class III HD-ZIP and KANADI genes. *Curr. Biol.* **13**, 1768–1774.

Endrizzi, K., Moussian, B., Haecker, A., Levin, J. Z., and Laux, T. (1996). The SHOOT MERISTEMLESS gene is required for maintenance of undifferentiated cells in Arabidopsis shoot and floral meristems and acts at a different regulatory level than the meristem genes WUSCHEL and ZWILLE. *Plant J.* **10**, 967–979.

Fiers, M., Hause, G., Boutilier, K., Casamitjana-Martinez, E., Weijers, D., Offringa, R., van der Geest, L., van Lookeren Campagne, M., and Liu, C. M. (2004). Mis-expression of the CLV3/ESR-like gene CLE19 in Arabidopsis leads to a consumption of root meristem. *Gene* **327**, 37–49.

Fischer, C., Speth, V., Fleig-Eberenz, S., and Neuhaus, G. (1997). Induction of zygotic polyembryos in wheat: influence of auxin polar transport. *Plant Cell* **9**, 1767–1780.

Friml, J. (2003). Auxin transport—shaping the plant. *Curr. Opin. Plant Biol.* **6**, 7–12.

Friml, J., Vieten, A., Sauer, M., Weijers, D., Schwarz, H., Hamann, T., Offringa, R., and Jürgens, G. (2003). Efflux-dependent auxin gradients establish the apical-basal axis of Arabidopsis. *Nature* **426**, 147–153.

Friml, J., Yang, X., Michniewicz, M., Weijers, D., Quint, A., Tietz, O., Benjamins, R., Ouwerkerk, P. B., Ljung, K., Sandberg, G., Hooykaas, P. J., Palme, K., *et al.* (2004). A PINOID-dependent binary switch in apical-basal PIN polar targeting directs auxin efflux. *Science* **306**, 862–865.

Galinha, C., Hofhuis, H., Luijten, M., Willemsen, V., Blilou, I., Heidstra, R., and Scheres, B. (2007). PLETHORA proteins as dose-dependent master regulators of Arabidopsis root development. *Nature* **449**, 1053–1057.

Gallois, J. L., Woodward, C., Reddy, G. V., and Sablowski, R. (2002). Combined SHOOT MERISTEMLESS and WUSCHEL trigger ectopic organogenesis in Arabidopsis. *Development* **129**, 3207–3217.

Hadfi, K., Speth, V., and Neuhaus, G. (1998). Auxin-induced developmental patterns in Brassica juncea embryos. *Development* **125**, 879–887.

Haecker, A., Gross-Hardt, R., Geiges, B., Sarkar, A., Breuninger, H., Herrmann, M., and Laux, T. (2004). Expression dynamics of *WOX* genes mark cell fate decisions during early embryonic patterning in *Arabidopsis thaliana*. *Development* **131**, 657–668.

Hakman, I., Hallberg, H., and Palovaara, J. (2009). The polar auxin transport inhibitor NPA impairs embryo morphology and increases the expression of an auxin efflux facilitator protein PIN during Picea abies somatic embryo development. *Tree Physiol.* **29**, 483–496.

Hamann, T., Mayer, U., and Jürgens, G. (1999). The auxin-insensitive *bodenlos* mutation affects primary root formation and apical-basal patterning in the *Arabidopsis* embryo. *Development* **126**, 1387–1395.

Hardtke, C. S., and Berleth, T. (1998). The *Arabidopsis* gene *MONOPTEROS* encodes a transcription factor mediating embryo axis formation and vascular development. *EMBO J.* **17**, 1405–1411.

Hong, S.-K., Aoki, T., Kitano, H., Satoh, H., and Nagato, Y. (1995). Phenotypic diversity of 188 rice embryo mutants. *Dev. Genet.* **16**, 298–310.

Husbands, A. Y., Chitwood, D. H., Plavskin, Y., and Timmermans, M. C. (2009). Signals and prepatterns: new insights into organ polarity in plants. *Genes Dev.* **23**, 1986–1997.

Jiang, K., and Feldman, L. J. (2005). Regulation of root apical meristem development. *Annu. Rev. Cell Dev. Biol.* **21**, 485–509.

Johnson, K. L., Degnan, K. A., Ross Walker, J., and Ingram, G. C. (2005). AtDEK1 is essential for specification of embryonic epidermal cell fate. *Plant J.* **44**, 114–127.

Johri, B. M., Ambegaokar, K. B., and Srivastava, P. S. (1992). *Comparative Embryology of Angiosperms.* Springer-Verlag, Berlin.

Jonsson, H., Heisler, M. G., Shapiro, B. E., Meyerowitz, E. M., and Mjolsness, E. (2006). An auxin-driven polarized transport model for phyllotaxis. *Proc. Natl. Acad. Sci. U.S.A.* **103**, 1633–1638.

Jürgens, G., and Mayer, U. (1994). Arabidopsis. In: Bard, J. B. L. (Ed.), *Embryos, Color Atlas of Development.* Wolffe Publishing, London, pp. 7–22.

Kamiya, N., Itoh, J., Morikami, A., Nagato, Y., and Matsuoka, M. (2003a). The SCARECROW gene's role in asymmetric cell divisions in rice plants. *Plant J.* **36**, 45–54.

Kamiya, N., Nagasaki, H., Morikami, A., Sato, Y., and Matsuoka, M. (2003b). Isolation and characterization of a rice *WUSCHEL*-type homeobox gene that is specifically expressed in the central cells of a quiescent center in the root apical meristem. *Plant J.* **35**, 429–441.

Kesseler, R., and Stuppy, W. (2006). *Seeds: Time Capsules of Life.* Papadakis, London.

Koltunow, A. M., and Grossniklaus, U. (2003). Apomixis: a developmental perspective. *Annu. Rev. Plant Biol.* **54**, 547–574.

Lakshmanan, K. K., and Ambegaokar, K. B. (1984). Polyembryony. In: Johri, B. M. (Ed.), *Embryology of Angiosperms.* Springer-Verlag, Berlin, pp. 445–474.

Lampard, G. R., Macalister, C. A., and Bergmann, D. C. (2008). Arabidopsis stomatal initiation is controlled by MAPK-mediated regulation of the bHLH SPEECHLESS. *Science* **322**, 1113–1116.

Lau, S., Jürgens, G., and De Smet, I. (2008). The evolving complexity of the auxin pathway. *Plant Cell* **20**, 1738–1746.

Lenhard, M., and Laux, T. (2003). Stem cell homeostasis in the Arabidopsis shoot meristem is regulated by intercellular movement of CLAVATA3 and its sequestration by CLAVATA1. *Development* **130**, 3163–3173.

Lersten, N. R. (2004). *Flowering Plant Embryology: With Emphasis on Economic Species.* Blackwell, Ames, IA.

Liu, C., Xu, Z., and Chua, N. H. (1993). Auxin polar transport is essential for the establishment of bilateral symmetry during early plant embryogenesis. *Plant Cell* **5**, 621–630.

Long, J. A., and Barton, M. K. (1998). The development of apical embryonic pattern in Arabidopsis. *Development* **125**, 3027–3035.

Long, J. A., Moan, E. I., Medford, J. I., and Barton, M. K. (1996). A member of the KNOTTED class of homeodomain proteins encoded by the STM gene of Arabidopsis. *Nature* **379**, 66–69.

Lukowitz, W., Roeder, A., Parmenter, D., and Somerville, C. (2004). A MAPKK kinase gene regulates extra-embryonic cell fate in Arabidopsis. *Cell* **116**, 109–119.

Mayer, K. F., Schoof, H., Haecker, A., Lenhard, M., Jürgens, G., and Laux, T. (1998). Role of WUSCHEL in regulating stem cell fate in the Arabidopsis shoot meristem. *Cell* **95**, 805–815.

Möller, B., and Weijers, D. (2009). Auxin control of embryo patterning. *Cold Spring Harb. Perspect. Biol.* **1**, a001545.

Nadeau, J. A. (2009). Stomatal development: new signals and fate determinants. *Curr. Opin. Plant Biol.* **12**, 29–35.

Nardmann, J., and Werr, W. (2007a). The evolution of plant regulatory networks: what Arabidopsis cannot say for itself. *Curr. Opin. Plant Biol.* **10**, 653–659.

Nardmann, J., Zimmermann, R., Durantini, D., Kranz, E., and Werr, W. (2007b). WOX gene phylogeny in Poaceae: a comparative approach addressing leaf and embryo development. *Mol. Biol. Evol.* **24**, 2474–2484.

Nodine, M. D., Yadegari, R., and Tax, F. E. (2007). RPK1 and TOAD2 are two receptor-like kinases redundantly required for Arabidopsis embryonic pattern formation. *Dev. Cell* **12**, 943–956.

Ottenschlager, I., Wolff, P., Wolverton, C., Bhalerao, R. P., Sandberg, G., Ishikawa, H., Evans, M., and Palme, K. (2003). Gravity-regulated differential auxin transport from columella to lateral root cap cells. *Proc. Natl. Acad. Sci. U.S.A.* **100**, 2987–2991.

Petrasek, J., and Friml, J. (2009). Auxin transport routes in plant development. *Development* **136**, 2675–2688.

Petricka, J. J., and Benfey, P. N. (2008). Root layers: complex regulation of developmental patterning. *Curr. Opin. Genet. Dev.* **18**, 354–361.

Prigge, M. J., Otsuga, D., Alonso, J. M., Ecker, J. R., Drews, G. N., and Clark, S. E. (2005). Class III homeodomain-leucine zipper gene family members have overlapping, antagonistic, and distinct roles in Arabidopsis development. *Plant Cell* **17**, 61–76.

Sarkar, A. K., Luijten, M., Miyashima, S., Lenhard, M., Hashimoto, T., Nakajima, K., Scheres, B., Heidstra, R., and Laux, T. (2007). Conserved factors regulate signalling in Arabidopsis thaliana shoot and root stem cell organizers. *Nature* **446**, 811–814.

Scanlon, M. J., Stinard, P. S., James, M. G., Myers, A. M., and Robertson, D. S. (1994). Genetic analysis of 63 mutations affecting maize kernel development isolated from Mutator stocks. *Genetics* **136**, 281–294.

Scheres, B., Wolkenfelt, H., Willemsen, V., Terlouw, M., Lawson, E., Dean, C., and Weisbeek, P. (1994). Embryonic origin of the Arabidopsis primary root and root meristem initials. *Development* **120**, 2475–2487.

Schlereth, A., Möller, B., Liu, W. M. K., Flipse, J, Rademacher, E. H., Schmid, M., Jürgens, G., and Weijers, D. (2010). MONOPTEROS controls embryonic root initiation by regulating a mobile transcription factor. *Nature* **464**, 913–916.

Schoof, H., Lenhard, M., Haecker, A., Mayer, K. F., Jürgens, G., and Laux, T. (2000). The stem cell population of Arabidopsis shoot meristems in maintained by a regulatory loop between the CLAVATA and WUSCHEL genes. *Cell* **100**, 635–644.

Seguí-Simarro, J. M., and Nuez, F. (2008). How microspores transform into haploid embryos: changes associated with embryogenesis induction and microspore-derived embryogenesis. *Physiol. Plant* **134**, 1–12.

Stahl, Y., Wink, R. H., Ingram, G. C., and Simon, R. (2009). A signaling module controlling the stem cell niche in Arabidopsis root meristems. *Curr. Biol.* **19**, 909–914.

Stepanova, A. N., Robertson-Hoyt, J., Yun, J., Benavente, L. M., Xie, D. Y., Dolezal, K., Schlereth, A., Jürgens, G., and Alonso, J. M. (2008). TAA1-mediated auxin biosynthesis is essential for hormone crosstalk and plant development. *Cell* **133**, 177–191.

Sundaresan, V., and Alandete-Saez, M. (2010). Pattern formation in miniature: the female gametophyte of flowering plants. *Development* **137**, 179–189

Taiz, L., and Zeiger, E. (2002). *Plant Physiology*. Sinauer, Sunderland, MA.

Takada, S., and Jürgens, G. (2007). Transcriptional regulation of epidermal cell fate in the Arabidopsis embryo. *Development* **134**, 1141–1150.

Tanaka, H., Onouchi, H., Kondo, M., Hara-Nishimura, I., Nishimura, M., Machida, C., and Machida, Y. (2001). A subtilisin-like serine protease is required for epidermal surface formation in Arabidopsis embryos and juvenile plants. *Development* **128**, 4681–4689.

Tanaka, H., Watanabe, M., Sasabe, M., Hiroe, T., Tanaka, T., Tsukaya, H., Ikezaki, M., Machida, C., and Machida, Y. (2007). Novel receptor-like kinase ALE2 controls shoot development by specifying epidermis in Arabidopsis. *Development* **134**, 1643–1652.

Tao, Y., Ferrer, J. L., Ljung, K., Pojer, F., Hong, F., Long, J. A., Li, L., Moreno, J. E., Bowman, M. E., Ivans, L. J., Cheng, Y., Lim, J., *et al.* (2008). Rapid synthesis of auxin via a new tryptophan-dependent pathway is required for shade avoidance in plants. *Cell* **133**, 164–176.

Ten Hove, C. A., and Heidstra, R. (2008). Who begets whom? Plant cell fate determination by asymmetric cell division. *Curr. Opin. Plant Biol.* **11**, 34–41.

Toonen, M. A. J., and de Vries, S. C. (1996). Initiation of somatic embryos from single cells. In: Wang, T. L., and Cuming, A. (Eds.), *Embryogenesis: The Generation of a Plant.* BIOS Scientific Publishers Ltd., Oxford, pp. 173–189.

Treml, B. S., Winderl, S., Radykewicz, R., Herz, M., Schweizer, G., Hutzler, P., Glawischnig, E., and Ruiz, R. A. (2005). The gene ENHANCER OF PINOID controls cotyledon development in the Arabidopsis embryo. *Development* **132**, 4063–4074.

Tucker, M. R., Hinze, A., Tucker, E. J., Takada, S., Jurgens, G., and Laux, T. (2008). Vascular signalling mediated by ZWILLE potentiates WUSCHEL function during shoot meristem stem cell development in the Arabidopsis embryo. *Development* **135**, 2839–2843.

Tucker, M. R., and Laux, T. (2007). Connecting the paths in plant stem cell regulation. *Trends Cell Biol.* **17**, 403–410.

Ulmasov, T., Murfett, J., Hagen, G., and Guilfoyle, T. J. (1997). Aux/IAA proteins repress expression of reporter genes containing natural and highly active synthetic auxin response elements. *Plant Cell* **9**, 1963–1971.

Wang, H., Ngwenyama, N., Liu, Y., Walker, J. C., and Zhang, S. (2007). Stomatal development and patterning are regulated by environmentally responsive mitogen-activated protein kinases in *Arabidopsis*. *Plant Cell* **19**, 63–73.

Weijers, D., Sauer, M., Meurette, O., Friml, J., Ljung, K., Sandberg, G., Hooykaas, P., and Offringa, R. (2005). Maintenance of embryonic auxin distribution for apical-basal patterning by PIN-FORMED-dependent auxin transport in Arabidopsis. *Plant Cell,* **17**, 2517–2526.

Weijers, D., Schlereth, A., Ehrismann, J. S., Schwank, G., Kientz, M., and Jürgens, G. (2006). Auxin triggers transient local signaling for cell specification in *Arabidopsis* embryogenesis. *Dev Cell* **10**, 265–270.

Wisniewska, J., Xu, J., Seifertova, D., Brewer, P. B., Ruzicka, K., Blilou, I., Rouquie, D., Benkova, E., Scheres, B., and Friml, J. (2006). Polar PIN localization directs auxin flow in plants. *Science* **312**, 883.

Woodward, A. W., and Bartel, B. (2005). Auxin: regulation, action, and interaction. *Ann. Bot. (Lond)* **95**, 707–735.

Yang, S., Johnston, N., Talideh, E., Mitchell, S., Jeffree, C., Goodrich, J., and Ingram, G. (2008). The endosperm-specific ZHOUPI gene of Arabidopsis thaliana regulates endosperm breakdown and embryonic epidermal development. *Development* **135**, 3501–3509.

Yeung, E. C., and Meinke, D. W. (1993). Embryogenesis in angiosperms: development of the suspensor. *Plant Cell* **5**, 1371–1381.

Zhao, Y., Christensen, S. K., Fankhauser, C., Cashman, J. R., Cohen, J. D., Weigel, D., and Chory, J. (2001). A role for flavin monooxygenase-like enzymes in auxin biosynthesis. *Science* **291**, 306–309.

CHAPTER TWO

Light-Regulated Plant Growth and Development

Chitose Kami, Séverine Lorrain, Patricia Hornitschek, *and* Christian Fankhauser

Contents

Abstract

Plants are sessile and photo-autotrophic; their entire life cycle is thus strongly influenced by the ever-changing light environment. In order to sense and respond to those fluctuating conditions higher plants possess several families of photoreceptors that can monitor light from UV-B to the near infrared (far-red). The molecular nature of UV-B sensors remains unknown, red (R) and far-red (FR) light is sensed by the phytochromes (phyA–phyE in Arabidopsis) while three classes of UV-A/blue photoreceptors have been identified:

Center for Integrative Genomics, University of Lausanne, Lausanne, Switzerland

Current Topics in Developmental Biology, Volume 91
ISSN 0070-2153, DOI 10.1016/S0070-2153(10)91002-8

cryptochromes, phototropins, and members of the Zeitlupe family (cry1, cry2, phot1, phot2, ZTL, FKF1, and LKP2 in Arabidopsis). Functional specialization within photoreceptor families gave rise to members optimized for a wide range of light intensities. Genetic and photobiological studies performed in Arabidopsis have shown that these light sensors mediate numerous adaptive responses (e.g., phototropism and shade avoidance) and developmental transitions (e.g., germination and flowering). Some physiological responses are specifically triggered by a single photoreceptor but in many cases multiple light sensors ensure a coordinated response. Recent studies also provide examples of cross-talk between the responses of Arabidopsis to different external factors, in particular among light, temperature, and pathogens. Although the different photoreceptors are unrelated in structure, in many cases they trigger similar signaling mechanisms including light-regulated protein–protein interactions or light-regulated stability of several transcription factors. The breath and complexity of this topic forced us to concentrate on specific aspects of photomorphogenesis and we point the readers to recent reviews for some aspects of light-mediated signaling (e.g., transition to flowering).

1. Multiple Photoreceptors to Sense a Variety of Light Colors and Intensities

Higher plants not only transform solar energy into chemical energy through the process of photosynthesis but also use light as an informational cue to control a multitude of physiological responses throughout their life cycle. Collectively these responses are known as photomorphogenesis (Kendrick and Kronenberg, 1994). Such responses can be reversible such as stomata opening or irreversible such as seed germination. The light quality (spectral composition), quantity, direction, and duration change depending on the season, latitude (magnitude of day-length variable), and local condition (weather, position within plant communities). For instance, light under a plant canopy has a typical signature with a strong reduction of blue and red light absorbed by the photosynthetic pigments while levels of green and in particular far red (FR) light (near infra-red $\lambda = 700–750$ nm) remain relatively high (Fig. 2.1A). To sense such a diversity of light conditions, higher plants possess multiple light sensors (Chen *et al.*, 2004; Quail, 2002; Somers and Fujiwara, 2009). Plant photobiology has a long history with detailed descriptions of photomorphogenic responses dating back to the 19th century (e.g., Sage, 1992; Whippo and Hangarter, 2006). In the second half of the 20th century light responses were analyzed in many plant species using primarily physiological, photobiological, and biochemical approaches (Kendrick and Kronenberg, 1994). With the development of molecular genetics, Arabidopsis became the primary model to discover photoreceptors and signaling factors (Chen *et al.*, 2004; Quail, 2002; Somers

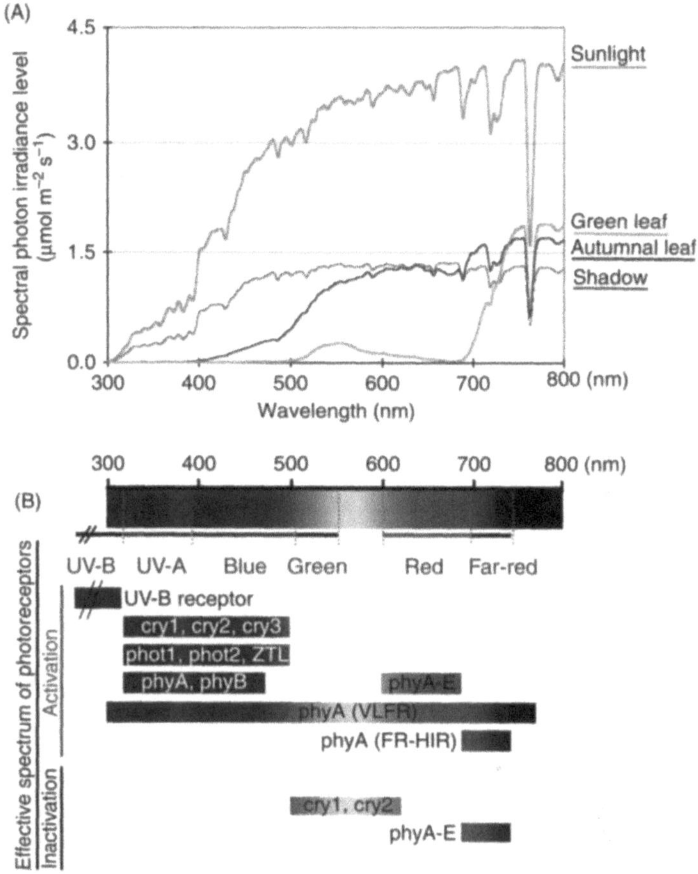

Figure 2.1 The spectral photon irradiance of natural-light environments and effective spectrum of photoreceptors. (A) All of spectra were measured by spectroradiometer (LI-1800; Li-Cor, Lincoln, NE) in Nara, Japan (May 2003, 15:00, fine weather). Unfiltered sunlight (Sunlight), the shadow of building (Shadow), sunlight transmitted through a green *Pueraria lobata* leaf and light brown *P. lobata* leaf (Autumnal leaf). (B) Effective spectrum of photoreceptors for activation and inactivation. (See Color Insert.)

and Fujiwara, 2009). In this review, we will describe how the light environment shapes most aspects of the Arabidopsis life cycle and present recent progress in light-regulated development.

Four classes of photoreceptors have been identified in Arabidopsis. These photoreceptor families are present in all sampled higher plants although the number of members in each family is somewhat variable. Arabidopsis possesses five phytochromes (phyA–phyE) maximally absorbing red and FR light (Fig. 2.1B; Franklin and Quail, 2010; Rockwell *et al.*, 2006). Three distinct classes of specific UV-A/blue light sensors are known:

cryptochromes (cry1 and cry2), phototropins (phot1 and phot2), and Zeitlupes (ZTL, FKF1, and LKP2) (Fig. 2.1B) (Christie, 2007; Demarsy and Fankhauser, 2009; Lin and Shalitin, 2003; Somers and Fujiwara, 2009). Cryptochromes that are related to DNA photolyases possess a third member in Arabidopsis known as cry3 (or cry-DASH). It is still unclear whether cry3 is a photosensory photoreceptor; however, it is involved in DNA repair mechanisms (Pokorny *et al.*, 2008). Higher plants possess a UV-B receptor of unknown molecular nature with broad roles in photomorphogenesis (Fig. 2.1B, Table 2.1; Jenkins, 2009). Finally, a number of green light responses in plants might be mediated by a yet-to-be-identified photoreceptor (Folta and Maruhnich, 2007).

Photoreceptors are chromoproteins composed of an apo-protein bound to a variety of chromophores (Christie *et al.*, 1998; Imaizumi *et al.*, 2003; Lin *et al.*, 1995; Rockwell *et al.*, 2006). The characteristic absorption spectra are determined by the chemical character of the chromophore and apo-protein. The chromophores and a simplified scheme of the light reactions of plant photoreceptors are presented in Fig. 2.2. Phytochromes are synthesized in their red light ($\lambda_{max} = 670$ nm) absorbing state known as Pr. Upon light absorption, the chromophore isomerizes leading to a FR-absorbing state ($\lambda_{max} = 730$ nm) known as Pfr. This is due to the isomerization of a double bond between the C and D rings of the tetrapyrrol (Fig. 2.2) (Rockwell *et al.*, 2006). This traditional view has however been challenged by recent nuclear magnetic resonance data of the Pr and Pfr states of a bacterial phytochrome indicating that the primary light reaction is an isomerization between the A and B rings (Ulijasz *et al.*, 2010). Whether this new finding is applicable to all phytochromes remains to be established. Phytochromes also possess a secondary absorption peak in the UV-A/blue range (Rockwell *et al.*, 2006). These spectral properties correlate with the action spectra of phytochrome-mediated responses showing maximal activity for red and FR light and also responses in the UV-A/blue range (Table 2.1). Light-induced changes of the chromophore lead to structural changes of the protein initiating photoreceptor-mediated signal transduction (e.g., Harper *et al.*, 2003; Pfeifer *et al.*, 2010). For the phototropins this primary light reaction leads to activation of the protein kinase activity (Tokutomi *et al.*, 2008). Light-regulated kinase activity has also been proposed for plant phytochromes and cryptochromes, but the physiological relevance of those activities remains to be firmly established (Bouly *et al.*, 2003; Shalitin *et al.*, 2003; Shen *et al.*, 2009; Yeh and Lagarias, 1998). By contrast, the light-induced interaction between phytochromes in their Pfr form and transcription factors of the PHYTOCHROME INTERACTING FAMILY (PIF) class is an established output of phytochrome light activation (see below) (Castillon *et al.*, 2007).

Although members of one photoreceptor family have similar absorption properties, each of the best-characterized plant photoreceptor families is,

Table 2.1 Photoreceptor-dependent responses in Arabidopsis

Physiological response	Light	Photoreceptor	Total fluence	References
Germination				
Induction of germination	Broad wavelength (320–780 nm)[a]	phyA	Very low-P	Shinomura *et al.* (1996)[b]
	Red	phyA, B	>Low-P	Shinomura *et al.* (1994), Shinomura *et al.* (1996),[b] and Botto *et al.* (1996)
		phyE	High-P	Hennig *et al.* (2002)
	Far-red	phyA	>Low-P, R:FR ratio	Shinomura *et al.* (1994), Shinomura *et al.* (1996),[b] and Botto *et al.* (1996)
		phyE	High-P	Hennig *et al.* (2002)
Inhibition of germination	Far-red	phyB	>Low-P	Shinomura *et al.* (1994)
De-etiolation				
Hypocotyl growth inhibition	UV-A & blue	cry1	>Low-C	Ahmad and Cashmore (1993) and Ahmad *et al.* (2002)[b]
		cry2	Low-C	Lin *et al.* (1998)
		phyA	High-C, >Low-P	Whitelam *et al.* (1993) and Shinomura *et al.* (2000)[b]
		phyB	>Low-C	Casal and Boccalandro (1995)
		phyC	>Low-C	Franklin *et al.* (2003b)
		phot1	>Low-C	Folta and Spalding (2001)
	Red	phyA	>Very low-P, >Low-C	Reed *et al.* (1994), Mazzella *et al.* (1997), and Shinomura *et al.* (2000)[b]

(*Continued*)

Table 2.1 (*Continued*)

Physiological response	Light	Photoreceptor	Total fluence	References
		phyB	>Low-P, >Low-C	McCormac *et al.* (1993) and Reed *et al.* (1993)
		phyC	High-C	Franklin *et al.* (2003b) and Monte *et al.* (2003)
	Far-red	phyA	High-C, High-P	Nagatani *et al.* (1993), Parks and Quail (1993), Whitelam *et al.* (1993), and Shinomura *et al.* (2000)[b]
Promotion of hypocotyl growth	Green	cry1	>Low-C	Bouly *et al.* (2007)
	Far-red	phyB	>Low-C	McCormac *et al.* (1993) and Nagatani *et al.* (1991)
Cotyledon opening	UV-A & blue	cry1	>Low-C	Casal and Boccalandro (1995)
		cry2	Low-C	Lin *et al.* (1998)
	Red	phyA	High-C	Neff and Chory (1998)
		phyB	High-C	Parks and Quail (1993)
		phyA, B, D, E	High-C	Franklin *et al.* (2003a)
		phyD	High-C	Aukerman *et al.* (1997)
	Far-red	phyA	High-C	Parks and Quail (1993)
Cotyledon expansion	UV-A and blue	cry1	High-C	Jackson and Jenkins (1995)
		cry2	High-C	Ohgishi *et al.* (2004)
		phot1,2	>Low-C	Ohgishi *et al.* (2004)
		phyA	High-C	Neff and Chory (1998)
		phyB	High-C	Neff and Van Volkenburgh (1994)

	Red	phyB	>Low-C	Neff and Van Volkenburgh (1994)
		phyC	High-C	Monte *et al.* (2003)
		phyA, D, E	High-C	Franklin *et al.* (2003a)
Anthocyanin accumulation	UV-A and blue	cry1	>Low-C	Jackson and Jenkins (1995) and Ahmad and Cashmore (1995)
		phyA, B	High-C	Poppe *et al.* (1998)
	Far-red	phyA	High-C	Kunkel *et al.* (1996)
Inhibition of gravitropism	Red	phyA	High-C	Poppe *et al.* (1996) and Robson and Smith (1996),
		phyB	High-P, High-C	Liscum and Hangarter (1993a)
	Far-red	phyA	High-P, High-C	Poppe *et al.* (1996)
Phototropism				
Hypocotyl phototropism	UV-A and blue	phot1	>Very low-P, > Low-C	Liscum and Briggs (1995) and Lascève *et al.* (1999)
		phot2	High-C	Sakai *et al.* (2001)
Modulation of phototropism	UV-A and blue	cry1, 2	>Low-C	Lascève *et al.* (1999) and Whippo and Hangarter (2003)
		phy	>Low-C	Janoudi and Poff (1992)[b]
		phyA	Low-C	Whippo and Hangerter (2004) and Lariguet and Fankhauser (2004)
		phyB, D	Low-C	Whippo and Hangerter (2004)
	Red	phy	>Low-C	Janoudi and Poff (1992)[b]
		phyA, B	>Low-P	Parks *et al.* (1996) and Janoudi *et al.* (1997)
Phototropism of inflorescence stem and petioles	UV-A and blue	phot1	>Low-C	Kagawa *et al.* (2009)
		phot2	High-C	Kagawa *et al.* (2009)

(*Continued*)

Table 2.1 (*Continued*)

Physiological response	Light	Photoreceptor	Total fluence	References
Negative root phototropism	UV-A and blue	phot1	>Low-C	Liscum and Briggs (1995) and Boccalandro *et al.* (2008)
Positive root phototropism	Red	phyA, B	High-C	Kiss *et al.* (2003)
Vegetative growth				
Leaf expansion	UV-A and blue	phot1, 2	High-C	Sakamoto and Briggs (2002)
	Red	phyA, B, C, D	Wc, R:FR ratio	Franklin *et al.* (2003b) and Devlin *et al.* (1999)
Leaf movement, positioning	UV-A and blue	phot1	>Low-C	Inoue *et al.* (2008)
	Far-red	phyB	R:FR ratio	Ballare and Scopel (1997)
Stomata development	UV-A and blue	cry1, 2	High-C	Kang *et al.* (2009)
	Red	phyB	High-C	Boccalandro *et al.* (2009) and Casson *et al.* (2009)
	Far-red	phyA	High-C	Kang *et al.* (2009)
Stomata opening	UV-A and blue	phot1, 2	High-C	Kinoshita (2001)
		cry1, 2	High-C	Mao *et al.* (2005)
	Red	phyB	High-C	Wang *et al.* (2010)
Chloroplast accumulation	UV-A and blue	phot1	High-P, >Low-C	Kagawa and Wada (2000) and Sakai *et al.* (2001)
		phot2	High-C	Sakai *et al.* (2001)
Chloroplast avoidance	UV-A and blue	phot2	High-C	Kagawa *et al.* (2001) and Jarillo *et al.* (2001)

Inhibition of petiole growth	UV-A and blue	cry1	High-C	Jackson and Jenkins (1995)
	Red	phyB	LD	Reed *et al.* (1993)
		phyA, B	SD	Devlin *et al.* (1996)
		phyC	Wc	Monte *et al.* (2003)
		phyD	SD	Devlin *et al.* (1999)
		phyE	SD	Devlin *et al.* (1998)
Promotion of petiole growth	Far-red	phyB	EOD	Nagatani *et al.* (1991)
		phyD	EOD	Devlin *et al.* (1999)
		phyE	EOD	Devlin *et al.* (1998)
Stem and internode elongation	Far-red	phyB	R:FR ratio	Nagatani *et al.* (1991)
		phyD	R:FR ratio	Devlin *et al.* (1999)
		phyA, B, E	LD	Devlin *et al.* (1998)
Root development	UV-A and blue	phot1	High-C	Galen *et al.* (2007)
	Red	phyB	High-C	Reed *et al.* (1993)
Root greening	UV-A and blue	phyA, B, cry1	High-C	Usami *et al.* (2004)
Fresh weight accumulation	UV-A and blue	phot1	>Low-C	Takemiya *et al.* (2005)
		phot2	High-C	Takemiya *et al.* (2005)
	Red	phyA, B	EOD	Devlin *et al.* (1996)

(Continued)

Table 2.1 *(Continued)*

Physiological response	Light	Photoreceptor	Total fluence	References
Control of flowering time				
Acceleration of flowering	UV-A and blue	cry1	High-C	Bagnall *et al.* (1996)
		cry2	High-C	Guo *et al.* (1998)
		ZTL	LD	Somers *et al.* (2000)
		FKF1	LD	Imaizumi *et al.* (2003)
	Far-red	phyA	low R:FR	Johnson *et al.* (1994)
Inhibition of flowering	Green	cry2	>Low-C	Banerjee *et al.* (2007)
	Red	phyB	High-C, R:FR ratio, EOD	Reed *et al.* (1993), Devlin *et al.* (1999), and Franklin *et al.* (2003a)
		phyC	Wc, LD	Balasubramanian *et al.* (2006) and Monte *et al.* (2003)
		phyD	R:FR ratio, EOD	Devlin *et al.* (1999) and Franklin *et al.* (2003a)
		phyE	R:FR ratio, EOD	Franklin *et al.* (2003a)

Summary of the photomorphogenic responses with the identified photoreceptor triggering these reactions. The type of light treatments leading to these responses is indicated. P stands for Pulse with Very low corresponding to a fluence of $10^{-6} \sim 10^{-2}$ μmol m^{-2}, Low to $10^{-2} \sim 10^{3}$ μmol m^{-2}, and High to $>10^{3}$ μmol m^{-2}. C stands for continuous light with Low corresponding to a fluence rate of $10^{-2} \sim 10^{1}$ μmol m^{-2} s^{-1} and High to $>10^{1}$ μmol m^{-2} s^{-1}. Wc = continuous white light, LD = long days (16 h white light/8 h dark), SD = short days (8 h white light/16 h dark), EOD = end-of-day treatment, which is done with FR light and leads to inactivation of the phytochromes.

[a] Multiple narrowband monochromatic lights between 320 and 780 nm were used for this experiment.

[b] Publication showing action spectra in Arabidopsis.

broadly speaking, composed of members specialized for high and low light responses. This difference in activity can be traced back to both changes of the biochemical properties of the photoreceptor and expression patterns (Christie, 2007; Demarsy and Fankhauser, 2009; Lin and Shalitin, 2003; Rockwell *et al.*, 2006). For example, phyA, cry2, and phot1 function in low light while phyB-E, cry1, and phot2 are more specialized for high light responses (Lin *et al.*, 1998; Sakai *et al.*, 2001; Smith and Whitelam, 1990). This correlates with the high levels of phyA and phot1 in etiolated seedlings allowing them to perceive minute amounts of light as the seedling grows toward the upper layers of the soil (Christie, 2007; Rockwell *et al.*, 2006). Moreover, phyA, cry2, and to a lower extent phot1 are unstable and degraded soon after activation (Clough and Viestra, 1997; Sakamoto and Briggs, 2002; Shalitin *et al.*, 2002). By contrast, a photoreceptor-like phot2 is transcriptionally induced by light, which correlates with the need for this photoreceptor under high light (Christie, 2007). Finally, there are differences in the photochemistry of the members of a given photoreceptor family. This has been extensively characterized for the phytochromes with phyA having much slower dark-reversion (thermal relaxation of the activated Pfr to the Pr ground state) (Rockwell *et al.*, 2006). Similarly for the phototropins, differences in LOV2 (light, oxygen, voltage domain) photochemistry between phot1 and phot2 may also contribute to the functional specialization of those photoreceptors (Aihara *et al.*, 2008).

The analysis of phytochrome evolution in the green lineage has shown that members of this gene family have diverged very early in the evolution of seed plants. The repertoire of phytochromes is somewhat species specific with most having at least three types (phyA, phyB, and phyC) and all seed plants sampled to date having a phyA and a phyB (Mathews, 2006). The functions and molecular properties of phyA and phyB are quite distinct, and it has been argued that phyA has evolved novel properties in order for seedlings to better cope with the higher density of large plants (Table 2.1) (Mathews, 2006; Rockwell *et al.*, 2006). Indeed, phyA has the ability to signal in FR-rich light conditions, which leads to a very low ratio of Pfr/Ptot (due to the overlapping spectra of Pfr and Pr even canopy light leads to some Pfr production). This ability of phyA allows seedling establishment (de-etiolation) under dense plant cover and is thus a competitive advantage for angiosperms, which all possess a phyA (Mathews, 2006). Studies comparing Arabidopsis accessions have also demonstrated that the phytochromes are subjects of natural variation that may enhance the fitness of those plants in diverse environments (Balasubramanian *et al.*, 2006; Filiault *et al.*, 2008; Samis *et al.*, 2008). Thus, plants can respond to a wide range of distinct and variable light conditions and natural genetic variation at the level of the photoreceptors participates to their adaptation to their environment.

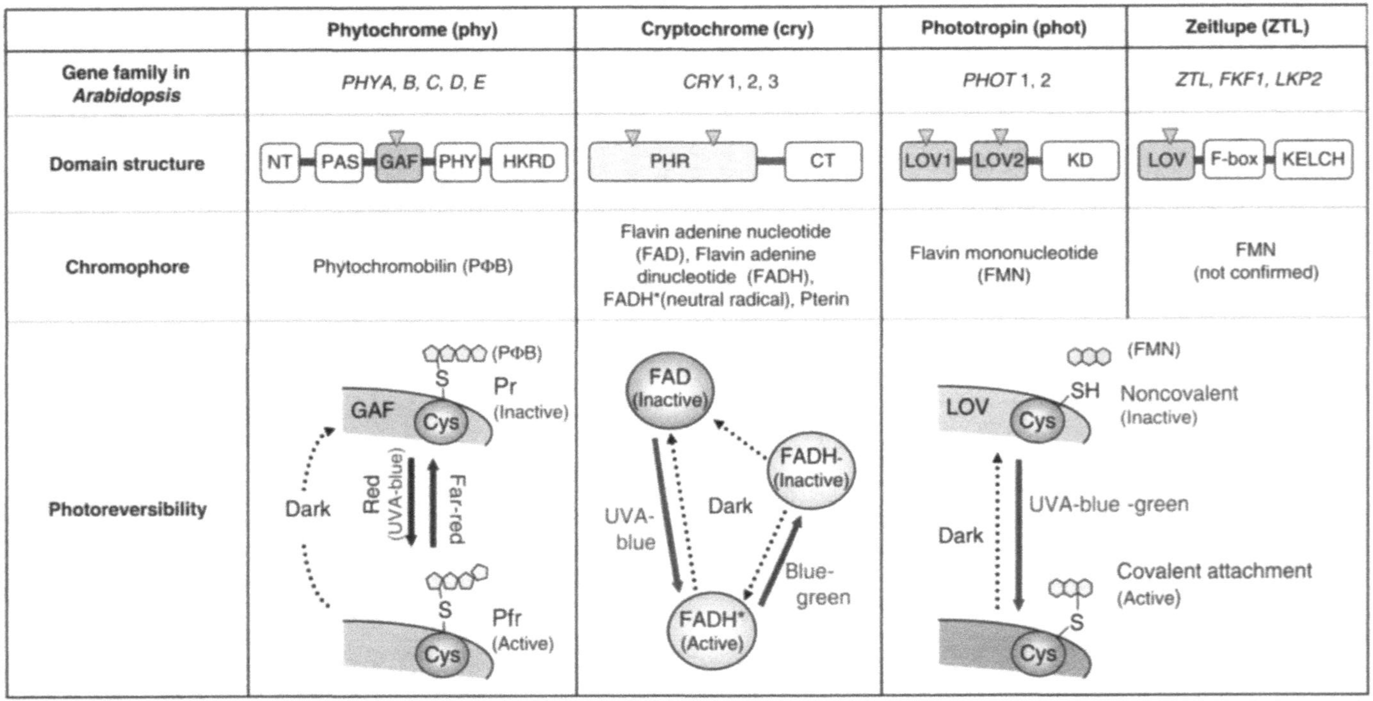

Figure 2.2 *(Continued)*

2. Physiological Responses Mediated by Plant Photoreceptors

Photoreceptors modulate plant growth and development throughout their life cycle; moreover, by monitoring the light environment they contribute to the timing of key developmental transitions such as seed germination and initiation of flowering (Fig. 2.3). We have summarized this information in Table 2.1, which lists the light-dependent physiological responses for which the photoreceptor has been determined. We will describe many of these responses quite briefly and would like to point the readers to more specific reviews in particular for photoperiodic induction of flowering (Imaizumi *et al.*, 2006; Kobayashi and Weigel, 2007; Turck *et al.*, 2008). Traditionally, light responses are subdivided depending on the amount of light needed and whether the response requires continuous irradiation or whether it is efficiently triggered by light pulses (Kendrick and Kronenberg, 1994). The so-called very low fluence responses (VLFRs) are initiated in response to as little as

---

Figure 2.2 Primary light reactions in the different classes of plant photoreceptors. Arabidopsis has five phytochrome-encoding genes (*PHYA–E*), three cryptochrome genes (*CRY1-3*), two phototropin genes (*PHOT1* and *PHOT2*), and three Zeitlupe family genes (*ZTL, FKF1, LKP2*). The protein domain organization of the different photoreceptors is schematized with the position of chromophore attachment marked with an arrowhead. Phytochromes have an N-terminal extension of unknown fold (NT) followed by a PAS (Per, ARNT, Sim) domain, a GAF (cGMP phosphodiesterase/adenyl cyclase/FhlA) domain that binds the chromophore, a PHY domain (related to PAS domains), and a C-terminus that is composed of two PAS domains and a histidine-kinase-related domain (HKRD). Cryptochromes have a photolyase homology region (PHR) and a C-terminus of unknown structure (CT). Phototropins are composed of two LOV (light, oxygen, voltage) domains in their N-terminus (LOV1 and LOV2) and a Ser/Thr protein kinase domain (KD). Members of the ZTL family have an N-terminal LOV domain, an F-box, and KELCH repeats. Phytochromes have phytochromobilin (PΦB) as a chromophore that is covalently bound to an invariant Cys residue in a GAF domain and photoreversibly switches between the Pr and the Pfr conformers upon isomerization of a double bond between the (A) and (B) rings of the tetrapyrrol. Cryptochromes have two chromophores; flavin adenine nucleotide (FAD) and a pterin acting as an antenna pigment. The light reactions from FAD to flavin adenine dinucleotide ($FADH^-$) or neutral radical form of FADH (FADH*) are depicted on the figure (adapted from Bouly et al. (2007)). Phototropins use flavin mononucleotide (FMN) as a chromophore. In darkness, each of the LOV domains noncovalently binds to FMN. After absorbing UV-A/blue light, an invariant Cys in the LOV domain covalently binds to FMN. This activated state rapidly returns to the dark state. Zeitlupe family light sensors also have a LOV photosensory domain. In contrast to the phototropins, these LOV domains remain in the light-activated state for a long time (hours).

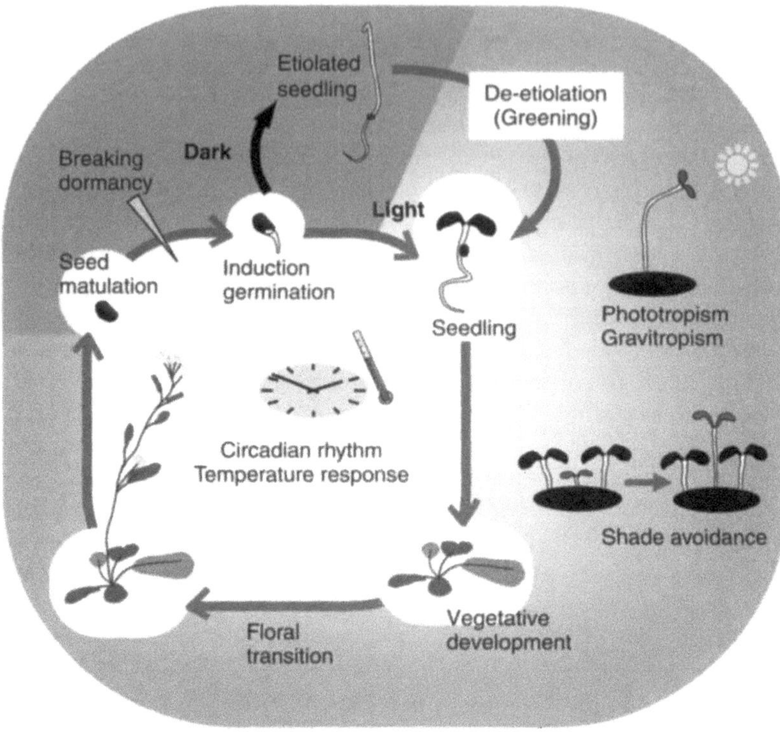

Figure 2.3 Photomorphogenesis in Arabidopsis. After germination, the seedling undergoes etiolated development in darkness or develops as a photosynthetically active seedling in the light. As the etiolated seedling emerges from the soil it will undergo de-etiolation. Light affects growth and development throughout the life cycle of plants.

100 pmol m^{-2}. Low fluence responses (LFRs) occur in the range of 10–1000 μmol m^{-2} while high irradiance responses (HIRs) require continuous light with a total fluence typically in excess to 10 mmol m^{-2}. This classification has been particularly useful to characterize different types of phytochrome-mediated responses (Rockwell *et al.*, 2006; Shinomura *et al.*, 1996; 2000). PhyA is the only phytochrome mediating both the VLFRs and the FR-HIRs (Shinomura *et al.*, 1996; 2000). In addition, all phytochromes are also capable of mediating the classic R/FR-reversible LFR (Rockwell *et al.*, 2006). The exact difference in signaling of the phytochromes under these different light conditions is not fully understood; however, a number of studies have indicated that phyA uses at least partially distinct signaling mechanisms when acting in the VLFRs and FR-HIRs (Casal *et al.*, 2000; Kneissl *et al.*, 2009; Lariguet *et al.*, 2003; Staneloni *et al.*, 2009)

2.1. Germination

In the seed, the embryo is in a much protected environment while the young seedling is very vulnerable. Germination is thus under strong environmental control, including effects of water, oxygen, temperature, and light (Penfield and King, 2009). Seeds only become competent to respond to those environmental signals once they have broken dormancy (Penfield and King, 2009). Among all photoreceptors the phytochromes play the most predominant role to promote germination under favorable light conditions and to prevent it when the light conditions are suboptimal as, for example, under a canopy (Franklin and Quail, 2010). In order to promote germination the phytochromes primarily act on gibberellin (GA) synthesis and signaling (Oh *et al.*, 2007; Piskurewicz *et al.*, 2009). The study of germination illustrates the distinct roles of phyA and phyB, with phyA acting as a broadband sensor for low light and phyB controlling R/FR-reversible induction of germination (Shinomura *et al.*, 1996). In addition, phyD and phyE have also been shown to promote germination (Dechaine *et al.*, 2009; Hennig *et al.*, 2002). It should be pointed out that the germination potential of seeds is conditioned by light-stable phytochrome that is transmitted in the dry seed in a Pfr/Ptot ratio depending on the growth conditions of the mother plant (Casal and Sánchez, 1998). By contrast, phyA is only synthesized during seed imbibition (Casal and Sánchez, 1998; Shinomura *et al.*, 1996). This may explain the recent findings showing that the phytochromes in the mother plant also condition the germination potential of their seeds by modulating seed maturation and dormancy (Dechaine *et al.*, 2009; Donohue *et al.*, 2008; Heschel *et al.*, 2007). Finally, as discussed in a later section, the predominance of a given phytochrome to control germination is modulated by temperature.

2.2. Young seedling development

Following germination the young seedling may encounter a variety of environments. The most extreme case is darkness, and higher plants have evolved a strategy to survive for a few days in this situation by living from their seed reserves (etiolated development). This developmental strategy is characterized by fast elongation of the hypocotyl growing against the gravity vector, maintenance of an apical hook, inhibition of cotyledon expansion, and inhibition of leaf initiation (Chen *et al.*, 2004). This strategy maximizes the chances of the seedling to rapidly reach the soil surface where the process of de-etiolation will be initiated. Phytochromes and cryptochromes predominantly control the de-etiolation phase by inhibiting hypocotyl elongation, initiating chloroplast development, promoting cotyledon expansion, and initiating leaf growth to enable the seedling to start its photosynthetic life (Chen *et al.*, 2004). The predominant photoreceptor controlling de-etiolation depends on the light environment, with phyA single-handedly mediating this response in FR light (as encountered under a canopy), whereas multiple phytochromes are involved in red light (with phyB playing a predominant role) and the cryptochromes with contributions from the phytochromes in blue light (Chen *et al.*, 2004; Franklin and Quail, 2010; Lin and Shalitin, 2003; Quail, 2002). UV-B also mediates this developmental transition and does not require any of the known photoreceptors. This UV-B response critically depends on UV RESISTANCE LOCUS8 (UVRB), the ubiquitin E3 ligase CONSTITUTIVE PHOTOMORPHOGENIC1 (COP1), and the transcription factor ELONGATED HYPOCOTYL5 (HY5; Jenkins, 2009). The phototropins also play an important role during this phase of development because once emerging from the soil the seedling might have to direct its growth toward a more favorable place (with better light) (Christie, 2007). The phototropins mediate a fast but transient inhibition of hypocotyl elongation and later are crucial to guide growth toward a directional light cue (Christie, 2007; Folta and Spalding, 2001). Although directional growth depends on the phototropins, the magnitude of this response is modulated by both the phytochromes and the cryptochromes (Nagashima *et al.*, 2008). By modulating the hypocotyl growth rate and inhibiting the hypocotyl gravitropic response, the latter two classes of photoreceptors affect the phototropic potential of young seedlings (Iino, 2006; Lariguet and Fankhauser, 2004; Whippo and Hangarter, 2003; 2004).

2.3. Vegetative development

Broadly speaking, throughout vegetative development, the different photoreceptors allow the optimization of photosynthesis according to the prevalent light conditions (Table 2.1). The shade avoidance response (SAR) is a good illustration of this concept (Franklin, 2008; Franklin and Quail, 2010; Vandenbussche *et al.*, 2005). Shade from the vegetation has a distinct spectral signature (Fig. 2.1), which has a strong influence on the phytochrome photoequilibrium due to alterations of the R:FR ratio (Franklin, 2008). Moreover, the green : blue ratio is also modified, which is predicted to alter the proportion of the cryptochromes in their signaling state (Figs. 2.1 and 2.2; Banerjee *et al.*, 2007; Bouly *et al.*, 2007). A plant response related to shade avoidance is neighbor detection, which occurs in response to a decrease of the R:FR ratio due to the reflexion of FR light from neighboring plants but does not necessarily lead to a reduction of photosynthetically active radiation (PAR) (Franklin, 2008). In shade-intolerant plants such as Arabidopsis, this reduction in the R:FR ratio has a number of striking effects on plant growth and development. The SAR is characterized by increased hypocotyl, stem, and petiole elongation; a more erect leaf position; increased apical dominance; and early flowering (Franklin, 2008;

Vandenbussche *et al.*, 2005). As for the other light responses discussed above, these morphological changes are accompanied by rapid and extensive alterations of the gene expression profile (Franklin, 2008; Vandenbussche *et al.*, 2005). The light-stable phytochromes (phyB–phyE) with phyB playing a prevalent function repress the SAR in direct sunlight (Franklin, 2008; Franklin and Quail, 2010). By contrast, phyA due to its ability to limit hypocotyl growth in FR light counteracts the activity of the other phytochromes. However, the light-labile nature of phyA limits its role in de-etiolated plants (Franklin and Quail, 2010; Salter *et al.*, 2003). A role for the cryptochromes in this adaptive response has also been demonstrated; this can be explained by the overall reduction of blue light and the low blue : green ratio under a canopy, which inhibits cryptochromes activity (Banerjee *et al.*, 2007; Bouly *et al.*, 2007; Vandenbussche *et al.*, 2005; Yanovsky *et al.*, 1995). Upward positioning of leaves, which is typical of the SAR (leaf epinasty), is also triggered by low-light environments and has been shown to depend on the combined action of phyA, phyB, cry1, and cry2 (Millenaar *et al.*, 2009; Vandenbussche *et al.*, 2005). This growth response, which presumably depends on asymmetric growth of the two sides of the petioles, is also under the control of the phototropins (Inoue *et al.*, 2008b). Interestingly, the phototropins also ensure that leaves are flat, which, combined with their effect on leaf and stem positioning, maximizes the photosynthetic potential in low-light environments (Christie, 2007; Inoue *et al.*, 2008b; Takemiya *et al.*, 2005).

Controlled gas exchange by the stomata represents another physiological parameter that is directly related to photosynthetic capacity. A tight regulation of this activity is essential to properly control CO_2 uptake, release of O_2, plant cooling, and water loss depending on the external conditions (Casson and Hetherington, 2010). It has long been known that blue light promotes stomata opening. This response primarily depends on phot1 and phot2 with cry1 and cry2 playing a more limited role (Kinoshita *et al.*, 2001; Mao *et al.*, 2005). While the relatively weak effect of red light on stomata opening was regarded as a photosynthetic effect, recent evidence suggests a role for the phytochromes in this response (Sharkey and Raschke, 1981; Wang *et al.*, 2010). In addition to light's short-term effect on stomata opening, several photoreceptors have also been shown to control the stomata index (ratio of guard cells over total cells). A decrease in both irradiance and the R:FR ratio typical of shading is perceived in mature leaves and will decrease the stomata index in the newly emerging leaves (Boccalandro *et al.*, 2009; Casson *et al.*, 2009; Lake *et al.*, 2001). By analyzing stomatal pattern in monochromatic lights and in different mutant backgrounds, it was shown that phyA, phyB, and cryptochromes are the principal regulators of stomata formation and pattern in FR, red, and blue lights, respectively (Boccalandro *et al.*, 2009; Casson *et al.*, 2009; Kang *et al.*, 2009). Interestingly, this developmental response involves elements of light signaling that were initially identified in the context of seedling de-etiolation such as COP1 and PIF4 (Casson *et al.*, 2009; Kang *et al.*, 2009). However, while *pif4* and *phyB* have opposite phenotypes during seedling de-etiolation in red light (see below), their stomata development phenotypes are similar suggesting a developmentally regulated mode of interaction of these two factors.

Also of great importance for the optimization of photosynthesis is a proper control of chloroplast positioning (Suetsugu and Wada, 2007). In low-light conditions, the chloroplasts display an accumulation response toward the source of blue light presumably in order to maximize light capture. This response is controlled by phot1 and phot2 (Sakai *et al.*, 2001). In contrast, high-light conditions that are typical of a sunny day trigger the phot2-mediated chloroplast avoidance response preventing photooxydative damage and allowing plant survival (Jarillo *et al.*, 2001; Kagawa *et al.*, 2001; Kasahara *et al.*, 2002). As for other responses primarily controlled by the phototropins the phytochromes modulate the extent of the response (DeBlasio *et al.*, 2003). Moreover, both the phytochromes and the cryptochromes play a central role in the development of etioplasts into chloroplasts (Rockwell *et al.*, 2006; Ruckle *et al.*, 2007).

2.4. Transition to flowering

Both light and temperature influence the transition from vegetative to reproductive growth. The phytochromes, cryptochromes, and members of the Zeitlupe family all contribute to this complex regulatory network, either by acting directly on key regulators of floral transition such as CONSTANS (CO) or by modulating the circadian clock, which has a profound influence on photoperiodic flowering. A detailed description of daylength-regulated flowering is beyond the scope of this review; we recommend the following recent review articles: Imaizumi and Kay, 2006; Kobayashi and Weigel, 2007; Turck *et al.*, 2008. Light quality and in particular shading also leads to accelerated flowering (Franklin, 2008; Vandenbussche *et al.*, 2005). The early flowering in short days of mutants such as *phyB* presumably reflects the constitutive shade avoidance phenotype of these plants (Franklin, 2008; Franklin and Quail, 2010). This specific aspect of the SAR requires PHYTOCHROME AND FLOWERING TIME (PFT1) a component of the mediator complex that has global roles in transcriptional regulation by acting as an adaptor between transcription factors and RNA polymerase II (Backstrom *et al.*, 2007; Cerdan and Chory, 2003). Interestingly, PFT1 has also been implicated in disease resistance providing an example of the possible crosstalk between different environmental factors affecting plant development (Kidd *et al.*, 2009 (see below).

3. Photomorphogenesis in a Changing Environment

Phenotypes developed by plants in response to changes in light quality/quantity can also be observed in other situations. For instance, the shade-avoidance phenotype described above is similar to the phenotype of plants grown under high temperatures (28°C vs. 22°C) or in flood conditions (Franklin, 2008; 2009; Koini *et al.*, 2009; Millenaar *et al.*, 2005; Pierik *et al.*, 2005; van Zanten *et al.*, 2009; Vandenbussche *et al.*, 2003). Taking the leaves away from the warm/submerged soil presumably optimizes plant fitness by preserving their "power supply" (leaves). More surprising was the identification of a gene-controlling shade avoidance that encodes a TIR-NBS-LRR protein typically involved in plant–pathogen interaction (Faigon-Soverna *et al.*, 2006). Altogether these data suggest that multiple signaling pathways may control the same "core-genetic program" to modulate plant growth according to the environmental cues. This could be achieved through independent and parallel signaling pathways or by using shared signaling components. Several pieces of evidence suggest that common regulators of growth are used in a variety of conditions. For instance, the DELLA proteins that negatively regulate GA pathways are involved in repression of growth responses to shade, pathogen, salt stress, or cold treatment (Achard *et al.*, 2006; 2008a; Achard *et al.*, 2008b; Navarro *et al.*, 2008). Similarly, PIF4 is involved in elongation-growth responses in a variety of light responses and also in response to increased temperatures (Huq and Quail, 2002; Koini *et al.*, 2009; Lorrain *et al.*, 2008; 2009; Stavang *et al.*, 2009). The variety of conditions in which PIF4 modulates growth is paralleled by a great complexity of PIF4 regulation. The control of PIF4 activity includes interaction with the DELLA proteins and LONG HYPOCOTYL in FR (HFR1) to prevent it from binding to DNA, transcriptional regulation by the clock and temperature, and proteolytic degradation upon interaction with the phytochromes (de Lucas *et al.*, 2008; Feng *et al.*, 2008; Hornitschek *et al.*, 2009; Koini *et al.*, 2009; Nozue *et al.*, 2007; Stavang *et al.*, 2009). Although PIF4 was identified as a light-signaling component, it also acts independently of photoreceptor activity indicating that light-signaling components also work in other pathways (Leivar *et al.*, 2008b; Koini *et al.*, 2009). As a consequence, light sensing can influence responses induced by different stimuli and vice versa. Examples of such interactions are presented in the following sections.

3.1. Light sensing and pathogen defense

Light is required to mount an efficient response to pathogens in Arabidopsis (Chandra-Shekara *et al.*, 2006; Genoud *et al.*, 2002; Griebel and Zeier,

2008; Roden and Ingle, 2009; Zeier *et al.*, 2004). Especially, light is important during the first hours following inoculation as dark-infected plants support more bacterial growth and mount less-efficient defense mechanisms than light-infected plants (Griebel and Zeier, 2008). Interestingly, the appearance of spontaneous necrotic lesions in the so-called lesion-mimic mutants is also light dependent (Brodersen *et al.*, 2002; Dietrich *et al.*, 1994; Lorrain *et al.*, 2004; Lu *et al.*, 2003; Mach *et al.*, 2001). This may reflect the requirement of photosynthetic energy for an efficient resistance or a crosstalk between chloroplast-derived molecules and defense mechanisms (Genoud *et al.*, 2002). In few cases participation of the photoreceptors in the defense mechanisms was studied (Chandra-Shekara *et al.*, 2006; Genoud *et al.*, 2002; Griebel and Zeier, 2008; Wu and Yang, 2010). PhyA, phyB, and cry1 are involved in the resistance response to *Pseudomonas syringae pv. tomato* carrying the *AvrRpt2* gene while they do not seem to play any role in response to *P. syringae pv. maculicola* (*AvrRpm1*) or to Turnip crinkle virus except in the set up of the systemic acquired resistance (Chandra-Shekara *et al.*, 2006; Genoud *et al.*, 2002; Griebel and Zeier, 2008; Wu and Yang, 2010). This does not necessarily mean that light is differentially required in response to various pathogens but rather that different mechanisms are involved in response to different pathogens. In both cases, light seems to act through salicylic acid perception/signaling.

The response of plants to pathogens and unfavorable light conditions may also compete with each other. This situation is encountered when plants are grown close to each other, as in agriculture. Plants competing for light resources may at the same time face pathogens such as attack by herbivores. Carbon resources have to be reallocated either to growth responses to reach the light or to defense mechanisms against pathogens: this is the "plant dilemma" (Ballare, 2009; Roberts and Paul, 2006). Priority is given to shade avoidance as it has been shown that shade increases the leaf area eaten by herbivores/herbivore fitness or the leaf area infected by pathogens (Izaguirre *et al.*, 2006; Moreno *et al.*, 2009; Roberts and Paul, 2006). This is not a passive phenomenon due to the modified plant architecture under shade conditions. Indeed the response to pathogens is also reduced by shade in *sav3* although this mutant does not respond to low R:FR by elongation growth (Moreno *et al.*, 2009). Shade conditions inhibit biosynthetic pathways linked to the production of defense molecules such as phenolic compounds, a response that is phytochrome dependent in tomato (Izaguirre *et al.*, 2006; Moreno *et al.*, 2009). One hypothesis is that low R:FR ratio reduces plant sensitivity to jasmonates (JA) (Moreno *et al.*, 2009). These phytohormones not only are important for defense against herbivores but were also shown to inhibit cell division and elongation (Bari and Jones, 2009; Zhang and Turner, 2008). Thus by reducing JA action, shade may weaken defense mechanisms but allow full growth of the plants (Moreno *et al.*, 2009).

3.2. Crosstalk between light and temperature

Important development transitions such as germination and flowering need to be synchronized with the environment to maximize the chances of species survival. For instance, most Arabidopsis accessions are typical "winter annual" that germinate in the autumn, spend all the winter as rosette, and flower in the spring. In addition to light, a cold period called vernalization is required to induce flowering. Furthermore, if light is necessary to induce germination in fresh seeds, it is not sufficient: a cold and wet treatment called stratification is required to break seed dormancy. These mechanisms are believed to prevent seed maturation and germination during the winter, a period that can be deleterious for the newly developing organisms. Phytochrome functions are well described in these mechanisms (Franklin, 2009; Franklin and Whitelam, 2004; Penfield, 2008; Samach and Wigge, 2005; Seo *et al.*, 2009). Depending on the temperature, the different phytochromes perform different functions (Dechaine *et al.*, 2009; Donohue *et al.*, 2008; 2007; Franklin, 2009; Halliday and Whitelam, 2003; Heschel *et al.*, 2007). For instance, while phyB plays a major role in germination under a wide range of temperatures, phyE and phyA are particularly important at cooler and warmer temperatures, respectively (Heschel *et al.*, 2007). Phytochrome function is not limited to germination: they are also involved in the perception of the temperature and light environment of the maternal plant during seed maturation (Dechaine *et al.*, 2009; Donohue *et al.*, 2007; 2008). Indeed, seeds from plants that experienced a cool environment are heavier but germinate less efficiently than those maturated under a warm environment (Dechaine *et al.*, 2009). PhyA prevents germination/favors dormancy when seeds are matured under a cold and shaded environment while phyB is important to promote germination under cold environments (Dechaine *et al.*, 2009). Integration of light- and cold-induced pathways will thus allow plants to anticipate changes in their environment. Another good example is the freezing tolerance conferred to Arabidopsis by low R:FR ratio when plants are grown at 16°C (Franklin and Whitelam, 2007). This is associated with the fact that plants experience longer periods of low R:FR ratio as winter approaches. This tolerance is conferred by the increased expression at 16°C of CRT/DRE-binding factors and COLD-REGULATED (COR) genes in low R:FR compared with high R:FR and depends on the inactivation of phyB and phyD (Franklin and Whitelam, 2007). As a consequence, *phyD* mutant grown in high R:FR at 16°C presents a higher expression of *COR15a* gene compared with the wild type, which is correlated with a higher freezing tolerance (Franklin and Whitelam, 2007).

The way phytochromes and temperature-signaling pathways interact is not well known yet. Temperature can influence phytochrome photoequilibrium, which may explain why the phytochromes have different roles depending on the temperature (Kristie and Fielding, 1994; Pons, 1986). The two signals can also share signaling components such as PIF4 (Koini *et al.*, 2009; Stavang *et al.*, 2009). Interestingly, temperature changes lead to alterations of the nucleosome composition (Kumar and Wigge, 2010). Thus temperature changes can modify the expression of different signaling components such as PIF4 whose activity could further be modulated by the phytochromes. Interestingly, different signaling pathways are involved in response to temperature: decreasing temperature can suppress the *phyB* early-flowering phenotype but not *phyB*-induced elongation responses (Halliday *et al.*, 2003). This suggests a developmentally regulated crosstalk between light and temperature pathways.

4. Sites of Perception and Action of a Light Signal

4.1. Tissue-specific considerations

Photoreceptors are present in most if not all plant tissues (Goosey *et al.*, 1997; Sakamoto and Briggs, 2002); however, light responses are tissue specific as illustrated during seedling de-etiolation when light inhibits growth in the hypocotyl while promoting it in the cotyledons. Such organ-specific light responses can also be observed at the gene expression level (Lopez-Juez *et al.*, 2008; Ma *et al.*, 2005). This could reflect tissue and/or developmentally regulated expression of signaling components. Furthermore, photoreceptors in one organ/tissue can influence responses in distant parts of the plant in a noncell-autonomous way. The best-known example of such systemic signal induced by light is "florigen" that is induced in the leaves and moves to apical meristems to regulate the formation of flowers (Kobayashi and Weigel, 2007; Turck *et al.*, 2008). Other examples include irradiation of mature leaves by low light, which leads to decreased stomatal index in newly emerging leaves that grow under high irradiance (Lake *et al.*, 2001). More recently, using tissue-specific phytochrome inactivation, Warnasooriya *et al.* showed that a FR light signal perceived by the cotyledons controls hypocotyl elongation (Warnasooriya and Montgomery, 2009). This complements the data of Tanaka *et al.* (2002) where FR light irradiation of the cotyledons induced the specific expression of a reporter gene in the hypocotyl while irradiation of the hypocotyl did not. This indicates that systemic signals are rather prevalent in the control of plant responses to light. We recommend two recent reviews extensively covering this aspect of light signaling (Bou-Torrent *et al.*, 2008; Montgomery, 2008).

4.2. Sites of action of the photoreceptors within the cells

The cellular site of action of several photoreceptors has received considerable attention over the past few years and has been recently reviewed for the phytochromes (Chen, 2008; Fankhauser and Chen, 2008). The bulk of all phytochromes is cytoplasmic in their inactive Pr state. However, light triggers rapid translocation of the activated phytochromes into the nucleus. Interestingly, only phyA has the ability to enter the nucleus under light conditions triggering a very low Pfr/Ptot ratio correlating with the ability of phyA to trigger the FR-HIRs and the VLFRs (Fankhauser and Chen, 2008). This special property is enabled by a dedicated phyA nuclear import system depending on FHY1 and FHL, two related proteins that interact with the light-activated phyA (Genoud *et al.*, 2008; Hiltbrunner *et al.*, 2006; Pfeiffer *et al.*, 2009; Rosler *et al.*, 2007). In addition, the transcription of those phyA importers depends on two transposase-derived transcription factors, which therefore indirectly control phyA nuclear accumulation (Lin *et al.*, 2007). Two recent publications indicate that FHY1 and FHL may play additional functions in addition to regulating phyA nuclear import (Shen *et al.*, 2009; Yang *et al.*, 2009). The light regulation of phyB nuclear import appears to depend on the unmasking of a nuclear localization signal (NLS) specifically in the light-activated Pfr conformation (Chen, 2008; Chen *et al.*, 2005). The phenotypic analysis of a variety of mutants (e.g., *fhy1fhl*) and transgenic lines expressing phyA or phyB fused to nuclear export signal (NES) or NLS sequences in phytochrome mutant backgrounds has determined that the major site of action of the phytochromes is the nucleus (Genoud *et al.*, 2008; Huq *et al.*, 2003; Matsushita *et al.*, 2003; Rosler *et al.*, 2007). This fits well with the rapid phytochrome-dependent effects on gene regulation and with their direct control of the activity of transcription factors of the PIF family (see below) (Castillon *et al.*, 2007; Leivar *et al.*, 2009; Tepperman *et al.*, 2006). However, phyA in particular may also have roles in the cytoplasm (Rosler *et al.*, 2007).

While for the phytochromes light-induced nuclear import is an important regulatory mechanism, both cry1 and cry2 are already nuclear in dark-grown seedlings and blue light does not alter the localization of those photoreceptors (Wu and Spalding, 2007). Cry1 is found in both the nucleus and the cytoplasm contrasting with the exclusively nuclear localization of cry2 (Kleiner *et al.*, 1999; Lin and Shalitin, 2003; Wu and Spalding, 2007). Complementation studies with GFP-cry1 fused either to an NES or an NLS were used to determine that the major site of cry1 action is the nucleus (Wu and Spalding, 2007). However, this study highlighted that different subcellular pools of cry1 have different functions with, for example, cytoplasmic cry1 promoting cotyledon expansion in blue light (Wu and Spalding, 2007).

In contrast to the aforementioned photoreceptors, the phototropins are primarily found at the plasma membrane (Kong *et al.*, 2006; Wan *et al.*,

2008). However, a fraction of both phot1 and phot2 leaves the plasma membrane upon blue light perception. Phot1 relocalizes to the cytoplasm while phot2 is found on the Golgi (Kong *et al.*, 2006; Wan *et al.*, 2008). The significance of this light-induced relocalization remains poorly understood. Interestingly, however, phyA modulates the blue-light-induced phot1 relocalization, and it has been proposed that this activity explains the phytochrome enhancement of phototropism (Han *et al.*, 2008). This model fits with the previously proposed role of phyA in the cytoplasm to promote phototropism (Rosler *et al.*, 2007).

The subcellular localization and cellular site of action of the members of the Zeitlupe family (ZTL, FKF1, and LKP2) have been analyzed less extensively (Somers and Fujiwara, 2009). Nevertheless, the currently available data on FKF1 and ZTL reveal some interesting differences between these two related photoreceptors (Kim *et al.*, 2007; Sawa *et al.*, 2007). ZTL controls the stability of the central circadian clock component TIMING OF CAB1 EXPRESSION1 (TOC1) and the related protein PSEUDO RESPONSE REGULATORS (PRR5) by interacting with those proteins in a time-of-day-specific manner in the cytoplasm (Kiba *et al.*, 2007; Kim *et al.*, 2007). In contrast, FKF1 appears to be nuclear where this F-box protein controls the stability of a repressor of *CO* expression thereby contributing to daylength-regulated flowering (Fornara *et al.*, 2009; Sawa *et al.*, 2007). The precise function of LKP2 is less well understood but its localization is also nuclear (Yasuhara *et al.*, 2004).

5. Signal Transduction

A large number of signaling components acting downstream of the photoreceptors have been identified. However, in many cases the events leading from photoreceptor activation to the specific function of a given signaling component is poorly understood. We will thus not attempt an exhaustive coverage of photoreceptor-mediated signal transduction, but rather focus on specific facets of the better understood signaling events. The analysis of these signaling pathways reveals a number of communalities such as the importance of light-regulated protein–protein interaction, protein stability, and kinase activity that are briefly reviewed below. Finally, we will present the PIF branch of phytochrome signaling in more detail.

5.1. Light-regulated degradation, protein–protein interactions, and kinase activity

The importance of light-regulated protein abundance was revealed with the identification of COP1, a ubiquitin E3 ligase that is required to maintain the de-etiolation program in the dark (reviewed in Jiao *et al.*, 2007; Yi and

Deng, 2005). COP1 in conjunction with members of the SPA family control the abundance of several light-signaling components, including HY5, LONG AFTER FR LIGHT (LAF1), HFR1, CO, and phyA (Jang *et al.*, 2008; Jiao *et al.*, 2007; Zhu *et al.*, 2008). This list includes numerous transcription factors that need to be maintained at low levels in the dark and accumulate in the light to mediate multiple aspects of photomorphogenesis (Fig. 2.4). Both the phytochromes and the cryptochromes inhibit COP1 activity, but the underlying mechanism is still not fully understood (reviewed in Jiao *et al.*, 2007). The cryptochromes directly interact with COP1 both in the light and in the dark and a yet-to-be-identified light-induced event in the cryptochromes leads to the inhibition of COP1 (Wang *et al.*, 2001; Yang *et al.*, 2001). How the phytochrome limits COP1 activity remains an open question but it is worth pointing out that by inhibiting COP1 and leading to the degradation of several PIFs (see below) the phytochromes profoundly influence transcription factor composition in a light-dependent fashion (Fig. 2.4; Jiao *et al.*, 2007). Importantly, COP1 and HY5 also play essential roles during UV-B signaling. These exciting new developments are beyond the scope of this review and we point the interested readers to these excellent recent publications (Favory *et al.*, 2009; Jenkins, 2009). Thus with the exception of phototropin signaling, all other plant photoreceptors described here mediate light responses at least in part by controlling the abundance of key regulators in a light-dependent fashion. Interestingly, NON-PHOTOTROPIC HYPOCOTYL 3 (NPH3), a central component of phototropin signaling, codes for a BTB/POZ-containing protein (Pedmale and Liscum, 2007). Such protein domains typically interact with cullin3 leading to the speculation that regulated proteolysis may also take center stage during phototropin signaling.

Light-regulated protein–protein interactions are also a recurrent theme in light signaling, as exemplified by the interaction between phytochromes and the PIFs (see below) and during ZTL and FKF1 signaling (Somers and Fujiwara, 2009). A light-regulated interaction between photoactivated cry2 and the bHLH protein CRYPTOCHROME-INTERACTING BHLH1 (CIB1) has also recently been described (Liu *et al.*, 2008). This interaction requires the cryptochrome chromophore and blue light. CIB1 and related bHLHs play a role in cryptochrome-regulated flowering; however, they do not regulate seedling de-etiolation (Liu *et al.*, 2008). Although more work is needed to fully understand the consequences of this interaction and the role of CIB1 and its family members, this work suggests an analogous signaling mechanism for the phytochromes and the cryptochromes as both photoreceptor families display light-regulated interactions with bHLH factors. UV-B signaling provides another striking example of light-regulated protein–protein interactions with the discovery that the UV-B-specific signaling component UVR8 interacts with COP1 in a UV-B-dependent manner (Favory *et al.*, 2009).

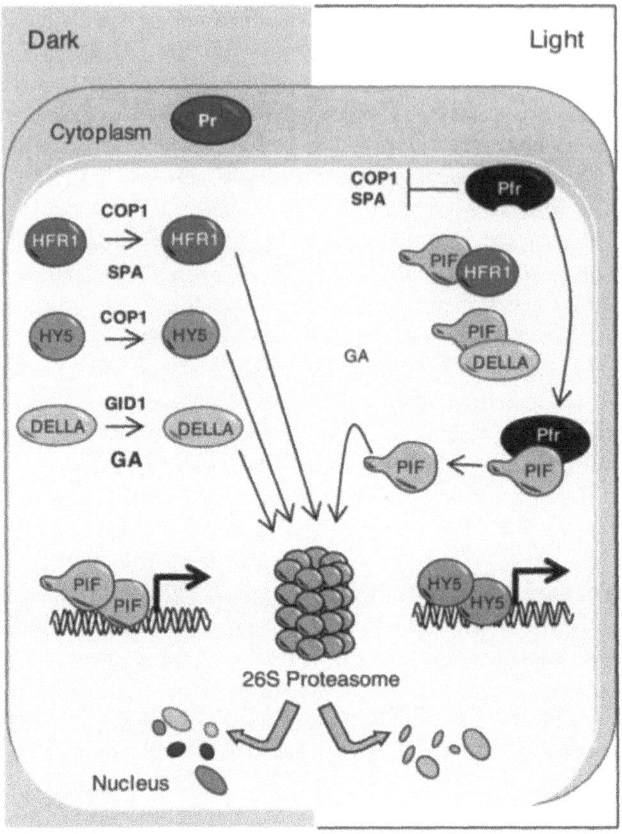

Figure 2.4 Phytochromes modulate the activity of multiple transcription factors. In the dark, the phytochromes are present in the cytoplasm in their ground Pr state. COP1 in association with SPA proteins leads to the degradation of multiple transcription factors, including HY5 and HFR1. Multiple PIF proteins remain stable and contribute to the etiolated form of development of the seedling. Upon light perception, the phytochromes in their Pfr conformation translocate into the nucleus where they lead to the inactivation of COP1 and the degradation of several PIFs. In addition to degradation, the PIFs are also inactivated by dimerization with HFR1 and the DELLA proteins. The primary mechanism leading to PIF inactivation depends on the light condition with dimerization with HFR1 playing an important role in FR-rich light because under these conditions the PIFs are relatively stable and HFR1 accumulates to high levels. The abundance of DELLA proteins is also regulated by the environment with, for example, higher levels of GA in the etiolated seedlings leading to more DELLA degradation.

Light-regulated kinase activity has long been proposed as a primary signaling mechanism for the phytochromes; a hypothesis that received new impetus with the discovery that cyanobacterial phytochromes are light-regulated histidine kinases (Yeh *et al.*, 1997). However, uncovering the significance of the reported protein kinase activity of plant phytochromes requires more work. Similarly, plant cryptochromes have been reported to possess Ser/Thr kinase activity but the physiological relevance of this finding remains to be established (Bouly *et al.*, 2003; Shalitin *et al.*, 2003). One problem for both classes of photoreceptors is that these enzymes have no homology with the well-characterized Ser/Thr protein kinase family and thus the characterization of mutants specifically inhibiting protein kinase activity has not be performed. On the other hand, the importance of the light-regulated protein kinase activity of the phototropins has strong experimental support (Christie, 2007; Tokutomi *et al.*, 2008). Ironically, while there are potential targets of the debated phytochrome kinase activity (e.g., PIF and FHY1), there is still no known target of the phototropins except the photoreceptor itself (Inoue *et al.*, 2008a; Shen *et al.*, 2009).

5.2. The PIF branch of phytochrome signaling

Much attention was focused on the members of the PIF class of bHLH transcription factor since the discovery that PIF3, the founding member of this family, specifically interacts with the light-activated Pfr form of phyB (Ni *et al.*, 1999). This distinguishing feature suggests a mechanism by which the phytochromes directly control light-regulated gene expression (Castillon *et al.*, 2007; Monte *et al.*, 2007). The PIFs are part of a 15-member clade of bHLHs, many of which have been shown to act in light-regulated morphogenesis (Castillon *et al.*, 2007; Heim *et al.*, 2003; Toledo-Ortiz *et al.*, 2003). Some of them are known as PIFs or PILs (PIF3-like). We will use the PIF nomenclature for all members of this group that interact with phyB (Khanna *et al.*, 2004). This group comprises PIF1 (PIL5), PIF3, PIF4, PIF5 (PIL6), PIF6 (PIL2), and PIF7. Several other members of this group such as PIL1, HFR1, and SPT do not interact with the light-activated phytochrome but nevertheless play functions in phytochrome-mediated morphogenesis (Castillon *et al.*, 2007; Khanna *et al.*, 2004). Interestingly, PIF orthologs were identified in rice suggesting a similar mode of phytochrome-regulated transcriptional regulation in monocots and dicots (*Oryza sativa*) (Nakamura *et al.*, 2007).

Most members of this subfamily contain two characteristic domains: the active phytochrome binding (APB) domain and the bHLH domain. The HLH domain allows the formation of homo- and heterodimers and the basic domain is responsible for DNA binding. PIF1, PIF3, PIF4, PIF5, and PIF7 were shown to bind to a specific *cis*-element the E-box (5′-CANNTG-3′) and more precisely the G-box (5′-CACGTG-3′) frequently found in light-regulated promoter sequences (Martinez-Garcia *et al.*, 2000; Huq *et al.*, 2004; Huq and Quail, 2002; Oh *et al.*, 2007; Shen *et al.*, 2007; Hornitschek *et al.*, 2009). HFR1 has an atypical basic domain, lacking two conserved amino acids, resulting in the incapacity to bind to the G-box sequence (Fairchild *et al.*, 2000; Hornitschek *et al.*, 2009). PIFs contain an amino-terminal-located APB domain that is necessary and sufficient for interaction with light-activated phyB (Khanna *et al.*, 2004). *In vitro* interaction studies indicate that PIF3 and PIF1 have a higher affinity for phyB than other PIFs (Huq *et al.*, 2004; Khanna *et al.*, 2004). How the conformational change of phyB from Pr to Pfr leads to conformer-specific interaction with the PIFs is not fully understood. A combination of homology modeling based on the structure of a prokaryotic phytochrome and the analysis of phyB point mutants indicates that several residues present on the so-called light-sensing knot are crucial to mediate this interaction (Kikis *et al.*, 2009; Oka *et al.*, 2008). This is noteworthy given that this corresponds to a region of phyB that is predicted to change its conformation upon light excitation. In addition, PIF1 and PIF3 also bind to the Pfr conformer of phyA through the active phytochrome A binding site domain, which surprisingly is not well conserved between PIF1 and PIF3 (Al-Sady *et al.*, 2006; Shen *et al.*, 2008).

The activity of the PIFs is regulated at multiple levels most prominently via interaction with the light-activated phytochromes leading to phosphorylation and subsequent degradation of PIF1, PIF3, PIF4, and PIF5 (Al-Sady *et al.*, 2006; Bauer *et al.*, 2004; de Lucas *et al.*, 2008; Lorrain *et al.*, 2008; Park *et al.*, 2004; Shen *et al.*, 2005; 2007). For PIF1, PIF3, and PIF5, rapid light-induced ubiquitylation has been reported suggesting a common mechanism for the regulated abundance of all those PIFs (Al-Sady *et al.*, 2006; Shen *et al.*, 2008; 2007). This results in a relatively high level of those proteins in the dark and a rapid decline in their abundance in light with a high R:FR ratio (typical of sunlight). In addition and correlating with phyA interaction assays, PIF1 and PIF3 levels also significantly decline in FR light while the levels of PIF4 and PIF5 remain much higher under such light conditions (Bauer *et al.*, 2004; Lorrain *et al.*, 2009; Shen *et al.*, 2008). PIF7 represents an exception because despite its ability to interact with phyB, PIF7 protein levels are not light regulated (Leivar *et al.*, 2008a). The activity of several PIFs is also inhibited by dimerization with transcriptional regulators leading to the formation of non-DNA-binding heteromers (de Lucas *et al.*, 2008; Feng *et al.*, 2008; Hornitschek *et al.*, 2009). PIF3 and PIF4 heterodimerize with the members of the DELLA family, leading to a crosstalk between hormone- and light-regulated growth (Alabadi *et al.*, 2008; de Lucas *et al.*, 2008; Feng *et al.*, 2008). The activity of PIF4 and PIF5 is inhibited by HFR1, which accumulates to high levels in low R:FR conditions typical of shade (Hornitschek *et al.*, 2009; Sessa *et al.*, 2005). The regulation of HFR1 levels results from the combination of transcript upregulation in FR-rich environments and light-regulated COP1-mediated protein stability (Fig. 2.4; Fairchild *et al.*, 2000; Hornitschek *et al.*, 2009; Sessa *et al.*, 2005). Finally, transcriptional regulation of *PIF4* and *PIF5* represents an additional level of regulation that determines when these factors promote elongation growth. Both genes are expressed under circadian control; in addition, *PIF4* levels are strongly temperature dependent (Koini *et al.*, 2009; Nozue *et al.*, 2007; Stavang *et al.*, 2009).

Members of the PIF family play both specific and overlapping roles in the control of a variety of phytochrome responses. For instance, germination is primarily controlled by PIF1 with SPT and PIF6 having a much more limited role for this response (Oh *et al.*, 2004; Penfield *et al.*, 2005, 2010; Shin *et al.*, 2009). Remarkably, none of the 2031 genes that are regulated by a germination-inducing red light treatment in wild-type seeds are light regulated in a *pif1* mutant demonstrating large-scale effect of PIF1 in the control of seed germination (Oh *et al.*, 2009). The picture emerging from these studies is that PIF1 is an inhibitor of seed germination that directly acts on ABA and GA, the two principal hormones regulating seed germination (Oh *et al.*, 2009; Piskurewicz *et al.*, 2009). By triggering the degradation of PIF1 the light-activated phytochromes release this break and promote germination.

In contrast to the specific role of PIF1 in the control of seed germination, at least four PIFs control the morphology of etiolated seedlings in a partially redundant manner (Leivar *et al.*, 2008b; 2009; Shin *et al.*, 2009). An etiolated quadruple *pif1pif3pif4pif5* mutant displays numerous features of light-grown seedlings, including morphology and the global gene expression pattern (Leivar *et al.*, 2008b; 2009; Shin *et al.*, 2009). In etiolated seedlings, the PIFs play a particularly important function by controlling chlorophyll biosynthesis, with PIF1, PIF3, and PIF5 being the most important for this response (Huq *et al.*, 2004; Shin *et al.*, 2009; Stephenson *et al.*, 2009). Thus the PIFs are required for the etiolated mode of seedling development, and the phytochromes promote de-etiolation (at least in part) by leading to their inactivation.

The analysis of *pif* mutants during seedling development in red and FR light leads to similar conclusions. It should however be pointed out that the interpretation of the results obtained in red light is somewhat complicated by the finding that PIFs regulate the abundance of phyB, which is the primary photoreceptor controlling de-etiolation in red light (Leivar *et al.*, 2008a). PhyB levels are increased in *pif* loss-of-function mutants while they are reduced in PIF-overexpressing plants correlating with enhanced de-etiolation in *pif* mutants and inhibited light responses in PIF overexpressers (Huq and Quail, 2002; Leivar *et al.*, 2008a). The mechanism by which the PIFs control phyB abundance remains unclear, but as for the light-regulated PIF degradation it requires interaction between the transcription factor and the photoreceptor (Al-Sady *et al.*, 2008). In contrast, in FR light, PIF1, PIF4, and PIF5 also negatively regulate the de-etiolation response but without affecting the levels of phyA, which controls this light response (Lorrain *et al.*, 2009; Oh *et al.*, 2004). Moreover, PIF4 and PIF5 also promote elongation growth in more mature plants during the shade-avoidance response (Hornitschek *et al.*, 2009; Lorrain *et al.*, 2008). Phenotypic analysis and gene expression studies show that PIF4 and PIF5 regulate a subset of SAR and more studies are needed to understand the relationship between the PIFs and other regulators of the SAR (Hornitschek *et al.*, 2009; Lorrain *et al.*, 2008; Roig-Villanova *et al.*, 2007; Sorin *et al.*, 2009; Tao *et al.*, 2008).

6. Concluding Remarks

Despite considerable progress during the last two decades on the elucidation of molecular events underlying photomorphogenesis, there are still a large number of unresolved issues. What is the molecular nature of the elusive UV-B receptor and are there really green light sensors in higher plants? Although extremely sophisticated biophysical approaches have been applied to analyze plant photoreceptors, our biochemical understanding of the events triggered by activation of these light sensors remains in its infancy (with the possible exception of the PIF branch of phytochrome signaling). The application of more biochemical and biophysical approaches should allow us to better address questions such as: How do both the phytochromes and cryptochromes inhibit COP1 in a light-dependent fashion? Which photoreceptors are really light-regulated kinases and what are their substrates? Photoreceptor-induced signaling mechanisms influence numerous aspects of plant development; however, in most cases we do not understand how the photoreceptors modulate development at the molecular level. Significant progress on these important issues require the combined approaches of developmental and photo-biologists. In many cases, phenotypes of relevant mutants will have to be analyzed with greater temporal and spatial resolution in order to understand the sequence of events and order of action of molecular players underlying a light response. A final fascinating avenue of research is to examine how all this knowledge gathered in Arabidopsis grown in the laboratory can be used to understand plant adaptation to their local environment.

REFERENCES

Achard, P., Cheng, H., De Grauwe, L., Decat, J., Schoutteten, H., Moritz, T., Van Der Straeten, D., Peng, J., and Harberd, N. P. (2006). *Science* **311**, 91–94.

Achard, P., Gong, F., Cheminant, S., Alioua, M., Hedden, P., and Genschik, P. (2008a). *Plant Cell* **20**, 2117–2129.

Achard, P., Renou, J. P., Berthome, R., Harberd, N. P., and Genschik, P. (2008b). *Curr. Biol.* **18**, 656–660.

Ahmad, M., and Cashmore, A. R. (1993). *Nature* **366**, 162–166.

Ahmad, M., Grancher, N., Heil, M., Black, R. C., Giovani, B., Galland, P., and Lardemer, D. (2002). *Plant Physiol.* **129**, 774–785.
Ahmad, M., Lin, C., and Cashmore, A. R. (1995). *Plant J.* **8**, 653–658.
Aihara, Y., Tabata, R., Suzuki, T., Shimazaki, K., and Nagatani, A. (2008). *Plant J.* **56**, 364–375.
Al-Sady, B., Kikis, E. A., Monte, E., and Quail, P. H. (2008). *Proc. Natl. Acad. Sci. U.S.A.* **105**, 2232–2237.
Al-Sady, B., Ni, W., Kircher, S., Schafer, E., and Quail, P. H. (2006). *Mol. Cell.* **23**, 439–446.
Alabadi, D., Gallego-Bartolome, J., Orlando, L., Garcia-Carcel, L., Rubio, V., Martinez, C., Frigerio, M., Iglesias-Pedraz, J. M., Espinosa, A., Deng, X. W., and Blazquez, M. A. (2008). *Plant J.* **53**, 324–335.
Aukerman, M. J., Hirschfeld, M., Wester, L., Weaver, M., Clack, T., Amasino, R. M., and Sharrock, R. A. (1997). *Plant Cell* **9**, 1317–1326.
Backstrom, S., Elfving, N., Nilsson, R., Wingsle, G., and Bjorklund, S. (2007). *Mol. Cell.* **26**, 717–729.
Bagnall, D. J., King, R. W., and Hangarter, R. P. (1996). *Planta* **200**, 278–280.
Balasubramanian, S., Sureshkumar, S., Agrawal, M., Michael, T. P., Wessinger, C., Maloof, J. N., Clark, R., Warthmann, N., Chory, J., and Weigel, D. (2006). *Nat. Genet.* **38**, 711–715.
Ballare, C. L. (2009). *Plant Cell Environ.* **32**, 713–725.
Ballare, C. L., and Scopel, A. L. (1997). *Funct. Ecol.* **11**, 441–450.
Banerjee, R., Schleicher, E., Meier, S., Viana, R. M., Pokorny, R., Ahmad, M., Bittl, R., and Batschauer, A. (2007). *J. Biol. Chem.* **282**, 14916–14922.
Bari, R., and Jones, J. D. (2009). *Plant Mol. Biol.* **69**, 473–488.
Bauer, D., Viczian, A., Kircher, S., Nobis, T., Nitschke, R., Kunkel, T., Panigrahi, K. C., Adam, E., Fejes, E., Schafer, E., and Nagy, F. (2004). *Plant Cell* **16**, 1433–1445.
Boccalandro, H. E., De Simone, S. N., Bergmann-Honsberger, A., Schepens, I., Fankhauser, C., and Casal, J. J. (2008). *Plant Physiol.* **146**, 108–115.
Boccalandro, H. E., Rugnone, M. L., Moreno, J. E., Ploschuk, E. L., Serna, L., Yanovsky, M. J., and Casal, J. J. (2009). *Plant Physiol.* **150**, 1083–1092.
Botto, J. F., Sanchez, R. A., Whitelam, G. C., and Casal, J. J. (1996). *Plant Physiol.* **110**, 439–444.
Bou-Torrent, J., Roig-Villanova, I., and Martinez-Garcia, J. F. (2008). *Trends Plant Sci.* **13**, 108–114.
Bouly, J. P., Giovani, B., Djamei, A., Mueller, M., Zeugner, A., Dudkin, E. A., Batschauer, A., and Ahmad, M. (2003). *Eur. J. Biochem.* **270**, 2921–2928.
Bouly, J. P., Schleicher, E., Dionisio-Sese, M., Vandenbussche, F., Van Der Straeten, D., Bakrim, N., Meier, S., Batschauer, A., Galland, P., Bittl, R., and Ahmad, M. (2007). *J. Biol. Chem.* **282**, 9383–9391.
Brodersen, P., Petersen, M., Pike, H. M., Olszak, B., Skov, S., Odum, N., Jorgensen, L. B., Brown, R. E., and Mundy, J. (2002). *Genes Dev.* **16**, 490–502.
Casal, J. J., and Boccalandro, H. (1995). *Planta* **197**, 213–218.
Casal, J. J., and Sánchez, R. A. (1998). *Seed Sci. Res.* **8**, 317–329.
Casal, J. J., Yanovsky, M. J., and Luppi, J. P. (2000). *Photochem. Photobiol.* **71**, 481–486.
Casson, S. A., Franklin, K. A., Gray, J. E., Grierson, C. S., Whitelam, G. C., and Hetherington, A. M. (2009). *Curr. Biol.* **19**, 229–234.
Casson, S. A., and Hetherington, A. M. (2010). *Curr. Opin. Plant Biol.* **13**, 90–95.
Castillon, A., Shen, H., and Huq, E. (2007). *Trends Plant Sci.* **12**, 514–521.
Cerdan, P. D., and Chory, J. (2003). *Nature* **423**, 881–885.
Chandra-Shekara, A. C., Gupte, M., Navarre, D., Raina, S., Raina, R., Klessig, D., and Kachroo, P. (2006). *Plant J.* **45**, 320–334.

Chen, M. (2008). *Curr. Opin. Plant Biol.* **11**, 503–508.
Chen, M., Chory, J., and Fankhauser, C. (2004). *Annu. Rev. Genet.* **38**, 87–117.
Chen, M., Tao, Y., Lim, J., Shaw, A., and Chory, J. (2005). *Curr. Biol.* **15**, 637–642.
Christie, J. M. (2007). *Annu. Rev. Plant Biol.* **58**, 21–45.
Christie, J. M., Reymond, P., Powell, G. K., Bernasconi, P., Raibekas, A. A., Liscum, E., and Briggs, W. R. (1998). *Science* **282**, 1698–1701.
Clough, R. C., and Viestra, R. D. (1997). *Plant Cell Environ.* **20**, 657–844.
DeBlasio, S. L., Mullen, J. L., Luesse, D. R., and Hangarter, R. P. (2003). *Plant Physiol.* **133**, 1471–1479.
Dechaine, J. M., Gardner, G., and Weinig, C. (2009). *Plant Cell Environ.* **32**, 1297–1309.
de Lucas, M., Daviere, J. M., Rodriguez-Falcon, M., Pontin, M., Iglesias-Pedraz, J. M., Lorrain, S., Fankhauser, C., Blazquez, M. A., Titarenko, E., and Prat, S. (2008). *Nature* **451**, 480–484.
Demarsy, E., and Fankhauser, C. (2009). *Curr. Opin. Plant Biol.* **12**, 69–74.
Devlin, P. F., Halliday, K. J., Harberd, N. P., and Whitelam, G. C. (1996). *Plant J.* **10**, 1127–1134.
Devlin, P. F., Patel, S. R., and Whitelam, G. C. (1998). *Plant Cell* **10**, 1479–1487.
Devlin, P. F., Robson, P. R., Patel, S. R., Goosey, L., Sharrock, R. A., and Whitelam, G. C. (1999). *Plant Physiol.* **119**, 909–915.
Dietrich, R. A., Delaney, T. P., Uknes, S. J., Ward, E. R., Ryals, J. A., and Dangl, J. L. (1994). *Cell* **77**, 565–577.
Donohue, K., Heschel, M. S., Butler, C. M., Barua, D., Sharrock, R. A., Whitelam, G. C., and Chiang, G. C. (2008). *New Phytol.* **177**, 367–379.
Donohue, K., Heschel, M. S., Chiang, G. C., Butler, C. M., and Barua, D. (2007). *Plant Cell Environ.* **30**, 202–212.
Faigon-Soverna, A., Harmon, F. G., Storani, L., Karayekov, E., Staneloni, R. J., Gassmann, W., Mas, P., Casal, J. J., Kay, S. A., and Yanovsky, M. J. (2006). *Plant Cell* **18**, 2919–2928.
Fairchild, C. D., Schumaker, M. A., and Quail, P. H. (2000). *Genes Dev.* **14**, 2377–2391.
Fankhauser, C., and Chen, M. (2008). *Trends Plant Sci.* **13**, 596–601.
Favory, J. J., Stec, A., Gruber, H., Rizzini, L., Oravecz, A., Funk, M., Albert, A., Cloix, C., Jenkins, G. I., Oakeley, E. J., Seidlitz, H. K., Nagy, F., *et al.* (2009). *EMBO J.* **28**, 591–601.
Feng, S., Martinez, C., Gusmaroli, G., Wang, Y., Zhou, J., Wang, F., Chen, L., Yu, L., Iglesias-Pedraz, J. M., Kircher, S., Schafer, E., Fu, X., *et al.* (2008). *Nature* **451**, 475–479.
Filiault, D. L., Wessinger, C. A., Dinneny, J. R., Lutes, J., Borevitz, J. O., Weigel, D., Chory, J., and Maloof, J. N. (2008). *Proc. Natl. Acad. Sci. U.S.A.* **105**, 3157–3162.
Folta, K. M., and Maruhnich, S. A. (2007). *J. Exp. Bot.* **58**, 3099–3111.
Folta, K. M., and Spalding, E. P. (2001). *Plant J.* **26**, 471–478.
Fornara, F., Panigrahi, K. C., Gissot, L., Sauerbrunn, N., Ruhl, M., Jarillo, J. A., and Coupland, G. (2009). *Dev. Cell* **17**, 75–86.
Franklin, K. A. (2008). *New Phytol.* **179**, 930–944.
Franklin, K. A. (2009). *Curr. Opin. Plant Biol.* **12**, 63–68.
Franklin, K. A., Davis, S. J., Stoddart, W. M., Vierstra, R. D., and Whitelam, G. C. (2003a). *Plant Cell* **15**, 1981–1989.
Franklin, K. A., Praekelt, U., Stoddart, W. M., Billingham, O. E., Halliday, K. J., and Whitelam, G. C. (2003b). *Plant Physiol.* **131**, 1340–1346.
Franklin, K. A., and Quail, P. H. (2010). *J. Exp. Bot.* **61**, 11–24.
Franklin, K. A., and Whitelam, G. C. (2004). *J. Exp. Bot.* **55**, 271–276.
Franklin, K. A., and Whitelam, G. C. (2007). *Nat. Genet.* **39**, 1410–1413.
Galen, C., Rabenold, J. J., and Liscum, E. (2007). *New Phytol.* **173**, 91–99.
Genoud, T., Buchala, A. J., Chua, N. H., and Metraux, J. P. (2002). *Plant J.* **31**, 87–95.

Genoud, T., Schweizer, F., Tscheuschler, A., Debrieux, D., Casal, J. J., Schafer, E., Hiltbrunner, A., and Fankhauser, C. (2008). *PLoS Genet.* **4**, e1000143.
Goosey, L., Palecanda, L., and Sharrock, R. A. (1997). *Plant Physiol.* **115**, 959–969.
Griebel, T., and Zeier, J. (2008). *Plant Physiol.* **147**, 790–801.
Guo, H., Yang, H., Mockler, T. C., and Lin, C. (1998). *Science* **279**, 1360–1633.
Halliday, K. J., Salter, M. G., Thingnaes, E., and Whitelam, G. C. (2003). *Plant J.* **33**, 875–885.
Halliday, K. J., and Whitelam, G. C. (2003). *Plant Physiol.* **131**, 1913–1920.
Han, I. S., Tseng, T. S., Eisinger, W., and Briggs, W. R. (2008). *Plant Cell* **20**, 2835–2847.
Harper, S. M., Neil, L. C., and Gardner, K. H. (2003). *Science* **301**, 1541–1544.
Heim, M. A., Jakoby, M., Werber, M., Martin, C., Weisshaar, B., and Bailey, P. C. (2003). *Mol. Biol. Evol.* **20**, 735–747.
Hennig, L., Stoddart, W. M., Dieterle, M., Whitelam, G. C., and Schafer, E. (2002). *Plant Physiol.* **128**, 194–200.
Heschel, M. S., Selby, J., Butler, C., Whitelam, G. C., Sharrock, R. A., and Donohue, K. (2007). *New Phytol.* **174**, 735–741.
Hiltbrunner, A., Tscheuschler, A., Viczian, A., Kunkel, T., Kircher, S., and Schafer, E. (2006). *Plant Cell Physiol.* **47**, 1023–1034.
Hornitschek, P., Lorrain, S., Zoete, V., Michielin, O., and Fankhauser, C. (2009). *EMBO J.* **28**, 3893–3902.
Huq, E., Al-Sady, B., Hudson, M., Kim, C., Apel, K., and Quail, P. H. (2004). *Science* **305**, 1937–1941.
Huq, E., Al-Sady, B., and Quail, P. H. (2003). *Plant J.* **35**, 660–664.
Huq, E., and Quail, P. H. (2002). *EMBO J.* **21**, 2441–2450.
Iino, M. (2006). *Curr. Opin. Plant Biol.* **9**, 89–93.
Imaizumi, T., and Kay, S. A. (2006). *Trends Plant Sci.* **11**, 550–558.
Imaizumi, T., Tran, H. G., Swartz, T. E., Briggs, W. R., and Kay, S. A. (2003). *Nature* **426**, 302–306.
Inoue, S., Kinoshita, T., Matsumoto, M., Nakayama, K. I., Doi, M., and Shimazaki, K. (2008a). *Proc. Natl. Acad. Sci. U.S.A.* **105**, 5626–5631.
Inoue, S., Kinoshita, T., Takemiya, A., Doi, M., and Shimazaki, K. (2008b). *Mol. Plant.* **1**, 15–26.
Izaguirre, M. M., Mazza, C. A., Biondini, M., Baldwin, I. T., and Ballare, C. L. (2006). *Proc. Natl. Acad. Sci. U.S.A.* **103**, 7170–7174.
Jackson, J. A., and Jenkins, G. I. (1995). *Planta* **197**, 233–239.
Jang, S., Marchal, V., Panigrahi, K. C., Wenkel, S., Soppe, W., Deng, X. W., Valverde, F., and Coupland, G. (2008). *EMBO J.* **27**, 1277–1288.
Janoudi, A., and Poff, K. L. (1992). *Photochem. Photobiol.* **56**, 655–659.
Jarillo, J. A., Gabrys, H., Capel, J., Alonso, J. M., Ecker, J. R., and Cashmore, A. R. (2001). *Nature* **410**, 952–954.
Jenkins, G. I. (2009). *Annu. Rev. Plant Biol.* **60**, 407–431.
Jiao, Y., Lau, O. S., and Deng, X. W. (2007). *Nat. Rev. Genet.* **8**, 217–230.
Johnson, E., Bradley, M., Harberd, N. P., and Whitelam, G. C. (1994). *Plant Physiol.* **105**, 141–149.
Kagawa, T., Kimura, M., and Wada, M. (2009). *Plant Cell Physiol.* **50**, 1774–1785.
Kagawa, T., Sakai, T., Suetsugu, N., Oikawa, K., Ishiguro, S., Kato, T., Tabata, S., Okada, K., and Wada, M. (2001). *Science* **291**, 2138–2141.
Kagawa, T., and Wada, M. (2000). *Plant Cell Physiol.* **41**, 84–93.
Kang, C. Y., Lian, H. L., Wang, F. F., Huang, J. R., and Yang, H. Q. (2009). *Plant Cell* **21**, 2624–2641.
Kasahara, M., Kagawa, T., Oikawa, K., Suetsugu, N., Miyao, M., and Wada, M. (2002). *Nature* **420**, 829–832.

Kendrick, R. E., and Kronenberg, G. H. M. (1994). *Photomorphogenesis in Plants*, 2nd ed. K. A. (Publishers, ed.), Kluwer, Dordrecht.
Khanna, R., Huq, E., Kikis, E. A., Al-Sady, B., Lanzatella, C., and Quail, P. H. (2004). *Plant Cell* **16**, 3033–3044.
Kiba, T., Henriques, R., Sakakibara, H., and Chua, N. H. (2007). *Plant Cell* **19**, 2516–2530.
Kidd, B. N., Edgar, C. I., Kumar, K. K., Aitken, E. A., Schenk, P. M., Manners, J. M., and Kazan, K. (2009). *Plant Cell* **21**, 2237–2252.
Kikis, E. A., Oka, Y., Hudson, M. E., Nagatani, A., and Quail, P. H. (2009). *PLoS Genet.* **5**, e1000352.
Kim, W. Y., Fujiwara, S., Suh, S. S., Kim, J., Kim, Y., Han, L., David, K., Putterill, J., Nam, H. G., and Somers, D. E. (2007). *Nature* **449**, 356–360.
Kinoshita, T., Doi, M., Suetsugu, N., Kagawa, T., Wada, M., and Shimazaki, K. (2001). *Nature* **414**, 656–660.
Kiss, J. Z., Mullen, J. L., Correll, M. J., and Hangarter, R. P. (2003). *Plant Physiol.* **131**, 1411–1417.
Kleiner, O., Kircher, S., Harter, K., and Batschauer, A. (1999). *Plant J.* **19**, 289–296.
Kneissl, J., Wachtler, V., Chua, N. H., and Bolle, C. (2009). *Plant Cell* **21**, 3212–3225.
Kobayashi, Y., and Weigel, D. (2007). *Genes Dev.* **21**, 2371–2384.
Koini, M. A., Alvey, L., Allen, T., Tilley, C. A., Harberd, N. P., Whitelam, G. C., and Franklin, K. A. (2009). *Curr. Biol.* **19**, 408–413.
Kong, S. G., Suzuki, T., Tamura, K., Mochizuki, N., Hara-Nishimura, I., and Nagatani, A. (2006). *Plant J.* **45**, 994–1005.
Kristie, D. N., and Fielding, A. (1994). *Seed Sci. Res.* **4**, 19–25.
Kumar, S. V., and Wigge, P. A. (2010). *Cell* **140**, 136–147.
Lake, J. A., Quick, W. P., Beerling, D. J., and Woodward, F. I. (2001). *Nature* **411**, 154.
Lariguet, P., Boccalandro, H. E., Alonso, J. M., Ecker, J. R., Chory, J., Casal, J. J., and Fankhauser, C. (2003). *Plant Cell* **15**, 2966–2978.
Lariguet, P., and Fankhauser, C. (2004). *Plant J.* **40**, 826–834.
Lasceve, G., Leymarie, J., Olney, M. A., Liscum, E., Christie, J. M., Vavasseur, A., and Briggs, W. R. (1999). *Plant Physiol.* **120**, 605–614.
Leivar, P., Monte, E., Al-Sady, B., Carle, C., Storer, A., Alonso, J. M., Ecker, J. R., and Quail, P. H. (2008a). *Plant Cell* **20**, 337–352.
Leivar, P., Monte, E., Oka, Y., Liu, T., Carle, C., Castillon, A., Huq, E., and Quail, P. H. (2008b). *Curr. Biol.* **18**, 1815–1823.
Leivar, P., Tepperman, J. M., Monte, E., Calderon, R. H., Liu, T. L., and Quail, P. H. (2009). *Plant Cell* **21**, 3535–3553.
Lin, R., Ding, L., Casola, C., Ripoll, D. R., Feschotte, C., and Wang, H. (2007). *Science* **318**, 1302–1305.
Lin, C., Robertson, D. E., Ahmad, M., Raibekas, A. A., Jorns, M. S., Dutton, P. L., and Cashmore, A. R. (1995). *Science* **269**, 968–970.
Lin, C., and Shalitin, D. (2003). *Annu. Rev. Plant Biol.* **54**, 469–496.
Lin, C., Yang, H., Guo, H., Mockler, T., Chen, J., and Cashmore, A. R. (1998). *Proc. Natl. Acad. Sci. U.S.A.* **95**, 2686–2690.
Liscum, E., and Briggs, W. R. (1995). *Plant Cell* **7**, 473–485.
Liscum, E., and Hangarter, R. P. (1993). *Plant Physiol.* **103**, 15–19.
Liu, H., Yu, X., Li, K., Klejnot, J., Yang, H., Lisiero, D., and Lin, C. (2008). *Science* **322**, 1535–1539.
Lopez-Juez, E., Dillon, E., Magyar, Z., Khan, S., Hazeldine, S., de Jager, S. M., Murray, J. A., Beemster, G. T., Bogre, L., and Shanahan, H. (2008). *Plant Cell* **20**, 947–968.
Lorrain, S., Allen, T., Duek, P. D., Whitelam, G. C., and Fankhauser, C. (2008). *Plant J.* **53**, 312–323.
Lorrain, S., Lin, B., Auriac, M. C., Kroj, T., Saindrenan, P., Nicole, M., Balague, C., and Roby, D. (2004). *Plant Cell* **16**, 2217–2232.

Lorrain, S., Trevisan, M., Pradervand, S., and Fankhauser, C. (2009). *Plant J.* **60**, 449–461.
Lu, H., Rate, D. N., Song, J. T., and Greenberg, J. T. (2003). *Plant Cell* **15**, 2408–2420.
Ma, L., Sun, N., Liu, X., Jiao, Y., Zhao, H., and Deng, X. W. (2005). *Plant Physiol.* **138**, 80–91.
Mach, J. M., Castillo, A. R., Hoogstraten, R., and Greenberg, J. T. (2001). *Proc. Natl. Acad. Sci. U.S.A.* **98**, 771–776.
Mao, J., Zhang, Y. C., Sang, Y., Li, Q. H., and Yang, H. Q. (2005). *Proc. Natl. Acad. Sci. U.S.A.* **102**, 12270–12275.
Martinez-Garcia, J. F., Huq, E., and Quail, P. H. (2000). *Science* **288**, 859–863.
Mathews, S. (2006). *Mol. Ecol.* **15**, 3483–3503.
Matsushita, T., Mochizuki, N., and Nagatani, A. (2003). *Nature* **424**, 571–574.
Mazzella, M. A., Alconda-Magliano, T. M., and Casal, J. J. (1997). *Plant Cell Environ.* **20**, 261–267.
McCormac, A. C., Wagner, D., Boylan, M. T., Quail, P. H., Smith, H., and Whitelam, G. C. (1993). *Plant J.* **4**, 19–27.
Millenaar, F. F., Cox, M. C., van Berkel, Y. E., Welschen, R. A., Pierik, R., Voesenek, L. A., and Peeters, A. J. (2005). *Plant Physiol.* **137**, 998–1008.
Millenaar, F. F., van Zanten, M., Cox, M. C., Pierik, R., Voesenek, L. A., and Peeters, A. J. (2009). *New Phytol.* **184**, 141–152.
Monte, E., Al-Sady, B., Leivar, P., and Quail, P. H. (2007). *J. Exp. Bot.* **58**, 3125–3133.
Monte, E., Alonso, J. M., Ecker, J. R., Zhang, Y., Li, X., Young, J., Austin-Phillips, S., and Quail, P. H. (2003). *Plant Cell* **15**, 1962–1980.
Montgomery, B. L. (2008). *Plant Signal. Behav.* **3**, 1053–1060.
Moreno, J. E., Tao, Y., Chory, J., and Ballare, C. L. (2009). *Proc. Natl. Acad. Sci. U.S.A.* **106**, 4935–4940.
Nagashima, A., Suzuki, G., Uehara, Y., Saji, K., Furukawa, T., Koshiba, T., Sekimoto, M., Fujioka, S., Kuroha, T., Kojima, M., Sakakibara, H., Fujisawa, N., *et al.* (2008). *Plant J.* **53**, 516–529.
Nagatani, A., Chory, J., and Furuya, M. (1991). *Plant Cell Physiol.* **32**, 1119–1122.
Nagatani, A., Reed, J. W., and Chory, J. (1993). *Plant Physiol.* **102**, 269–277.
Nakamura, Y., Kato, T., Yamashino, T., Murakami, M., and Mizuno, T. (2007). *Biosci. Biotechnol. Biochem.* **71**, 1183–1191.
Navarro, L., Bari, R., Achard, P., Lison, P., Nemri, A., Harberd, N. P., and Jones, J. D. (2008). *Curr. Biol.* **18**, 650–655.
Neff, M. M., and Chory, J. (1998). *Plant Physiol.* **118**, 27–35.
Neff, M. M., and Van Volkenburgh, E. (1994). *Plant Physiol.* **104**, 1027–1032.
Ni, M., Tepperman, J. M., and Quail, P. H. (1999). *Nature* **400**, 781–784.
Nozue, K., Covington, M. F., Duek, P. D., Lorrain, S., Fankhauser, C., Harmer, S. L., and Maloof, J. N. (2007). *Nature* **448**, 358–361.
Oh, E., Kang, H., Yamaguchi, S., Park, J., Lee, D., Kamiya, Y., and Choi, G. (2009). *Plant Cell* **21**, 403–419.
Oh, E., Kim, J., Park, E., Kim, J. I., Kang, C., and Choi, G. (2004). *Plant Cell* **16**, 3045–3058.
Oh, E., Yamaguchi, S., Hu, J., Yusuke, J., Jung, B., Paik, I., Lee, H. S., Sun, T. P., Kamiya, Y., and Choi, G. (2007). *Plant Cell* **19**, 1192–1208.
Ohgishi, M., Saji, K., Okada, K., and Sakai, T. (2004). *Proc. Natl. Acad. Sci. U.S.A.* **101**, 2223–2228.
Oka, Y., Matsushita, T., Mochizuki, N., Quail, P. H., and Nagatani, A. (2008). *PLoS Genet.* **4**, e1000158.
Park, E., Kim, J., Lee, Y., Shin, J., Oh, E., Chung, W. I., Liu, J. R., and Choi, G. (2004). *Plant Cell Physiol.* **45**, 968–975.
Parks, B. M., and Quail, P. H. (1993). *Plant Cell* **5**, 39–48.

Parks, B. M., Quail, P. H., and Hangarter, R. P. (1996). *Plant Physiol.* **110**, 155–162.
Pedmale, U. V., and Liscum, E. (2007). *J. Biol. Chem.* **282**, 19992–20001.
Penfield, S. (2008). *New Phytol.* **179**, 615–628.
Penfield, S., Josse, E. M., and Halliday, K. J. (2010). *Plant Mol. Biol.* **73**, 89–95.
Penfield, S., Josse, E. M., Kannangara, R., Gilday, A. D., Halliday, K. J., and Graham, I. A. (2005). *Curr. Biol.* **15**, 1998–2006.
Penfield, S., and King, J. (2009). *Proc. Biol. Sci.* **276**, 3561–3569.
Pfeifer, A., Mathes, T., Lu, Y., Hegemann, P., and Kottke, T. (2010). *Biochemistry.* **49**, 1024–1032.
Pfeiffer, A., Kunkel, T., Hiltbrunner, A., Neuhaus, G., Wolf, I., Speth, V., Adam, E., Nagy, F., and Schafer, E. (2009). *Plant J.* **57**, 680–689.
Pierik, R., Millenaar, F. F., Peeters, A. J., and Voesenek, L. A. (2005). *Ann. Bot.* **96**, 533–540.
Piskurewicz, U., Tureckova, V., Lacombe, E., and Lopez-Molina, L. (2009). *EMBO J.* **28**, 2259–2271.
Pokorny, R., Klar, T., Hennecke, U., Carell, T., Batschauer, A., and Essen, L. O. (2008). *Proc. Natl. Acad. Sci. U.S.A.* **105**, 21023–21027.
Pons, T. (1986). *Physiol. Plant.* **68**, 252–258.
Poppe, C., Hangarter, R. P., Sharrock, R. A., Nagy, F., and Schafer, E. (1996). *Planta* **199**, 511–514.
Quail, P. H. (2002). *Curr. Opin. Cell Biol.* **14**, 180–188.
Reed, J. W., Nagatani, A., Elich, T. D., Fagan, M., and Chory, J. (1994). *Plant Physiol.* **104**, 1139–1149.
Reed, J. W., Nagpal, P., Poole, D. S., Furuya, M., and Chory, J. (1993). *Plant Cell* **5**, 147–157.
Roberts, M. R., and Paul, N. D. (2006). *New Phytol.* **170**, 677–699.
Robson, P. R., and Smith, H. (1996). *Plant Physiol.* **110**, 211–216.
Rockwell, N. C., Su, Y. S., and Lagarias, J. C. (2006). *Annu. Rev. Plant Biol.* **57**, 837–858.
Roden, L. C., and Ingle, R. A. (2009). *Plant Cell* **21**, 2546–2552.
Roig-Villanova, I., Bou-Torrent, J., Galstyan, A., Carretero-Paulet, L., Portoles, S., Rodriguez-Concepcion, M., and Martinez-Garcia, J. F. (2007). *EMBO J.* **26**, 4756–4767.
Rosler, J., Klein, I., and Zeidler, M. (2007). *Proc. Natl. Acad. Sci. U.S.A.* **104**, 10737–10742.
Ruckle, M. E., DeMarco, S. M., and Larkin, R. M. (2007). *Plant Cell* **19**, 3944–3960.
Sage, L. C. (1992). *Pigment of the Imagination: A History of Phytochrome Research* (Academic Press, inc.), San Diego.
Sakai, T., Kagawa, T., Kasahara, M., Swartz, T. E., Christie, J. M., Briggs, W. R., Wada, M., and Okada, K. (2001). *Proc. Natl. Acad. Sci. U.S.A.* **98**, 6969–6974.
Sakamoto, K., and Briggs, W. R. (2002). *Plant Cell* **14**, 1723–1735.
Salter, M. G., Franklin, K. A., and Whitelam, G. C. (2003). *Nature* **426**, 680–683.
Samach, A., and Wigge, P. A. (2005). *Curr. Opin. Plant Biol.* **8**, 483–486.
Samis, K. E., Heath, K. D., and Stinchcombe, J. R. (2008). *Evolution* **62**, 2971–2983.
Sawa, M., Nusinow, D. A., Kay, S. A., and Imaizumi, T. (2007). *Science* **318**, 261–265.
Seo, M., Nambara, E., Choi, G., and Yamaguchi, S. (2009). *Plant Mol. Biol.* **69**, 463–472.
Sessa, G., Carabelli, M., Sassi, M., Ciolfi, A., Possenti, M., Mittempergher, F., Becker, J., Morelli, G., and Ruberti, I. (2005). *Genes Dev.* **19**, 2811–2815.
Shalitin, D., Yang, H., Mockler, T. C., Maymon, M., Guo, H., Whitelam, G. C., and Lin, C. (2002). *Nature* **417**, 763–767.
Shalitin, D., Yu, X., Maymon, M., Mockler, T., and Lin, C. (2003). *Plant Cell* **15**, 2421–2429.
Sharkey, T. D., and Raschke, K. (1981). *Plant Physiol.* **68**, 1170–1174.
Shen, Y., Khanna, R., Carle, C. M., and Quail, P. H. (2007). *Plant Physiol.* **145**, 1043–1051.
Shen, H., Moon, J., and Huq, E. (2005). *Plant J.* **44**, 1023–1035.
Shen, Y., Zhou, Z., Feng, S., Li, J., Tan-Wilson, A., Qu, L. J., Wang, H., and Deng, X. W. (2009). *Plant Cell* **21**, 494–506.

Shen, H., Zhu, L., Castillon, A., Majee, M., Downie, B., and Huq, E. (2008). *Plant Cell* **20**, 1586–1602.

Shin, J., Kim, K., Kang, H., Zulfugarov, I. S., Bae, G., Lee, C. H., Lee, D., and Choi, G. (2009). *Proc. Natl. Acad. Sci. U.S.A.* **106**, 7660–7665.

Shinomura, T., Nagatani, A., Chory, J., and Furuya, M. (1994). *Plant Physiol.* **104**, 363–371.

Shinomura, T., Nagatani, A., Hanzawa, H., Kubota, M., Watanabe, M., and Furuya, M. (1996). *Proc. Natl. Acad. Sci. U.S.A.* **93**, 8129–8133.

Shinomura, T., Uchida, K., and Furuya, M. (2000). *Plant Physiol.* **122**, 147–156.

Smith, H., and Whitelam, G. C. (1990). *Plant Cell Environ.* **13**, 695–707.

Somers, D. E., and Fujiwara, S. (2009). *Trends Plant Sci.* **14**, 206–213.

Somers, D. E., Schultz, T. F., Milnamow, M., and Kay, S. A. (2000). *Cell* **101**, 319–329.

Sorin, C., Salla-Martret, M., Bou-Torrent, J., Roig-Villanova, I., and Martinez-Garcia, J. F. (2009). *Plant J.* **59**, 266–277.

Staneloni, R. J., Rodriguez-Batiller, M. J., Legisa, D., Scarpin, M. R., Agalou, A., Cerdan, P. D., Meijer, A. H., Ouwerkerk, P. B., and Casal, J. J. (2009). *Proc. Natl. Acad. Sci. U.S. A.* **106**, 13624–13629.

Stavang, J. A., Gallego-Bartolome, J., Gomez, M. D., Yoshida, S., Asami, T., Olsen, J. E., Garcia-Martinez, J. L., Alabadi, D., and Blazquez, M. A. (2009). *Plant J.* **60**, 589–601.

Stephenson, P. G., Fankhauser, C., and Terry, M. J. (2009). *Proc. Natl. Acad. Sci. U.S.A.* **106**, 7654–7659.

Suetsugu, N., and Wada, M. (2007). *Biol. Chem.* **388**, 927–935.

Takemiya, A., Inoue, S., Doi, M., Kinoshita, T., and Shimazaki, K. (2005). *Plant Cell* **17**, 1120–1127.

Tanaka, S., Nakamura, S., Mochizuki, N., and Nagatani, A. (2002). *Plant Cell Physiol.* **43**, 1171–1181.

Tao, Y., Ferrer, J. L., Ljung, K., Pojer, F., Hong, F., Long, J. A., Li, L., Moreno, J. E., Bowman, M. E., Ivans, L. J., Cheng, Y., Lim, J., *et al.* (2008). *Cell* **133**, 164–176.

Tepperman, J. M., Hwang, Y. S., and Quail, P. H. (2006). *Plant J.* **48**, 728–742.

Tokutomi, S., Matsuoka, D., and Zikihara, K. (2008). *Biochim. Biophys. Acta* **1784**, 133–142.

Toledo-Ortiz, G., Huq, E., and Quail, P. H. (2003). *Plant Cell* **15**, 1749–1770.

Turck, F., Fornara, F., and Coupland, G. (2008). *Annu. Rev. Plant Biol.* **59**, 573–594.

Ulijasz, A. T., Cornilescu, G., Cornilescu, C. C., Zhang, J., Rivera, M., Markley, J. L., and Vierstra, R. D. (2010). *Nature* **463**, 250–254.

Usami, T., Mochizuki, N., Kondo, M., Nishimura, M., and Nagatani, A. (2004). *Plant Cell Physiol.* **45**, 1798–1808.

Vandenbussche, F., Pierik, R., Millenaar, F. F., Voesenek, L. A., and Van Der Straeten, D. (2005). *Curr. Opin. Plant Biol.* **8**, 462–468.

Vandenbussche, F., Vriezen, W. H., Smalle, J., Laarhoven, L. J., Harren, F. J., and Van Der Straeten, D. (2003). *Plant Physiol.* **133**, 517–527.

van Zanten, M., Voesenek, L. A., Peeters, A. J., and Millenaar, F. F. (2009). *Plant Physiol.* **151**, 1446–1458.

Wan, Y. L., Eisinger, W., Ehrhardt, D., Kubitscheck, U., Baluska, F., and Briggs, W. (2008). *Mol. Plant* **1**, 103–117.

Wang, F. F., Lian, H. L., Kang, C. Y., and Yang, H. Q. (2010). *Mol. Plant.* **3**, 246–259.

Wang, H., Ma, L. G., Li, J. M., Zhao, H. Y., and Deng, X. W. (2001). *Science* **294**, 154–158.

Warnasooriya, S. N., and Montgomery, B. L. (2009). *Plant Physiol.* **149**, 424–433.

Whippo, C. W., and Hangarter, R. P. (2003). *Plant Physiol.* **132**, 1499–1507.

Whippo, C. W., and Hangarter, R. P. (2004). *Plant Cell Environ.* **27**, 1223–1228.

Whippo, C. W., and Hangarter, R. P. (2006). *Plant Cell* **18**, 1110–1119.

Whitelam, G. C., Johnson, E., Peng, J., Carol, P., Anderson, M. L., Cowl, J. S., and Harberd, N. P. (1993). *Plant Cell* **5**, 757–768.

Wu, G., and Spalding, E. P. (2007). *Proc. Natl. Acad. Sci. U.S.A.* **104**, 18813–18818.

Wu, L., and Yang, H. Q. (2010). *Mol. Plant.* **3**, 539–548.
Yang, S. W., Jang, I. C., Henriques, R., and Chua, N. H. (2009). *Plant Cell* **21**, 1341–1359.
Yang, H. Q., Tang, R. H., and Cashmore, A. R. (2001). *Plant Cell* **13**, 2573–2587.
Yanovsky, M. J., Casal, J. J., and Whitelam, G. C. (1995). *Plant Cell Environ.* **18**, 788–794.
Yasuhara, M., Mitsui, S., Hirano, H., Takanabe, R., Tokioka, Y., Ihara, N., Komatsu, A., Seki, M., Shinozaki, K., and Kiyosue, T. (2004). *J. Exp. Bot.* **55**, 2015–2027.
Yeh, K. C., and Lagarias, J. C. (1998). *Proc. Natl. Acad. Sci. U.S.A.* **95**, 13976–13981.
Yeh, K. C., Wu, S. H., Murphy, J. T., and Lagarias, J. C. (1997). *Science* **277**, 1505–1508.
Yi, C., and Deng, X. W. (2005). *Trends Cell Biol.* **15**, 618–625.
Zeier, J., Pink, B., Mueller, M. J., and Berger, S. (2004). *Planta* **219**, 673–683.
Zhang, Y., and Turner, J. G. (2008). *PLoS One* **3**, e3699.
Zhu, D., Maier, A., Lee, J. H., Laubinger, S., Saijo, Y., Wang, H., Qu, L. J., Hoecker, U., and Deng, X. W. (2008). *Plant Cell* **20**, 2307–2323.

CHAPTER THREE

Root Development–Two Meristems for the Price of One?

Tom Bennett *and* Ben Scheres

Contents

Abstract

In this review, we analyze progress in understanding the mechanisms of root meristem development and function. The formation of embryonic and lateral roots, together with the remarkable regenerative ability of roots, seems to be

Department of Molecular Genetics, University of Utrecht, Utrecht, The Netherlands

Current Topics in Developmental Biology, Volume 91

ISSN 0070-2153, DOI 10.1016/S0070-2153(10)91003-X

linked to an auxin-dependent patterning mechanism, the "reflux loop," that can act at least partly independently of cellular context. A major feature of root formation is the production of the "structural initials," the center of the developing root. These cells form an organizing center (OC), the quiescent center (QC), which is needed for meristem activity. The exact role of the QC remains somewhat unclear, though it maintains a stem cell (SC) state in adjacent cells and acts as a long-term SC pool itself. SCs in the root can be defined on an operational basis, but a molecular definition for SC identity remains elusive. Instead, the behavior of cells in the proximal root might better be understood as the result of a "potential" gradient in the meristem, which confers cellular characteristics with respect to proximity to the QC. This potential gradient also seems to be auxin-dependent, possibly as a result of the effect of auxin on the expression of *PLETHORA* genes, key regulators of meristem function. Only in the root cap (RC) has distinct SC identity been proposed; but increasingly, evidence suggests that regulation of RC development is rather different from that in the proximal meristem; interestingly, a similar dichotomy can also be observed in the shoot meristem. Cell cycle progression must lie at the core of meristematic activity, and recent work has begun to uncover how hormonal regulation feeds forward into various aspects of the cell cycle. The emergent picture is one of coordinate regulation of cell division and elongation by a hormonal signaling network that is integrated by the auxin reflux loop to control root growth.

1. Introduction

A typical root system consists of a primary root bearing a repetitively branched system of lateral roots. Since each lateral root effectively behaves like the primary root, much of the analysis of root development falls under one of two major problems: how *de novo* root meristems form and how root meristems produce new tissues. In this review, we focus on recent advances in understanding how the different functional zones of the root meristem operate to produce new, functional tissues.

2. Structure of the Root

The root has a simple structure established during embryogenesis (Fig. 3.1; Dolan *et al.*, 1993). At the distal tip of the root is the root cap (RC), consisting of the central columella (COL) and lateral root cap (LRC). Located just above this is the quiescent center (QC), and beyond that the main, proximal body of the root. The main body of the root consists of concentric tissue layers with different functional properties; from outside-in, these are the epidermis (Epi), cortex (Cor), endodermis (Endo), pericycle,

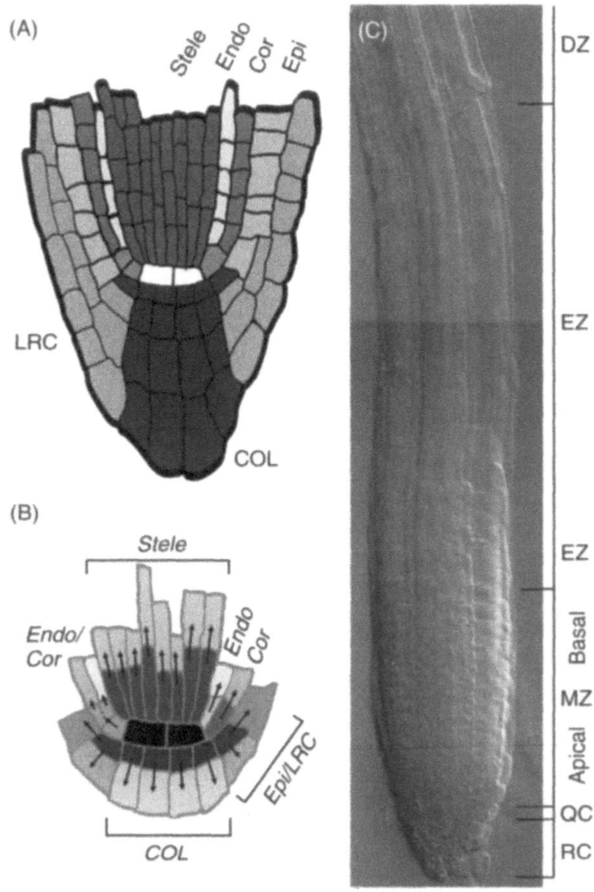

Figure 3.1 (A) Schematic diagram showing tissues in the *Arabidopsis* root meristem; red, stele (pericycle and vasculature); yellow, endodermis (Endo); green, cortex (Cor); pink, epidermis (Epi); light purple, lateral root cap (LRC); blue, columella (COL); white, quiescent center (QC). (B) Schematic diagram showing stem cell types and division planes in the *Arabidopsis* root meristem. On the left hand side, the diagram shows the organization in young *Arabidopsis* roots, with a common endodermal–cortical stem cell (Endo/Cor); on the right hand side, the organization in older roots is shown, with separate Endo and Cor stem cells. Stem cell daughters are shown in lighter shading; red, stele; mustard, Endo/Cor; yellow, Endo; green, Cor; dark purple, epidermal/lateral root cap stem cells (Epi/LRC); pink, Epi; light purple, LRC; blue, COL; black, QC. (C) Differential contrast interference light micrograph showing longitudinal organization of zones in the *Arabidopsis* root meristem; DZ, differentiation zone; EZ, elongation zone; MZ, meristem zone; QC, quiescent center; RC, root cap. (See Color Insert.)

and the centrally located stele (Fig. 3.1A). The stele contains cambial cells competent to form vascular elements; of these cells, a subset is specified as xylem cells, and others as phloem (see chapter 8). In young *Arabidopsis* roots, each of the outer layers is one cell wide, and the stele is approximately five cells in diameter; in other developmental stages, and in other species, the diameter of the cortex and stele layers can vary quite widely.

Above the QC lies a zone of cells which undergo mitotic divisions, often called the meristem zone (MZ). From an operational point of view, the MZ has been divided in two: the apical MZ, consisting of the most actively dividing cells, and a basal MZ, in which cells become larger, divide less frequently, and begin to acquire distinct fates (de Smet *et al.*, 2007 and references therein). The layer of RC cells directly adjacent to the QC is also mitotically competent but, as discussed below, can be considered to form a separate meristem (conceptually and genetically, if not physically). We thus distinguish between the RC meristem (RCM) and the main, proximal meristem (PM). As cells exit mitosis, they leave the MZ and undergo a period of elongation, thereby defining the elongation zone (EZ). Finally, cells cease elongation and enter the differentiation zone (DZ), which is typically defined by the appearance of root hairs and vascular cells with visible secondary cell wall formation. These proximo-distally arranged functional domains (Fig. 3.1C) are the main focus of this review, and for the purposes of this review, we consider all of these zones as part of the "greater" meristem. Radial patterning of the root tissue layers, together with patterning within the vasculature and patterning of the epidermal cell fate, has been recently reviewed elsewhere (Iyer-Pascuzzi and Benfey, 2009). We will begin, however, by examining the formation of roots.

3. Ontogeny of Roots

3.1. Embryonic root formation in *Arabidopsis*

Much of our knowledge of embryonic root formation comes from work on *Arabidopsis*. However, it should be noted that *Arabidopsis* embryogenesis is far from typical of dicot embryogenesis, let alone that of angiosperms or spermatophytes in general. In *Arabidopsis*, the first embryonic division gives rise to an apical cell and a basal cell. The basal cell gives rise to the suspensor, whilst the apical cell will go onto to form the entire embryo proper, with one small exception (reviewed in chapter 1; Laux *et al.*, 2004). A critical point in embryonic root development in *Arabidopsis* is the recruitment of the uppermost suspensor cell, the hypophysis, into the embryo proper, where it will give rise to the QC and COL of the root (Fig. 3.2; Mansfield and Briarty, 1991; Scheres *et al.*, 1994). The formation of a QC (whose function is discussed in greater detail below) can be taken as a marker for the initiation of a functional root. Mutants that fail to recruit the hypophysis completely lack an embryonic root (Berleth and Jurgens, 1993; Hamann *et al.*, 1999). A substantial body of work has strongly implicated the phytohormone auxin (primarily in the form of indole-3-acetic acid, IAA) in the

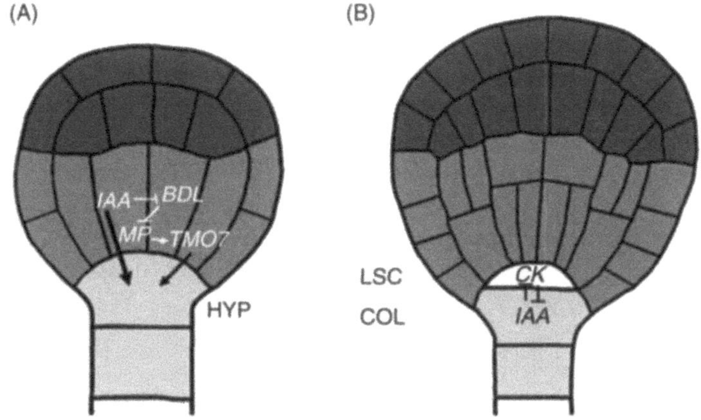

Figure 3.2 (A) Hypophyseal specification in the early *Arabidopsis* embryo. Auxin (IAA) relieves repression of the MONOPTEROS (MP) transcription factor by targeting the Aux/IAA protein BODENLOS (BDL) for destruction. MP then activates its target TMO7, which moves into the presumptive hypophysis (HYP). Together with auxin transported into the hypophysis, TMO7 regulates the recruitment and division of the hypophysis. (B) Separation of quiescent center (QC) and columella fates in the *Arabidopsis* embryo by mutually repressive auxin (IAA) and cytokinin (CK) signaling. After hypophyseal division, cytokinin signaling becomes predominant in the lens-shaped cell (LSC) and specifies it as the future QC, whilst auxin dominates in the columella precursor cell (COL).

recruitment of the hypophysis (reviewed in Laux *et al.*, 2004). Prior to the recruitment event, auxin is transported from the apical region into the basal pole of the embryo (Friml *et al.*, 2003), where it targets the repressive Aux/IAA transcription factor BODENLOS (BDL) for degradation. Aux/IAA proteins block the transcriptional activity of auxin response factor (ARF) proteins (mechanism reviewed in Mockaitis and Estelle, 2008), and BDL degradation therefore allows the MONOPTEROS (MP) ARF protein to activate transcription pathways leading to hypophyseal recruitment. However, MP and BDL are not expressed in the hypophysis itself, and hypophyseal specification therefore occurs non-cell autonomously (Weijers *et al.*, 2006). At least one component of this mechanism is movement of the TARGET OF MONOPTEROS7 (TMO7) transcription factor from the embryo proper into the hypophysis; it remains to be established whether there are also other aspects to this mechanism (Schlereth *et al.*, 2010).

Once specified, the hypophysis divides periclinally; the upper, lens-shaped cell will go on to form the cells of the embryonic QC, whilst the lower cell will form the COL region of the RC (Scheres *et al.*, 1994). Recent work has shown that interactions between auxin and another plant growth factor, cytokinin, are required for specification of these two cells. Auxin downregulates cytokinin signaling in the COL precursor, by upregulating the transcription factors ARR7 and ARR17, whilst cytokinin signaling in the lens-shaped cell remains unaffected; this asymmetry seems to arise from higher auxin levels in the COL precursor after hypophyseal division (Müller and Sheen, 2008). This distinction between these lineages is required for the formation of the embryonic root and expression of key root patterning genes (discussed further below; Müller and Sheen, 2008). This stereotyped pattern of division in the hypophyseal descendents produces the two major "distal" root tissues. Specification of the embryonic QC seems to be necessary to maintain specific division patterns in adjacent cells during embryogenesis and prevent differentiation of those cells when post-embryonic growth starts (Willemsen *et al.*, 1998). Although the root in the mature embryo is very small compared to the hypocotyl and cotyledons, all the specification and patterning events needed to form the primary root occur during embryogenesis in *Arabidopsis*, and post-embryonic root growth is, in essence, an enlargement of this pattern.

3.2. Root regeneration

One of the more fascinating properties of the root is its ability to regenerate a new root meristem after excision of the apical MZ of the existing root (Feldman, 1976; Sena *et al.*, 2009). Similar regeneration can also be achieved by ablation of the QC cells in *Arabidopsis*; a new QC is then formed within the existing meristem (Xu *et al.*, 2006). Recent work has demonstrated just how quickly this occurs after either treatment (Xu *et al.*, 2006; Sena *et al.*, 2009). In both cases, regeneration only requires mitotic cells and centers around the formation of a new QC, as is the case in *Arabidopsis* embryonic root formation. This dynamic re-specification of the meristem and QC is likely to be the result of the action of the plant hormone auxin (again, comparable to embryonic root formation). An auxin concentration maximum exists in the QC (Petersson *et al.*, 2009), and various genetic and chemical manipulations to auxin signaling or transport alter the positioning of QC-like cells in conjunction with alterations to the auxin maximum (Sabatini *et al.*, 1999). Furthermore, a shift in the auxin maximum occurs within 3 h of QC ablation, preceding re-specification of the QC at that new position (Xu *et al.*, 2006).

What mechanism lies behind these dynamic shifts of the auxin maximum? It is now well-established that a network of auxin efflux transporters, the extensively studied PIN proteins, is required to maintain the characteristic auxin distribution found in the root meristem (Blilou *et al.*, 2005). PIN proteins are polarly localized within cells and determine the directional transport of auxin within plant tissues (reviewed in Petrásek and Friml, 2009). In brief, five PIN proteins coordinate transport in the *Arabidopsis* root meristem (Fig. 3.3A–D). PIN1 is largely responsible for the primary, basally directed transport, which transports auxin toward the root tip. PIN2 is required for the apically directed transport which occurs in the epidermal and LRC layers. PIN3 and PIN7 link these two opposing flows together; in

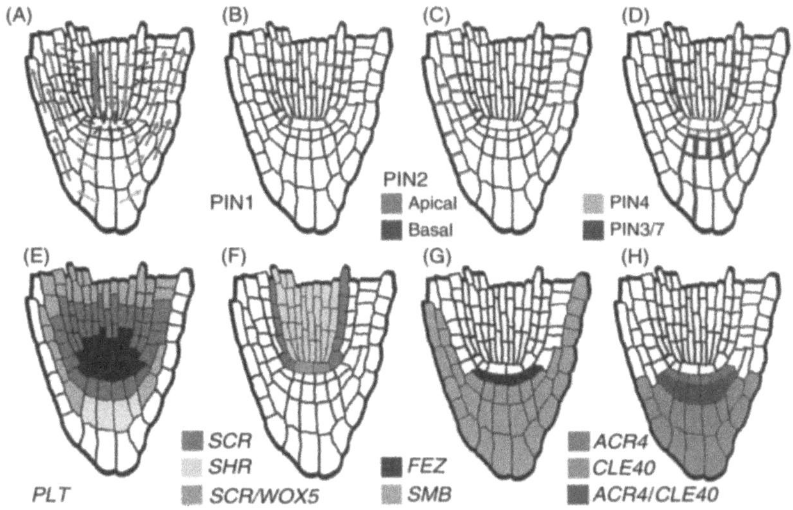

Figure 3.3 (A) Simplified model of auxin flux within the *Arabidopsis* root meristem. Basal auxin transport (green arrows) from the shoot enters the root meristem and is concentrated in the quiescent center (QC) area (purple arrows). In the columella, auxin is transported laterally (gold arrows) and then transported apically in the epidermis and lateral root cap (red arrows). Finally, apically transported auxin is recycled into the basal auxin stream by inward lateral transport (blue arrows). (B) Schematic diagram showing expression domain and protein localization of PIN1 in the root meristem. (C) Schematic diagram showing expression domain and protein localization of PIN2 in the root meristem; apical localization in epidermal and lateral root cap cells, and basal localization in cortical cells. (D) Simplified schematic diagram showing expression domains and protein localizations of PIN3, PIN4, and PIN7 in the root meristem. (E) Simplified schematic diagram showing graded expression of PLETHORA (PLT) proteins in the root meristem corresponds with cell division activity. (F) Schematic diagram showing expression of *SHORT-ROOT* (*SHR*), *SCARECROW* (*SCR*), and *WOX5* genes in the root meristem. (G) Schematic diagram showing expression of the FEZ protein in stem cells and SMB protein in mature root cap cells. (H) Schematic diagram showing expression of the CLE40 and ACR4 proteins in the root cap. (See Color Insert.)

the RC, they are responsible for *outward* lateral transport of auxin, whilst in the MZ, they are responsible for *inward* lateral transport of auxin (Blilou *et al.*, 2005). PIN4, which has a more specialized role, is expressed around and localized toward the QC and contributes to the concentration of auxin in the QC (Friml *et al.*, 2002). Computer modeling shows that this network is, in principle, able to maintain auxin distribution in the absence of any exogenous auxin, by creating an auxin "reflux" loop in the root meristem (Fig. 3.3A; Grieneisen *et al.*, 2007). How can this distribution network dynamically alter the auxin maximum? A possible hint comes from experiments showing that excision of the RC leads to the specification of a new QC in much the same way as if the QC had been ablated (Feldman, 1977).

The auxin concentration maximum in the root can be seen as a type of standing wave formed by fluxes of auxin centered on the presumptive QC area (Grieneisen *et al.*, 2007). It is probable that such a standing wave cannot be formed at boundary positions (i.e., next to dead QC cells, or at the outside of the root), and therefore, as an emergent property of the transport system, the auxin maximum will automatically shift proximally under such circumstances, leading to re-specification of the QC, even if the original QC was undamaged. This auxin-dependent mechanism of QC re-specification is very likely similar if not identical to that which specifies the position of the root during embryogenesis.

The AP2-domain PLETHORA (PLT) proteins are key factors in the initial specification, and re-specification of the root meristem and QC; evidence so far accumulated suggests that these proteins act downstream of auxin in these processes, though this link remains to be established conclusively. In the *plt1 plt2* double mutant, the root meristem gradually "collapses," and the meristem cells have all (apparently) differentiated a few days after germination (Aida *et al.*, 2004). Moreover, *plt1 plt2 plt3* triple mutants have no post-embryonic root activity; further addition of a single mutant allele of *plt4/bbm* removes the embryonic root leading to a rootless phenotype similar to *mp* (Galinha *et al.*, 2007). Conversely, *PLT* overexpression causes the production of larger MZs, or even ectopic root meristems (Aida *et al.*, 2004; Galinha *et al.*, 2007). Although *PLT* expression has been shown to react to altered auxin levels, the response is slow and may not precede alterations in development (Aida *et al.*, 2004). Similarly, while *PLT* expression is reduced in the ARF double mutant *mp non-phototropic hypocotyl4* (*nph4*), which is dramatically impaired in embryonic auxin response, it is not completely abolished (Aida *et al.*, 2004); the lack of the basal embryonic region in this mutant also complicates interpretation. Conversely, PLT proteins induce transcription of *PIN* genes and other auxin-related genes, and indeed do so on faster time scales than auxin induces *PLT* transcription (Blilou *et al.*, 2005; Galinha *et al.*, 2007). Which of these two factors is primarily responsible for specifying the root meristem then? The cyclical nature of feedback loops like this makes it difficult to disentangle the root initiation pathway; it remains likely that auxin is the primary signal, but a definitive answer awaits further experiments.

3.3. Lateral root formation

Lateral roots form as completely *de novo* organs, from divisions in the mature pericycle layer, although it has been suggested that certain pericycle cells are preconditioned to do this while residing in the basal MZ (de Smet *et al.*, 2007). Auxin is well established as acting in several steps during lateral root formation, including the initial specification of the new primordium (recently reviewed in Péret *et al.*, 2009). Once a new lateral root is specified, the characteristic pattern of auxin transport seen in primary roots is set-up in the developing primordium and this drives formation and growth of the new root (Benková *et al.*, 2003). Recent attempts to model lateral root formation suggest that small differences in auxin levels in the pericycle could be sufficient to trigger formation of a lateral auxin reflux loop, which could couple lateral root specification, formation and growth (Laskowski *et al.*, 2008).

3.4. Specification of the embryonic root in other species

In general, the embryonic root of spermatophytes (seed plants) develops at the basal pole of the embryo (the point where the embryo attaches to the maternal gametophyte), opposite the shoot apical meristem (SAM). In species such as maize, the embryonic root is established in the ground tissue at the basal pole of the embryo and does not seem to require recruitment of a specific cell type (e.g., the hypophysis). This rather more flexible specification process is qualitatively similar to the regeneration of the root which can occur in post-embryonic *Arabidopsis* roots. As has previously been suggested, it is unlikely that the general mechanism of embryonic root specification in other species is fundamentally different from that of *Arabidopsis*, despite the apparent morphological differences (Jiang and Feldman, 2005). We therefore suggest that the hypophyseal origin of the root in *Arabidopsis* is a special, restricted case of the more general mechanism which generates roots embryonically and post-embryonically in other species. It is plausible that a self-organizing auxin network is responsible for the formation of all root meristems in higher plants.

The basal location of the embryonic root meristem is probably a derived evolutionary characteristic; in "lower" plants, the embryonic origin of the root is rather different, for instance subtending the single embryonic leaf in ferns (reviewed in Cooke *et al.*, 2004). This raises some important questions. If there is a root developmental program which is not inherently tied to any particular location, what determines exactly where the embryonic root will form? Furthermore, can an embryonic root develop anywhere, or are there certain preconditions, for instance at the basal embryonic pole, that allow the root developmental program to occur? An obvious hypothesis is that, in order to specify a root, auxin must be transported into the appropriate region during early embryogenesis. However, in *Arabidopsis*, auxin accumulation in the basal pole of the embryo is not by itself sufficient for root specification, demonstrating that there is a deeper answer to this question. A recent report shows that the failure to repress the activity of the class III HD-ZIP genes *PHABULOSA* (*PHB*) and *PHAVOLUTA* (*PHV*) in the basal pole of *serrate* (*se*) mutant embryos leads to aberrant hypophyseal divisions followed by exaggerated cell proliferation, and a loss of the embryonic root; auxin, however, accumulates correctly at the basal pole of these embryos (Grigg *et al.*, 2009). Moreover, direct expression of class III HD-ZIPs in the presumptive embryonic root can cause conversion of the root to a shoot fate (Smith and Long, 2010). These data suggest that there are other activities, independent of auxin, necessary for embryonic root specification, which include the absence of class III HD-ZIP transcription factors (Grigg *et al.*, 2009; Smith and Long, 2010).

4. The Root Stem Cell Niche

4.1. Features of the stem cell niche

As was touched upon above, the very heart of the root meristem, both functionally and structurally, is the QC. The QC was initially identified as a region of low mitotic activity at the tip of the root, at the point where the different cell files that make up the root converge. These observations initially led to the idea that the QC a region of "initial" cells, each of which gives rise to a whole cell file lineage (reviewed in Jiang and Feldman, 2005). The demonstration of much-reduced (but not absent) mitotic activity in the region of convergence somewhat modified this theory, because the "functional" initials, the actual cells that undergo regular divisions to extend the cells files, must lie outside the point of convergence. The quiescent cells in the region of convergence have since been described as "structural" initials (Barlow, 1997). Conceptually then, the QC is intimately linked to the concepts of initials, and work over the last decade has shown that there is also an intimate functional link between the QC and the activity of the functional initials, similar to animal stem cell (SC) niches (Scheres, 2007).

More recently, the functional initials have been described as SCs, since they retain the capacity to create new cells while remaining undifferentiated themselves (Barlow, 1997). The structural initials/QC has been described as an OC (van den Berg, 1997), since cells in this region divide only rarely, and seem to be required for activity in the functional initials. Hereafter, we refer to the structural initials as the QC and the functional initials as SCs; together they have been characterized as forming a "stem cell niche" (Aida *et al.*, 2004). This concept is particularly pertinent in species with large meristems, where the distinction between the structural and functional initials is not especially clear, and in which there may be a gradient of mitotic activity, rather than a sharp boundary. In the very small *Arabidopsis* root meristem, on the other hand, the distinction between the QC and the functional initials is generally very distinct (Dolan *et al.*, 1993).

4.2. Features of the QC

Although the QC has been extensively analyzed over the past decade, in some ways it remains as mysterious as ever; while the role of the QC in root development is now relatively well understood at the level of the whole organ, there are still many unsolved questions regarding the QC. It is clear at least that the QC is required for the indeterminate growth of roots; in *Arabidopsis* mutants which do not maintain a QC, the root meristem is eventually exhausted and all cells cease to divide (Aida *et al.*, 2004; Sabatini *et al.*, 2003). One proposed explanation for this observation is that the QC is required to maintain SCs in an "immortal" state; ablation of the QC causes rapid differentiation of the RC SCs (van den Berg *et al.*, 1997).

In *Arabidopsis*, the QC will initiate infrequent divisions after about 6–7 days after germination (Baum *et al.*, 2002) and can be induced to do so at higher rates under stressful conditions (Kidner *et al.*, 2000). These divisions in the QC allow replacement of the SC pool and also regenerate the QC itself (Kidner *et al.*, 2000). It has recently been shown that mutants which overproduce ethylene have a somewhat higher rate of QC division but that these divisions do not affect the function of the QC (Ortega-Martínez *et al.*, 2007). This demonstrates that quiescence is not necessary *per se* for the function of the QC; indeed, in species with larger meristems, the QC is only quiescent in a relative sense. Thus, the QC, far from being a fixed entity, is a much more dynamic structure than it outwardly appears. Indeed, so dynamically can the QC be re-specified that it therefore appears to be not so much an irreversibly specialized group of cells, but a group of cells temporarily exposed to special conditions.

4.3. QC specification

What are the conditions that specify the QC? The QC is normally associated with the cell file initials established during embryogenesis or during lateral root formation. However, as root regeneration experiments show, these initials are not actually required for re-specification of the QC. Indeed, newly regenerated QCs are actually sufficient to re-pattern the meristem so as to "re-initialize" each cell file, demonstrating the potential of a QC as an OC. These data also provide further evidence for the theory that positional information is the key factor in determining fate in the root meristem, and not cell lineage (van den Berg *et al.*, 1995). As discussed above, root formation is very probably induced by an auxin-dependent patterning system; the auxin maximum in this system specifies the position of the presumptive QC. Auxin therefore seems to be a critical input into QC specification (Sabatini *et al.*, 1999).

In addition to auxin, two distinct transcription factor activities have previously been associated with QC specification, on the basis of mutant phenotypes. In both *plt1 plt2* and *scarecrow* (*scr*) mutants, a functional QC is not maintained, and the root meristem collapses post germination (Aida *et al.*, 2004; Sabatini *et al.*, 2003). This suggests that the PLT proteins, discussed above, are required for QC specification, along with the SCR and SHORT-ROOT (SHR) proteins; importantly, PLT and SHR/SCR appear required for different aspects of QC specification (Aida *et al.*, 2004). SCR and SHR are also known for their role in creating and specifying the Endo layer during embryogenesis, and maintaining Endo identity in later root development (see below). In both QC and Endo specification, the same mechanism prevails; *SHR* is transcribed in the vasculature, but SHR protein moves outwards into the ground tissue or QC, where it activates transcription of *SCR* (Fig. 3.3F). SCR then acts to restrict further movement of SHR by confining it to the nucleus, where both proteins also act to drive specification of the QC or Endo layer (Di Laurenzio *et al.*, 1996; Helariutta *et al.*, 2000; Nakajima *et al.*, 2001).

Recent work suggests that root regeneration (including formation of a presumptive QC) can occur in both *plt1 plt2* and *scr* mutants, even after the root meristem has collapsed (Sena *et al.*, 2009). This rather striking finding suggests that the former meristematic cells in *plt1 plt2*/*scr* are not completely differentiated and retain the capacity to re-enter the cell cycle. Furthermore, the data suggest that neither SCR nor PLT activity alone is necessary for regeneration of roots, in live with the idea that these genes are needed for partially overlapping aspects of QC specification. We suggest that these genes might be dispensable for formation of new structural initials *per se* but are probably required to specify those initials as a functional QC. In turn, these data also suggest that a functional QC is not required for the formation of a root, even though subsequent root development generally centers around the QC. It may also be mentioned that the extensive redundancy of PLT proteins means that a strict requirement for these proteins in root regeneration has not yet been excluded.

4.4. QC activity

Regardless of its exact role in QC specification, SCR is certainly required for QC maintenance. SCR is required to maintain expression of the homeodomain transcription factor WOX5, since in *scr* mutants, *WOX5* is not expressed (Fig. 3.3F; Sarkar *et al.*, 2007). *WOX5* expression does not require PLT1 or PLT2 but does seem to act downstream of ARF transcription factors such as MP (Sarkar *et al.*, 2007). WOX5 is required for the ongoing activity of the QC but apparently not its specification. The *Arabidopsis wox5* QC has reduced function and is unable to maintain the undifferentiated state of the RCM. However, the wox5 mutant root continues to grow indeterminately, which suggests that the QC maintains at least some activity, and that the RC SCs are continually re-specified during development (Sarkar *et al.*, 2007). The phenotype of *wox5* is comparable to the phenotype seen in *wuschel* (*wus*) mutants, which cannot maintain an undifferentiated SC pool in the SAM (Laux *et al.*, 1996). WOX5 and WUS are very closely related paralogues (Nardmann *et al.*, 2009) that are functionally equivalent; driving expression of *WOX5* under the *WUS* promoter is sufficient to rescue *wus* mutants, and *vice versa* (Sarkar *et al.*, 2007). Overexpression of *WOX5* leads to an accumulation of cells in the distal root meristem which superficially resemble COL SCs; however, these cells do not differentiate into COL cells upon QC ablation, unlike wild-type COL SCs. This phenotype therefore does not seem to be a result of "potentiating" the QC, and increasing the range of its signaling, since the effect is cell autonomous (and does not need the true QC at all). One interpretation of the phenotype would be that ectopic WOX5 activity converts functional SCs to a structural QC-like fate, and SC daughters to a functional SC fate, thereby increasing the size of the SC niche. As discussed above, in other species, the distinction between structural and functional initials is less clear-cut, and the overall pool of cells in the SC niche is more dynamic. By extension, a plausible hypothesis is that differences in the size of SC niches, both between species and between developmental stages, is the result of different patterns of WOX5 activity.

In the SAM OC, WUS is required for activating short-range signaling which keeps shoot SCs from differentiating; the nature of this signal is unknown (Schoof *et al.*, 2000). Laser ablation of a single QC cell in *Arabidopsis* causes differentiation of the RC SCs that are directly adjoined to that cell, but not the other RC SCs (van den Berg, 1997). It is therefore likely that a signal from the QC maintains the SC state of adjacent RC cells. Analysis of differentiation and cell division in RC SCs, in various mutant backgrounds, suggests that the QC signal inhibits differentiation rather than promoting division (van den Berg *et al.*, 1997). However, accurate tools for assessing differentiation in the PM are lacking, and the effect of QC ablation on the PM remains difficult to establish. QC-to-SC signaling has been proposed as one of the central functions of the QC, especially since short-range signaling between OC and SCs also exists. It is a reasonable hypothesis that the short-range signals in the root and shoot meristems are of a similar nature and are produced under the control of the WUS/WOX5 paralogues. However, despite much effort, both signals have remained elusive.

The analysis of cellular redox status and plasmodesmatal (PD) connectivity suggests new ideas on the mode of action of the QC signal. The QC has a more oxidizing environment than the neighboring cells, with lower levels of the reduced form of glutathione (GSH), a tripeptide that acts as cellular redox buffer (reviewed in Jiang and Feldman, 2005). This has led to the suggestion that this oxidizing environment is required for QC function, possibly by causing cell cycle arrest (Jiang *et al.*, 2003). It has also been suggested that auxin accumulation in the QC is responsible for this oxidizing environment, possibly directly as a result of catabolic breakdown of auxin, or by regulation of redox-related gene expression (Jiang and Feldman, 2005), which might explain the requirement for a PIN4-dependent auxin-concentrating mechanism surrounding the QC. Interestingly, recent work has shown that an oxidizing cellular environment causes restrictions in PD connectivity in the root meristem (and elsewhere; Benitez-Alfonso and Jackson *et al.*, 2009; Benitez-Alfonso *et al.*,2009). Could the oxidizing environment in the QC therefore be connected with signal transmission? The apparent dependence on direct cellular connections makes transport through PD an obvious mechanism for the spread of the QC signal. An intriguing possibility is therefore that the oxidizing environment of the QC restricts the PD connectivity of the adjacent initials, and thereby limits the intercellular movement and therefore range of a QC signal. Clearly, more work is required to examine the link between redox status and PD connections in the root meristem, but it remains an interesting possibility, especially since PD trafficking has also been demonstrated in the shoot meristem (Kim *et al.*, 2003).

4.5. The stem cell concept

The concept of "stem cells" has truly caught the imagination of both scientists and the public, at least partly because of the theoretical potential of SCs in medicine and tissue engineering. Most SC activity can only be defined from an operational point of view; generalized molecular definitions of SCs have been difficult to produce and might indeed be something of a red herring (Lander, 2009). The two general operational properties of mammalian SCs can be described as "potency" (the ability to produce other cell types) and "self-renewal" (maintenance of SC numbers; Lander, 2009). This operational SC concept has also been widely applied to plant meristems.

Both the root and shoot meristems produce new cells which extend the growth of the existing organ indeterminately, clearly indicating the existence of an *ipso facto* SC activity. In the PM of *Arabidopsis*, almost all cell divisions are anticlinal, that is, *within* a tissue layer. The principal exception to this is in the formation of the new ground tissue cells (Endo and Cor) during early root development. In young *Arabidopsis* roots, there is a single type of ground tissue initial (Fig. 3.1B—Gro) that divides anticlinally to give rise to the endodermal/cortical (Endo/Cor) sub-initial, which in turn divides periclinally to form new Endo and Cor cells (under the influence of the SCR/SHR pathway). Later in development, however, the Gro initial also divides periclinally, giving rise to separate Endo and Cor initials, which thereafter divide only anticlinally, as in other tissue layers. Furthermore, in some *Arabidopsis* ecotypes, and in other species, the Endo and Cor initials are present from embryogenesis onwards. The post-embryonic endodermal–cortical divisions in the *Arabidopsis* PM are therefore not a fundamental aspect of root development, and may arise due to topological constraints in the small *Arabidopsis* root.

In general, the anticlinal direction of division means that each tissue layer in the PM must by definition have its own SCs, which act to extend the layer by production of new cells (Fig. 3.1B). In *Arabidopsis*, clonal analysis shows that, in the PM, only those cells adjacent to the QC can be operationally defined as SCs. Large clonal sectors in the root always terminate in one of these cells; sectors induced in all other dividing cells, even immediate SC daughters, are much smaller (Scheres *et al.*, 1994). Thus, after a SC division, the cell adjacent the QC will remain a SC, and the other cell will undergo approximately four further rounds of division before exiting the cell cycle (Scheres *et al.*, 1994). These different developmental fates mean that the cell divisions in these SCs can be defined as asymmetric, a property often associated with SCs (Lander, 2009). However, they are only asymmetrical by operational criteria; there is currently no evidence in the PM for inherent asymmetry (i.e., molecular or physical differences between the two daughter cells).

In contrast to the PM, all SC divisions in the RCM are periclinal, that is, they form new tissue layers. In *Arabidopsis*, the COL SCs divide periclinally; the cell adjacent to the QC remains a SC, while the other daughter cell immediately exits the cell cycle (Dolan *et al.*, 1993). The *Arabidopsis* LRC initials are also the epidermal initials; these Epi/LRC SCs undergo both anti- and periclinal divisions, which give rise to Epi and LRC daughter cells, respectively (Fig. 3.1B; Dolan *et al.*, 1993). This juxtaposition may also be due to topological constraints caused by the small size of the *Arabidopsis* meristem. In other species, such as maize, there are distinct LRC SCs (Bengough *et al.*, 2001). LRC daughter cells may undergo a few anticlinal divisions in order to extend the cell file, so that the LRC can ensheath the MZ, but there are no further periclinal divisions (Willemsen *et al.*, 2008).

Thus, in *Arabidopsis*, there is a group of SCs directly adjacent to the QC, one-cell deep on all faces. Indeed, it could scarcely be otherwise; the fixed position of plant cells relative to each other means that almost all cells in the MZ will eventually be forced out of the meristem, by divisions in cells nearer the center of the meristem. The only exceptions are the QC cells and those cells directly adjacent to the QC; these cells are therefore, by an operational definition, the only possible SCs in the root meristem. This raises three important points regarding SC characteristics. Firstly, it suggests that the group of functional SCs surrounding the QC can only be one-cell deep, no matter which species is considered. Secondly, it suggests that SC status in any given cell is transient (even if long-lived); while the functional/structural SC pool is self-renewing as a whole, each functional SC will eventually be displaced by the rare divisions in the QC cells. Thirdly, it demonstrates that central location in the meristem is necessary for ongoing SC activity. It is therefore worth asking whether the SC concept is really a useful one in root development, as has recently been done for mammalian SC systems (Lander, 2009). For instance, do the SCs actually have specific properties which set them apart from other meristematic cells, or are they only SCs by simple fact of their central location and the positional cues that they therefore receive? Are root SC divisions inherently asymmetrical, or are they simply symmetrical divisions, in which the different positions of the daughter cells results in an asymmetry in their eventual fate? And if there are indeed distinct "stemness" characteristics, are they inherent to the cells, or to the environment (niche) the cells are located in?

4.6. The root cap and the stem cell concept

The simplest clues to the above questions can be found in the RCM. Since COL daughter cells exit the cell cycle immediately, and since they have an obvious physical marker for differentiation, in the form of starch-rich, gravity-sensing statocytes, it is clear that COL SCs have properties different from those of their daughter cells; proximity to the QC is certainly not their only distinguishing feature. In LRC daughter cells, the change in division plane indicates that they also have properties different from those of the SCs. Moreover, factors that control SC behavior in the RC have recently been identified in *Arabidopsis*, which shows that, in the RC at least, there is probably a distinct SC "identity." The NAC domain transcription factor FEZ is required for periclinal cell divisions in the COL and LRC SCs; *fez* mutants have much reduced SC activity and fewer RC layers (Willemsen *et al.*, 2008). Lack of FEZ activity does not prevent anticlinal divisions in the Epi/LRC initial, or in the LRC daughter cells, meaning its activity is specific to the RC SC divisions (Willemsen *et al.*, 2008). Conversely, SOMBRERO (SMB), a NAC protein from a different sub-family, is required to repress SC-like activity in the RC daughter cells, and *smb* mutants have an extra layer of SC-like cells (Willemsen *et al.*, 2008). FEZ and SMB appear to act in a feedback loop, in which FEZ promotes *SMB* expression in daughter cells and SMB in turn downregulates *FEZ* to prevent ectopic SC-like activity (Fig. 3.3G; Willemsen *et al.*, 2008). Thus, in the RC at least, there is a genetic network which regulates SC activity and the asymmetry between SC daughter cells.

4.7. Proximal meristem stem cells

It has been more difficult to identify or define "stemness" in the PM SCs. The properties of the PM SCs have been inferred primarily from the behavior of RC SCs (e.g., van den Berg *et al.*, 1997). Similarly, the activity (or lack thereof) of the RCM is often used a as proxy to assess activity in the SC niche as a whole. However, this approach is problematic, since it is increasingly apparent that the behavior of RC SCs is rather idiosyncratic and both qualitatively and quantitatively different from that of PM SCs. Firstly, all RC SC divisions are periclinal, whereas PM SC divisions are mostly anticlinal. Furthermore, in the *Arabidopsis* Epi/LRC initials, the two types of division are genetically separable by mutations in *FEZ*, suggesting that this SC type effectively has two different activities juxtaposed, and operating according to different genetic programs. RC SC activity can therefore be reduced without affecting PM SC function. Secondly, the QC has a clear function in maintaining the SC state of RCM cells (van den Berg, 1997), but, with hindsight, no clear effect on the PM SCs. Is the short-range QC signal really important in maintaining PM SC activity? Consistent with the idea that it is not, *wox5* mutations only cause differentiation in the RC SCs, and not PM SCs (Sarkar *et al.*, 2007). Similarly, root meristem-specific inhibition of the activity of RETINOBLASTOMA-RELATED (RBR), a probable master regulator of differentiation, causes a large increase in the number of undifferentiated SC-like cells in the RM, but has no obvious effect on PM SCs (Wildwater *et al.*, 2005). Taken together, these data suggest that RC SCs have characteristics definitively different from those in the PM, are regulated differently, and therefore should not be treated as equivalent to PM SCs. In view of these data, and evidence from phylogenetic and comparative studies, we suggest that the root meristem should be treated as two separate meristems, which may have evolved separately from one another.

The body plan of primitive trachaeophytes (for instance, those seen in the Rhynie Chert fossils) seems to have been based around horizontally growing stem axes ("telomes"). These telomes seem to have had primitive meristems at their apices; in most species these meristems did not produce distinct lateral organs (e.g., leaves) or have a structure analogous to the RC. One major theory for the evolution of modern trachaeophyte body plans is that both shoot and root systems are derived from these telomic growth axes; this idea finds support in the similarity of regulatory networks which regulate SAM and RM function in *Arabidopsis* (reviewed in Friedman *et al.*, 2004). We therefore propose that the PM of higher plants is descended from a generic telomic meristem, driving axial growth of an organ, but incapable of producing new organs (Fig. 3.4). The more advanced fossil species *Asteroxylon mackiei* had upright shoot-like telomes in which the meristem produced very simple leaf-like "enations" as lateral organs, and root-like axes which grew into the soil, but lacked a RC; *Asteroxylon* therefore suggests that generic telomic meristems later diversified and specialized. There is little direct evidence for the evolution of the RC in the fossil record, but we suggest, based on the differences between the RCM and PM noted above, that the RCM was one such specialization, evolving in the context of pre-existing, root-like telomic meristem. Given its functions in facilitating and guiding root growth, the RC may plausibly have been the evolutionary innovation which allowed the penetrative growth into the soil that is characteristic of "true" roots. Taking this idea a step further, we also suggest that SAM can also be thought of as having two separate meristems activities: a generic meristem activity in the rib zone (RZ) of the

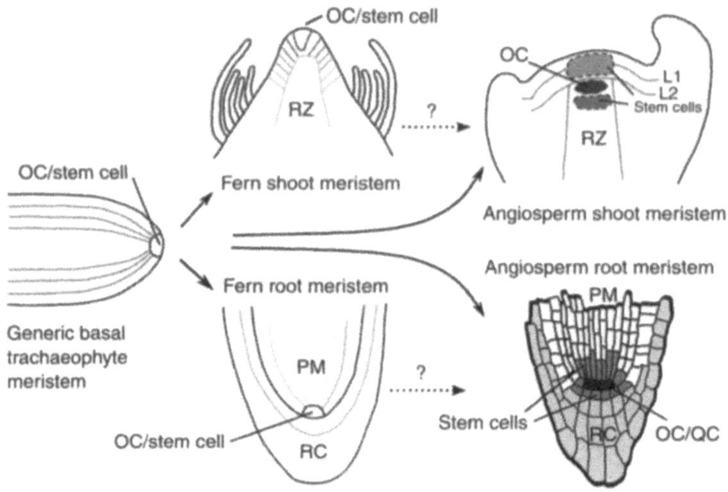

Figure 3.4 Hypothetical scheme for meristem evolution. Basal trachaeophytes, now extinct but known from fossilized remains, grew primarily as a network of horizontal stems, or telomes, and had primitive meristematic activity at the growing points of these stem axes. Root and shoot meristems in modern plants (such as ferns and angiosperms) are probably descended from these basic meristems. The proximal meristem (PM) of roots and the rib zone (RZ) meristem of shoots, in both ferns and angiosperms, are conceptually similar to these simple meristems, primarily driving the axial growth of an existing organ. The root cap (RC), a structure not observed in these fossils, but present in all living trachaeophytes, seems to be a separate and probably later evolutionary innovation in root meristems. Organogenesis differs markedly in the shoot meristems of monilophytes (including ferns) and embryophytes (including angiosperms); it is generally thought that leaf development evolved separately in these lineages, suggesting organogenesis is also a later evolutionary innovation than simple axial growth in the shoot meristem.

shoot, which drives axial growth of the organ (and corresponds to the PM in the root), and a specialized, organogenic meristem activity in the L1, L2, and L3 layers (see chapter 4; Fig. 3.4). Organogenesis in the SAM is thought to have evolved separately at least 3 times in trachaeophytes (Friedman *et al.*, 2004), suggesting that this activity is also a later innovation, and not inherent to generic shoot meristem function. Mutants such as *pin1*, in which organogenic activity is abolished, but axial growth of the shoot continues, may also support the idea of two separable meristematic activities in the shoot (Okada *et al.*, 1991).

4.8. The search for stemness

Where are the implications of these data for the role of the QC and the concept of stemness in the PM? Regarding the QC, new experiments are obviously required to probe its relationship with the PM. It is plausible that the QC has a more general effect of maintaining mitosis in the PM, rather than specifically maintaining the SCs, especially if the SCs do not have any inherent stemness factors. In this regard, it should be noted that *wox5* mutations and RBR inhibition do enhance the collapse of the PM in *plt1 plt2* and *scr* mutants, even though alone they have no apparent effect on the PM (Sarkar *et al.*, 2007; Wildwater *et al.*, 2005).

The operational definition of root SCs presented here does not allow ectopic SCs to exist. However, it is also clear that RC cells can ectopically acquire SC-like characteristics. Is the same true of PM cells, and if so, how can we tell? The appearance of PM SCs is very similar to their daughter cells, making potential changes in SC-like activity hard to identify; one possible marker for PM SC-like activity is the periclinal division which occurs in the Endo/Cor initials. Overexpression of PLT2 in the MZ (in an RBR RNAi background) leads to ectopic SC-like periclinal divisions in the Endo layer (Galinha *et al.*, 2007). Conversely, re-introduction of PLT2 into the *plt1 plt2* double mutant under the control of the *RCH2* promoter (which leads to PLT expression at approximately normal levels in MZ cells, but no PLT expression in stem cells) can maintain meristematic activity for up to 12 days (Galinha *et al.*, 2007), but does not restore indeterminate, SC like-activity to the meristem. A gradient of PLT protein, highest in the SC niche, appears to extend up the basal MZ (Fig. 3.3E; Galinha *et al.*, 2007), and given the above data, it is therefore probable that high PLT protein levels confer SC-like properties. However, the graded expression of PLT proteins could imply that they are not specific enough to define stemness in the PM SCs; it might be truer to say that high PLT expression confers "division potency" in cells near the QC. The QC has the highest PLT expression but rarely divides; it is therefore probable that other factors, acting downstream of auxin, impose a division block on these high potency cells (discussed below). Whole genome transcriptional analysis in roots has promise for indentifying candidate stemness factors, since factors specifically up- or downregulated in the SC area can now be identified (Brady *et al.*, 2007). Ultimately though, it remains possible that there is no requirement for the plant to define its PM SCs (transcriptionally or otherwise), if their SC properties anyway emerge as a higher level property of the developmental system.

4.9. Feedback signaling in the stem cell niche

In the SAM feedback loop mentioned above, SCs negatively feed back on the production of more SCs, thus helping to maintain a consistent pool of cells. The CLAVATA3 (CLV3) peptide ligand, which is secreted by SCs in the SAM, acts through the receptor-like kinase CLV1 (and the receptor-like protein CLV2) to downregulate *WUS* expression in the OC of the SAM, thereby restricting the SC promoting activity downstream of WUS (Schoof *et al.*, 2000; reviewed in Dodsworth, 2009). Such a pathway has also been suggested to exist in the root meristem, since over- or misexpression of *CLV3*, or its homologues *CLE19* or *CLE40*, leads to reduced meristem activity (Brand *et al.*, 2000; Casamitjana-Martínez *et al.*, 2003; Hobe *et al.*, 2003). This idea is strengthened by the demonstration that WOX5 acts in the root OC, as discussed above (Sarkar *et al.*, 2007). The effect of *CLE19* overexpression on the root meristem can be suppressed by mutations in at least two genes, *SUPPRESSOR OF LLP1 1* (*SOL1*) and *SOL2*; *SOL1* encodes a zinc-carboxypeptidase that might be involved in processing the CLV/CLE ligands (Casamitjana-Martínez *et al.*, 2003). A recent report has shown that *SOL2* is identical to *CORYNE* (*CRN*; Miwa *et al.*, 2008), a gene encoding a membrane-associated serine/threonine kinase (Müller *et al.*, 2008). Intriguingly, mutations in *CRN* also block the effects of *CLV3* overexpression in the shoot meristem and result in *clv*-like phenotypes (Müller *et al.*, 2008). It has been demonstrated that CRN and CLV2 physically interact, both at the plasma membrane and in the endoplasmic reticulum, and that this interaction is required for both of them to be recruited to the membrane (Bleckmann *et al.*, 2010; Zhu *et al.*, 2009); CRN/CLV2 multimers would, in effect, form a receptor-like kinase protein similar to CLV1. Müller *et al.* (2008) suggested that CLV1 and CLV2/CRN complexes can independently perceive CLV3/CLE ligands, but that CLV2/CRN complexes can also heterodimerize with CLV1 complexes, suggesting that there is crosstalk between these receptors (Zhu *et al.*, 2009; Bleckmann *et al.*, 2010). CRN and CLV2 also act to perceive CLE signals in the root meristem (Miwa *et al.*, 2008; Müller *et al.*, 2008), which further strengthens the argument that a CLV/WUS-like pathway acts in the root meristem.

Is any CLE peptide actually endogenously involved in signaling in the root meristem? Furthermore, is there a CLV1-like activity in the root meristem, or does the CLV2/CRN complex suffice for signaling in the root meristem? Recent work from Stahl and colleagues (2009) suggests possible answers to both these questions. They show that mutations in CLE40, which is expressed in COL daughter cells, leads to their delayed differentiation; conversely, addition of CLE40 peptide causes differentiation in the COL SCs of wild-type roots. These data suggest that CLE40 endogenously regulates proliferation of COL SCs. They further show that mutations in the ARABIDOPSIS CRINKLY4 (ACR4) receptor-like kinase (but not CLV2) block the effect of CLE40 peptide, demonstrating that ACR4 is a novel target of CLE signaling, active in RCM development. However, it is clear that this signaling cascade is not a canonical CLV–WUS type feedback loop, since it possesses some unusual features. Firstly, *ACR4* is usually only expressed in the COL SCs, and the CLE40 signal therefore appears to be a signal from differentiating cells to SCs, rather than from SCs to an OC (Fig. 3.3H). Secondly, addition of CLE40 peptide to wild type causes a proximal shift in expression of both *ACR4* (to the QC) and *WOX5* (to the PM SCs). The exact role of the

CLE4–ACR4 pathway is therefore not entirely clear; for instance, does it target *WOX5* expression directly, and if so, how (given the expression pattern of *ACR4*)? It is also interesting that this noncanonical pathway controls a feedback loop in the RCM given the above hypothesis that the RC evolved separately to the rest of the root meristem. An intriguing possibility is that the CLE40–ACR4 pathway represents a RC-specific evolutionary innovation, which was able to utilize parts of a pre-existing, canonical CLV/WUS regulatory network in the root. It remains a distinct possibility that there is a canonical CLV/WUS pathway in the PM, since many CLE peptides also strongly promote differentiation in the PM, which can be seen both in the overexpression of CLE-encoding genes (see above) and in the exogenous addition of CLE peptides (Ito *et al.*, 2006).

Intriguingly, further components of the CLV/WUS feedback loop from the SAM control SC function in the root meristem. POLTERGEIST (POL) and its close homologue POLTERGEIST-LIKE1 (PLL1) are type 2C protein phosphatases, which act to promote *WUS* transcription in the OC and which are deactivated by CLV signaling (Song *et al.*, 2006). Double mutants of *pol* and *pll1* are embryo lethal and do not specify a root SC niche, or maintain *WOX5* expression, during embryogenesis (Song *et al.*, 2008). In these embryos, both the hypophyseal and pro-cambial lineages show defective cell divisions, resulting in a failure to specify both the embryonic stele and QC. The early and severe embryonic defects in these mutants makes it difficult to pinpoint the exact function of POL/PLL1 in the root meristem, but the failure to initiate *WOX5* expression indicates that they may play a similar role in the root as in the shoot.

5. The Meristematic Zone

The majority of the PM consists of so-called "transit amplifying cells" (TACs). These are SC daughters, which remain in mitosis, and which divide at a faster rate than the SCs, thereby increasing the overall number of new cells produced by the meristem during a given period. Unlike SCs, their divisions are operationally symmetrical; both daughter cells will eventually leave the meristem and differentiate. A new TAC will undergo approximately four rounds of cell division before finally exiting mitosis and leaving the meristem (Scheres *et al.*, 1994).

The size of the MZ is controlled at the level of the whole organ, with a sharp boundary between the MZ and EZ; once TACs pass this boundary, they have exited the cell cycle and lose much of their competence to re-enter mitosis (Sena *et al.*, 2009). Although the boundary between the MZ and EZ (sometimes called the transition zone) can be clearly defined, it is also true that cells in the basal MZ show much less mitotic activity than those in the apical part, so there is also a gradient of transition between MZ and EZ. TACs leave the meristem not because of any inherent characteristic of the cell then, but because divisions in cells nearer the QC gradually force TACs further away from the QC, until they cross this meristem boundary. It is therefore probable that there is no specific TA "identity," and that TACs arise as a simple consequence of the higher-order properties of the meristem as a whole. Understanding these higher order properties, and how root MZ size is controlled, is therefore very important in understanding root development, since the rate of root growth is a direct function of MZ size. Over the past decade, many factors have been found to control MZ size and root growth rate; in particular, most of the major plant hormones seem to have an input into controlling MZ size (discussed below).

5.1. Cell cycle and meristem size

Several of the factors that influence MZ size which were identified through genetic screens seem to be part of either the core or peripheral cell cycle machinery. For instance, mutations in HOBBIT/CDC27B, part of the core APC (anaphase promoting complex) that targets cyclin proteins for destruction, cause defects in cell division in the MZ (Blilou *et al.*, 2002; Pérez-Pérez *et al.*, 2008; Serralbo *et al.*, 2006). It is unsurprising that defects in the cell cycle lead to reduced MZ size; cell cycle factors are clearly necessary for meristem function, and the cell cycle machinery should be expected to be the ultimate target for regulation of MZ size. The interest in these factors is therefore not because they themselves regulate MZ size, but in identifying potential downstream nodes in the network, upon which regulation might act. To give a recent example, two *Arabidopsis* genes encoding CCS52A proteins, which act as activators of the APC, have been shown to regulate cell cycle in the root. CCS52A1 is expressed, very precisely, in the EZ, whilst CCS52A2 is expressed in the SC niche and RC (Vanstraelen *et al.*, 2009). Mutations in CCS52A1 lead to an increased root length and MZ size, suggesting that it is required for the prompt exit from the cell cycle that occurs at the boundary of the MZ (Vanstraelen *et al.*, 2009). Conversely, the *ccs52a2* mutant has a much shorter root, caused by collapse of the meristem; the organization of the SC niche is lost, the QC divides, the RC SCs differentiate, and markers for QC fate are lost (Vanstraelen *et al.*, 2009). CCS52A2 therefore seems to be required, post-embryonically, to maintain quiescence in the QC, and perhaps also to maintain the low division rates seen in the SCs. As might be expected, CCS52A2 acts downstream of auxin, and probably also PLT activity (Vanstraelen *et al.*, 2009), suggesting that it is a downstream effector of known meristematic regulators. In respect of the control of MZ size, CCS52A1 is perhaps more interesting as a potential downstream effector, since it could easily be a target of signals which promote exit from division; whether this is the case, is currently unknown.

In a similar vein, the SUMO E3 ligase MMS21/HIGH PLOIDY2 (HPY2), which is required to prevent the onset of endoreduplication, is necessary for correct root meristem development (Ishida *et al.*, 2009). MMS21/HPY2 positively regulates expression of *CDKB1;1* and *CDKB2;1*; in turn, the CDKB proteins prevent cells exiting mitosis and entering the endocycle (Ishida *et al.*, 2009). As a consequence, *mms21/hpy2* mutants have very short roots and small MZs (Huang *et al.*, 2009; Ishida *et al.*, 2009). MMS21/HPY2 appears to be transcriptionally regulated in response to PLT activity, suggesting MMS21/HPY2 is another downstream effector of MZ size regulation. It has also recently demonstrated that high levels of auxin signaling are needed to prevent plant cells leaving mitosis and entering endoreduplication (Ishida *et al.*, 2010); it is likely that this occurs at least partially through PLT and HPY2-dependent regulation of the cell cycle. We can now therefore begin to understand how the control of meristematic activity arises by the interaction of developmental regulators, hormones, and the cell cycle (Fig. 3.5).

5.2. Hormonal control of root meristem size

The ability of auxin to regulate root growth is long-established; in recent years, effects of other hormones have become apparent. The effect of auxin on MZ size appears to depend primarily on basipetal auxin transport through the LRC and epidermal layers. Blocking this transport, by either ablation of the COL or the LRC (which prevents redistribution of auxin to the basipetal transport stream), or via mutation of multiple PIN auxin efflux transporters, leads to a strong reduction in the number of meristematic cells in all files (Blilou *et al.*, 2005). It is likely that the effect on other cell files occurs by disruption of the auxin "reflux" loop; cells not provided with auxin by reflux from the epidermis also exit mitosis. However, it is not clear why these cells need to be provided by auxin specifically from the reflux loop, if auxin is still being transported acropetally from the shoot. Perhaps the reflux loop prevents over-canalization of auxin transport and ensures that all cells are exposed to auxin. The requirement for auxin in order for cells to remain in mitosis is probably connected to *PLT* expression; certainly, as discussed above, *PLT* expression in TA cells is sufficient to keep them in mitosis (Galinha *et al.*, 2007). The size of the predicted reflux loop (based on PIN expression patterns) approximately matches the size of the MZ; *PLT* expression levels also correspond neatly with the pattern of cell division in the MZ (i.e., higher in the apical MZ) (Fig. 3.3E). Auxin-dependent transcription of *PLT* genes may therefore define the size of the MZ. It is a long-standing hypothesis that auxin is directly required for cell cycle progression (see for instance, den Boer and Murray, 2000), so it is possible that the auxin reflux loop also has PLT-independent roles in the control of MZ size. Recent work has shown that the enigmatic AUXIN-BINDING PROTEIN1 (ABP1), which has

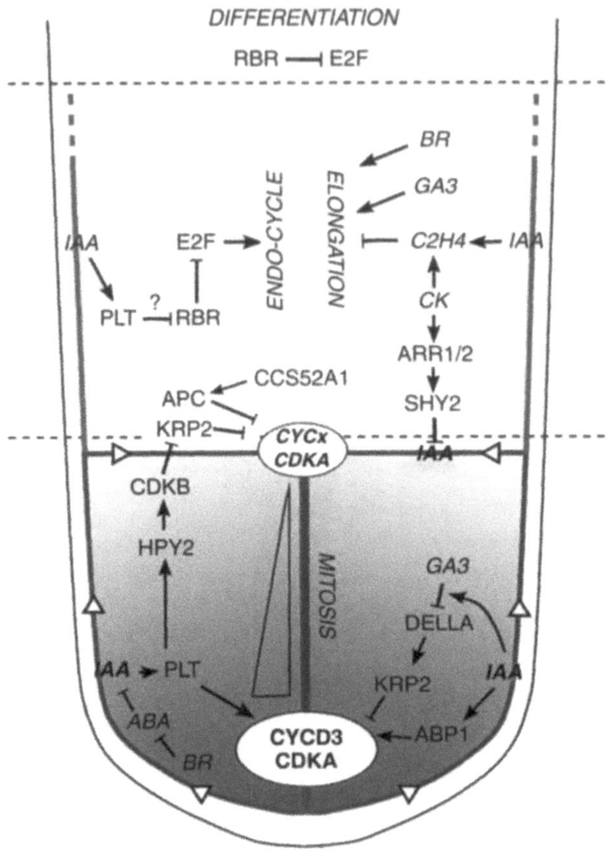

Figure 3.5 An auxin-dependent network controls root growth. Characteristic "reflux" patterns of auxin transport within the root meristem (gray lines and arrows) lead to a graded distribution of auxin (shading). The distribution of auxin is connected to the gradient of mitosis in the MZ and to endocycling and elongation in the EZ, while the absence of auxin correlates with differentiation in the DZ. Hormonal networks, including those acting through the auxin network, ultimately converge on regulation of the cell cycle. For full explanation of abbreviations, see accompanying text.

long been proposed as a probable auxin receptor, is required to maintain cell divisions in the MZ of *Arabidopsis* (Tromas *et al.*, 2009). How signaling may occur downstream of ABP1 is not clear; it was previously thought that ABP1 responses would be nontranscriptional. However, the results of Tromas and colleagues show that ABP1 is needed for upregulation of primary auxin response genes, such as Aux/IAAs, and also for transcription of D-type cyclins (Tromas *et al.*, 2009); *PLT* expression is essentially unchanged however. This suggests that auxin can indeed promote cell cycle progression in roots in a PLT-independent fashion, although perhaps not as directly as traditionally thought.

Computer modeling of auxin transport in the root suggests that the reflux loop should continue to operate in the absence of exogenous auxin, although the size of the meristem would gradually diminish (Grieneisen *et al.*, 2007). Thus, when coupled with auxin transported from the shoot, auxin transport in the root is probably sufficient to explain root growth. However, it remains highly likely that, in "real" roots, auxin levels are constantly modulated by synthesis and degradation of auxin. For instance, two tryptophan amino-transferases (TAA1 and TAR2), which are required for synthesis of auxin, are expressed in the QC/RC area, and double mutants in these genes cannot maintain meristematic activity (Stepanova *et al.*, 2008). Furthermore, six (out of eight) GH3 auxin-conjugating enzymes are expressed in the root, particularly in the epidermis and RC (Brady *et al.*, 2007; Staswick *et al.*, 2005; T.B. unpublished observations); mutants in at least one of these genes seem to affect root growth (Khan and Stone, 2007). A plausible hypothesis is therefore that the QC and RC act together as a kind of "clearing house" for auxin, receiving auxin from the acropetal transport stream, then adding or removing auxin as appropriate, before dispatching the auxin laterally for reflux around the meristem. This also raises the possibility that the QC can have longer-range effects on the root meristem, through regulation of auxin levels. Although, given the large quantities of auxin normally present in the root, these effects would be relatively mild, they might be enough to, for example, fine-tune the position of the MZ boundary.

Recently, it has been shown that cytokinin negatively regulates MZ size by upregulating expression of the transcription factors ARR1 and ARR2 (Dello Ioio *et al.*, 2007). The majority of the effect of cytokinin on MZ size seems to be mediated through transcriptional upregulation of the Aux/IAA protein SHY2 by ARR1/ARR12, since *shy2* loss-of-function mutants completely negate ARR1 overexpression (Dello Ioio *et al.*, 2008). SHY2 is a negative regulator of auxin signaling and appears to regulate *PIN* gene expression in the MZ; it is unclear which ARF proteins are targeted by SHY2, however. It is therefore likely that cytokinin ultimately acts by downregulating *PIN* expression in the MZ, weakening the reflux loop, and therefore MZ size. Cytokinin has traditionally been seen as a growth-promoting factor, signaling the presence of adequate nutrient levels (in particular nitrate; see for instance Takei *et al.*, 2002); it is therefore somewhat surprising that cytokinin should inhibit root growth. One plausible explanation for this would be that cytokinin is involved not in nutrient signaling *per se*, but rather in nutrient allocation. Under high nitrate conditions (=high cytokinin), resources can be shifted away from root growth toward development of the shoot system. Conversely, under low nitrate conditions, it is imperative that roots grow more strongly in order to locate new sources of soil nitrate, and therefore that resources are instead allocated to root growth. Thus, the effect of cytokinin on root growth could be part of mechanism that prioritizes resource allocation in response to environmental stimuli.

Recently, gibberellin (GA) has also been shown to regulate MZ size in *Arabidopsis*, and specifically by its activity in the endodermis. The principle mechanism of GA action is to target DELLA proteins for degradation via ubiquitination (reviewed in Harberd *et al.*, 2009). DELLA proteins act as repressors of growth, and GA therefore promotes plant growth by removing growth repressors. Mutants defective in GA biosynthesis have a smaller MZ, a phenotype which can be suppressed by mutation of multiple *DELLA* genes (Achard *et al.*, 2009; Ubeda-Tomás *et al.*, 2009). Furthermore, expression of a GA-resistant version of the DELLA protein GAI (*gai*) in the root meristem causes a large reduction in MZ size (Ubeda-Tomás *et al.*, 2009). A similar effect can be observed if *gai* is expressed in the endodermis alone, but not in other tissue layers, suggesting that GA response in the root meristem is only required in the endodermis (Ubeda-Tomás *et al.*, 2009). It is a well-established idea that certain tissue layers provide the driving force (or constraining force) for the growth of plant organs; for instance, in the leaf, the epidermis seems to drive growth, at least in response to brassinosteroids (BRs) (Savaldi-Goldstein *et al.*, 2007). It therefore seems that the endodermis is a critical tissue for determining MZ size, since in lines where growth is inhibited by endodermal *gai*, the adjacent cell files adjust their division rates to match the endodermis (but not *vice versa*). How does GA signaling fit into the framework of MZ size control? Achard and colleagues show that cyclin-dependent kinase inhibitors (CKIs) such as KRP2, which inhibit the action of the cyclin/cyclin-dependent kinase (CYC/CDK) complexes necessary for cell cycle progression, are upregulated by DELLA proteins, and therefore downregulated by GA signaling (Achard *et al.*, 2009); GA therefore seems to target the cell-cycle in TACs to control MZ size. It has previously been proposed that auxin signaling in the root enhances GA response (Fu and Harberd, 2003), suggesting that the GA effect is also mechanistically linked to auxin.

BRs themselves also have an important input into root growth. Mutants in the *BREVIS RADIX* (*BRX*) gene have, as the name suggests, shorter roots, partly due to reduced to decreased MZ size (Mouchel *et al.*, 2004). BRX is required for transcription of BR synthesis genes, and mutants can be rescued by addition of BR; furthermore, mutants in BR perception also show defects in root growth (Mouchel *et al.*, 2006). The BR effect seems to be mediated by potentiating auxin signaling in the root meristem; BR does not seem to directly modulate expression of the auxin signaling machinery, but instead BR-dependent transcription factors probably co-regulate genes with auxin-dependent transcription factors (Mouchel *et al.*, 2006). BRX is strongly positively regulated by auxin, at the level of transcription, protein stability, and protein localization (Scacchi *et al.*, 2009), but is mildly downregulated by

BR, suggesting it acts to fine-tune BR and auxin interactions in the root meristem (Mouchel *et al.*, 2006). Interestingly, BRX also seems to regulate the response of root growth to abscisic acid (ABA). ABA inhibits root growth, although through what process is not entirely clear, and the sensitivity to ABA inhibition is increased in *brx* mutants (Rodrigues *et al.*, 2009). Introduction of ABA insensitivity by transgenic methods into *brx* allows more effective rescue of the *brx* phenotype by BR than in control lines (Rodrigues *et al.*, 2009). Since other BR-deficient/insensitive mutants are also ABA-hypersensitive in root growth, it seems that one of the main functions of BR in controlling root growth may be to relieve ABA inhibition.

Most of the described hormonal inputs into MZ size ultimately seem to act by modulating auxin levels, transport, or signaling in the root, probably thereby altering the "strength" of the auxin reflux loop. Auxin "in the loop" in turn feeds through to GA response and cell cycle activity, to drive meristematic activity and cell cycle progression (Fig. 3.5).

6. Elongation

The elongation of cells after leaving the MZ is an active process, which makes the largest contribution to actual growth of the root. Plants with small MZs invariably have low root growth, which implies that each cell has a definite capacity to elongate and that a deficiency in cell production cannot be overcome by extra elongation. Cell expansion in leaves is correlated with ploidy, and repeated cycles of endo-reduplication are needed to boost chromosomal copy number as high as 32C (reviewed in Inzé and de Veylder, 2006). The same process may therefore be required for elongation of the cells in the EZ, as illustrated by the data discussed above with regard to MMS21/HPY2. The endocycle is certainly induced in post-meristematic cells, although it still remains to be proven that it is required for elongation.

6.1. Hormones and elongation

As with control of MZ size, hormonal signaling is strongly implicated in the control of elongation; indeed, the ability of auxin to modulate elongation was the first hormonal response discovered in plants. Auxin very rapidly promotes cell expansion in shoots; it is plausible that this response is not transcriptional, and acts through ABP1 (Christian *et al.*, 2006; Schenck *et al.*, 2010). However, in the root, auxin has always been associated with repression of cell elongation, except possibly at very low concentrations (Fu *et al.*, 2003). This inhibition of root elongation forms part of a growth system which allows roots to flexibly respond to stimuli such as gravity. When such stimuli are detected by the RC (invariably the point of signal detection or integration in root tropic responses), auxin is laterally redistributed (mediated by PIN3 and PIN7) to one side of the RC, and asymmetric reflux of auxin occurs through the epidermis. As a consequence, more auxin accumulates on one side of the EZ than the other, leading to asymmetric elongation, bending the root either toward or away from the stimulus (as appropriate). A plausible hypothesis is that the inhibition of elongation by auxin at least partially represents the flip-side of its promotion of MZ fate; perhaps auxin causes cells on the inhibited side to remain in the cell cycle for longer, leading to greater elongation on the other side. The regulation of elongation may also therefore be seen as part of a higher-order meristematic system deriving from auxin reflux (Fig. 3.5).

Ethylene is another hormone which has long been known to inhibit elongation in the root. Recent work has demonstrated that a major part of this ethylene response is mediated through the auxin reflux system. Ethylene upregulates auxin biosynthesis genes in the RC/QC area, including *TAA1* and *TAR2*, (discussed above); extra auxin thus produced is then transported apically to the EZ, where it acts to inhibit elongation (Růzicka *et al.*, 2007; Stepanova *et al.*, 2007; Swarup *et al.*, 2007). The interaction between auxin and ethylene has added layers of complexity, since auxin upregulates ethylene synthesis in the EZ, where it acts to increase the inhibition of elongation caused by auxin (Stepanova *et al.*, 2007). This demonstrates that the effect of auxin on elongation is not just due to its effects on cell proliferation. Cytokinin also seems to negatively regulate elongation in root cells and also acts to do this by upregulating ethylene biosynthesis in the EZ (Růzicka *et al.*, 2009).

At least two other hormones have been shown to regulate cell elongation in the root. In addition to its role in controlling MZ size, GA is also required in the EZ to promote cell elongation (Ubeda-Tomás *et al.*, 2008). Again, the endodermis appears to be the critical tissue; blocking GA response in the endodermis alone prevents elongation in all cell files, although cells in other files can still expand in other directions (Ubeda-Tomás *et al.*, 2008). DELLA proteins also therefore prevent elongation; based on their inhibition of cell cycle progression in the MZ, this could feasibly be by blocking endo-reduplication in the EZ. BRs are well-characterized regulators of cell expansion (Clouse and Sasse, 1998), and BRX, which has reduced BR levels (as discussed above), also has smaller final root cell size; the cells in the EZ still elongate, but to a lesser degree (Mouchel *et al.*, 2004). It is therefore probable that BR also regulates cell elongation in the root, although whether this occurs in an auxin-dependent manner is not clear.

7. Differentiation

When root cells reach their final size, they begin to differentiate; again, this is an active process, and not just the final exit from the cell cycle; differentiation in some root cells, such as root hair cells, requires substantial re-modeling of the cell, and cells usually only acquire their final functional characteristics when they enter the DZ. Several differentiation processes in the root, typically those accompanied by obvious visual markers, such as growth of root hairs (reviewed in Cárdenas, 2009), secondary cell wall deposition in xylem, or statocyte differentiation in the RC, are relatively well characterized in *Arabidopsis*. However, the general mechanism by which cells entering the DZ undergo generic cellular differentiation remains poorly characterized.

The aforementioned RBR protein probably plays a significant role in the transition to differentiation. Certainly, in mammalian systems, RETINOBLASTOMA proteins are implicated in cellular differentiation. Canonical RBR function involves inhibiting the activity of E2F transcription factors, which are themselves required to drive entry into the DNA replication phase of the cell cycle (reviewed in Inzé and de Veylder, 2006). Reduction in RBR expression delays the differentiation of RC daughter cells and also causes extra divisions in the MZ (Wildwater *et al.*, 2005); as discussed above, the sensitivity of RC and PM cells to RBR seems to be different. Conversely, overexpression of RBR rapidly leads to loss of cell division in the MZ, premature elongation, and differentiation (Wildwater *et al.*, 2005). RBR is expressed ubiquitously, but its activity is repressed by the action of several of the proteins previously discussed. SCR seems to oppose RBR activity in SCs, since in RBR RNAi lines, *scr* mutations no longer cause collapse of the meristem (Wildwater *et al.*, 2005). It is likely too that PLT activity in SCs, TACs, and perhaps elongating cells somehow opposes RBR activity, though inhibition of RBR cannot rescue *plt1 plt2* mutants, demonstrating the PLT proteins certainly have other roles besides inhibiting RBR. Can RBR be fitted elsewhere into the emerging framework of root meristem function? The answer seems to be yes, on several counts. CYCD3;1 overexpression results in a phenotype similar to RBR inhibition, and genetic evidence suggests that CYCD3;1 (and probably other D-type cyclins) inhibits RBR action in the root (den Boer and Murray, 2000). CYCD3;1 seems to be a downstream target of ABP1 (and therefore auxin), and both overexpression of CYCD3;1 and inhibition of RBR expression can rescue the defects seen in ABP1-deficient roots (Tromas *et al.*, 2009). It therefore seems likely that control of differentiation, in a general sense, is also integrated into the overarching, auxin-based regulatory network that controls activity of the root meristem (Fig. 3.5).

8. Concluding Remarks

Our understanding of root meristem function continues to be dominated by auxin. In *Arabidopsis*, an auxin reflux loop directly connects SC niche specification, cell division, and cell elongation in a flexible yet robust network, which determines many of the properties of the root meristem, and therefore growth of the root. It seems probable that this system has self-organising properties, making it independent of cell topology (yet capable of altering it), which would go some way to explaining the diversity in embryonic root specification amongst species, and the wide variety of situations in which *de novo* root meristems can form; including, for instance, adventitious rooting and root regeneration. This auxin network seems to be the target of many signaling inputs, providing a mechanism by which different environmental or developmental stimuli can be integrated with each other, and together modulate all aspects of root growth. Much progress has been made into understanding how regulatory mechanisms interact with the cell cycle machinery to drive or inhibit meristematic activity; elucidating exactly how the auxin network impacts on cellular behavior is an important future challenge in understanding the operation of the root meristem.

With regard to the functional zones of the meristem, it rather seems that the more closely the problem is analyzed, the less distinct the answer becomes. Although SCs in the PM can be defined operationally by the observer, it is not clear that plants make such a distinction; although an MZ–EZ boundary can be externally defined, it seems that cells are rather more oblivious to it. The overall impression formed is of a gradient of "potential" within the meristem, highest in the SC niche, and declining toward the DZ. Cells in the apical MZ still have the potential to become SCs, or to regenerate the root; cells in the basal MZ do not, but still have the potential to divide, even when they usually have stopped; and so on. Only the distinction between the RCM and PM seems clearer on the basis of recent data; indeed, the differences in appearance and behavior between the two make it tempting to speculate, as we have done, that they may have evolved separately from each other. Addressing such hypotheses will require analysis of root meristem development in a wide range of species, something that we anticipate will become a major aspect of studies of root development in the next decade.

REFERENCES

Achard, P., Gusti, A., Cheminant, S., Alioua, M., Dhondt, S., Coppens, F., Beemster, G. T., and Genschik, P. (2009). Gibberellin signaling controls cell proliferation rate in Arabidopsis. *Curr. Biol.* **19**, 1188–1193.

Aida, M., Beis, D., Heidstra, R., Willemsen, V., Blilou, I., Galinha, C., Nussaume, L., Noh, Y. S., Amasino, R., and Scheres, B. (2004). The *PLETHORA* genes mediate patterning of the Arabidopsis root stem cell niche. *Cell* **119**, 109–120.

Barlow, P. W. (1997). Stem cells and founder zones in plants, particularly their roots. In: Potten, C. S. (Ed.), *Stem Cells*. Academic, London, pp. 29–57.

Baum, S., Dubrovsky, J. F., and Rost, T. L. (2002). Apical organization and maturation of the cortex and vascular tissue in *Arabidopsis thaliana* (Brassicaceae) roots. *Am. J. Bot.* **89**, 908–920.

Bengough, A. G., Iijima, M., and Barlow, P. W. (2001). Image analysis of maize root caps—estimating cell numbers from 2-D longitudinal sections. *Ann. Bot.* **87**, 693–698.

Benitez-Alfonso, Y., Cilia, M., San Roman, A., Thomas, C., Maule, A., Hearn, S., and Jackson, D. (2009). Control of Arabidopsis meristem development by thioredoxin-dependent regulation of intercellular transport. *Proc. Natl. Acad. Sci. U.S.A.* **106**, 3615–3620.

Benitez-Alfonso, Y., and Jackson, D. (2009). Redox homeostasis regulates plasmodesmal communication in Arabidopsis meristems. *Plant Signal. Behav.* **4**, 655–659.

Benková, E., Michniewicz, M., Sauer, M., Teichmann, T, Seifertová, D., Jürgens, G., and Friml, J. (2003). Local, efflux-dependent auxin gradients as a common module for plant organ formation. *Cell* **115**, 591–602.

Berleth, T., and Jurgens, G. (1993). The role of the *MONOPTEROS* gene in organizing the basal body region of the *Arabidopsis* embryo. *Development* **118**, 575–587.

Bleckmann, A., Weidtkamp-Peters, S., Seidel, C. A., and Simon, R. (2010). Stem cell signaling in Arabidopsis requires CRN to localize CLV2 to the plasma membrane. *Plant Physiol.* **152**, 166–176.

Blilou, I., Frugier, F., Folmer, S., Serralbo, O., Willemsen, V., Wolkenfelt, H., Eloy, N. B., Ferreira, P. C., Weisbeek, P., and Scheres, B. (2002). The Arabidopsis *HOBBIT* gene encodes a CDC27 homolog that links the plant cell cycle to progression of cell differentiation. *Genes Dev.* **16**, 2566–2575.

Blilou, I., Xu, J., Wildwater, M., Willemsen, V., Paponov, I., Friml, J., Heidstra, R., Aida, M., Palme, K., and Scheres, B. (2005). The PIN auxin efflux facilitator network controls growth and patterning in Arabidopsis roots. *Nature* **433**, 39–44.

Brady, S. M., Orlando, D. A., Lee, J. Y., Wang, J. Y., Koch, J., Dinneny, J. R., Mace, D., Ohler, U., and Benfey, P. N. (2007). A high-resolution root spatiotemporal map reveals dominant expression patterns. *Science* **318**, 801–806.

Brand, U., Fletcher, J. C., Hobe, M., Meyerowitz, E. M., and Simon, R. (2000). Dependence of stem cell fate in Arabidopsis on a feedback loop regulated by CLV3 activity. *Science* **289**, 617–619.

Cárdenas L. (2009). New findings in the mechanisms regulating polar growth in root hair cells. *Plant Signal. Behav.* **4**, 4–8.

Casamitjana-Martínez, E., Hofhuis, H. F., Xu, J., Liu, C. M., Heidstra, R., and Scheres, B. (2003). Root-specific CLE19 overexpression and the sol1/2 suppressors implicate a CLV-like pathway in the control of Arabidopsis root meristem maintenance. *Curr. Biol.* **13**, 1435–1441.

Christian, M., Steffens, B., Schenck, D., Burmester, S., Böttger, M., and Lüthen, H. (2006). How does auxin enhance cell elongation? Roles of auxin-binding proteins and potassium channels in growth control. *Plant Biol. (Stuttg)* **8**, 346–352.

Clouse, S. D., and Sasse, J. M. (1998). Brassinosteroids: essential regulators of plant growth and development. *Annu. Rev. Plant Phys. Mol. Biol.* **49**, 427–451.

Cooke, T. J., Poli, D., and Cohen, J. D. (2004). Did auxin play a crucial role in the evolution of novel body plans during the Late Silurian–Early Devonian radiation of land plants? In: Hemsley, A., and Poole, I. (Eds.), *The Evolution of Plant Physiology*. Academic, London, pp. 85–107.

Dello Ioio, R., Linhares, F. S., Scacchi, E., Casamitjana-Martinez, E., Heidstra, R., Costantino, P., and Sabatini, S. (2007). Cytokinins determine Arabidopsis root-meristem size by controlling cell differentiation. *Curr. Biol.* **17**, 678–682.

Dello Ioio, R., Nakamura, K., Moubayidin, L., Perilli, S., Taniguchi, M., Morita, M. T., Aoyama, T., Costantino, P., and Sabatini, S. (2008). A genetic framework for the control of cell division and differentiation in the root meristem. *Science* **322**, 1380–1384.

den Boer, B. G., and Murray, J. A. (2000). Triggering the cell cycle in plants. *Trends Cell Biol.* 10, 245–250.

de Smet, I., Tetsumura, T., de Rybel, B., Frey, N. F., Laplaze, L., Casimiro, I., Swarup, R., Naudts, M,, Vanneste, S., Audenaert, D., Inzé, D., Bennett, M. J., *et al.* (2007). Auxin-dependent regulation of lateral root positioning in the basal meristem of Arabidopsis. *Development* **134**, 681–690.

Di Laurenzio, L., Wysocka-Diller, J., Malamy, J. E., Pysh, L., Helariutta, Y., Freshour, G., Hahn, M. G., Feldmann, K. A., and Benfey, P. N. (1996). The *SCARECROW* gene regulates an asymmetric cell division that is essential for generating the radial organization of the Arabidopsis root. *Cell* **86**, 423–433.

Dodsworth, S. (2009). A diverse and intricate signalling network regulates stem cell fate in the shoot apical meristem. *Dev. Biol.* **336**, 1–9.

Dolan, L., Janmaat, K., Willemsen, V., Linstead, P., Poethig, S., Roberts, K., and Scheres, B. (1993). Cellular organisation of the Arabidopsis thaliana root. *Development* **119**, 71–84.

Feldman, L. J. (1976). The de novo origin of the quiescent center in regenerating root apices of *Zea mays*. *Planta* **128**, 207–212.

Feldman L. J. (1977). The generation and elaboration of primary vascular tissue patterns in roots of *Zea*. *Bot. Gaz.* **138**, 393–402.

Friedman, W. E., Moore, R. C., and Purugganan M. D. (2004). The evolution of plant development. *Am. J. Bot.* **91**, 1726–1741.

Friml, J., Benková, E., Blilou, I., Wisniewska, J., Hamann, T., Ljung, K., Woody, S., Sandberg, G., Scheres, B., Jürgens, G., and Palme, K. (2002). AtPIN4 mediates sink-driven auxin gradients and root patterning in Arabidopsis. *Cell* **108**, 661–673.

Friml, J., Vieten, A., Sauer, M., Weijers, D., Schwarz, H., Hamann, T., Offringa, R., and Jürgens, G. (2003). Efflux-dependent auxin gradients establish the apical-basal axis of Arabidopsis. *Nature* **426**, 147–153.

Fu, X., and Harberd, N. P. (2003). Auxin promotes Arabidopsis root growth by modulating gibberellin response. *Nature* **421**, 740–743.

Galinha, C., Hofhuis, H., Luijten, M., Willemsen, V., Blilou, I., Heidstra, R., and Scheres, B. (2007). PLETHORA proteins as dose-dependent master regulators of Arabidopsis root development. *Nature* **449**, 1053–1057.

Grieneisen, V. A., Xu, J., Marée, A. F., Hogeweg, P., and Scheres, B. (2007). Auxin transport is sufficient to generate a maximum and gradient guiding root growth. *Nature* **449**, 1008–1013.

Grigg, S. P., Galinha, C., Kornet, N., Canales, C., Scheres, B., and Tsiantis, M. (2009). Repression of apical homeobox genes is required for embryonic root development in Arabidopsis. *Curr. Biol.* **19**, 1485–1490.

Hamann, T., Mayer, U., and Jürgens, G. (1999).The auxin-insensitive *bodenlos* mutation affects primary root formation and apical-basal patterning in the Arabidopsis embryo. *Development* **126**, 1387–1395.

Harberd, N. P., Belfield, E., and Yasumura, Y. (2009). The angiosperm gibberellin-GID1-DELLA growth regulatory mechanism: how an "inhibitor of an inhibitor" enables flexible response to fluctuating environments. *Plant Cell* **21**, 1328–1339.

Helariutta, Y., Fukaki, H., Wysocka-Diller, J., Nakajima, K., Jung, J., Sena, G., Hauser, M. T. and Benfey, P.N. (2000). The *SHORT-ROOT* gene controls radial patterning of the Arabidopsis root through radial signaling. *Cell* **101**, 555–567.

Hobe, M., Müller, R., Grünewald, M., Brand, U., and Simon, R. (2003). Loss of CLE40, a protein functionally equivalent to the stem cell restricting signal CLV3, enhances root waving in Arabidopsis. *Dev. Genes Evol.* **213**, 371–381.

Huang, L., Yang, S., Zhang, S., Liu, M., Lai, J., Qi, Y., Shi, S., Wang, J., Wang, Y., Xie, Q., and Yang, C. (2009). The Arabidopsis SUMO E3 ligase AtMMS21, a homologue of NSE2/MMS21, regulates cell proliferation in the root. *Plant J.* **60**, 666–678.

Inzé, D., and de Veylder, L. (2006). Cell cycle regulation in plant development. *Annu. Rev. Genet.* **40**, 77–105.

Ishida, T., Adachi, S., Yoshimura, M., Shimizu, K., Umeda, M., and Sugimoto, K. (2010). Auxin modulates the transition from the mitotic cycle to the endocycle in Arabidopsis. *Development* **137**, 63–71.

Ishida, T., Fujiwara, S., Miura, K., Stacey, N., Yoshimura, M., Schneider, K., Adachi, S., Minamisawa, K., Umeda, M., and Sugimoto, K. (2009). SUMO E3 ligase HIGH PLOIDY2 regulates endocycle onset and meristem maintenance in Arabidopsis. *Plant Cell* **21**, 2284–2297.

Ito, Y., Nakanomyo, I., Motose, H., Iwamoto, K., Sawa, S., Dohmae, N., and Fukuda, H. (2006). Dodeca-CLE peptides as suppressors of plant stem cell differentiation. *Science* **313**, 842–845.

Iyer-Pascuzzi, A. S., and Benfey, P. N. (2009). Transcriptional networks in root cell fate specification. *Biochim. Biophys. Acta* **1789**, 315–325.

Jiang, K., and Feldman, L. J. (2005). Regulation of root apical meristem development. *Annu. Rev. Cell Dev. Biol.* **21**, 485–509.

Jiang, K., Meng, Y. L., and Feldman, L. J. (2003). Quiescent center formation in maize roots is associated with an auxin-regulated oxidizing environment. *Development* **130**, 1429–1438.

Khan, S., and Stone, J. M. (2007). *Arabidopsis thaliana* GH3.9 influences primary root growth. *Planta* **226**, 21–34.

Kidner, C., Sundaresan, V., Roberts, K., and Dolan, L. (2000). Clonal analysis of the *Arabidopsis* root confirms that position, not lineage, determines cell fate. *Planta* **211**, 191–199.

Kim, J. Y., Yuan, Z., and Jackson, D. (2003). Developmental regulation and significance of KNOX protein trafficking in Arabidopsis. *Development* 130, 4351–4362.

Lander, A. D. (2009). The "stem cell" concept: is it holding us back? *J. Biol.* **21**, 70–74.

Laskowski, M., Grieneisen, V. A., Hofhuis, H., Hove, C. A., Hogeweg, P., Marée, A. F., and Scheres, B. (2008). Root system architecture from coupling cell shape to auxin transport. *PLoS Biol.* **6**, e307.

Laux, T., Mayer, K. F., Berger, J., and Jürgens, G. (1996). The *WUSCHEL* gene is required for shoot and floral meristem integrity in Arabidopsis. *Development* **122**, 87–96.

Laux, T., Würschum, T., and Breuninger, H. (2004). Genetic regulation of embryonic pattern formation. *Plant Cell* **16**, S190–202.

Mansfield, S. G., and Briarty, L. G. (1991). Early embryogenesis in *Arabidopsis thaliana*. II. The developing embryo. *Can. J. Bot.* **69**, 461–476.

Miwa, H., Betsuyaku, S., Iwamoto, K., Kinoshita, A., Fukuda, H., and Sawa, S. (2008). The receptor-like kinase SOL2 mediates CLE signaling in Arabidopsis. *Plant Cell Physiol.* **49**, 1752–1757.

Mockaitis, K., and Estelle, M. (2008). Auxin receptors and plant development: a new signaling paradigm. *Annu. Rev. Cell Dev. Biol.* **24**, 55–80.

Mouchel, C. F., Briggs, G. C., and Hardtke, C. S. (2004). Natural genetic variation in Arabidopsis identifies BREVIS RADIX, a novel regulator of cell proliferation and elongation in the root. *Genes Dev.* **18**, 700–714.

Mouchel, C. F., Osmont, K. S., and Hardtke, C. S. (2006). BRX mediates feedback between brassinosteroid levels and auxin signalling in root growth. *Nature* **443**, 458–461.

Müller, R., Bleckmann, A, and Simon, R. (2008). The receptor kinase CORYNE of Arabidopsis transmits the stem cell-limiting signal CLAVATA3 independently of CLAVATA1. *Plant Cell* **20**, 934–946.
Müller, B., and Sheen, J. (2008). Cytokinin and auxin interaction in root stem-cell specification during early embryogenesis. *Nature* **453**, 1094–1097.
Nakajima, K., Sena, G., Nawy, T. and Benfey, P.N. (2001). Intercellular movement of the putative transcription factor SHR in root patterning. *Nature* **413**, 307–311.
Nardmann, J., Reisewitz, P., and Werr, W. (2009). Discrete shoot and root stem cell-promoting WUS/WOX5 functions are an evolutionary innovation of angiosperms. *Mol. Biol. Evol.* **26**, 1745–1755.
Okada, K., Ueda, J., Komaki, M. K., Bell, C. J., and Shimura, Y. (1991). Requirement of the auxin polar transport system in early stages of Arabidopsis floral bud formation. *Plant Cell* **3**, 677–684.
Ortega-Martínez, O., Pernas, M., Carol, R. J., and Dolan, L. (2007). Ethylene modulates stem cell division in the Arabidopsis thaliana root. *Science* **317**, 507–510.
Péret, B., De Rybel, B., Casimiro, I., Benková, E., Swarup, R., Laplaze, L., Beeckman, T., and Bennett, M. J. (2009). Arabidopsis lateral root development: an emerging story. *Trends Plant Sci.* **14**, 399–408.
Pérez-Pérez, J. M., Serralbo, O., Vanstraelen, M., González, C., Criqui, M. C., Genschik, P., Kondorosi, E., and Scheres, B. (2008). Specialization of CDC27 function in the *Arabidopsis thaliana* anaphase-promoting complex (APC/C). *Plant J.* **53**, 78–89.
Petersson, S. V., Johansson, A. I., Kowalczyk, M., Makoveychuk, A., Wang, J. Y., Moritz, T., Grebe, M., Benfey, P. N., Sandberg, G., and Ljung, K. (2009). An auxin gradient and maximum in the Arabidopsis root apex shown by high-resolution cell-specific analysis of IAA distribution and synthesis. *Plant Cell* **21**, 1659–1668.
Petrásek, J., and Friml, J. (2009). Auxin transport routes in plant development. *Development* **136**, 2675–2688.
Rodrigues, A., Santiago, J., Rubio, S., Saez, A., Osmont, K. S., Gadea, J., Hardtke, C. S., and Rodriguez, P. L. (2009). The short-rooted phenotype of the *brevis radix* mutant partly reflects root abscisic acid hypersensitivity. *Plant Physiol.* **149**, 1917–1928.
Růzicka, K., Ljung, K., Vanneste, S., Podhorská, R., Beeckman, T., Friml, J., and Benková, E. (2007). Ethylene regulates root growth through effects on auxin biosynthesis and transport-dependent auxin distribution. *Plant Cell* **19**, 2197–2212.
Růzicka, K., Simásková, M., Duclercq, J., Petrásek, J., Zazímalová, E., Simon, S., Friml, J., Van Montagu, M. C., and Benková, E. (2009). Cytokinin regulates root meristem activity via modulation of the polar auxin transport. *Proc. Natl. Acad. Sci. U.S.A.* **106**, 4284–4289.
Sabatini, S., Beis, D., Wolkenfelt, H., Murfett, J., Guilfoyle, T., Malamy, J., Benfey, P., Leyser, O., Bechtold, N., Weisbeek, P., and Scheres, B. (1999). An auxin-dependent distal organizer of pattern and polarity in the Arabidopsis root. *Cell* **99**, 463–472.
Sabatini, S., Heidstra, R., Wildwater, M., and Scheres, B. (2003). SCARECROW is involved in positioning the stem cell niche in the Arabidopsis root meristem. *Genes Dev.* **17**, 354–358.
Sarkar, A. K., Luijten, M., Miyashima, S., Lenhard, M., Hashimoto, T., Nakajima, K., Scheres, B., Heidstra, R., and Laux, T. (2007). Conserved factors regulate signalling in Arabidopsis thaliana shoot and root stem cell organizers. *Nature* **446**, 811–814.
Savaldi-Goldstein, S., Peto, C., and Chory, J. (2007). The epidermis both drives and restricts plant shoot growth. *Nature* **446**, 199–202.
Scacchi, E., Osmont, K. S., Beuchat, J., Salinas, P., Navarrete-Gómez, M., Trigueros, M., Ferrándiz, C., and Hardtke, C. S. (2009). Dynamic, auxin-responsive plasma membrane-to-nucleus movement of Arabidopsis BRX. *Development* **136**, 2059–2067.
Schenck, D., Christian, M., Jones, A., and Lüthen, H. (2010). Rapid auxin-induced cell expansion and gene expression: a four-decade old question revisited. *Plant Physiol.* **152**, 1183–1185.

Scheres, B. (2007). Stem-cell niches: nursery rhymes across kingdoms. *Nat. Rev. Mol. Cell Biol.* **8**, 345–354.

Scheres, B., Wolkenfelt, H., Willemsen, V., Terlouw, M., Lawson, E., Dean, C., and Weisbeek, P. (1994). Embryonic origin of the *Arabidopsis* primary root and root-meristem initials. *Development* **120**, 2475–2487.

Schlereth, A., Möller, B., Liu, W., Kientz, M., Flipse, J., Rademacher, E. H., Schmid, M., Jürgens, G., and Weijers, D. (2010). MONOPTEROS controls embryonic root initiation by regulating a mobile transcription factor. *Nature* **464**, 913–916.

Schoof, H., Lenhard, M., Haecker, A., Mayer, K. F., Jürgens, G., and Laux, T. (2000). The stem cell population of Arabidopsis shoot meristems in maintained by a regulatory loop between the *CLAVATA* and *WUSCHEL* genes. *Cell* **100**, 635–644.

Sena, G., Wang, X., Liu, H. Y., Hofhuis, H., and Birnbaum, K. D. (2009). Organ regeneration does not require a functional stem cell niche in plants. *Nature* **457**, 1150–1153.

Serralbo, O., Pérez-Pérez, J. M., Heidstra, R., and Scheres B. (2006). Non-cell-autonomous rescue of anaphase-promoting complex function revealed by mosaic analysis of HOBBIT, an Arabidopsis CDC27 homolog. *Proc. Natl. Acad. Sci. U.S.A.* **103**, 13250–13255.

Smith, Z. R., and Long, J. A. (2010). Control of Arabidopsis apical–basal embryo polarity by antagonistic transcription factors. *Nature* **464**, 423–426.

Song, S. K., Hofhuis, H., Lee, M. M., and Clark, S. E. (2008). Key divisions in the early Arabidopsis embryo require POL and PLL1 phosphatases to establish the root stem cell organizer and vascular axis. *Dev. Cell* **15**, 98–109.

Song, S. K., Lee, M. M., and Clark, S. E. (2006). POL and PLL1 phosphatases are CLAVATA1 signaling intermediates required for Arabidopsis shoot and floral stem cells.*Development* **133**, 4691–4698.

Stahl, Y., Wink, R. H., Ingram, G. C., and Simon, R. (2009). A signaling module controlling the stem cell niche in Arabidopsis root meristems. *Curr. Biol.* **19**, 909–914.

Staswick, P. E., Serban, B., Rowe, M., Tiryaki, I., Maldonado, M. T., Maldonado, M. C., and Suza, W. (2005). Characterization of an Arabidopsis enzyme family that conjugates amino acids to indole-3-acetic acid. *Plant Cell* **17**, 616–627.

Stepanova, A. N., Robertson-Hoyt, J., Yun, J., Benavente, L. M., Xie, D. Y., Dolezal, K., Schlereth, A., Jürgens, G., and Alonso, J. M. (2008). TAA1-mediated auxin biosynthesis is essential for hormone crosstalk and plant development. *Cell* **133**, 177–191.

Stepanova, A. N., Yun, J., Likhacheva, A. V., and Alonso, J. M. (2007). Multilevel interactions between ethylene and auxin in Arabidopsis roots. *Plant Cell* **19**, 2169–2185.

Swarup, R., Perry, P., Hagenbeek, D., van der Straeten, D., Beemster, G. T., Sandberg, G., Bhalerao, R., Ljung, K., and Bennett, M. J. (2007). Ethylene upregulates auxin biosynthesis in Arabidopsis seedlings to enhance inhibition of root cell elongation. *Plant Cell* **19**, 2186–2196.

Takei, K., Takahashi, T., Sugiyama, T., Yamaya, T., and Sakakibara, H. (2002). Multiple routes communicating nitrogen availability from roots to shoots: a signal transduction pathway mediated by cytokinin. *J. Exp. Bot.* **53**, 971–977.

Tromas, A., Braun, N., Muller, P., Khodus, T., Paponov, I. A., Palme, K., Ljung, K., Lee, J. Y., Benfey, P., Murray, J. A., Scheres, B., and Perrot-Rechenmann, C. (2009). The AUXIN BINDING PROTEIN 1 is required for differential auxin responses mediating root growth. *PLoS One* **4**, e6648.

Ubeda-Tomás, S., Federici, F., Casimiro, I., Beemster, G. T., Bhalerao, R., Swarup, R., Doerner, P., Haseloff, J., and Bennett, M. J. (2009). Gibberellin signaling in the endodermis controls Arabidopsis root meristem size. *Curr. Biol.* **19**, 1194–1199.

Ubeda-Tomás, S., Swarup, R., Coates, J., Swarup, K., Laplaze, L., Beemster, G. T., Hedden, P., Bhalerao, R., and Bennett, M. J. (2008). Root growth in Arabidopsis requires gibberellin/DELLA signalling in the endodermis. *Nat. Cell Biol.* **10**, 625–628.

van den Berg, C., Willemsen, V., Hage, W., Weisbeek, P., and Scheres, B. (1995). Cell fate in the Arabidopsis root meristem determined by directional signalling. *Nature* **378**, 62–65.

van den Berg, C., Willemsen, V., Hendriks, G., Weisbeek, P., and Scheres, B. (1997). Short-range control of cell differentiation in the Arabidopsis root meristem. *Nature* **390**, 287–289.

Vanstraelen, M., Baloban, M., Da Ines, O., Cultrone, A., Lammens, T., Boudolf, V., Brown, S. C., De Veylder, L., Mergaert, P., and Kondorosi, E. (2009). APC/C-CCS52A complexes control meristem maintenance in the Arabidopsis root. *Proc. Natl. Acad. Sci. U.S.A.* **106**, 11806–11811.

Weijers, D., Schlereth, A., Ehrismann, J. S., Schwank, G., Kientz, M., and Jürgens, G. (2006). Auxin triggers transient local signaling for cell specification in Arabidopsis embryogenesis. *Dev. Cell* **10**, 265–270.

Wildwater, M., Campilho, A., Perez-Perez, J. M., Heidstra, R., Blilou, I., Korthout, H., Chatterjee, J., Mariconti, L., Gruissem, W., and Scheres, B. (2005). The *RETINOBLASTOMA-RELATED* gene regulates stem cell maintenance in Arabidopsis roots. *Cell* **123**, 1337–1349.

Willemsen, V., Bauch, M., Bennett, T., Campilho, A., Wolkenfelt, H., Xu, J., Haseloff, J., and Scheres, B. (2008). The NAC domain transcription factors FEZ and SOMBRERO control the orientation of cell division plane in Arabidopsis root stem cells. *Dev. Cell* **15**, 913–922.

Willemsen, V., Wolkenfelt, H., de Vrieze, G., Weisbeek, P., and Scheres, B. (1998). The *HOBBIT* gene is required for formation of the root meristem in the Arabidopsis embryo. *Development* **125**, 521–531.

Xu, J., Hofhuis, H., Heidstra, R., Sauer, M., Friml, J., and Scheres, B. (2006). A molecular framework for plant regeneration. *Science* **311**, 385–388.

Zhu, Y., Wang, Y., Li, R., Song, X., Wang, Q., Huang, S., Jin, J. B., Liu, C. M., and Lin, J. (2009). Analysis of interactions among the CLAVATA3 receptors reveals a direct interaction between CLAVATA2 and CORYNE in Arabidopsis. *Plant J.* **61**, 223–233.

CHAPTER FOUR

Shoot Apical Meristem Form and Function

Chan Man Ha,[*,†] Ji Hyung Jun,[*,†] *and* Jennifer C. Fletcher[*,†]

Contents

Abstract

The shoot apical meristem (SAM) generates above-ground aerial organs throughout the lifespan of higher plants. In order to fulfill this function, the meristem must maintain a balance between the self-renewal of a reservoir of central stem cells and organ initiation from peripheral cells. The activity of the pluripotent stem cell population in the SAM is dynamically controlled by complex, overlapping signaling networks that include the feedback regulation of meristem maintenance genes and the signaling of plant hormones. Organ initiation likewise requires the function of multifactor gene regulatory networks, as well as instructive cues from the plant hormone auxin and reciprocal signals from the shoot meristem. Floral meristems (FMs) are products of the

* Plant Gene Expression Center, USDA/UC Berkeley, Albany, California, USA
† Department of Plant and Microbial Biology, University of California, Berkeley, California, USA
Chan Man Ha and Ji Hyung Jun have contributed equally to this work

Current Topics in Developmental Biology, Volume 91
ISSN 0070-2153, DOI 10.1016/S0070-2153(10)91004-1

reproductive SAM that sustains a transient stem cell reservoir for flower formation. Regulation of FM activity involves both feedback loops shared with the SAM and floral-specific factors. Recent studies have rapidly advanced our understanding of SAM function by adopting newly developed molecular and computational techniques. These advances are becoming integrated with data from traditional molecular genetics methodologies to develop a framework for understanding the central principles of SAM function.

1. Introduction

Plants continuously generate organs from stem cells located at their growing tips. During plant development, the stem cells function as a reservoir of undifferentiated cells sustained through continuous cell divisions. In the shoot, the stem cells are located within a specialized structure called the shoot apical meristem (SAM) (Fig. 4.1A) and are maintained in a pluripotent state by signals from the neighboring cells. As the stem cells divide, some daughter cells are displaced toward the periphery of the SAM to be consumed in the production of lateral organs, while others are retained at the shoot apex to replenish the reservoir. During vegetative development the SAM produces leaves, whereas during reproductive development the main products of the SAM are flowers, which are derived from floral meristems (FMs) that contain their own transient stem cell population (Fig. 4.1A, B). Whatever the developmental stage, keeping the delicate balance between self-renewal of meristem cells and continuous organ development needs to be guaranteed. Because the SAM is a dynamic structure regulated by complicated signaling networks consisting of the activities of many genes and plant hormones, coordination among them is a prerequisite for maintaining the SAM. Much progress has been achieved in studying SAM establishment and maintenance via molecular genetic methods, which have been more recently coupled with live imaging technologies and computational modeling. Here we discuss the molecular mechanisms that organize and control SAM function with an emphasis on the dicotyledonous model plant *Arabidopsis thaliana*.

2. Shoot Apical Meristem Organization

2.1. SAM structure

Classical studies indicate that the angiosperm SAM is organized into overlapping cell layers and domains (Fig. 4.1A). The Arabidopsis SAM consists of three cell layers, L1–L3, which remain clonally distinct from one another due to their specific cell division patterns (Irish and Sussex, 1992; Leyser and

Furner, 1992; Steeves and Sussex, 1989). The single-layered L1 and L2 cells comprise the tunica and divide anticlinally, perpendicular to the plane of the meristem. These cell layers generate the epidermis and subepidermis, respectively, of the shoots, leaves, and flowers. The multitiered L3 cells comprise the corpus and divide in all planes, forming the internal tissues such as the stem vasculature and pith.

The SAM also consists of three distinct functional domains (Steeves and Sussex, 1989). The central zone (CZ) is located at the summit of the SAM and contains the slowly dividing stem cell population. This region does not consist of permanent initial cells but rather acts as a reservoir through which individual stem cells pass (Ruth *et al.*, 1985). Stem cell division replenishes the CZ and displaces the daughter cells outward into the surrounding peripheral zone (PZ), where cells are rapidly dividing and new organs are initiated, as well as downward into the rib zone (RZ) that provides cells for the internal tissues of the stem and lateral organs. At the top of the RZ is positioned a small group of cells called the organizing center (OC), which acts as a niche that sustains the overlying stem cell reservoir (Carles and Fletcher, 2003). Although clear boundaries between the three domains have not been observed, expression studies reveal that the cells in each zone have different gene expression patterns (Nishimura *et al.*, 1999; Yadav *et al.*, 2009) and thus distinct molecular as well as functional characteristics.

Clonal analysis demonstrated that the identity of any given shoot meristematic cell depends on its position rather than its lineage (Poethig, 1987; Stewart and Dermen, 1975). Beyond the L1 and L2 anticlinal cell division patterns that maintain the clonal layers, no fixed cell lineage patterns occur in the SAM (Furner and Pumfrey, 1992; Poethig *et al.*, 1986). Because plant cells cannot move freely, they must therefore receive and interpret their positional information from their neighbors in the SAM. The L1 cells control the division patterns of the subtending L2 cells and prevent their differentiation, whereas the CZ has an inhibitory effect on the PZ that prevents it from generating new stem cell reservoirs (Reinhardt *et al.*, 2003a). Thus, the dynamic properties of the SAM are maintained by the coordination of cellular behavior between different zones and layers via intercellular communication (Fletcher, 2002).

2.2. SAM formation during embryogenesis

The SAM forms gradually during embryogenesis. Patterning of the developing plant embryo occurs primarily through the control of cell division rates, locations, and planes, coupled with cell elongation. In eight-cell octant-stage Arabidopsis embryos, transverse cell divisions partition the embryo into an apical and a basal domain (Takada and Tasaka, 2002; West *et al.*, 1993). Additional rounds of asymmetric cell division yield the 16-cell-stage embryo and then the globular stage embryo, the upper hemisphere of which now consists of an outer epidermal cell layer and an internal hypodermal layer. Beginning at this stage the cells that form the SAM become confined to the most apical, central portion of the embryo (chapter 1; Barton and Poethig, 1993). Further division of the apical cells produces the heart-stage embryo, wherein the embryonic leaves, called cotyledons, emerge as two distinct bulges of cells from the apical, peripheral region of the embryo. Stem cells first become detectable at this stage in the apical cells that lie between the cotyledon primordia (Fletcher *et al.*, 1999). The embryo then transitions to the torpedo stage, in which the cells in the hypodermal layer divide periclinally to generate the characteristic three-layered SAM structure. As the embryo matures, further anticlinal divisions in the upper two layers and mixed-plane divisions in the underlying cells beneath culminate in the formation of a distinct dome of SAM cells at the shoot apex.

The SAM is gradually patterned during embryogenesis via the successive activation of meristem-restricted genes. *WUSCHEL* (*WUS*), which encodes a WOX family homeodomain transcription factor, is one of the earliest genes expressed during embryonic SAM initiation (Mayer *et al.*, 1998). Mutations in *WUS* lead to misspecification of stem cells and premature termination of the SAM (Laux *et al.*, 1996). *WUS* is initially expressed at the 16-cell stage of embryogenesis in the four subepidermal cells of the apical domain (Mayer *et al.*, 1998). Through subsequent rounds of asymmetric cell division, *WUS* expression becomes confined to the deeper OC region of the developing SAM. It is still unclear how *WUS* is initially activated during embryogenesis. Because the plant hormone cytokinin (CK) acts to specifically localize induction of *WUS* transcription during postembryonic SAM development (Gordon *et al.*, 2009), it is possible that CK may contribute to the initial establishment of the OC by inducing *WUS* expression. However, the synthetic CK response reporter *TCS : GFP* is not expressed in the apical tier during early embryogenesis (Muller and Sheen, 2008), indicating that apical CK levels are low during this critical developmental window.

The *CUP-SHAPED COTYLEDON* (*CUC*) and *SHOOT MERISTEMLESS* (*STM*) genes are also expressed early during embryo development, and are required for SAM establishment and maintenance. The closely related *CUC1, CUC2*, and *CUC3* genes encode NAC domain transcription factors homologous to the petunia NO APICAL MERISTEM protein (Aida *et al.*, 1999; Takada *et al.*, 2001; Vroemen *et al.*, 2003). Loss-of-function *cuc* single mutants display mild phenotypic defects, whereas *cuc* double mutants fail to form a SAM and exhibit cotyledon fusion (Aida *et al.*, 1999; Takada *et al.*, 2001; Vroemen *et al.*, 2003). The *CUC* genes are first expressed at the apex of the globular stage embryo between the presumptive cotyledon primordia and are involved in the induction of *STM* expression (Aida *et al.*, 1999). *STM* encodes a class 1 KNOTTED-LIKE HOMEOBOX (KNOX1) transcription factor that is required to establish and maintain the SAM (Barton and Poethig, 1993; Long *et al.*, 1996). *STM* is initially expressed during the globular stage in the presumptive SAM cells at the embryo apex (Long *et al.*, 1996), such that the *STM* and *CUC* expression domains overlap during early embryogenesis. However, as embryogenesis proceeds, the *STM* and *CUC* expression domains resolve into complementary patterns, with *STM* expressed throughout the SAM and the *CUC* genes confined to the boundary between the SAM and the cotyledons to allow the separation of the two cotyledon primordia (Aida *et al.*, 1999; Takada *et al.*, 2001; Vroemen *et al.*, 2003). *PIN-FORMED1* (*PIN1*) and *PINOID* (*PID*), which mediate auxin-dependent morphogenesis, as well as *TCP* transcription factor genes are expressed in the presumptive cotyledon primordia from the torpedo stage and appear to regulate the initial spatial restriction of *CUC* gene expression during early embryo development (Aida *et al.*, 2002; Furutani *et al.*, 2004; Koyama *et al.*, 2007; Palatnik *et al.*, 2003).

By the early heart stage, genes that restrict SAM activity, *CLAVATA1* (*CLV1*) and *CLV3*, are induced in the presumptive SAM between the developing cotyledons. As embryogenesis proceeds, their expression domains become restricted to the interior and superficial regions of the SAM, respectively (Clark *et al.*, 1996; Fletcher *et al.*, 1999). Maintenance of *CLV1* expression requires STM function (Long and Barton, 1998), whereas maintenance of high-level *CLV3* expression depends on both WUS and STM (Brand *et al.*, 2002). In sum, patterning of the embryonic SAM is a gradual and dynamic process that is essential for the successful establishment of the stem cells and the OC.

3. Shoot Apical Meristem Maintenance

3.1. The CLV–WUS pathway

A fundamental feature of the SAM (Fig. 4.1C) is that it maintains a stable stem cell population throughout plant development despite the constant displacement of their daughter cells for lateral organ initiation. Much of our understanding about the molecular mechanisms that maintain a functional SAM is derived from molecular genetic studies. *WUS* has been shown to be a key regulator of shoot stem cell maintenance. Loss-of-function *wus* mutants fail to organize a functional embryonic SAM (Fig. 4.1E), with the CZ occupied by cells that are larger and more vacuolated than normal CZ cells (Mayer *et al.*, 1998; Schoof *et al.*, 2000). After germination, *wus* plants develop multiple transient meristems randomly across the shoot apex that prematurely terminate after forming a few organs (Laux *et al.*, 1996). *WUS* is expressed in the OC and specifies the identity of the overlying stem cells (Mayer *et al.*, 1998), suggesting that WUS acts in a noncell-autonomous manner to promote stem cell renewal. Not only is WUS

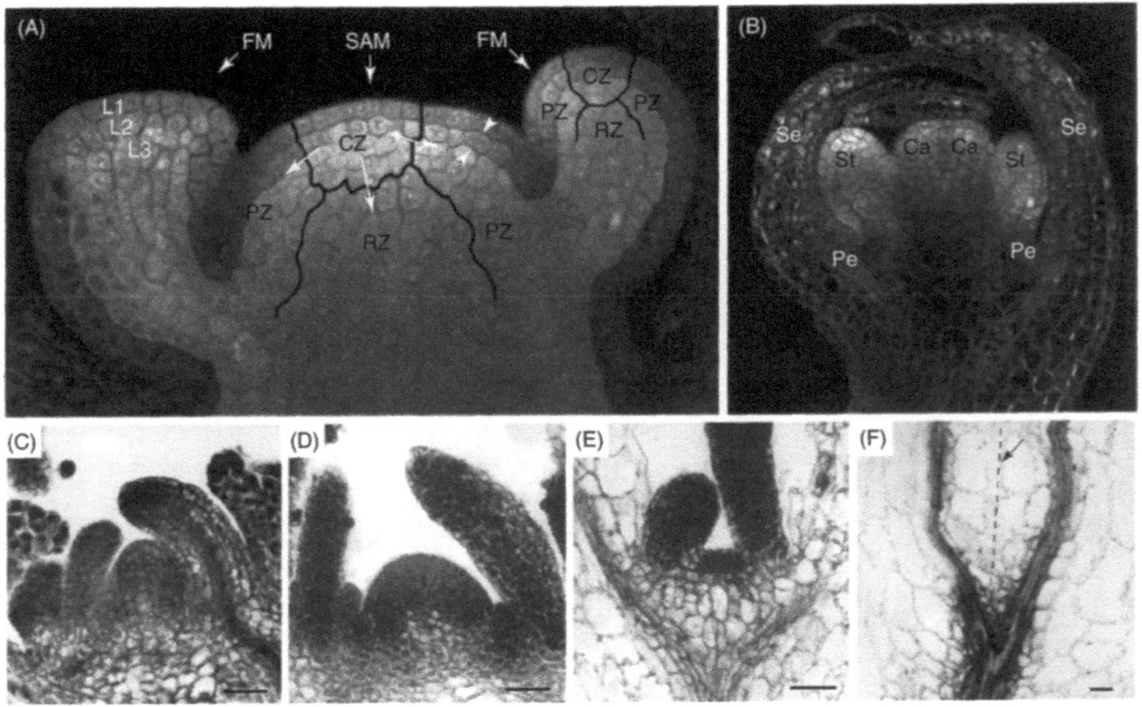

Figure 4.1 *(Continued)*

necessary to maintain stem cell identity in the SAM, but ectopic expression of *WUS* is sufficient to induce shoot stem cell identity in organ primordia and root tissues (Gallois *et al.*, 2004).

Conversely, the *CLV* signal transduction pathway is required throughout development to restrict stem cell accumulation (Fig. 4.1D) by limiting the size of the *WUS* expression domain (reviewed in Baurle and Laux, 2003; Fletcher, 2002). CLV3 is a member of the CLV3/ESR (CLE) family of proteins that contains a short conserved CLE domain (Cock and McCormick, 2001), and encodes a secreted 13 amino-acid signaling peptide (Fletcher *et al.*, 1999) derived from the CLE domain by proteolytic processing (Ohyama *et al.*, 2009). *CLV3* is transcribed in the stem cells in a *WUS*-dependent manner (Fletcher *et al.*, 1999; Schoof *et al.*, 2000), and its secretion through the secretary pathway into the extracellular space is critical for its function (Rojo *et al.*, 2002). Imaging experiments have shown that CLV3 protein moves outward from the stem cell reservoir and can act over a distance of several cell diameters (Lenhard and Laux, 2003), confirming its role as an intercellular signal in the SAM.

The CLV3 signal is perceived by two parallel receptor systems that converge on the regulation of *WUS* expression. One receptor module consists of the leucine-rich repeat (LRR) receptor-like serine/threonine kinase CLV1 (Clark *et al.*, 1997). *CLV1* is specifically transcribed in the interior region of the SAM, underlying the *CLV3* domain and encompassing the *WUS* expression domain in the OC (Clark *et al.*, 1997; Fletcher *et al.*, 1999). CLV1 proteins form plasma membrane-localized homodimers (Bleckmann *et al.*, 2010), and the CLV3 polypeptide directly interacts with the CLV1 extracellular LRR domain (Ogawa *et al.*, 2008). The range of

Figure 4.1 Organization of the Arabidopsis shoot apical meristem (SAM). (A) Confocal laser-scanning micrograph through an inflorescence SAM and its adjacent floral meristems (FMs). Superimposed on the micrograph are the meristem layers and zones. The shaded domains depict the three cell layers: the epidermal (L1) layer, the subepidermal layer (L2), and the internal corpus layers (L3). The two arrowheads denote an anticlinal cell division occurring in the L2. The black outlines represent the approximate boundaries between the meristematic zones: the central zone (CZ), the peripheral zone (PZ), and the rib zone (RZ). (B) Confocal laser-scanning micrograph of a stage 6 Arabidopsis FM. The floral meristem terminates after the production of cells that are incorporated into the four whorls of organs—sepals (Se), petals (Pe), stamens (St), and carpels (Ca). Longitudinal sections through the Arabidopsis vegetative SAMs of 7-day-old (C) wild type, (D) *clavata1*-4 (*clv1*-4), (E) *wuschel*-1 (*wus-1*), and (F) *shoot meristemless*-11 (*stm*-11) plants. (D) The SAM of *clv1*-4 plants is broader and taller than the wild-type SAM (C). (E) The shoot apex of *wus*-1 plants is small and flat, and will terminate after the initiation of the two leaf primordia. (F) Plants carrying the *stm*-11 mutation lack a SAM and the two cotyledons are fused at their base. The arrow indicates the boundary between the two cotyledons. Scale bars = 50 μm. Reproduced from *Trends in Plant Science* (Carles and Fletcher, 2003).

CLV3 action is limited due to its sequestration by the CLV1 receptor complex in cells that lie beneath the stem cell reservoir, which ensures continued SAM activity by preventing the CLV3 signal from spreading into the OC and repressing *WUS* transcription (Lenhard and Laux, 2003). The CLV1-related BARELY ANY MERISTEM (BAM1), BAM2, and BAM3 receptor-like kinases appear to act in parallel with CLV1 to restrict stem cell accumulation in the center of the SAM (DeYoung and Clark 2008), but are proposed to additionally play a role in stem cell maintenance on the SAM periphery (DeYoung and Clark, 2008; DeYoung *et al.*, 2006).

The second receptor module consists of CLV2 and CORYNE (CRN). *CLV2* encodes a LRR receptor-like protein that is similar to CLV1 but lacks a kinase domain (Jeong *et al.*, 1999). CRN is a plasma membrane-localized serine/threonine protein kinase that contains only a short extracellular domain (Muller and Sheen, 2008; Zhu *et al.*, 2010). CRN forms homodimers and also directly interacts with CLV2 through its transmembrane domain, thus constituting a complete transmembrane receptor kinase module (Bleckmann *et al.*, 2010; Zhu *et al.*, 2010). Genetic evidence indicates that CLV2 and CRN act together but independently of the CLV1 receptor pathway (Clark *et al.*, 1995; Kayes and Clark, 1998; Muller and Sheen, 2008). These two CLV3-mediated signaling pathways negatively regulate the size of the *WUS* expression domain in the underlying OC (Brand *et al.*, 2000; Schoof *et al.*, 2000). Thus, stem cell homeostasis in the SAM is achieved by a negative feedback regulatory loop composed of *WUS* and the CLV3-mediated signaling pathways (Fig. 4.2), whereby departure of cells from the stem cell reservoir causes reduced *CLV3* expression levels, which leads to reduced negative signaling through the parallel receptor pathways. This in turn permits the elevation of *WUS* expression in the OC and results in an enhancement of the WUS-mediated signal to induce the production of more stem cells.

Several intracellular components of the CLV1 signal transduction pathway have been identified. The type-2C kinase-associated protein phosphatase KAPP negatively regulates CLV1 signal transduction by directly interacting with and dephosphorylating the CLV1 kinase domain (Stone *et al.*, 1998; Williams *et al.*, 1997). Two closely related members of a novel type of PP2C family, *POLTERGEIST* (*POL*) and *PLL1*, act as rate-limiting modulators of stem cell specification (Song and Clark, 2005) that are intermediates of the CLV1 complex in the regulation of *WUS* expression and stem cell maintenance (Song *et al.*, 2006). A Rho GTPase-related protein, which is related to Ras small GTPase superfamily members in animals and is typically associated with cytosolic mitogen-activated protein (MAP) kinase cascades, is an additional component of the CLV complex (Trotochaud *et al.*, 1999). However, whether a MAP kinase cascade is involved in intracellular CLV signal transduction remains unknown.

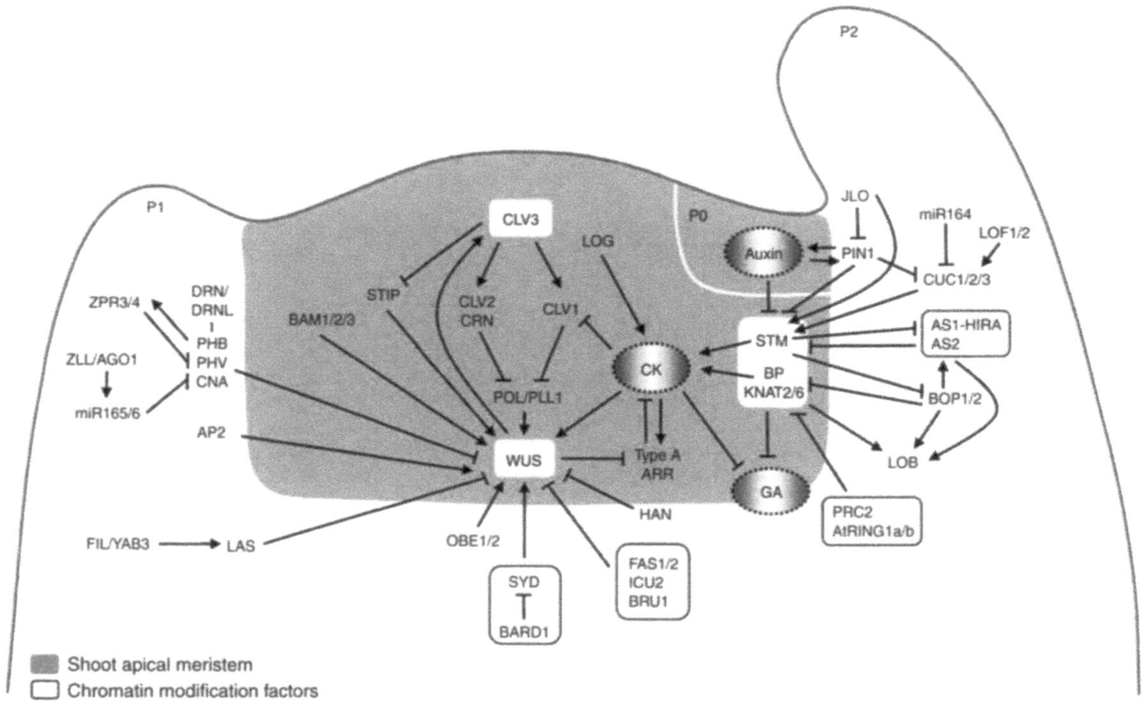

Figure 4.2 Regulatory relationships between genes acting in the shoot apical meristem and lateral organ primordia. The SAM region is shaded. Chromatin modification factors are boxed, and key meristem regulators are depicted in white squares. Hormones are depicted in dashed ovals. Positive (arrow) and negative (bar) regulatory interactions between genes and hormones are shown irrespective of whether the interaction is indirect or direct. P0, incipient organ primordium; P1, newly arising organ primordium; P2, outgrowing organ primordium.

3.2. The KNOX1 pathway

Genetic screens also identified a parallel pathway to the WUS–CLV pathway that is necessary for SAM establishment and maintenance. This pathway is mediated by the activity of *KNOX1* genes, principally *STM* in Arabidopsis and *KN1* in maize. These genes suppress differentiation throughout the SAM to permit stem cell proliferation (Long *et al.*, 1996). Loss-of-function *stm* mutants fail to establish a SAM (Fig. 4.1F) and form fused cotyledons but no other elements of the shoot system (Long *et al.*, 1996). The suppression of meristem cell differentiation by STM occurs mainly via repression of the MYB-related gene *ASYMMETRIC LEAVES1* (*AS1*) (Byrne *et al.*, 2002), and the *KNOX1* gene *BREVIPEDICILLUS* (*BP*) acts redundantly with *STM* in this process (see below). STM and WUS act in largely independent pathways in SAM regulation, although increased *WUS* expression can partially compensate for the loss of *STM* function (Lenhard *et al.*, 2002).

3.3. Meristem regulation by phytohormones

Although information about cellular mechanisms controlled by phytohormones in the SAM is limited, studies suggest that the heterogenous distribution of hormone activity contributes to patterning processes in the SAM. The spatial regulation of CK biosynthesis and cellular responses to CK in the central domain are important for SAM function, whereas localization of high gibberellin (GA) and auxin activities is closely associated with lateral organ development (Fig. 4.2).

CKs are adenine-derived phytohormones that are positive regulators of cell proliferation and play a key role in SAM function. CKs stimulate the formation of new SAMs in plant tissue culture (Skoog and Miller, 1957). Plants with mutations in multiple CK receptor genes have smaller SAMs than wild-type plants (Higuchi *et al.*, 2004; Nishimura *et al.*, 2004), and the loss of CK perception causes a failure in SAM establishment and consequently vegetative growth arrest (Skylar *et al.*, 2010), indicating that CK is essential for SAM activity. CK action in the SAM is regulated at multiple levels, including biosynthesis, activation, and cellular response.

KNOX1 genes mediate CK biosynthesis in the SAM by directly activating several *ISOPENTENYL TRANSFERASE* (*IPT*) genes, which encode enzymes that catalyze the initial step of CK production (Sakakibara, 2006). Activation of *IPT* gene expression and exogenous application of CK partially rescues the *stm* phenotype, whereas the *wooden leg* (*wol*) mutation in a CK receptor gene enhances the phenotype of a weak *stm* allele (Jasinski *et al.*, 2005; Yanai *et al.*, 2005). Thus, one mechanism through which *KNOX1* genes sustain SAM activity is via their promotion of CK accumulation at the shoot apex. In addition the *LONELY GUY* (*LOG*) gene, encoding a phosphoribohydrolase that acts at the final step of bioactive CK synthesis, is required to maintain SAM activity in rice (Kurakawa *et al.*, 2007). The *log* plants display premature SAM termination resulting from reduced levels of active CK in the SAM. *LOG* expression is confined to the upper part of the SAM encompassing the stem cell domain, indicating that, at least in the shoot tip, CK functions as a local paracrine signal (Kurakawa *et al.*, 2007) and that region-specific modulation of CK activity is important for maintaining full SAM function.

CK signaling in Arabidopsis is perceived by a family of HISTIDINE KINASE (AHK) receptors, which transduce the signal to two classes of Arabidopsis RESPONSE REGULATOR (ARR) transcription factors. The type B ARRs activate CK-induced genes, whereas the type A ARRs down-regulate CK signaling through a negative feedback loop. In Arabidopsis, four type-A *ARR* genes are direct targets of *WUS* suppression, establishing a mechanistic connection between the CLV/WUS network and hormonal control of meristem maintenance (Leibfried *et al.*, 2005). Over-expression of a constitutively active form of the WUS target gene *ARR7* induces meristem arrest in transgenic plants (Leibfried *et al.*, 2005), whereas a mutation in the maize type-A ARR homolog ABPHYL1 causes SAM enlargement (Giulini *et al.*, 2004). Thus, type-A *ARR* genes negatively influence meristem size and, by negatively regulating these genes, WUS promotes CK-dependent cell division and stem cell proliferation.

Conversely, CK induces the transcription of meristem-promoting genes. CK signaling plays a role in activating the *WUS*-related homeobox gene *STIMPY* (*STIP/WOX9*) in SAM tissues (Skylar *et al.*, 2010). *STIP* is required for maintaining meristem cell division and growth, and functions downstream of CLV3 in positively regulating *WUS* expression (Wu *et al.*, 2005). CK treatment likewise promotes *CLV3* and *WUS* transcription (Buechel *et al.*, 2010; Lindsay *et al.*, 2006), and computational analysis suggests that positive regulation of *WUS* expression by CK occurs via both CLV-dependent and CLV-independent mechanisms (Gordon *et al.*, 2009). Thus, the WUS and CK-signaling pathways reinforce each other through multiple positive feedback loops to maintain the stem cell niche. The expression patterns of *WUS* and the CK receptor gene *AHK4* overlap in the OC, whereas type-A *ARR* gene expression occurs in a complementary region of the SAM. Localized perception of CK, perhaps generated from the overlying stem cells through the activity of *LOG*, via its receptors in the interior tissues, can therefore lead to the induction of *WUS* in the OC. WUS, in turn, promotes stem cell fate in the overlying cells, resulting in a positive reinforcement mechanism to ensure the continuous repatterning of the *WUS*-expressing OC that is required as meristem cells are displaced during growth (Gordon *et al.*, 2009).

In contrast to CK, the growth regulator gibberellin promotes cell differentiation by inducing longitudinal cell expansion and cytoskeletal rearrangement (Shibaoka and Nagai, 1994). As such its activity is antagonistic to SAM function, which perpetuates populations of dividing, undifferentiated cells, and indeed CK has been shown to act antagonistically to GA (Greenboim-Wainberg *et al.*, 2005). KNOX1 proteins maintain low GA levels in the Arabidopsis SAM by suppressing the expression of *GA20-oxidase1*, which encodes a rate-limiting enzyme for GA biosynthesis (Hay *et al.*, 2002). Conversely, maize KN1 directly induces expression of the *ga2ox1* gene encoding a GA-inactivating enzyme by binding to an intronic *cis*-regulatory element containing two TGAC motifs (Bolduc and Hake, 2009). Thus, KNOX1 proteins also modulate GA accumulation by activating genes involved in GA catabolism (Kessler *et al.*, 2006). Like *GA20ox1*, Arabidopsis *GA2ox2* mRNA levels increase in response to *STM* induction, although it is not clear if this modulation is direct or mediated via CK biosynthesis (Jasinski *et al.*, 2005). Because *GA2ox* expression is detected at the base of the SAM, activation of these genes by KNOX1 proteins may prevent the influx of active GAs from the leaf primordia into the SAM. *KNOX1* gene function therefore promotes SAM activity by repressing GA biosynthesis as well as inducing CK biosynthesis, coordinating these functions to establish an environment that favors cell proliferation.

3.4. Regulation of meristem gene activity

In animal systems, bivalent states of both repressive H3K27 methylation and active H3K4 methylation at genes involved in tissue differentiation are responsible for keeping the loci silent in embryonic stem (ES) cells while simultaneously priming them for rapid activation (Bernstein *et al.*, 2006), indicating an important role for epigenetic factors in the control of stem cell homeostasis. Although the chromatin states of plant stem cells remain to be analyzed, chromatin remodeling factors in several processes—including nucleosome assembly/disassembly, ATP-dependent chromatin remodeling, and histone modification—play important roles in stem cell homeostasis and SAM function in plants.

Polycomb group (PcG) proteins represent one mechanism for mediating stem cell homeostasis that has been well described in animal systems (Francis and Kingston, 2001). PcG proteins are found in at least two complexes, PRC1 and PRC2, which maintain stable transcriptional repression at their target loci. Multiple PRC2 complexes exist in plants, and one complex containing CURLY LEAF/SWINGER (CLF/SWN), EMBRYONIC FLOWER2 (EMF2), VERNALIZATION2 (VRN2), and FERTILIZATION INDEPENDENT ENDOSPERM (FIE) (Chanvivattana *et al.*, 2004) regulates *KNOX1* gene expression by repressing *STM* and *KNAT2* in leaf primordia, and by down-regulating *BP* expression (Katz *et al.*, 2004). CLF and SWN encode closely related histone methyltransferases that direct the trimethylation of histone 3 lysine 27 (H3K27me3), a chromatin mark that is recognized by PRC1 in animals and an LIKE-HETEROCHROMATIN PROTEIN1 (LHP1)-containing complex in plants. CLF protein is directly recruited to the *STM* locus, and *STM* mis-expression in *clf* plants correlates with reduced levels of repressive H3K27me3 marks at the locus (Schubert *et al.*, 2006). De-repression of *KNOX1* gene expression and ectopic meristem formation were also observed in plants carrying mutations in both AtRING1a and AtRING1b, which encode homologs of the animal PRC1 core component RING1 (Xu and Shen, 2008). The interaction of AtRING1a with CLF suggests that a PRC1-like complex acts to suppress *KNOX1* gene expression in conjunction with the PRC2 complex.

The activities of several chromatin-remodeling factors converge on *WUS* regulation. *WUS* is a direct target of induction by *SPLAYED* (*SYD*), which encodes a SWI/SNF-type chromatin-remodeling ATPase involved in transcriptional regulation (Kwon *et al.*, 2005). The closest homolog of SYD, BRAHMA (AtBRM), also acts in SAM maintenance (Farrona *et al.*, 2004). As a counteracting factor of SYD, the BRCA1-associated RING domain 1 protein BARD1 represses *WUS* transcription by interfering with SYD function (Han *et al.*, 2008). The *FASCIATA1* (*FAS1*) and *FAS2* genes, which encode the p150 and p60 subunits of chromatin assembly factor-1 (CAF-1) that mediates nucleosome assembly during DNA replication, are required for proper SAM organization and for stable, OC-restricted *WUS* expression (Kaya *et al.*, 2001). A similar function is proposed for BRUSHY1 (BRU1)/MGOUN3/TONSOKU, which encodes a novel protein containing LRRs and potential DNA-binding motifs (Guyomarc'h *et al.*, 2004; Suzuki *et al.*, 2004; Takeda *et al.*, 2004). BRU1 appears to be involved in the structural and functional stabilization of heterochromatin, and BRU1 and CAF-1 are proposed to contribute additively to post-replication stabilization of chromatin structure (Takeda *et al.*, 2004). Activity of the DNA polymerase α-subunit INCURVATA2 (ICU2) is also required to restrict the *WUS* expression domain (Turck *et al.*, 2007), and the synergistic interaction between *icu2* and *fas1* suggests a functional relationship between the DNA replication and chromatin-remodeling machineries for *WUS* regulation (Barrero *et al.*, 2007).

The *OBERON1* (*OBE1*), *OBE2*, *APETALA2* (*AP2*), and *HANABA TARANU* (*HAN*) genes regulate SAM activity by controlling *WUS* expression at the transcriptional level (Casson *et al.*, 2009; Saiga *et al.*, 2008; Wurschum *et al.*, 2006; Zhao *et al.*, 2004). The OBE1 and OBE2 proteins contain a PHD finger domain and coiled-coil domain with a predicted role in transcriptional control. Premature termination of the shoot meristem after the generation of several leaf primordia and reduced *WUS* expression in *obe1 obe2* plants indicates a redundant role for OBE1 and OBE2 in SAM maintenance (Casson *et al.*, 2009; Saiga *et al.*, 2008). The onset of *WUS* and *CLV3* induction is unchanged in young *obe1 obe2* embryos, but during later stages of embryo development their expression is altered resulting in the terminal differentiation of SAM stem cells (Saiga *et al.*, 2008; Tucker *et al.*, 2008). Thus, *OBE1* and *OBE2* maintain shoot meristem stem cells via the regulation of WUS and CLV3 activities. The study of a dominant-negative allele of the *AP2* floral homeotic gene identifies a novel pathway in the SAM regulated by the redundant function of *AP2* and unknown members of the AP2 subfamily. The I28 allele suppresses both *WUS* and *CLV3* expression in the SAM, and wild-type AP2 protein promotes stem cell maintenance either by negatively regulating CLV signaling or by independently antagonizing its affects (Wurschum *et al.*, 2006). In contrast, the GATA-3-like transcription factor HAN restricts SAM size by repressing *WUS* expression in the L1 and L2 cell layers through a CLV-independent pathway (Zhao *et al.*, 2004). Because *HAN* is transcribed at the boundary between the SAM and initiating organ primordia, HAN appears to function in a noncell-autonomous fashion to delimit the boundary that confines *WUS* expression to the OC.

It is interesting to note that, although the constant and precise *WUS* expression pattern seems to be controlled by many regulators, a 57-bp *cis*-regulatory sequence is sufficient to direct proper *WUS* transcription *in planta* (Baurle and Laux, 2005). This result suggests that regulation of the *WUS* locus may occur at the level of integrating multiple inputs as trans-acting complexes.

4. Lateral Organ Initiation

In higher plants, lateral organs emerge from the PZ of the SAM. One of the earliest indications of a developmental switch from SAM cell fate to lateral organ cell fate is the down-regulation of *KNOX1* gene expression (Jackson *et al.*, 1994; Long *et al.*, 1996). In Arabidopsis, the *KNOX1* genes *STM, BP, KNAT2*, and *KNAT6* are expressed in overlapping patterns within or adjacent to the SAM but are down-regulated in initiating lateral organ primordia (Belles-Boix *et al.*, 2006; Lincoln *et al.*, 1994; Long *et al.*, 1996). Ectopic *KNOX1* gene expression in Arabidopsis organ primordia leads to the formation of ectopic meristems on the adaxial leaf surface (Chuck *et al.*, 1996; Long *et al.*, 1996). Thus, repression of *KNOX1* expression outside the SAM is essential for proper organ differentiation.

Several genes are involved in repressing *KNOX1* gene expression during lateral organ formation (Fig. 4.2). *AS1*, which encodes a MYB-domain transcription factor, is transcribed in organ primordia cells in a complementary pattern to the *KNOX1* genes (Byrne *et al.*, 2000). *STM* is required to repress *AS1* expression in the SAM to maintain its integrity (Byrne *et al.*, 2000). In turn, *AS1* down-regulates *BP, KNAT2*, and *KNAT6* in the developing leaf primordia (Byrne *et al.*, 2000; Ori *et al.*, 2000). *AS2*, which encodes a plant-specific ASL/LOB-domain (LBD) protein, has a similar genetic and molecular function to *AS1* and the two proteins physically interact *in vitro* (Iwakawa *et al.*, 2002; Ori *et al.*, 2000; Semiarti *et al.*, 2001; Shuai *et al.*, 2002; Xu *et al.*, 2003). AS1 and AS2 also interact with the histone chaperone protein HIRA to form an AS1–AS2–HIRA repressive protein complex (Phelps-Durr *et al.*, 2005). This complex binds directly to two sites in the *BP* and *KNAT2* promoters that flank an enhancer element required for *KNOX1* expression in the leaf (Guo *et al.*, 2008). The protein complexes at these two sites are proposed to interact to establish a DNA loop that can sequester the enhancer elements and prevent their function at the promoters, creating an inactive chromatin state at the *BP* and *KNAT2* loci in developing organ primordia (Guo *et al.*, 2008). The chromatin-remodeling genes *PICKLE* (*PKL*), *SERRATE* (*SE*), and RNA-DEPENDENT RNA POLYMERASE6 (*RDR6/SDE1/SGS2*) also suppress *KNOX1* ectopic expression in organ primordia (Li *et al.*, 2005; Ori *et al.*, 2000), providing additional evidence that the stable silencing of *KNOX1* genes in differentiated tissues is maintained by epigenetic mechanisms.

The *BLADE-ON-PETIOLE 1* (*BOP1*) and *BOP2* genes are likewise involved in the acquisition of proper cellular differentiation by leaf primordia cells through the regulation of *KNOX1* gene expression (Ha *et al.*, 2003; 2007). *BOP1* and *BOP2* encode BTB/POZ domain and ankyrin repeat-containing proteins that are expressed at the base of initiating leaf primordia (Ha *et al.*, 2004; Hepworth *et al.*, 2005; Norberg *et al.*, 2005). *BOP1* and *BOP2* negatively regulate *KNOX1* gene expression in developing leaves, while positively regulating three *ASL/LBD* genes, including *AS2* (Ha *et al.*, 2003; 2007). BOP1 functions as a transcriptional coactivator that binds directly to the promoter of its target gene *AS2* (Jun *et al.*, 2010). *BOP1* and *BOP2* activity is required for *AS2* activation specifically in the proximal region of the leaf to condition proper organ differentiation through the repression of *KNOX1* gene expression. Furthermore, repression of *BOP1* and *BOP2* expression in the SAM by *STM* is crucial for embryonic SAM development (Jun *et al.*, 2010).

The plant hormone auxin is a key factor in lateral organ initiation. Auxin concentration maxima are established at the flanks of the SAM where new organ primordia initiate through the activity of PIN1 auxin efflux carriers and AUXIN1 (AUX1) influx carriers (Benkova *et al.*, 2003; Reinhardt *et al.*, 2000; 2003b). Further, exogenous auxin application induces lateral organ initiation from the flanks of the SAM (Reinhardt *et al.*, 2000; 2003b), and many plants with defects in auxin biosynthesis, polar transport, or signaling display lateral organ formation phenotypes (Bennett *et al.*, 1995; Cheng *et al.*, 2007; 2008; Friml *et al.*, 2003; 2004; Furutani *et al.*, 2007; Mayer *et al.*, 1993; Mravec *et al.*, 2008; Okada *et al.*, 1991; Treml *et al.*, 2005; Vieten *et al.*, 2005). *STM* expression in the SAM is complementary to that of the auxin reporter genes *PIN1* and *DR5*, with *STM* down-regulation in newly arising organ primordia occurring coincidently with up-regulation of *PIN1* (Heisler *et al.*, 2005). PIN1 acts together with AS1 in repressing *BP* expression to promote leaf and flower initiation from the SAM, and conversely PIN1-mediated auxin action is opposed by KNOX1 activity (Hay *et al.*, 2006). Thus, auxin and PIN1 activity function antagonistically to the *KNOX1* genes to establish lateral organ fate in groups of cells on the flanks of the SAM.

As lateral organs initiate, morphological boundaries are formed between the organ and the SAM. These boundary regions have distinctive growth characteristics and cell identities, and thus different molecular signatures, than the neighboring meristematic or organ primordia cells (Aida and Tasaka, 2006; Rast and Simon, 2008). Many transcription factor genes are involved in organ boundary formation. In addition to their involvement in SAM initiation, the *CUC* genes also function in the specification of the boundary between the meristem and the initiating organs (Aida *et al.*, 1997; 1999; Hibara *et al.*, 2006; Takada *et al.*, 2001; Vroemen *et al.*, 2003). The *CUC* genes are proposed to act redundantly to inhibit boundary cell proliferation, which is a prerequisite for organ separation from the SAM. *CUC* gene expression is restricted to boundary region by the activities of *STM* and auxin (Aida *et al.*, 1999; 2002; Friml *et al.*, 2004; Furutani *et al.*, 2004; Takada and Tasaka, 2002; Takada *et al.*, 2001; Vernoux *et al.*, 2000), and *CUC1* and *CUC2*, but not *CUC3*, are also regulated at the post-transcriptional level by miR164 (Laufs *et al.*, 2004; Mallory *et al.*, 2004). The *KNOX1* gene *KNAT6* is expressed at the boundary between the SAM and the cotyledons later than *CUC* and *STM*, and its expression is required for the activities of *CUC1* and *CUC2* (Belles-Boix *et al.*, 2006).

CUC activity in organ boundary specification overlaps with that of members of two other gene families. The *LATERAL ORGAN FUSION1* (*LOF1*) and *LOF2* MYB-domain transcription factor genes contribute redundantly to meristem–organ separation during development, and *LOF1* interacts genetically with *CUC2* and *CUC3* (Lee *et al.*, 2009). *LOF1* is required for maximal transcription of *LOF2* and all three *CUC* genes, as well as of *LATERAL SUPPRESSOR* (*LAS*), which encodes a GRAS family putative transcription factor expressed at the SAM–lateral organ boundary (Greb *et al.*, 2003). *LAS* acts partially redundantly with *CUC3* but downstream of *CUC1* and *CUC2* (Hibara *et al.*, 2006), indicating that the three *CUC* genes are not equivalent in their functions and participate in overlapping genetic pathways. These data place *LOF1* at the top of a genetic regulatory hierarchy for meristem–organ separation, with *LOF2* and *CUC* genes acting downstream and *LAS* acting downstream of *CUC1* and *CUC2*.

The *LATERAL ORGAN BOUNDARY* (*LOB*) gene, a founding member of the *LBD* family of plant-specific putative transcription factors, is expressed at the boundaries of the developing organs, suggesting it may play a role in boundary specification (Shuai *et al.*, 2002). Loss-of-function *lob* plants have no distinct phenotypes, possibly because other *LBD* family members function redundantly with *LOB*. *LOB* expression is induced by the activities of *AS1, AS2, BOP1*, and *BOP2* in the adjacent organ primordia (Byrne *et al.*, 2002; Ha *et al.*, 2007), and appears to function in a separate pathway from *LOF1* and *LOF2* (Lee *et al.*, 2009). Another *LBD* gene, *JAGGED LATERAL ORGANS* (*JLO*), is expressed at the SAM–organ border and is required for auxin-dependent patterning of the early embryo (Borghi *et al.*, 2007). During vegetative and floral development *JLO* activates *KNOX1* expression while repressing the expression of *PIN* genes, suggesting that *JLO* coordinates auxin accumulation with down-regulation of meristem-specific gene expression at the boundary between the SAM and the initiating organ primordia.

5. Phyllotaxis

Leaf primordia arise from the flanks of the SAM in a highly ordered spatial and temporal pattern, termed phyllotaxis. Phyllotaxis is a major determinant of plant architecture, and a common form of phyllotaxis, the spiral phyllotaxis found in Arabidopsis (Fig. 4.3), has been an attractive phenomenon for mathematicians as well as biologists to study. This is because consecutive leaves in the spirals are arranged at a divergence angle of 137.5°, the so-called golden ratio or Fibonacci angle, that has striking regularity and curious mathematical properties (Kuhlemeier, 2007). A consequence of the regularity of phyllotactic patterning is that the position where a new leaf will initiate can be predicted based on the sites of the existing leaf primordia. Surgical experiments provided evidence that a chemical and/or physical factor could affect the angle between organ primordia (Snow and Snow, 1932; Wardlaw, 1949), and modern molecular studies have demonstrated that auxin is the key instructive signal for phyllotactic pattern formation (reviewed in Kuhlemeier, 2007; Reinhardt *et al.*, 2000; 2003b).

In the SAM, expression peaks of the synthetic auxin-signaling reporter DR5 : GFP, which reflects the location of auxin concentration maxima, in the L1 layer correlates with the sites of primordia initiation on the meristem flanks (Benkova *et al.*, 2003; Heisler *et al.*, 2005). In the tomato SAM, inhibition of polar auxin transport specifically inhibits leaf formation, which can be restored by the micro-application of auxin to the flanks of the meristem (Reinhardt *et al.*, 2000). The initiation of the new leaf occurs at precisely the position where the exogenous auxin is applied. Arabidopsis *pinformed1* (*pin1*) plants with a defect in polar auxin transport develop pin-like inflorescence apices devoid of lateral organs (Okada *et al.*, 1991; Petrasek *et al.*, 2006). Local application of auxin likewise restores organ primordia formation to *pin1* inflorescences (Reinhardt *et al.*, 2000). These experiments indicate that localized auxin accumulation within the shoot apex determines the radial positioning of primordia initiation.

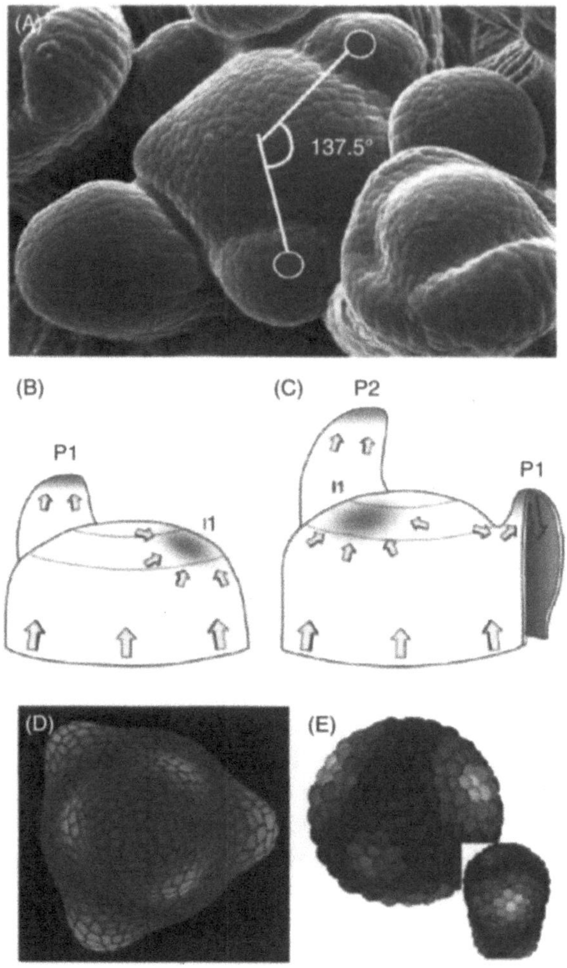

Figure 4.3 Phyllotaxis at the Arabidopsis inflorescence meristem. (A) The flower primordia initiate in a spiral fashion. The divergence angle between successive primordia is approximately 137.5°. Reproduced from Traas and Vernoux (2002). (B) Conceptual model of the regulation of phyllotaxis by polar auxin fluxes in the shoot apical meristem. PIN1 orientation directs auxin fluxes (arrows) in the L1 layer, leading to accumulation of auxin (dark shading) at the initiation site (I1) in the peripheral zone. This accumulation eventually results in organ induction. (C) Later, basipetal PIN1 polarization inside the bulging primordium (P1) drains auxin into the inner layers, depleting the neighboring L1 cells. As a consequence, another auxin maximum is created in the peripheral zone at position I1 removed from primordia P1 and P2. (D) Computer-generated output from an auxin transport model of phyllotaxis (Smith *et al.*, 2006) depicting the spiral pattern of auxin maxima (light shading) on a two-dimensional template mimicking the meristem epidermis. (E) Computer-generated output from an auxin transport model (Jonsson *et al.*, 2006) on a mechanically growing layer of cells that mimic the meristem epidermis. The lightest shading corresponds to the highest concentration of auxin. Inset shows side view. Reproduced from (Jonsson *et al.* 2006; Smith *et al.* 2006), copyright 2006 *National Academy of Sciences*, USA.

Analyses of the expression and subcellular localization of various auxin transporters show that the high intracellular auxin concentration required for leaf initiation is generated by directional auxin transport through the SAM (Reinhardt *et al.*, 2003b; Scarpella *et al.*, 2006). The *PIN1* gene, which encodes an auxin efflux carrier, is predominantly expressed in the L1 layer, and PIN1 protein is polarly localized in the plasma membrane of L1 cells toward the auxin convergence point corresponding to the site of auxin maxima at the center of the incipient primordium (Heisler *et al.*, 2005; Reinhardt *et al.*, 2003b). The available evidence suggests that auxin is transported toward the SAM via PIN1 activity in the L1 (Reinhardt *et al.*, 2000; Smith *et al.*, 2006). Once in the meristem, auxin is absorbed by developing organ primordia, which act as sinks that deplete auxin from the surrounding tissue. Auxin thus reaches the organogenic PZ of the SAM at a certain minimal distance from the youngest primordia, which in Arabidopsis corresponds to the Fibonacci angle (Fig. 4.3A–C). Accumulation of auxin at this location induces the formation of a new primordium, as well as the expression of *PIN1* (Reinhardt *et al.*, 2000). Mathematical modeling coupled with imaging analysis confirms that this molecular mechanism is sufficient to account for the generation of spiral phyllotaxis as well as other phyllotactic patterns observed in nature (Jonsson *et al.*, 2005; Smith *et al.*, 2006) (Fig. 4.3D, E).

PIN1 polarity and expression undergoes cyclic alterations during organ primordia formation, and PIN1 polarity changes are correlated with changes in auxin distribution within the developing primordia (Heisler *et al.*, 2005). As primordia develop, PIN1 protein becomes basally-oriented to evacuate auxin from the surface to the inner tissues, transporting it inward toward the provascular tissue (Scarpella *et al.*, 2006). Through this process the primordia becomes transformed into an auxin sink. In addition, PIN1 polarity reversal away from older and toward younger primordia is proposed to contribute to specifying new initiation sites (Heisler *et al.*, 2005). This process may be mediated by continuous GNOM ARF (ADP ribosylation factor) GEF (guanine nucleotide exchange factor)-dependent cycling of PIN1 protein between endosomes and the plasma membrane (Geldner *et al.*, 2003), and by the PINOID protein kinase which can act as a binary switch of PIN1 polarity (Friml *et al.*, 2004; Furutani *et al.*, 2004). The phyllotactic pattern therefore results from a dynamic interaction between existing and newly initiating organ primordia on the flanks of the SAM, mediated by actively transported auxin, such that new primordia emerge at sites of high auxin concentration and maintain a minimum distance from each other by depleting auxin in their vicinity.

Auxin influx carriers also contribute to phyllotactic patterning. Arabidopsis has four permease-like auxin influx carriers, AUX1 and its three paralogs, LAX1, LAX2, and LAX3 (Parry *et al.*, 2001). Like PIN1, AUX1 is mainly detected in the L1 layer of the SAM. Yet in contrast to PIN1, AUX1 protein is localized throughout the circumference of the plasma membrane and has been shown to actively transport auxin into cells (Yang *et al.*, 2006). Quadruple mutant plants carrying mutations in the *AUX1, LAX1, LAX2,* and *LAX3* genes show clustered or irregular phyllotactic patterns accompanying the loss of a well-defined auxin signal and reduced or altered PIN1 convergence points around the meristem (Bainbridge *et al.*, 2008), and it has been suggested that AUX1 works to retain auxin and coordinate PIN1 polarization within the L1 cells of the SAM (Heisler and Jonsson, 2007; Bainbridge *et al.*, 2008). Thus, AUX1 family-mediated auxin uptake seems to be involved in stabilizing auxin gradients and PIN1 polarization in the L1 layer, whereas the PIN1 auxin efflux carrier is the key element for coordinating the polar auxin transport necessary to generate regular phyllotactic patterns.

6. Communication Between the SAM and Differentiated Tissues

Despite its central role in plant development as an engine for continuous organogenesis, the SAM does not function in isolation but rather is in constant communication with its neighboring organs and tissues. The related *ARGONAUTE* (*AGO*) genes *AGO1* and *ZWILLE* (*ZLL*), aka *PINHEAD* and *AGO10*, are involved in noncell-autonomous SAM maintenance by more differentiated tissues. AGO proteins are central components of RNA-induced silencing complexes (RISC) that bind small regulatory RNA molecules to mediate gene regulation by RNA silencing (reviewed by Voinnet, 2009). *AGO1* and *ZLL* have overlapping functions to maintain the SAM during embryo development (Lynn *et al.*, 1999) and promote *STM* expression in the SAM (Kidner and Martienssen, 2005) while also restricting the size of the *WUS* expression domain (Tucker *et al.*, 2008). In addition AGO1 is required to initiate and ZLL to sustain *CLV3* expression, suggesting that AGO1 and ZLL maintain stem cell activity by potentiating *WUS* signaling from the OC. It has been proposed that AGO1 and ZLL act sequentially in regulating ES cell activity, such that *WUS*-dependent signaling from the OC to the stem cells is promoted by *AGO1* and subsequently sustained by a *ZLL*-mediated signaling pathway (Tucker *et al.*, 2008). Notably, *ZLL* function in the embryonic provascular tissue appears to be sufficient for stem cell maintenance, although ZLL protein itself does not move (Tucker *et al.*, 2008). Thus, ZLL is likely to be involved in the production or transmission of a signal from the provascular cells to the SAM.

Surgical experiments indicate that the SAM produces a signal that specifies the cells on the top (adaxial) side of the organ primordia nearest to the meristem as different from the cells on the bottom (abaxial) side (Steeves and Sussex, 1989). Experiments in which an incision was made between the shoot apex and the site of incipient leaf primordia initiation resulted in the production of a leaf that was radially symmetric, consisting entirely of abaxial cell types, whereas the next primordium that initiates away from the incision site develops into a normally polarized leaf (Snow and Snow, 1959; Sussex, 1951). Thus, the SAM sends a signal to the future organ primordium that orients its adaxial–abaxial axis and confers adaxial identity on the cells located nearest the SAM (see chapter 5). Primordia cells located away from the meristem are not exposed to the signal and develop abaxial fate, which appears to be the default state. Organ primordia that were isolated by incision after they had already begun their morphological differentiation also formed normal leaves (Sussex, 1951), indicating that the meristem-derived signal is perceived very early during the formation of the primordium.

Conversely, signals from the adaxial domain of organ primordia appear to be essential for SAM activity. Over-expression of the GARP transcription factor gene *KANADI1* (*KAN1*), which is expressed in abaxial tissues and transforms adaxial cell fate to abaxial cell fate, prevents the establishment of the embryonic SAM (Eshed *et al.*, 2001; Kerstetter *et al.*, 2001). Similarly loss-of-function Antirrhinum *phantastica* mutants form abaxialized leaves and lack a SAM (Waites and Hudson, 1995). On the other hand, semidominant *phabulosa-1d* (*phb-1d*) mutants that form radial, adaxialized organs develop ectopic SAMs on the undersides of their leaves (McConnell and Barton, 1998), indicating that adaxial cell identity promotes SAM fate. *PHB* is one of a family of class III HOMEODOMAIN-LEUCINE ZIPPER (HD-ZIPIII) proteins that include *PHAVOLUTA* (*PHV*), *REVOLUTA* (*REV*), and *CORONA* (*CNA*). *PHB, PHV*, and *REV* specify adaxial organ polarity and also act redundantly in embryonic SAM establishment (McConnell *et al.*, 2001; Prigge *et al.*, 2005). In contrast, *CNA* functions in parallel with the CLV pathway to regulate meristem size (Green *et al.*, 2005), and *PHB, PHV*, and *CNA* restrict SAM cell accumulation by down-regulating *WUS* transcription (Prigge *et al.*, 2005; Williams *et al.*, 2005). This function occurs independently of *REV* (Williams *et al.*, 2005). Thus, members of the HD-ZIPIII family have both overlapping and divergent functions in SAM regulation.

HD-ZIPIII activity in the SAM is regulated at multiple levels. *PHB, PHV*, and *CNA* expression levels are negatively regulated by members of the microRNA165/166 family of small regulatory RNAs, which act via a RISC that contains AGO1 (Williams *et al.*, 2005). HD-ZIPIII protein function is also modulated by their interaction with the LITTLE ZIPPER (ZPR) proteins. REV and PHB promote the transcription of the four *ZPR* genes, which encode small leucine zipper-containing proteins (Wenkel *et al.*, 2007). Among these, *ZPR3* and *ZPR4* are redundantly required to generate the proper shape and organization of the SAM, as well as to repress *CLV3* and *WUS* expression (Kim *et al.*, 2008). ZPR3 and ZPR4 negatively regulate HD-ZIPIII function during SAM development by directly interacting with the proteins via their leucine zipper motifs and preventing their dimerization, which is a prerequisite for their binding to DNA (Kim *et al.*, 2008; Wenkel *et al.*, 2007). The induction by *HD-ZIPIII* genes of ZPR proteins that interfere with their function generates a negative feedback loop that can serve to dampen the effects of fluctuations in HD-ZIPIII protein levels, thus contributing to SAM homeostasis.

The HD-ZIPIII proteins also physically interact with the plant-specific AP2/ERF domain protein DORNROSCHEN/ENHANCER OF SHOOT REGENERATION1 (DRN/ESR1). *DRN* and its paralog *DRNL* are expressed in the CZ of the SAM in a pattern that overlaps with *CLV3* (Kirch *et al.*, 2003). *DRN* and *DRNL* have partially redundant functions in patterning the early embryo and act upstream of auxin transport and response (Chandler *et al.*, 2007). PHB heterodimerizes with DRN *in planta*, and the two proteins act in a common pathway to control cell division in developing embryos (Chandler *et al.*, 2007). However, it remains to be determined whether any of the effects of the HD-ZIPIII proteins on SAM establishment and maintenance are mediated via auxin.

Activity of the *FILAMENTOUS FLOWER* (*FIL*) and *YABBY3* (*YAB3*) genes in lateral organs stimulates signals that maintain SAM organization and regular phyllotaxis (Goldshmidt *et al.*, 2008). *FIL* and *YAB3* encode YABBY family putative transcription factors (Sawa *et al.*, 1999; Siegfried *et al.*, 1999) that are expressed specifically in the abaxial domain of organ primordia and have multiple developmental functions. *FIL* and *YAB3* activity promotes abaxial cell identity and growth (Eshed *et al.*, 1999; 2004) and represses *KNOX1* gene expression in developing leaves (Kumaran *et al.*, 2002). In addition, *FIL* and *YAB3* have noncell-autonomous effects on the SAM, affecting the phyllotactic pattern of the inflorescence meristem and negatively regulating the *CLV3* and *WUS* expression domains (Goldshmidt *et al.*, 2008). The YAB-derived signal(s) from the organ primordia to the CZ of the SAM are mediated in part through the activity of the organ boundary gene *LATERAL SUPPRESSOR* (*LAS*), prompting the postulation that concentric expression domains of factors such as LAS may mediate between the activities of the central stem cells and the peripheral organ primordia by filtering or amplifying morphogenic signals to ensure proper SAM function (Goldshmidt *et al.*, 2008).

7. Floral Meristem Termination

FMs develop as primordia from the flanks of the SAM after environmental and endogenous signals induce the transition from vegetative to reproductive (inflorescence) growth. FM identity is conferred by the transcription factors LEAFY (LFY) and AP1, which are activated only in floral primordia (Blazquez *et al.*, 1997; Gustafson-Brown *et al.*, 1994). In Arabidopsis FMs, a transient stem cell reservoir is maintained by the CLV–WUS feedback loop to provide progeny cells for floral organ formation. However, unlike the primary SAM in which the stem cell population is continuously maintained, the FM ultimately terminates stem cell activity during the production of the innermost whorl of organs, the carpels. The timing of FM termination is tightly controlled so that sufficient cells are produced to initiate the correct number of floral organs.

Arabidopsis flowers consist of sepal, petal, stamen, and carpel organs arranged in four concentric whorls. The identity of each whorl of organs is specified by a specific combination of transcription factors that form complexes regulating a set of downstream target genes (Coen and Meyerowitz, 1991; Krizek and Fletcher, 2005). A key factor among these is the MADS domain protein AGAMOUS (AG), which specifies stamen and carpel identity and also acts as a central component of a temporal feedback loop that controls floral stem cell termination (Fig. 4.4). The *AG* locus is repressed throughout the embryonic and vegetative stages by the activity of the CLF PcG protein, which deposits H3K27me3 repressive histone

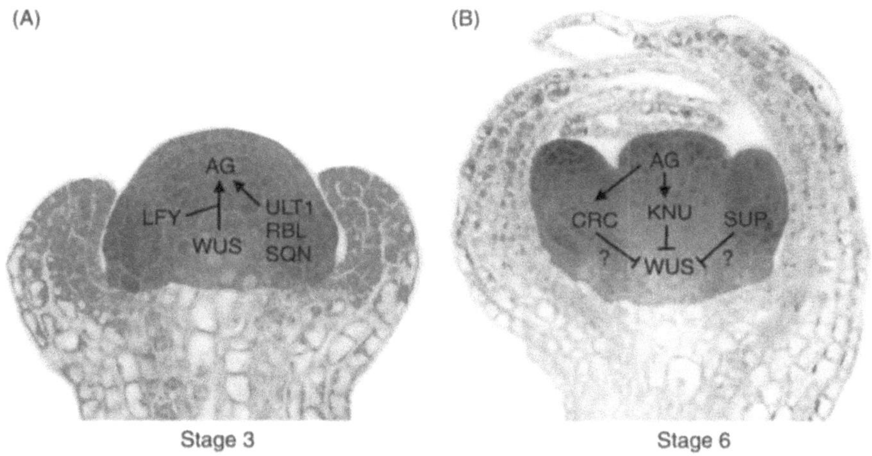

Figure 4.4 Floral meristem termination pathway. (A) At stage 3 of flower development, *AG* expression is induced in the center of the FM by the transcriptional activators LFY and WUS. The trxG factor ULT1 plays a role in timing *AG* activation by counteracting PcG-mediated deposition of repressive chromatin marks at the *AG* locus. RBL and SQN also contribute to *AG* induction through as yet unknown mechanisms. (B) At stage 6 of flower development, *KNU* transcription is directly activated by AG, whereupon KNU represses *WUS* transcription to terminate stem cell maintenance. *WUS* expression is also negatively regulated by *SUP* and by AG-induced *CRC* activity, although whether these interactions are direct or indirect is unknown (denoted by a ?). In both images the *AG* expression domain is shaded. Modified from *Trends in Plant Science* (Carles and Fletcher, 2003).

modifications on the *AG* chromatin for stable gene silencing (Schubert *et al.*, 2006). At stage 3 of flower development, when the sepal primordia emerge on the flanks of the FM (Smyth *et al.*, 1990), the transcription factors WUS and LFY bind independently to adjacent regions of the second intron of *AG* to induce its expression in the center of the FM (Lenhard *et al.*, 2001; Lohmann *et al.*, 2001).

The SAND-domain putative DNA-binding protein ULTRAPETALA1 (ULT1) also contributes to the correct temporal induction of *AG* expression (Carles *et al.*, 2005; 2004). *ULT1* promotes FM determinacy by inducing *AG* activation at stage 3 of FM formation (Fletcher, 2001) and limiting the duration of *WUS* expression in the center of the meristem (Carles *et al.*, 2004). ULT1 has trithorax Group (trxG) activity that counteracts the repressive function of CLF in controlling *AG* expression by limiting the deposition of H3K27me3 repressive marks at the *AG* locus, priming it for activation early in flower development (Carles and Fletcher, 2009). Physical interaction between ULT1 and the trxG protein ATX1, as well as the direct association of ULT1 with sequences in the promoter and second intron of *AG* (Carles and Fletcher, 2009), suggests a possible role for ULT1 in mediating the recruitment of a trxG complex to this key target locus.

In wild-type FMs *WUS* is initially expressed in a pattern similar to that of the SAM, but its transcription is repressed at stage 6 when the FM is consumed in the formation of the central carpels (Mayer *et al.*, 1998; Schoof *et al.*, 2000). *WUS* repression occurs via indirect feedback regulation by AG, mediated by the *KNUCKLES* (*KNU*) gene. *KNU* encodes a protein containing a C2H2-type zinc finger and an EAR-like transcriptional repression domain (Sun *et al.*, 2009). AG binds directly to the *KNU* promoter early in flower development, but *KNU* activation only occurs 2 days later after the high level of H3K27me3 repressive methylation marks on the *KNU* locus decrease. At stage 6 the AG-dependent removal of these repressive marks permits the induction of *KNU* transcription and consequent *WUS* repression (Sun *et al.*, 2009). Although the function of AG in chromatin remodeling and the mechanism through which KNU represses *WUS* are not known, these data demonstrate that KNU is a key component of the timing mechanism for floral stem cell termination. Thus, *AG* activation by WUS early in flower development induces KNU-mediated *WUS* repression at a later stage, forming a temporal feedback loop that is superimposed on the CLV/WUS spatial feedback loop to switch off stem cell activity at the time of carpel initiation in the center of the flower.

There is some evidence that floral stem cell proliferation is also negatively regulated by the C2H2-type zinc finger proteins CRABS CRAW (CRC) and SUPERMAN (SUP). Mutations in *CRC*, which encodes a YAB transcription factor, affect only carpel development, but *crc ag*-1/+ flowers generate supernumerary whorls inside the fourth whorl of carpels, indicating an additional role for *CRC* in promoting FM determinacy (Alvarez and

Smyth, 1999). *CRC* is a direct target of *AG* induction (Gomez-Mena *et al.*, 2005) that is initially expressed at stage 6 of flower formation (Bowman *et al.*, 1999), suggesting that it acts late in the FM determinacy process, potentially as part of the mechanism that represses *WUS* transcription. Two additional floral determinacy factors, *REBELOTE* (*RBL*) and *SQUINT* (*SQN*), were identified in a genetic screen for *crc* modifiers (Prunet *et al.*, 2008). At the genetic level both act redundantly with each other and *CRC* to control FM termination, and SQN promotes *AG* expression in an inner subdomain of whorl 4. *RBL* encodes a NOC2 domain-containing protein that in yeast is required for maturation and intranuclear transport of pre-ribosomes (Milkereit *et al.*, 2001). SQN is the Arabidopsis homolog of cyclophilin 40 (Berardini *et al.*, 2001), but their molecular roles in FM determinacy are still obscure.

The transcriptional repressor SUP is expressed at the boundary between the third and fourth whorls and appears to act synergistically with *AG* in FM determinacy (Jack *et al.*, 1997; Schultz *et al.*, 1991; Sun *et al.*, 2009). The *AG*-related floral homeotic MADS box genes *APETALA3* (*AP3*) and *PISTILLATA* (*PI*) are required to condition the *sup* phenotype (Bowman *et al.*, 1992; Schultz *et al.*, 1991), and their exclusion from the fourth whorl by SUP is necessary for FM termination (Bowman *et al.*, 1991). However, because ectopic expression of *AP3* and *PI* delays FM termination in wild-type but not *ag* flowers (Krizek and Meyerowitz, 1996), the effects of the *sup* mutation and of *AP3*/*PI* ectopic expression on FM termination are likely to be dependent on AG function, although they occur independently of *AG* transcription (Bowman *et al.*, 1992). The delay in FM termination in *sup* mutants is associated with prolonged *WUS* expression in the center of the FM (Prunet *et al.*, 2008), but whether the SUP pathway directly or indirectly affects *WUS* regulation remains to be determined.

8. Live Imaging and Computational Modeling Technology

Specification of cell fate within the SAM occurs dynamically throughout development, yet despite constant fluctuations in cell number, the overall size and organization of the SAM is maintained at homeostasis. Therefore, coordinated regulation of cell identity, cell division, and cell expansion in the SAM is crucial to keep meristem function intact and to coordinate proper organ initiation (Baurle and Laux, 2003; DeYoung and Clark, 2001; Williams and Fletcher, 2005). Molecular genetic studies have revealed diverse pathways that control SAM morphogenesis and growth, but have the limitation of being static and single-point analyses. Recent studies using live imaging techniques to visualize fluorescent reporter proteins *in vivo* have enabled more direct examination of the dynamic regulation of SAM function (Reddy *et al.*, 2004). In addition, computational modeling techniques are being applied to simulate meristem behavior and to devise experimentally testable predictions about the mechanisms of SAM activity.

Early live imaging studies focused on SAM growth determined that cell division rates vary across the SAM surface and that DNA synthesis is tightly coupled with specific cell division and expansion patterns (Grandjean *et al.*, 2004; Reddy *et al.*, 2004). Live imaging also revealed that *CLV3* restricts its own expression domain to the CZ by preventing the re-specification and/or de-differentiation of their nearest PZ cell neighbors into CZ cells (Reddy and Meyerowitz, 2005). Thus, the progeny of the stem cells retain their ability to respond to stem cell-promoting signals after their departure from the CZ, but this ability is gradually lost as the cells are displaced toward the periphery of the PZ. CLV3 signaling also positions the boundary between CZ and PZ identity, and is rapidly sensed by the *WUS*-expressing OC cells (Muller *et al.*, 2006). Thus, variability in CLV3 signal levels caused by transient fluctuations in stem cell number can be quickly detected and compensated for to maintain SAM function.

Although the CLV–WUS pathway represents a robust mechanism for sustaining SAM homeostasis as stem cell numbers fluctuate during growth, little is known about how meristem activity is maintained despite the variable environmental conditions that plants experience during their lives. A study combining experimental treatments with mathematical modeling showed that although the size of the entire SAM and of the OC varies greatly under different growth conditions, stem cell number remains relatively constant (Geier *et al.*, 2008). The authors were unable to predict the relative stability of the stem cell domain under different growth conditions using a single feedback model, suggesting that multiple feedback mechanisms exist to control stem cell proliferation and differentiation.

A second feedback-regulated system, one controlled by *WUS* activity, also appears to be necessary in order to compensate for fluctuations in CLV3 signal levels and guarantee stem cell homeostasis (Muller *et al.*, 2006). Indeed, mathematical modeling based on live imaging data predicts the existence of both the CLV3 repressive signal and an additional WUS-dependent activator that organizes the *WUS* expression domain within the SAM (Jonsson *et al.*, 2005). The type of reaction–diffusion mechanism utilized in that study has been applied more broadly to model the dynamics of gene regulation across the SAM, which showed that these combined feedback systems are sufficient to permit the establishment and maintenance of both the stem cell domain and the OC in the SAM (Hohm *et al.*, 2010).

Live imaging has also revealed that the process of organogenesis on the flanks of the SAM is associated with distinct cellular behavior. Initial primordium outgrowth is linked to oriented divisions of the primordium progenitor cells and their proximal neighbors, followed by a rapid burst of cell expansion and cell division that transforms a primordium into a three-dimensional organ

(Reddy *et al.*, 2004). Morphogenesis at the SAM also correlates with the supracellular alignment of microtubules along the boundary between the meristem and the primordium (Hamant *et al.*, 2008). Microtubule orientation occurs along the planes of predicted mechanical stresses in the SAM, as predicted by mathematical stress-feedback models (Hamant *et al.*, 2008). Thus, organ initiation at the SAM is mediated by the complex interplay between local developmental signals, mechanical stresses, and cell growth patterns.

9. Concluding Remarks

The SAM is the engine that drives continuous plant growth and development, a dynamic structure that produces new leaves, stems, and flowers while simultaneously sustaining a stem cell reservoir for future organ formation. Over a decade of research has demonstrated the existence of extensive gene regulatory networks that maintain the position and activity of the central stem cell domain, as well as those that organize the initiation of lateral organs on the periphery. These control networks consist of cell autonomous transcriptional and epigenetic regulators, short-range intercellular signals, and longer range hormone activities, and involve multiple positive and negative feedback mechanisms. Determining how all of these different factors are integrated to produce specific cellular outcomes represents a significant challenge for the future. The combination of traditional molecular genetic and biochemical studies with live imaging technology, computational modeling, and genome-wide transcriptome and epigenome analysis will enable us to decipher SAM behavior with increasing clarity, providing new insights into the complex, intricate functions of this unique structure.

REFERENCES

Aida, M., Ishida, T., Fukaki, H., Fujisawa, H., and Tasaka, M. (1997). Genes involved in organ separation in Arabidopsis: an analysis of the *cup-shaped cotyledon* mutant. *Plant Cell* **9**, 841–857.

Aida, M., Ishida, T., and Tasaka, M. (1999). Shoot apical meristem and cotyledon formation during Arabidopsis embryogenesis: interaction among the *CUP-SHAPED COTYLEDON* and *SHOOT MERISTEMLESS* genes. *Development* **126**, 1563–1570.

Aida, M., and Tasaka, M. (2006). Morphogenesis and patterning at the organ boundaries in the higher plant shoot apex. *Plant Mol. Biol.* **60**, 915–928.

Aida, M., Vernoux, T., Furutani, M., Traas, J., and Tasaka, M. (2002). Roles of PIN-FORMED1 and MONOPTEROS in pattern formation of the apical region of the Arabidopsis embryo. *Development* **129**, 3965–3974.

Alvarez, J., and Smyth, D. R. (1999). *CRABS CLAW* and *SPATULA*, two Arabidopsis genes that control carpel development in parallel with *AGAMOUS*. *Development* **126**, 2377–2386.

Bainbridge, K., Guyomarc'h, S., Bayer, E., Swarup, R., Bennett, M., Mandel, T., and Kuhlemeier, C. (2008). Auxin influx carriers stabilize phyllotactic patterning. *Genes Dev.* **22**, 810–823.

Barrero, J. M., Gonzalez-Bayon, R., del Pozo, J. C., Ponce, M. R., and Micol, J. L. (2007). *INCURVATA2* encodes the catalytic subunit of DNA Polymerase alpha and interacts with genes involved in chromatin-mediated cellular memory in *Arabidopsis thaliana. Plant Cell* **19**, 2822–2838.

Barton, M. K., and Poethig, R. S. (1993). Formation of the shoot apical meristem in *Arabidopsis thaliana*: an analysis of development in the wild type and in the *shoot meristemless* mutant. *Development* **119**, 823–831.

Baurle, I., and Laux, T. (2003). Apical meristems: the plant's fountain of youth. *Bioessays* **25**, 961–970.

Baurle, I., and Laux, T. (2005). Regulation of *WUSCHEL* transcription in the stem cell niche of the Arabidopsis shoot meristem. *Plant Cell* **17**, 2271–2280.

Belles-Boix, E., Hamant, O., Witiak, S. M., Morin, H., Traas, J., and Pautot, V. (2006). *KNAT6*: an Arabidopsis homeobox gene involved in meristem activity and organ separation. *Plant Cell* **18**, 1900–1907.

Benkova, E., Michniewicz, M., Sauer, M., Teichmann, T., Seifertova, D., Jurgens, G., and Friml, J. (2003). Local, efflux-dependent auxin gradients as a common module for plant organ formation. *Cell* **115**, 591–602.

Bennett, S.R.M., Alvarez, J., Bossinger, G., and Smyth, D. R. (1995). Morphogenesis in *pinoid* mutants of *Arabidopsis thaliana. Plant J.* **8**, 505–520.

Berardini, T. Z., Bollman, K., Sun, H., and Poethig, R. S. (2001). Regulation of vegetative phase change in *Arabidopsis thaliana* by cyclophilin 40. *Science* **291**, 2405–2407.

Bernstein, B. E., Mikkelsen, T. S., Xie, X., Kamal, M., Huebert, D. J., Cuff, J., Fry, B., Meissner, A., Wernig, M., Plath, K., Jaenisch, R., Wagschal, A., *et al.* (2006). A bivalent chromatin structure marks key developmental genes in embryonic stem cells. *Cell* **125**, 315–326.

Blazquez, M. A., Soowal, L. N., Lee, I., and Weigel, D. (1997). LEAFY expression and flower initiation in Arabidopsis. *Development* **124**, 3835–3844.

Bleckmann, A., Weidtkamp-Peters, S., Seidel, C.A.M., and Simon, R. (2010). Stem cell signaling in Arabidopsis requires CRN to localize CLV2 to the plasma membrane. *Plant Physiol.* **152**, 166–176.

Bolduc, N., and Hake, S. (2009). The maize transcription factor KNOTTED1 directly regulates the gibberellin catabolism gene *ga2ox1. Plant Cell* **21**, 1647–1658.

Borghi, L., Bureau, M., and Simon, R. (2007). Arabidopsis *JAGGED LATERAL ORGANS* is expressed in boundaries and coordinates KNOX and PIN activity. *Plant Cell* **19**, 1795–1808.

Bowman, J. L., Baum, S. F., Eshed, Y., Putterill, J., and Alvarez, J. (1999). Molecular genetics of gynoecium development in Arabidopsis. *Curr. Top. Dev. Biol.* **45**, 155–205.

Bowman, J. L., Sakai, H., Jack, T., Weigel, D., Mayer, U., and Meyerowitz, E. M. (1992). SUPERMAN, a regulator of floral homeotic genes in Arabidopsis. *Development* **114**, 599–615.

Bowman, J. L., Smyth, D. R., and Meyerowitz, E. M. (1991). Genetic interactions among floral homeotic genes of Arabidopsis. *Development* **112**, 1–20.

Brand, U., Fletcher, J. C., Hobe, M., Meyerowitz, E. M., and Simon, R. (2000). Dependence of stem cell fate in Arabidopsis on a feedback loop regulated by CLV3 activity. *Science* **289**, 617–619.

Brand, U., Grunewald, M., Hobe, M., and Simon, R. (2002). Regulation of *CLV3* expression by two homeobox genes in Arabidopsis. *Plant Physiol.* **129**, 565–575.

Buechel, S., Leibfried, A., To, J. P., Zhao, Z., Andersen, S. U., Kieber, J. J., and Lohmann, J. U. (2010). Role of A-type ARABIDOPSIS RESPONSE REGULATORS in meristem maintenance and regeneration. *Eur. J. Cell Biol.* **89**, 279–284.

Byrne, M. E., Barley, R., Curtis, M., Arroyo, J. M., Dunham, M., Hudson, A., and Martienssen, R. A. (2000). Asymmetric leaves1 mediates leaf patterning and stem cell function in Arabidopsis. *Nature* **408**, 967–971.

Byrne, M. E., Simorowski, J., and Martienssen, R. A. (2002). ASYMMETRIC LEAVES1 reveals *knox* gene redundancy in Arabidopsis. *Development* **129**, 1957–1965.

Carles, C. C., Choffnes-Inada, D., Reville, K., Lertpiriyapong, K., and Fletcher, J. C. (2005). *ULTRAPETALA1* encodes a SAND domain putative transcriptional regulator that controls shoot and floral meristem activity in Arabidopsis. *Development* **132**, 897–911.

Carles, C. C., and Fletcher, J. C. (2003). Shoot apical meristem maintenance: the art of a dynamic balance. *Trends Plant Sci.* **8**, 394–401.

Carles, C. C., and Fletcher, J. C. (2009). The SAND domain protein ULTRAPETALA1 acts as a trithorax group factor to regulate cell fate in plants. *Genes Dev.* **23**, 2723–2728.

Carles, C. C., Lertpiriyapong, K., Reville, K., and Fletcher, J. C. (2004). The *ULTRAPETALA1* gene functions early in Arabidopsis development to restrict shoot apical meristem activity and acts through *WUSCHEL* to regulate floral meristem determinacy. *Genetics* **167**, 1893–1903.

Casson, S. A., Topping, J. F., and Lindsey, K. (2009). MERISTEM-DEFECTIVE, an RS domain protein, is required for the correct meristem patterning and function in Arabidopsis. *Plant J.* **57**, 857–869.

Chandler, J. W., Cole, M., Flier, A., Grewe, B., and Werr, W. (2007). The AP2 transcription factors DORNROSCHEN and DORNROSCHEN-LIKE redundantly control Arabidopsis embryo patterning via interaction with PHAVOLUTA. *Development* **134**, 1653–1662.

Chanvivattana, Y., Bishopp, A., Schubert, D., Stock, C., Moon, Y. H., Sung, Z. R., and Goodrich, J. (2004). Interaction of Polycomb-group proteins controlling flowering in Arabidopsis. *Development* **131**, 5263–5276.

Cheng, Y., Qin, G., Dai, X., and Zhao, Y. (2008). NPY genes and AGC kinases define two key steps in auxin-mediated organogenesis in Arabidopsis. *Proc. Natl. Acad. Sci. U.S.A.* **105**, 21017–21022.

Cheng, C., Yun, K. Y., Ressom, H. W., Mohanty, B., Bajic, V. B., Jia, Y., Yun, S. J., and de los Reyes, B. G. (2007). An early response regulatory cluster induced by low temperature and hydrogen peroxide in seedlings of chilling-tolerant japonica rice. *BMC Genomics* **8**, 175.

Chuck, G., Lincoln, C., and Hake, S. (1996). KNAT1 induces lobed leaves with ectopic meristems when overexpressed in Arabidopsis. *Plant Cell* **8**, 1277–1289.

Clark, S. E., Jacobsen, S. E., Levin, J. Z., and Meyerowitz, E. M. (1996). The *CLAVATA* and *SHOOT MERISTEMLESS* loci competitively regulate meristem activity in Arabidopsis. *Development* **122**, 1567–1575.

Clark, S. E., Running, M. P., and Meyerowitz, E. M. (1995). CLAVATA3 is a specific regulator of shoot and floral meristem development affecting the same processes as CLAVATA1. *Development* **121**, 2057–2067.

Clark, S. E., Williams, R. W., and Meyerowitz, E. M. (1997). The *CLAVATA1* gene encodes a putative receptor kinase that controls shoot and floral meristem size in Arabidopsis. *Cell* **89**, 575–585.

Cock, J. M., and McCormick, S. (2001). A large family of genes that share homology with *CLAVATA3*. *Plant Physiol.* **126**, 939–942.

Coen, E. S., and Meyerowitz, E. M. (1991). The war of the whorls—genetic interactions controlling flower development. *Nature* **353**, 31–37.

DeYoung, B., Bickle, K. L., Schrage, K. J., Muskett, P., Patel, K., and Clark, S. E. (2006). The CLAVATA1-related BAM1, BAM2 and BAM3 receptor kinase-like proteins are required for meristem function in Arabidopsis. *Plant J.* **45**, 1–16.

DeYoung, B. and Clark, S. E. (2008). BAM receptors regulate stem cell specification and organ development through complex interactions with CLAVATA signaling. *Genetics* **180**, 895–904.

DeYoung, B. J., and Clark, S. E. (2001). Signaling through the CLAVATA1 receptor complex. *Plant Molecular Biology* **46**, 505–513.

Eshed, Y., Baum, S. F., and Bowman, J. L. (1999). Distinct mechanisms promote polarity establishment in carpels of Arabidopsis. *Cell* **99**, 199–209.

Eshed, Y., Baum, S. F., Perea, J. V., and Bowman, J. L. (2001). Establishment of polarity in lateral organs of plants. *Curr. Biol.* **11**, 1251–1260.

Eshed, Y., Izhaki, A., Baum, S. F., Floyd, S. K., and Bowman, J. L. (2004). Asymmetric leaf development and blade expansion in Arabidopsis are mediated by KANADI and YABBY activities. *Development* **131**, 2997–3006.

Farrona, S., Hurtado, L., Bowman, J. L., and Reyes, J. C. (2004). The *Arabidopsis thaliana* SNF2 homolog AtBRM controls shoot development and flowering. *Development* **131**, 4965–4975.

Fletcher, J. C. (2001). The *ULTRAPETALA* gene controls shoot and floral meristem size in Arabidopsis. *Development* **128**, 1323–1333.

Fletcher, J. C. (2002). Shoot and floral meristem maintenance in Arabidopsis. *Annu. Rev. Plant Biol.* **53**, 45–66.

Fletcher, J. C., Brand, U., Running, M. P., Simon, R., and Meyerowitz, E. M. (1999). Signaling of cell fate decisions by *CLAVATA3* in Arabidopsis shoot meristems. *Science* **283**, 1911–1914.

Francis, N. J., and Kingston, R. E. (2001). Mechanisms of transcriptional memory. *Nat. Rev. Mol. Cell Biol.* **2**, 409–421.

Friml, J., Vieten, A., Sauer, M., Weijers, D., Schwarz, H., Hamann, T., Offringa, R., and Jurgens, G. (2003). Efflux-dependent auxin gradients establish the apical–basal axis of Arabidopsis. *Nature* **426**, 147–153.

Friml, J., Yang, X., Michniewicz, M., Weijers, D., Quint, A., Tietz, O., Benjamins, R., Ouwerkerk, P. B., Ljung, K., Sandberg, G., Hooykaas, P. J., Palme, K., *et al.* (2004). A PINOID-dependent binary switch in apical–basal PIN polar targeting directs auxin efflux. *Science* **306**, 862–865.

Furutani, M., Kajiwara, T., Kato, T., Treml, B. S., Stockum, C., Torres-Ruiz, R. A., and Tasaka, M. (2007). The gene MACCHI-BOU 4/ENHANCER OF PINOID encodes a NPH3-like protein and reveals similarities between organogenesis and phototropism at the molecular level. *Development* **134**, 3849–3859.

Furutani, M., Vernoux, T., Traas, J., Kato, T., Tasaka, M., and Aida, M. (2004). PIN-FORMED1 and PINOID regulate boundary formation and cotyledon development in Arabidopsis embryogenesis. *Development* **131**, 5021–5030.

Gallois, J. L., Nora, F. R., Mizukami, Y., and Sablowski, R. (2004). WUSCHEL induces shoot stem cell activity and developmental plasticity in the root meristem. *Genes Dev.* **18**, 375–380.

Geier, F., Lohmann, J. U., Gerstung, M., Maier, A. T., Timmer, J., and Fleck, C. (2008). A quantitative and dynamic model for plant stem cell regulation. *PLoS One* **3**, e3553.

Geldner, N., Anders, N., Wolters, H., Keicher, J., Kornberger, W., Muller, P., Delbarre, A., Ueda, T., Nakano, A., and Jurgens, G. (2003). The Arabidopsis GNOM ARF-GEF mediates endosomal recycling, auxin transport, and auxin-dependent plant growth. *Cell* **112**, 219–230.

Giulini, A., Wang, J., and Jackson, D. (2004). Control of phyllotaxy by the cytokinin-inducible response regulator homologue ABPHYL1. *Nature* **430**, 1031–1034.

Goldshmidt, A., Alvarez, J. P., Bowman, J. L., and Eshed, Y. (2008). Signals derived from *YABBY* gene activities in organ primordia regulate growth and partitioning of Arabidopsis shoot apical meristems. *Plant Cell* **20**, 1217–1230.

Gomez-Mena, C., de Folter, S., Costa, M. M., Angenent, G. C., and Sablowski, R. (2005). Transcriptional program controlled by the floral homeotic gene *AGAMOUS* during early organogenesis. *Development* **132**, 429–438.

Gordon, S. P., Chickarmane, V. S., Ohno, C., and Meyerowitz, E. M. (2009). Multiple feedback loops through cytokinin signaling control stem cell number within the Arabidopsis shoot meristem. *Proc. Natl. Acad. Sci. U.S.A.* **106**, 16529–16534.

Grandjean, O., Vernoux, T., Laufs, P., Belcram, K., Mizukami, Y., and Traas, J. (2004). In vivo analysis of cell division, cell growth, and differentiation at the shoot apical meristem in Arabidopsis. *Plant Cell* **16**, 74–87.

Greb, T., Clarenz, O., Schafer, E., Muller, D., Herrero, R., Schmitz, G., and Theres, K. (2003). Molecular analysis of the *LATERAL SUPPRESSOR* gene in Arabidopsis reveals a conserved control mechanism for axillary meristem formation. *Genes Dev.* **17**, 1175–1187.

Green, K. A., Prigge, M. J., Katzman, R. B., and Clark, S. E. (2005). CORONA, a member of the class III homeodomain leucine zipper gene family in Arabidopsis, regulates stem cell specification and organogenesis. *Plant Cell* **17**, 691–704.

Greenboim-Wainberg, Y., Maymon, I., Borochov, R., Alvarez, J., Olszewski, N., Ori, N., Eshed, Y., and Weiss, D. (2005). Cross talk between gibberellin and cytokinin: the Arabidopsis GA response inhibitor SPINDLY plays a positive role in cytokinin signaling. *Plant Cell* **17**, 92–102.

Guo, M., Thomas, J., Collins, G., and Timmermans, M. C. (2008). Direct repression of *KNOX* loci by the ASYMMETRIC LEAVES1 complex of Arabidopsis. *Plant Cell* **20**, 48–58.

Gustafson-Brown, C., Savidge, B., and Yanofsky, M. F. (1994). Regulation of the Arabidopsis floral homeotic gene *APETALA1*. *Cell* **76**, 131–143.

Guyomarc'h, S., Vernoux, T., Traas, J., Zhou, D. X., and Delarue, M. (2004). *MGOUN3*, an Arabidopsis gene with TetratricoPeptide-Repeat-related motifs, regulates meristem cellular organization. *J. Exp. Bot.* **55**, 673–684.

Ha, C. M., Jun, J. H., Nam, H. G., and Fletcher, J. C. (2004). *BLADE-ON-PETIOLE1* encodes a BTB/POZ domain protein required for leaf morphogenesis in *Arabidopsis thaliana*. *Plant Cell Physiol.* **45**, 1361–1370.

Ha, C. M., Jun, J. H., Nam, H. G., and Fletcher, J. C. (2007). *BLADE-ON-PETIOLE 1* and *2* control Arabidopsis lateral organ fate through regulation of LOB domain and adaxial–abaxial polarity genes. *Plant Cell* **19**, 1809–1825.

Ha, C. M., Kim, G. T., Kim, B. C., Jun, J. H., Soh, M. S., Ueno, Y., Machida, Y., Tsukaya, H., and Nam, H. G. (2003). The *BLADE-ON-PETIOLE 1* gene controls leaf pattern formation through the modulation of meristematic activity in Arabidopsis. *Development* **130**, 161–172.

Hamant, O., Heisler, M. G., Jonsson, H., Krupinski, P., Uyttewaal, M., Bokov, P., Corson, F., Sahlin, P., Boudaoud, A., Meyerowitz, E. M., Couder, Y. and Traas, J. (2008). Developmental patterning by mechanical signals in Arabidopsis. *Science* **322**, 1650–1655.

Han, P., Li, Q., and Zhu, Y. X. (2008). Mutation of Arabidopsis BARD1 causes meristem defects by failing to confine *WUSCHEL* expression to the organizing center. *Plant Cell* **20**, 1482–1493.

Hay, A., Barkoulas, M., and Tsiantis, M. (2006). ASYMMETRIC LEAVES1 and auxin activities converge to repress *BREVIPEDICELLUS* expression and promote leaf development in Arabidopsis. *Development* **133**, 3955–3961.

Hay, A., Kaur, H., Phillips, A., Hedden, P., Hake, S., and Tsiantis, M. (2002). The gibberellin pathway mediates KNOTTED1-type homeobox function in plants with different body plans. *Curr Biol.* **12**, 1557–1565.

Heisler, M. G. and Jonsson, H. (2007). Modelling meristem development in plants. *Curr. Opin. Plant Biol.* **10**, 92–97.

Heisler, M. G., Ohno, C., Das, P., Sieber, P., Reddy, G. V., Long, J. A., and Meyerowitz, E. M. (2005). Patterns of auxin transport and gene expression during primordium development revealed by live imaging of the Arabidopsis inflorescence meristem. *Curr. Biol.* **15**, 1899–1911.

Hepworth, S. R., Zhang, Y., McKim, S., Li, X., and Haughn, G. W. (2005). BLADE-ON-PETIOLE-dependent signaling controls leaf and floral patterning in Arabidopsis. *Plant Cell* **17**, 1434–1448.

Hibara, K., Karim, M. R., Takada, S., Taoka, K. I., Furutani, M., Aida, M., and Tasaka, M. (2006). Arabidopsis CUP-SHAPED COTYLEDON3 regulates postembryonic shoot meristem and organ boundary formation. *Plant Cell* **18**, 2946–2957.

Higuchi, M., Pischke, M. S., Mahonen, A. P., Miyawaki, K., Hashimoto, Y., Seki, M., Kobayashi, M., Shinozaki, K., Kato, T., Tabata, S., Helariutta, Y., Sussman, M. R., *et al.* (2004). In planta functions of the Arabidopsis cytokinin receptor family. *Proc. Natl. Acad. Sci. U.S.A.* **101**, 8821–8826.

Hohm, T., Zitzler, E., and Simon, R. (2010). A dynamic model for stem cell homeostasis and patterning in Arabidopsis meristems. *PLoS One* **5**, e9189.

Iwakawa, H., Ueno, Y., Semiarti, E., Onouchi, H., Kojima, S., Tsukaya, H., Hasebe, M., Soma, T., Ikezaki, M., Machida, C., and Machida, Y. (2002). The *ASYMMETRIC LEAVES2* gene of *Arabidopsis thaliana*, required for formation of a symmetric flat leaf lamina, encodes a member of a novel family of proteins characterized by cysteine repeats and a leucine zipper. *Plant Cell Physiol.* **43**, 467–478.

Jack, T., Sieburth, L., and Meyerowitz, E. (1997). Targeted misexpression of AGAMOUS in whorl 2 of Arabidopsis flowers. *Plant J.* **11**, 825–839.

Jackson, D., Veit, B., and Hake, S. (1994). Expression of maize *KNOTTED1* related homeobox genes in the shoot apical meristem predicts patterns of morphogenesis in the vegetative shoot. *Development* **120**, 405–413.

Jasinski, S., Piazza, P., Craft, J., Hay, A., Woolley, L., Rieu, I., Phillips, A., Hedden, P., and Tsiantis, M. (2005). KNOX action in Arabidopsis is mediated by coordinate regulation of cytokinin and gibberellin activities. *Curr Biol.* **15**, 1560–1565.

Jeong, S., Trotochaud, A. E., and Clark, S. E. (1999). The Arabidopsis CLAVATA2 gene encodes a receptor-like protein required for the stability of the CLAVATA1 receptor-like kinase. *Plant Cell* **11**, 1925–1934.

Jonsson, H., Heisler, M., Reddy, G. V., Agrawal, V., Gor, V., Shapiro, B. E., Mjolsness, E., and Meyerowitz, E. M. (2005). Modeling the organization of the *WUSCHEL* expression domain in the shoot apical meristem. *Bioinformatics* **21**(Suppl 1), i232–i240.

Jonsson, H., Heisler, M. G., Shapiro, B. E., Meyerowitz, E. M., and Mjolsness, E. (2006). An auxin-driven polarized transport model for phyllotaxis. *Proc. Natl. Acad. Sci. U.S.A.* **103**, 1633–1638.

Jun, J. H., Ha, C. M., and Fletcher, J. C. (2010). BLADE-ON-PETIOLE1 coordinates organ determinacy and axial polarity in Arabidopsis by directly activating *ASYMMETRIC LEAVES2*. *Plant Cell* **22**, 62–76.

Katz, A., Oliva, M., Mosquna, A., Hakim, O., and Ohad, N. (2004). FIE and CURLY LEAF polycomb proteins interact in the regulation of homeobox gene expression during sporophyte development. *Plant J.* **37**, 707–719.

Kaya, H., Shibahara, K. I., Taoka, K. I., Iwabuchi, M., Stillman, B., and Araki, T. (2001). FASCIATA genes for chromatin assembly factor-1 in Arabidopsis maintain the cellular organization of apical meristems. *Cell* **104**, 131–142.

Kayes, J. M., and Clark, S. E. (1998). CLAVATA2, a regulator of meristem and organ development in Arabidopsis. *Development* **125**, 3843–3851.

Kerstetter, R. A., Bollman, K., Taylor, R. A., Bomblies, K., and Poethig, R. S. (2001). KANADI regulates organ polarity in Arabidopsis. *Nature* **411**, 706–709.

Kessler, S., Townsley, B., and Sinha, N. (2006). L1 division and differentiation patterns influence shoot apical meristem maintenance. *Plant Physiol.* **141**, 1349–1362.
Kidner, C. A., and Martienssen, R. A. (2005). The role of ARGONAUTE1 (AGO1) in meristem formation and identity. *Dev. Biol.* **280**, 504–517.
Kim, Y. S., Kim, S. G., Lee, M., Lee, I., Park, H. Y., Seo, P. J., Jung, J. H., Kwon, E. J., Suh, S. W., Paek, K. H., and Park, C. M. (2008). HD-ZIP III activity is modulated by competitive inhibitors via a feedback loop in Arabidopsis shoot apical meristem development. *Plant Cell* **20**, 920–933.
Kirch, T., Simon, R., Grunewald, M., and Werr, W. (2003). The DORNROSCHEN/ ENHANCER OF SHOOT REGENERATION1 gene of Arabidopsis acts in the control of meristem cell fate and lateral organ development. *Plant Cell* **15**, 694–705.
Koyama, T., Furutani, M., Tasaka, M., and Ohme-Takagi, M. (2007). TCP transcription factors control the morphology of shoot lateral organs via negative regulation of the expression of boundary-specific genes in Arabidopsis. *Plant Cell* **19**, 473–484.
Krizek, B. A., and Fletcher, J. C. (2005). Molecular mechanisms of flower development: An armchair guide. *Nat. Rev. Genet.* **6**, 688–698.
Krizek, B. A., and Meyerowitz, E. M. (1996). The Arabidopsis homeotic genes APETALA3 and PISTILLATA are sufficient to provide the B class organ identity function. *Development* **122**, 11–22.
Kuhlemeier, C. (2007). Phyllotaxis. *Trends Plant Sci.* **12**, 143–150.
Kumaran, M. K., Bowman, J. L., and Sundaresan, V. (2002). YABBY polarity genes mediate the repression of KNOX homeobox genes in Arabidopsis. *Plant Cell* **14**, 2761–2770.
Kurakawa, T., Ueda, N., Maekawa, M., Kobayashi, K., Kojima, M., Nagato, Y., Sakakibara, H., and Kyozuka, J. (2007). Direct control of shoot meristem activity by a cytokinin-activating enzyme. *Nature* **445**, 652–655.
Kwon, C. S., Chen, C., and Wagner, D. (2005). WUSCHEL is a primary target for transcriptional regulation by SPLAYED in dynamic control of stem cell fate in Arabidopsis. *Genes Dev.* **19**, 992–1003.
Laufs, P., Peaucelle, A., Morin, H., and Traas, J. (2004). MicroRNA regulation of the *CUC* genes is required for boundary size control in Arabidopsis meristems. *Development* **131**, 4311–4322.
Laux, T., Mayer, K. F., Berger, J., and Jurgens, G. (1996). The *WUSCHEL* gene is required for shoot and floral meristem integrity in Arabidopsis. *Development* **122**, 87–96.
Lee, D. K., Geisler, M., and Springer, P. S. (2009). LATERAL ORGAN FUSION1 and LATERAL ORGAN FUSION2 function in lateral organ separation and axillary meristem formation in Arabidopsis. *Development* **136**, 2423–2432.
Leibfried, A., To, J. P., Busch, W., Stehling, S., Kehle, A., Demar, M., Kieber, J. J., and Lohmann, J. U. (2005). WUSCHEL controls meristem function by direct regulation of cytokinin-inducible response regulators. *Nature* **438**, 1172–1175.
Lenhard, M., Bohnert, A., Jurgens, G., and Laux, T. (2001). Termination of stem cell maintenance in Arabidopsis floral meristems by interactions between WUSCHEL and AGAMOUS. *Cell* **105**, 805–814.
Lenhard, M., Jurgens, G., and Laux, T. (2002). The *WUSCHEL* and *SHOOTMERISTEMLESS* genes fulfill complementary roles in Arabidopsis shoot meristem regulation. *Development* **129**, 3195–3206.
Lenhard, M., and Laux, T. (2003). Stem cell homeostasis in the Arabidopsis shoot meristem is regulated by intercellular movement of CLAVATA3 and its sequestration by CLAVATA1. *Development* **130**, 3163–3173.
Li, H., Xu, L., Wang, H., Yuan, Z., Cao, X., Yang, Z., Zhang, D., Xu, Y., and Huang, H. (2005). The putative RNA-dependent RNA polymerase RDR6 acts synergistically with ASYMMETRIC LEAVES1 and 2 to repress BREVIPEDICELLUS and MicroRNA165/166 in Arabidopsis leaf development. *Plant Cell* **17**, 2157–2171.

Lincoln, C., Long, J., Yamaguchi, J., Serikawa, K., and Hake, S. (1994). A knotted1-like homeobox gene in Arabidopsis is expressed in the vegetative meristem and dramatically alters leaf morphology when overexpressed in transgenic plants. *Plant Cell* **6**, 1859–1876.

Lindsay, D. L., Sawhney, V. K., and Bonham-Smith, P. C. (2006). Cytokinin-induced changes in CLAVATA1 and WUSCHEL expression temporally coincide with altered floral development in Arabidopsis. *Plant Sci.* **170**, 1111–1117.

Lohmann, J. U., Hong, R. L., Hobe, M., Busch, M. A., Parcy, F., Simon, R., and Weigel, D. (2001). A molecular link between stem cell regulation and floral patterning in Arabidopsis. *Cell* **105**, 793–803.

Long, J. A., and Barton, M. K. (1998). The development of apical embryonic pattern in Arabidopsis. *Development* **125**, 3027–3035.

Long, J. A., Moan, E. I., Medford, J. I., and Barton, M. K. (1996). A member of the KNOTTED class of homeodomain proteins encoded by the *STM* gene of Arabidopsis. *Nature* **379**, 66–69.

Lynn, K., Fernandez, A., Aida, M., Sedbrook, J., Tasaka, M., Masson, P., and Barton, M. K. (1999). The PINHEAD/ZWILLE gene acts pleiotropically in Arabidopsis development and has overlapping functions with the ARGONAUTE1 gene. *Development* **126**, 469–481.

Mallory, A. C., Dugas, D. V., Bartel, D. P., and Bartel, B. (2004). MicroRNA regulation of NAC-domain targets is required for proper formation and separation of adjacent embryonic, vegetative, and floral organs. *Curr. Biol.* **14**, 1035–1046.

Mayer, U., Buttner, G., and Jurgens, G. (1993). Apical-basal pattern formation in the Arabidopsis embryo—studies on the role of the Gnom gene. *Development* **117**, 149–162.

Mayer, K. F., Schoof, H., Haecker, A., Lenhard, M., Jurgens, G., and Laux, T. (1998). Role of WUSCHEL in regulating stem cell fate in the Arabidopsis shoot meristem. *Cell* **95**, 805–815.

McConnell, J. R., and Barton, M. K. (1998). Leaf polarity and meristem formation in Arabidopsis. *Development* **125**, 2935–2942.

McConnell, J. R., Emery, J., Eshed, Y., Bao, N., Bowman, J., and Barton, M. K. (2001). Role of PHABULOSA and PHAVOLUTA in determining radial patterning in shoots. *Nature* **411**, 709–713.

Milkereit, P., Gadal, O., Podtelejnikov, A., Trumtel, S., Gas, N., Petfalski, E., Tollervey, D., Mann, M., Hurt, E., and Tschochner, H. (2001). Maturation and intranuclear transport of pre-ribosomes requires Noc proteins. *Cell* **105**, 499–509.

Mravec, J., Kubes, M., Bielach, A., Gaykova, V., Petrasek, J., Skupa, P., Chand, S., Benkova, E., Zazimalova, E., and Friml, J. (2008). Interaction of PIN and PGP transport mechanisms in auxin distribution-dependent development. *Development* **135**, 3345–3354.

Muller, R., Borghi, L., Kwiatkowska, D., Laufs, P., and Simon, R. (2006). Dynamic and compensatory responses of Arabidopsis shoot and floral meristems to CLV3 signaling. *Plant Cell* **18**, 1188–1198.

Muller, B., and Sheen, J. (2008). Cytokinin and auxin interaction in root stem-cell specification during early embryogenesis. *Nature* **453**, 1094–1097.

Nishimura, C., Ohashi, Y., Sato, S., Kato, T., Tabata, S., and Ueguchi, C. (2004). Histidine kinase homologs that act as cytokinin receptors possess overlapping functions in the regulation of shoot and root growth in Arabidopsis. *Plant Cell* **16**, 1365–1377.

Nishimura, A., Tamaoki, M., Sato, Y., and Matsuoka, M. (1999). The expression of tobacco knotted1-type class 1 homeobox genes corresponds to regions predicted by the cytohistological zonation model. *Plant J.* **18**, 337–347.

Norberg, M., Holmlund, M., and Nilsson, O. (2005). The *BLADE ON PETIOLE* genes act redundantly to control the growth and development of lateral organs. *Development* **132**, 2203–2213.

Ogawa, M., Shinohara, H., Sakagami, Y., and Matsubayashi, Y. (2008). Arabidopsis CLV3 peptide directly binds CLV1 ectodomain. *Science* **319**, 294.

Ohyama, K., Shinohara, H., Ogawa-Ohnishi, M., and Matsubayashi, Y. (2009). A glycopeptide regulating stem cell fate in *Arabidopsis thaliana*. *Nat. Chem. Biol.* **5**, 578–580.

Okada, K., Ueda, J., Komaki, M. K., Bell, C. J., and Shimura, Y. (1991). Requirement of the auxin polar transport system in early stages of Arabidopsis floral bud formation. *Plant Cell* **3**, 677–684.

Ori, N., Eshed, Y., Chuck, G., Bowman, J. L., and Hake, S. (2000). Mechanisms that control knox gene expression in the Arabidopsis shoot. *Development* **127**, 5523–5532.

Palatnik, J. F., Allen, E., Wu, X., Schommer, C., Schwab, R., Carrington, J. C., and Weigel, D. (2003). Control of leaf morphogenesis by microRNAs. *Nature* **425**, 257–263.

Parry, G., Marchant, A., and May, S. (2001). Quick on the uptake: characterization of a family of plant auxin influx carriers. *J. Plant Growth Reg.* **20**, 217–225.

Petrasek, J., Mravec, J., Bouchard, R., Blakeslee, J. J., Abas, M., Seifertova, D., Wisniewska, J., Tadele, Z., Kubes, M., Covanova, M., Dhonukshe, P., Skupa, P., *et al.* (2006). PIN proteins perform a rate-limiting function in cellular auxin efflux. *Science* **312**, 914–918.

Phelps-Durr, T. L., Thomas, J., Vahab, P., and Timmermans, M. C. (2005). Maize rough sheath2 and its Arabidopsis orthologue ASYMMETRIC LEAVES1 interact with HIRA, a predicted histone chaperone, to maintain knox gene silencing and determinacy during organogenesis. *Plant Cell* **17**, 2886–2898.

Poethig, R. S., Coen, E. H. J., and Johri, M. M. (1986). Cell lineage patterns in maize Zea mays embryogenesis: a clonal analysis. *Dev. Biol.* **117**, 392–404.

Poethig, R. S. (1987). Clonal analysis of cell lineage patterns in plant development. *Am. J. Bot.* **74**, 581–594.

Prigge, M. J., Otsuga, D., Alonso, J. M., Ecker, J. R., Drews, G. N., and Clark, S. E. (2005). Class III homeodomain-leucine zipper gene family members have overlapping, antagonistic, and distinct roles in Arabidopsis development. *Plant Cell* **17**, 61–76.

Prunet, N., Morel, P., Thierry, A. M., Eshed, Y., Bowman, J. L., Negrutiu, I., and Trehin, C. (2008). *REBELOTE, SQUINT*, and *ULTRAPETALA1* function redundantly in the temporal regulation of floral meristem termination in *Arabidopsis thaliana*. *Plant Cell* **20**, 901–919.

Rast, M. I., and Simon, R. (2008). The meristem-to-organ boundary: more than an extremity of anything. *Curr. Opin. Genet. Dev.* **18**, 287–294.

Reddy, G. V., Heisler, M. G., Ehrhardt, D. W., and Meyerowitz, E. M. (2004). Real-time lineage analysis reveals oriented cell divisions associated with morphogenesis at the shoot apex of *Arabidopsis thaliana*. *Development* **131**, 4225–4237.

Reddy, G. V., and Meyerowitz, E. M. (2005). Stem-cell homeostasis and growth dynamics can be uncoupled in the Arabidopsis shoot apex. *Science* **310**, 663–667.

Reinhardt, D., Frenz, M., Mandel, T., and Kuhlemeier, C. (2003a). Microsurgical and laser ablation analysis of interactions between the zones and layers of the tomato shoot apical meristem. *Development* **130**, 4073–4083.

Reinhardt, D., Mandel, T., and Kuhlemeier, C. (2000). Auxin regulates the initiation and radial position of plant lateral organs. *Plant Cell* **12**, 507–518.

Reinhardt, D., Pesce, E. R., Stieger, P., Mandel, T., Baltensperger, K., Bennett, M., Traas, J., Friml, J., and Kuhlemeier, C. (2003b). Regulation of phyllotaxis by polar auxin transport. *Nature* **426**, 255–260.

Rojo, E., Sharma, V. K., Kovaleva, V., Raikhel, N. V., and Fletcher, J. C. (2002). CLV3 is localized to the extracellular space, where it activates the Arabidopsis CLAVATA stem cell signaling pathway. *Plant Cell* **14**, 969–977.

Ruth, J., Klekowski, E. J., and Stein, O. L. (1985). Impermanent initials of the shoot apex and diplontic selection in a juniper-chimera. *Am. J. Bot.* **72**, 1127–1135.

Saiga, S., Furumizu, C., Yokoyama, R., Kurata, T., Sato, S., Kato, T., Tabata, S., Suzuki, M., and Komeda, Y. (2008). The Arabidopsis *OBERON1* and *OBERON2* genes encode plant homeodomain finger proteins and are required for apical meristem maintenance. *Development* **135**, 1751–1759.

Sakakibara, H. (2006). Cytokinins: activity, biosynthesis, and translocation. *Annu. Rev. Plant Biol.* **57**, 431–449.

Sawa, S., Watanabe, K., Goto, K., Kanaya, E., Morita, E. H., and Okada, K. (1999). *FILAMENTOUS FLOWER*, a meristem and organ identity gene of *Arabidopsis*, encodes a protein with a zinc finger and HMG-related domains. *Genes Dev.* **13**, 1079–1088.

Scarpella, E., Marcos, D., Friml, J., and Berleth, T. (2006). Control of leaf vascular patterning by polar auxin transport. *Genes Dev.* **20**, 1015–1027.

Schoof, H., Lenhard, M., Haecker, A., Mayer, K. F., Jurgens, G., and Laux, T. (2000). The stem cell population of Arabidopsis shoot meristems is maintained by a regulatory loop between the CLAVATA and WUSCHEL genes. *Cell* **100**, 635–644.

Schubert, D., Primavesi, L., Bishopp, A., Roberts, G., Doonan, J., Jenuwein, T., and Goodrich, J. (2006). Silencing by plant Polycomb-group genes requires dispersed trimethylation of histone H3 at lysine 27. *EMBO J.* **25**, 4638–4649.

Schultz, E. A., Pickett, F. B., and Haughn, G. W. (1991). The FLO10 gene product regulates the expression domain of homeotic genes AP3 and PI in Arabidopsis flowers. *Plant Cell* **3**, 1221–1237.

Semiarti, E., Ueno, Y., Tsukaya, H., Iwakawa, H., Machida, C., and Machida, Y. (2001). The ASYMMETRIC LEAVES2 gene of *Arabidopsis thaliana* regulates formation of a symmetric lamina, establishment of venation and repression of meristem-related homeobox genes in leaves. *Development* **128**, 1771–1783.

Shibaoka, H., and Nagai, R. (1994). The plant cytoskeleton. *Curr. Opin. Cell Biol.* **6**, 10–15.

Shuai, B., Reynaga-Pena, C. G., and Springer, P. S. (2002). The lateral organ boundaries gene defines a novel, plant-specific gene family. *Plant Physiol.* **129**, 747–761.

Siegfried, K. R., Eshed, Y., Baum, S. F., Otsuga, D., Drews, G. N., and Bowman, J. L. (1999). Members of the YABBY gene family specify abaxial cell fate in Arabidopsis. *Development* **126**, 4117–4128.

Skoog, F., and Miller, C. O. (1957). Chemical regulation of growth and organ formation in plant tissues cultured in vitro. *Symp. Soc. Exp. Biol.* **54**, 118–130.

Skylar, A., Hong, F. X., Chory, J., Weigel, D., and Wu, X. L. (2010). STIMPY mediates cytokinin signaling during shoot meristem establishment in Arabidopsis seedlings. *Development* **137**, 541–549.

Smith, R. S., Guyomarc'h, S., Mandel, T., Reinhardt, D., Kuhlemeier, C., and Prusinkiewicz, P. (2006). A plausible model of phyllotaxis. *Proc. Natl. Acad. Sci. U.S.A.* **103**, 1301–1306.

Smyth, D. R., Bowman, J. L., and Meyerowitz, E. M. (1990). Early flower development in Arabidopsis. *Plant Cell* **2**, 755–767.

Snow, M., and Snow, R. (1932). Experiments on phyllotaxis I—The effect of isolating a primordium. *Philos. Trans. R. Soc. Lond. B* **221**, 1–43.

Snow, M., and Snow, R. (1959). Regulation of sizes of leaf primordia by older leaves. *Proc. R. Soc. Lond. B. Biol. Sci.* **151**, 39–47.

Song, S. K., and Clark, S. E. (2005). POL and related phosphatases are dosage-sensitive regulators of meristem and organ development in Arabidopsis. *Dev Biol.* **285**, 272–284.

Song, S. K., Lee, M. M., and Clark, S. E. (2006). POL and PLL1 phosphatases are CLAVATA1 signaling intermediates required for Arabidopsis shoot and floral stem cells. *Development* **133**, 4691–4698.

Steeves, T., and Sussex, I. (1989). *Patterns in Plant Development.* Cambridge University Press, Cambridge, UK.

Stewart, R. N., and Dermen, H. (1975). Flexibility in ontogeny as shown by contribution of shoot apical layers to leaves of periclinal chimeras. *Am. J. Bot.* **62**, 935–947.

Stone, J. M., Trotochaud, A. E., Walker, J. C., and Clark, S. E. (1998). Control of meristem development by CLAVATA1 receptor kinase and kinase-associated protein phosphatase interactions. *Plant Physiol.* **117**, 1217–1225.

Sun, B., Xu, Y., Ng, K. H., and Ito, T. (2009). A timing mechanism for stem cell maintenance and differentiation in the Arabidopsis floral meristem. *Genes Dev.* **23**, 1791–1804.

Sussex, I. M. (1951). Experiments on the cause of dorsiventrality in leaves. *Nature* **167**, 651–652.

Suzuki, T., Inagaki, S., Nakajima, S., Akashi, T., Ohto, M. A., Kobayashi, M., Seki, M., Shinozaki, K., Kato, T., Tabata, S., Nakamura, K., and Morikami, A. (2004). A novel Arabidopsis gene TONSOKU is required for proper cell arrangement in root and shoot apical meristems. *Plant J.* **38**, 673–684.

Takada, S., Hibara, K., Ishida, T., and Tasaka, M. (2001). The CUP-SHAPED COTYLEDON1 gene of Arabidopsis regulates shoot apical meristem formation. *Development* **128**, 1127–1135.

Takada, S., and Tasaka, M. (2002). Embryonic shoot apical meristem formation in higher plants. *J. Plant Res.* **115**, 411–417.

Takeda, S., Tadele, Z., Hofmann, I., Probst, A. V., Angelis, K. J., Kaya, H., Araki, T., Mengiste, T., Mittelsten Scheid, O., Shibahara, K., Scheel, D., and Paszkowski, J. (2004). BRU1, a novel link between responses to DNA damage and epigenetic gene silencing in Arabidopsis. *Genes Dev.* **18**, 782–793.

Traas, J., and Vernoux, T. (2002). The shoot apical meristem: the dynamics of a stable structure. *Philos. Trans. R. Soc. Lond. B. Biol. Sci.* **357**, 737–747.

Treml, B. S., Winderl, S., Radykewicz, R., Herz, M., Schweizer, G., Hutzler, P., Glawischnig, E., and Ruiz, R. A. (2005). The gene ENHANCER OF PINOID controls cotyledon development in the Arabidopsis embryo. *Development* **132**, 4063–4074.

Trotochaud, A. E., Hao, T., Wu, G., Yang, Z., and Clark, S. E. (1999). The CLAVATA1 receptor-like kinase requires CLAVATA3 for its assembly into a signaling complex that includes KAPP and a Rho-related protein. *Plant Cell* **11**, 393–406.

Tucker, M. R., Hinze, A., Tucker, E. J., Takada, S., Jurgens, G., and Laux, T. (2008). Vascular signalling mediated by ZWILLE potentiates WUSCHEL function during shoot meristem stem cell development in the Arabidopsis embryo. *Development* **135**, 2839–2843.

Turck, F., Roudier, F., Farrona, S., Martin-Magniette, M. L., Guillaume, E., Buisine, N., Gagnot, S., Martienssen, R. A., Coupland, G., and Colot, V. (2007). Arabidopsis TFL2/LHP1 specifically associates with genes marked by trimethylation of histone H3 lysine 27. *PLoS Genet.* **3**, e86.

Vernoux, T., Kronenberger, J., Grandjean, O., Laufs, P., and Traas, J. (2000). PIN-FORMED 1 regulates cell fate at the periphery of the shoot apical meristem. *Development* **127**, 5157–5165.

Vieten, A., Vanneste, S., Wisniewska, J., Benkova, E., Benjamins, R., Beeckman, T., Luschnig, C., and Friml, J. (2005). Functional redundancy of PIN proteins is accompanied by auxin-dependent cross-regulation of PIN expression. *Development* **132**, 4521–4531.

Voinnet, O. (2009). Origin, biogenesis, and activity of plant microRNAs. *Cell* **136**, 669–687.

Vroemen, C. W., Mordhorst, A. P., Albrecht, C., Kwaaitaal, M.A.C.J., and de Vries, S. C. (2003). The CUP-SHAPED COTYLEDON3 gene is required for boundary and shoot meristem formation in Arabidopsis. *Plant Cell* **15**, 1563–1577.

Waites, R., and Hudson, A. (1995). Phantastica—a gene required for dorsoventrality of leaves in *Antirrhinum majus*. *Development* **121**, 2143–2154.

Wardlaw, C. W. (1949). Phyllotaxis and organogenesis in ferns. *Nature* **164**, 167–169.

Wenkel, S., Emery, J., Hou, B. H., Evans, M. M., and Barton, M. K. (2007). A feedback regulatory module formed by LITTLE ZIPPER and HD-ZIPIII genes. *Plant Cell* **19**, 3379–3390.

West, M. A. L., and Harada, J. J. (1993). Embryogenesis in higher plants: an overview. *Plant Cell* **5**, 1361–1369.

Williams, L., and Fletcher, J. C. (2005). Stem cell regulation in the Arabidopsis shoot apical meristem. *Curr Opin Plant Biol.* **8**, 582–586.

Williams, L., Grigg, S. P., Xie, M., Christensen, S., and Fletcher, J. C. (2005). Regulation of Arabidopsis shoot apical meristem and lateral organ formation by microRNA miR166g and its AtHD-ZIP target genes. *Development* **132**, 3657–3668.

Williams, R. W., Wilson, J. M., and Meyerowitz, E. M. (1997). A possible role for kinase-associated protein phosphatase in the Arabidopsis CLAVATA1 signaling pathway. *Proc. Nat. Acad. Sci. U.S.A.* **94**, 10467–10472.

Wu, X., Dabi, T., and Weigel, D. (2005). Requirement of homeobox gene STIMPY/WOX9 for Arabidopsis meristem growth and maintenance. *Curr. Biol.* **15**, 436–440.

Wurschum, T., Gross-Hardt, R., and Laux, T. (2006). APETALA2 regulates the stem cell niche in the Arabidopsis shoot meristem. *Plant Cell* **18**, 295–307.

Xu, L., and Shen, W. H. (2008). Polycomb silencing of *KNOX* genes confines shoot stem cell niches in *Arabidopsis*. *Curr. Biol.* **18**, 1966–1971.

Xu, L., Xu, Y., Dong, A., Sun, Y., Pi, L., and Huang, H. (2003). Novel as1 and as2 defects in leaf adaxial–abaxial polarity reveal the requirement for ASYMMETRIC LEAVES1 and 2 and ERECTA functions in specifying leaf adaxial identity. *Development* **130**, 4097–4107.

Yadav, R. K., Girke, T., Pasala, S., Xie, M., and Reddy, G. V. (2009). Gene expression map of the Arabidopsis shoot apical meristem stem cell niche. *Proc. Natl. Acad. Sci.* USA **106**, 4941–4946.

Yanai, O., Shani, E., Dolezal, K., Tarkowski, P., Sablowski, R., Sandberg, G., Samach, A., and Ori, N. (2005). Arabidopsis KNOXI proteins activate cytokinin biosynthesis. *Curr Biol.* **15**, 1566–1571.

Yang, Y., Hammes, U. Z., Taylor, C. G., Schachtman, D. P, and Nielsen, E. (2006) High-affinity auxin transport by the aux1 influx carrier protein. *Curr. Biol.* **16**, 1123–1127.

Zhao, Y., Medrano, L., Ohashi, K., Fletcher, J. C., Yu, H., Sakai, H., and Meyerowitz, E. M. (2004). HANABA TARANU is a GATA transcription factor that regulates shoot apical meristem and flower development in Arabidopsis. *Plant Cell* **16**, 2586–2600.

Zhu, Y., Wang, Y., Li, R., Song, X., Wang, Q., Huang, S., Jin, J. B., Liu, C.-M., and Lin, J. (2010). Analysis of interactions among the CLAVATA3 receptors reveals a direct interaction between CLAVATA2 and CORYNE in Arabidopsis. *Plant J.* **61**, 223–233.

CHAPTER FIVE

Signaling Sides: Adaxial–Abaxial Patterning in Leaves

Catherine A. Kidner* *and* Marja C. P. Timmermans†

Contents

Abstract

Most leaves are dorsiventrally flattened and develop clearly defined upper and lower surfaces. Light capturing is the specialization of the adaxial or upper surface and the abaxial or lower surface is specialized for gas exchange (Fig. 5.1). This division into adaxial and abaxial domains is also key for the outgrowth of the leaf blade or lamina, which occurs along the boundary between the upper and lower sides. How this polarity is set up is not clear but genetic analysis in a range of species suggests that several highly conserved interlocking pathways are involved. Positional information from the meristem is reinforced by

* Royal Botanic Garden Edinburgh, Edinburgh, UK, and Institute of Molecular Plant Sciences, University of Edinburgh, Edinburgh, UK
† Cold Spring Harbor Laboratory, Cold Spring Harbor, New York, USA

Current Topics in Developmental Biology, Volume 91
ISSN 0070-2153, DOI 10.1016/S0070-2153(10)91005-3

signaling through the epidermal layer as the meristem grows away from the leaf primordium. Opposing ta-siRNA and miRNA gradients help refine distinct adaxial and abaxial sides, and mutual inhibition between the genes expressed on each side stabilizes the boundary. In this review we consider how recent work in a range of species is clarifying our understanding of these processes.

1. Leaves Evolved Late in Land Plant History

Leaves seem such a key feature of plants that it is surprising that they evolved relatively late in land plant history. By the time leaves appear in the fossil record the land plant lineages were already diverse and at least four and possibly up to nine different lineages independently evolved the leaf lamina (Boyce, 2010). The earliest vascular land plants (dating from as early as 445 million years ago (Steemans *et al.*, 2009)) were simply branching axes without dorsiventral polarity or lamina outgrowth. The evolution of a flat lamina to intercept sunlight was constrained by high CO_2 levels. High levels of CO_2 meant plants needed a fairly low density of stomata to be able to take up as much CO_2 as they could fix. This low density of stomata was sufficient to cool an axis by evapotranspiration but was not enough to be able to cool a lamina, which absorbs much more heat as well as more light. During the late Paleozoic (360 million years ago), CO_2 levels began to drop. Plants responded to this by producing more stomata per square cm. This higher density of stomata was able to cool an expanse of lamina and large leaves duly appear in the fossil record at about this time (Beerling *et al.*, 2001). The diverse origins of leaves have two consequences. One is that the processes of leaf development vary between lineages. The other is that the genes that now regulate leaf polarity were playing different roles 360 million years ago and these roles may constrain leaf development.

2. Leaves Develop from a Patterned Meristem

The independent origin of leaves in multiple lineages has resulted in a range of different methods of forming a flattened, determinate lateral organ from an indeterminate growing tip. Some of the same genetic pathways are involved in leaf development in each lineage but the processes of leaf development are quite distinct and this affects how polarity and lamina outgrowth are controlled (Floyd and Bowman, 2006; Harrison *et al.*, 2005). In the lycophyte *Selaginella*, the meristem has a single pyramidical apical cell and leaves are formed from two adjacent epidermal daughter cells, which divide circumferentially and then tangentially to create a leaf primordia of 12–16 cells (Harrison *et al.*, 2007). In many ferns, leaves derive from a single cell, or a small number of cells in the shoot apex, one of which becomes the lens-shaped apical cell for the leaf primordia. This apical cell gives rise to two rows of marginal initials that produce abaxial and adaxial daughter cells, generating a lamina (White and Turner, 1995). Seed plants are distinctive among vascular plants in having layered meristems rather than single apical cells and leaf primordia develop as quite large populations of cells in a particular zone of the meristem.

The typical angiosperm meristem is comprised of distinct zones and cell layers that are defined cytologically, molecularly by gene expression patterns, and functionally (see chapter 4). The meristem contains a central zone of slowly dividing stem cells surrounded by a peripheral zone of daughter cells, which divide faster and are competent for organogenesis. Divisions in the outermost cell layers of the meristem are anticlinal. This gives rise to a discrete L1 layer, which generates the epidermis, and often at least one underlying layer that remains separate from the internal mass of cells. Leaf primordia develop from groups of peripheral zone cells (~25 in *Arabidopsis* and 100–200 in maize) encompassing all meristem layers that become determined in response to a phyllotactic signal. These incipient leaf cells lose expression of meristem-specific genes, switch on leaf-specific programs, and develop polarity defining the two sides of the leaf. Concomitantly, cell division patterns change and the leaf primordium begins to grow outward, defining proximo-distal and medio-lateral axes.

3. Maintaining Polarity beyond the Meristem

The population of cells that becomes defined as the incipient leaf primordium is thought to carry positional information with them from the meristem to pattern the developing leaf. Primordium founder cells positioned apically, adjacent to the central zone of the meristem, will become the adaxial side of the leaf, whereas the more basal cells of the primordium, at the outer edge of the peripheral zone, give rise to the abaxial side of the leaf (Fig. 5.1A). Genes required for adaxial cell fate are expressed in the central zone of the meristem and genes that promote abaxial cell fate are already expressed in the internode region below the peripheral zone (see Fig. 5.4 and below). Surgical experiments provide evidence that positional information from the zones of the meristem is required to produce a polarized mature leaf. Incisions that separate an incipient primordium (referred to as I1 or P0) from the meristem tip cause the formation of a radialized abaxial leaf. This

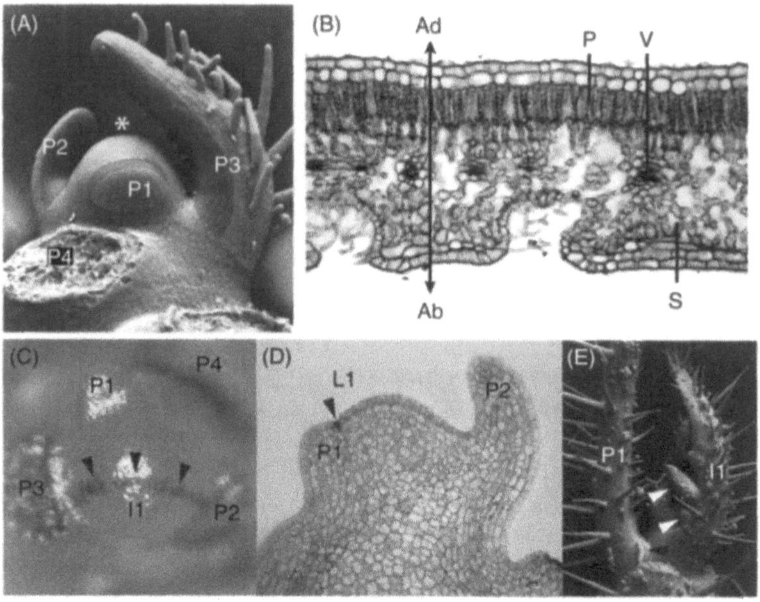

Figure 5.1 Meristem-regulated polarization of the leaf. (A) Scanning electron micrograph (SEM) of a tomato vegetative SAM showing the youngest (P1) and increasingly older (P2 and onward) leaf primordia surrounding the meristem (asterisks). Note that the adaxial/upper side of leaves (purple) develops in closer proximity to the SAM tip than the abaxial/lower side (yellow). (B) Transverse section through a *Nerium* leaf illustrating the differentiation of distinct cell types within the adaxial and abaxial domains. Palisade mesophyll (p) cells form beneath the adaxial epidermis, whereas spongy mesophyll (s) cells differentiate abaxially. Within the vasculature (v), water-bearing xylem cells differentiate adaxial to sugar-bearing phloem cells. (C) A vegetative tomato shoot apex showing ablation of the L1 layer (black arrowheads) separating the incipient leaf (I1) from the meristem. (D) Longitudinal section of a tomato apex showing an L1-restricted incision (black arrowhead) that separates the P1 primordium from the meristem. (E) SEM of a tomato apex a day after the P1 has been surgically separated from the SAM. The P1 develops as a radial, abaxialized primordium (black arrowheads show the scar from the ablation). In contrast, a primordium that developed in contact with the meristem (I1) is dorsiventrally flattened and develops leaflet pairs (white arrowheads). The SEM image in A was kindly provided by C. Kuhlemeier, and images C–E are reprinted with permission from Reinhardt et al. (Reinhardt *et al.*, 2005). (See Color Insert.)

shows that the polarization of the leaf primordium needs a signal from the meristem that promotes adaxial cell fate (Sussex, 1951, 1954; Fig. 5.1C–E).

Refinements of this surgical work have shown that the signal from the meristem travels through the L1 (Reinhardt *et al.*, 2005). The adaxial L1 has distinct gene expression patterns early in leaf development that are required for polarity (see below) and the meristem L1 is required to maintain cell division patterns (Reinhardt *et al.*, 2003). It could be argued that damage to the L1 by these ablations causes the loss of important L1-specific gene expression patterns; however, incisions that pass through the I1 epidermis and do not isolate it from the meristem result in leaves that are polarized normally (Reinhardt *et al.*, 2005). Even a small portion of intact L1 between the meristem and the I1 leaf primordium is sufficient to confer dorsiventrality. Primordia isolated at the I1 stage often initiated new meristems on the primordium side of the ablation. In these cases, the primordia developed normally, or with slight defects. This provides further support that an active meristem is the source of an adaxializing L1 signal.

This signal is required during a short developmental window. Primordia isolated at the P1 stage, when outgrowth from the meristem has been initiated, develop a dorsiventral distal region but are abaxialized in the proximal region (Reinhardt *et al.*, 2005). This suggests that the acquisition of organ polarity occurs gradually and depends on sustained signaling from the meristem. Incisions at the P2 stage of leaf development do not perturb adaxial–abaxial patterning, suggesting that mechanisms within the developing primordium are in place to maintain organ polarity. While the identity of the meristem-derived adaxializing signal remains unknown, much progress has been made toward understanding the genetic network that defines adaxial and abaxial cell fates.

4. Determinants of Adaxial Fate—*PHANTASTICA*

The isolated primordia in these surgical experiments not only lost adaxial cell fate, the leaf blade also failed to grow out along the medio-lateral axis. Several mutants have since been identified where lamina outgrowth is lost (see below). It was the phenotypic analysis of one of these mutants that generated the outgrowth-at-the-boundary hypothesis that has defined how adaxial–abaxial leaf polarity and lamina outgrowth are connected (Fig. 5.2). The *phantastica* (*phan*) mutant in *Antirrhinum* produces some fully radial leaves, some "trumpet-shaped" leaves, as well as leaves with broader, rumpled lamina that develop ridges of ectopic blade on their adaxial surface. Waites and Hudson (1995) interpreted these growth phenotypes to result from varying loss of adaxial cell fate. The radialized leaves were shown to contain abaxial cell types encircling central xylem tissue, whereas the other leaves were found to be partially abaxialized. The model proposes that complete loss of adaxial cell fate results in no polarity boundary and therefore no lamina; the limitation of adaxial cell fate to the tip of the leaf primordium results in limitation of lamina outgrowth to the displaced adaxial–abaxial boundary at the leaf tip; and that the ectopic patches of abaxial cell fate on the adaxial surface result in ectopic boundaries driving ectopic lamina outgrowth (Fig. 5.2).

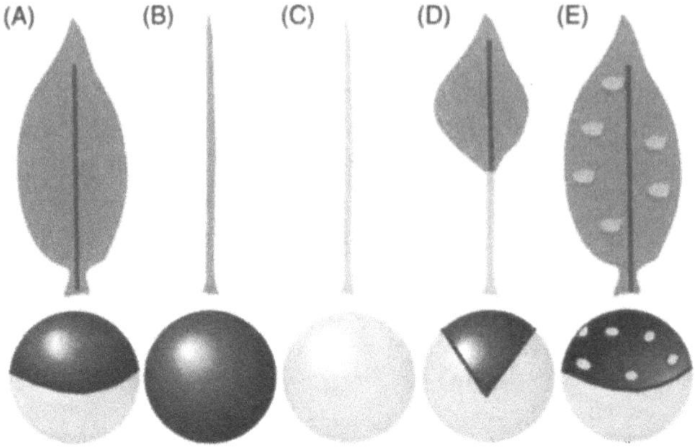

Figure 5.2 Lamina outgrowth occurs at the boundary between adaxial and abaxial regions. (A) In a normal leaf, the boundary between the adaxial and abaxial domains runs along the edge of the leaf primordium, driving lamina outgrowth along the length of the leaf. (B, C) On leaf primordia that are fully adaxialized (B) or abaxialized (C), no medio-lateral blade outgrowth occurs. (D) If the adaxial domain is reduced, blade outgrowth is limited to a smaller region of the leaf, in this example the distal tip, where adaxial–abaxial boundaries occur. (E) If cell fate is unstable and patches of abaxial cells form on the adaxial side of the primordium, outgrowths will develop at the ectopic adaxial–abaxial boundaries surrounding these patches. Redrawn with permission from Waites and Hudson (1995).

The *PHAN* gene from *Antirrhinum* and orthologs of other species (collectively referred to as the ARP family after the first three members to be cloned: *PHAN*, *rough sheath2* (*rs2*), and *ASYMMETRIC LEAVES1* (*AS1*)) encode Myb transcription factors, which promote adaxial cell fate and are further required to maintain the downregulation of meristem-promoting KNOX genes in the developing leaf (Byrne *et al.*, 2000; Kim *et al.*, 2003; Luo *et al.*, 2005; McHale and Koning, 2004a; Tattersall *et al.*, 2005; Timmermans *et al.*, 1999; Tsiantis *et al.*, 1999; Waites *et al.*, 1998). ARP genes are specific to vascular plants and their biochemical function has been conserved for over 300 million years as ARP from *Selaginella* can complement the *Arabidopsis as1* mutant (Harrison *et al.*, 2005). However, the ARP-KNOX regulatory module has been independently co-opted into leaf development in these two lineages as lycophytes (*Selaginella*) and seed plants (*Arabidopsis*) derive from a leafless common ancestor.

Despite their association with adaxial cell fate, ARP genes are expressed throughout young leaf primordia. This suggests that their role in adaxial cell fate is regulated at the protein level. The Lateral-Organ-Boundary (LOB) family member *ASYMMETRIC LEAVES2* (*AS2*) could provide one such mechanism. This Zn finger transcription factor interacts with AS1 and is expressed in the adaxial most cell layers of young leaf primordia, and ectopic expression leads to adaxialization of developing leaves (Evans, 2007; Guo *et al.*, 2008; Iwakawa *et al.*, 2007; Lin *et al.*, 2003; Phelps-Durr *et al.*, 2005). In *Arabidopsis* and maize, *arp* null mutants do not show a clear loss of adaxial cell identity (Byrne *et al.*, 2000; Schneeberger *et al.*, 1998; Timmermans *et al.*, 1999; Tsiantis *et al.*, 1999). This suggests that, in at least some lineages, their role in adaxial cell fate is redundant with other factors; a notion that is supported by enhancer–suppressor screens, which, among others, identified several polarity determinants (Garcia *et al.*, 2006; Li *et al.*, 2005; Pinon *et al.*, 2008; Ueno *et al.*, 2007).

Once the defects that characterize a polarity mutant had been recognized, additional genes that define the adaxial and abaxial domains were identified in a number of diverse plant species. These polarity mutants either fall into one of three distinct classes of highly conserved transcription factors (the class III HOMEODOMAIN LEUCINE ZIPPER (HD-ZIPIII), KANADI (KAN), and AUXIN RESPONSE FACTOR (ARF) proteins), or they perturb regulatory pathways that control the spatio-temporal accumulation of these transcription factors.

5. Determinants of Adaxial Fate—The HD-ZIPIII Genes

Dominant, gain-of-function mutations in the HD-ZIPIII genes have strong effects on leaf polarity. In *Arabidopsis*, dominant alleles of *PHABULOSA* (*PHB*), *PHAVOLUTA* (*PHV*), and *REVOLUTA* (*REV*) cause the formation of fully adaxialized filaments or trumpet-shaped leaves with adaxial tissue on the "stem" of the trumpet and abaxial in the "bell" (Emery *et al.*, 2003; McConnell and Barton, 1998). In addition, meristem size is increased in these adaxialized mutants, the inverse of the loss of meristem function seen in severely abaxialized *phan* mutants (McConnell and Barton, 1998; Waites and Hudson, 1995). In conjunction with data from surgical experiments, these phenotypes point to a mutually enhancing relationship between meristem activity and adaxial cell fate. But, just as the identity of the "Sussex" signal has remained elusive, the nature of the signal from lateral organs to the meristem remains unknown.

Individual loss-of-function mutants in HD-ZIPIII genes have weak or no discernable phenotypes; however, the importance of this family as a whole becomes evident in higher-order mutant combinations (Prigge *et al.*, 2005). *phb phv rev* triple mutants develop a single abaxialized cotyledon and have lost meristem function (Emery *et al.*, 2003; Prigge *et al.*, 2005). The HD-ZIPIII genes are expressed in the meristem tip, vasculature, and leaf primordia in patterns that support their roles in defining the adaxial domain of the developing leaf and the xylem pole of vasculature, as well as in maintaining stem cell function. Gain-of-function HD-ZIPIII alleles result in overexpression as well as in ectopic expression of functional transcripts on the abaxial side of developing primordia (Itoh *et al.*, 2008; Juarez *et al.*, 2004a; McConnell *et al.*, 2001). The fact that such alleles condition an adaxialized leaf phenotype indicates that HD-ZIPIII genes are not only necessary but also sufficient to specify adaxial fate.

6. Regulation of HD-ZIPIII Activity

Reflective of their prominent roles in plant development, HD-ZIPIII activity is subject to multiple levels of regulation (Fig. 5.3). The characterized gain-of-function HD-ZIPIII mutants are due to changes at a target site normally recognized by the 21 nucleotide microRNAs, miR165 and miR166 (Itoh *et al.*, 2008; Juarez *et al.*, 2004a; McConnell *et al.*, 2001; McHale and Koning, 2004b). These small RNAs direct the cleavage of HD-ZIPIII transcripts (Tang *et al.*, 2003). In both *Arabidopsis* and maize, miR165/166 accumulate on the abaxial side of young leaf primordia and thereby limit HD-ZIPIII expression to the adaxial side (Juarez *et al.*, 2004a; Kidner and Martienssen, 2004). In maize, miR166 accumulation is strongest immediately below the incipient leaf but a gradient of weaker miR166 expression extends into the abaxial domain of the young primordia (Juarez *et al.*, 2004a). This pattern of accumulation is particularly intriguing considering recent findings that small RNAs can traffic between cells (Carlsbecker *et al.*, 2010; Chitwood *et al.*, 2009; Dunoyer *et al.*, 2010). It presents the possibility that miR166 serves as a positional signal within the meristem that polarizes the incipient leaf (see below). The importance of miR156/166-mediated HD-ZIPIII regulation in creating adaxial–abaxial polarity is evident not only from the phenotypes caused by the dominant HD-ZIPIII mutations that prevent miRNA-directed transcript cleavage, but also from the severe abaxialization of leaves with uniform miR165/166 expression (Alvarez *et al.*, 2006; Nogueira *et al.*, 2007).

HD-ZIPIII activity in *Arabidopsis* is also modulated by the four LITTLE ZIPPER (ZPR1-4) proteins, which contain a leucine zipper motif that is similar to the leucine zipper found in the HD-ZIPIII proteins and mediates heterodimer formation. However, such HD-ZIPIII/ZPR heterodimers are unable to bind DNA. ZPR proteins thus effectively sequester the HD-ZIPIII transcription factors into inactive protein complexes (Kim *et al.*, 2008; Wenkel *et al.*, 2007). *ZPR* expression is induced, in part, by HD-ZIPIII activity suggesting that a negative feedback loop modulates HD-ZIPIII function. The greatly enlarged shoot apical meristem of *zpr3 zpr4* loss-of-function mutants illustrates the significance of this feedback loop in meristem regulation (Kim *et al.*, 2008). However, as *ZPR1* and

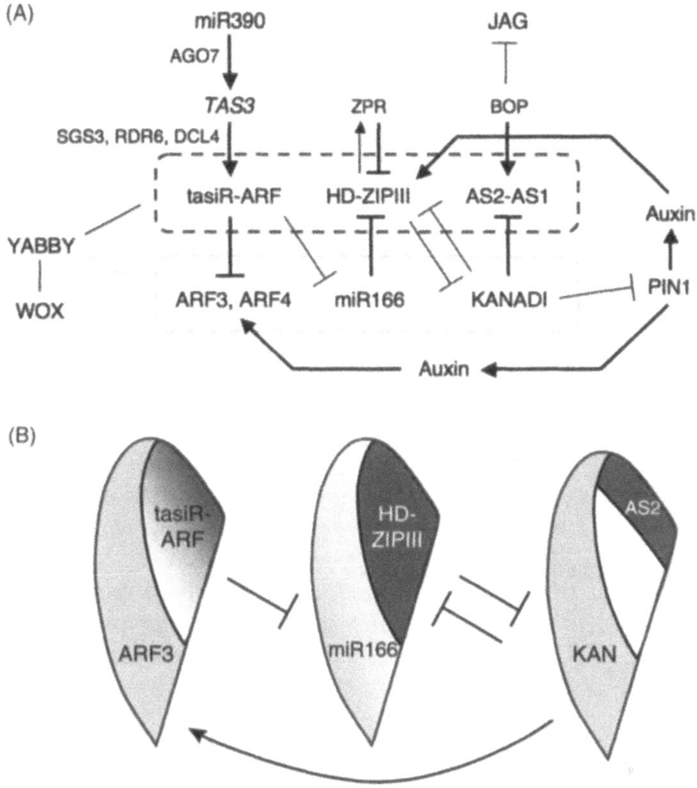

Figure 5.3 Genetic interactions in leaf polarity. (A) Specification of organ polarity involves three conserved interlocking genetic pathways: the *TAS3* ta-siRNA–ARF3/4, miR166–HD-ZIPIII, and AS1-AS2–KAN pathways. These pathways are distinguished by redundancies and by mutual antagonistic interactions between adaxial and abaxial components. Key determinants of adaxial identity are encircled in purple, and abaxial determinants in yellow. Heavy lines indicate direct relationships, light lines indirect. Colored lines show the general effects of each cell fate on YABBY expression, which varies between species. (B) Spatial arrangement of the key components in the three genetic pathways involved in leaf polarity in the P1 leaf. Gradients of tasiR-ARF and miR166 generate sharply defined expression domains of ARF and HD-ZIPIII genes, respectively. Further mutual inhibition between the abaxial and adaxial zones sharpens and defines the boundary. (See Color Insert.)

ZPR3 are expressed relatively late in leaf development (Wenkel *et al.*, 2007), their contribution to adaxial–abaxial patterning is less clear. Thus, although both miR165/166 and the ZPR proteins negatively regulate HD-ZIPIII activity, their contributions are quite distinct. Whereas miR165/166 limits the expression domain of the HD-ZIPIII genes to the adaxial side of developing leaves, the ZPR proteins act within the HD-ZIPIII expression domain to modulate their activity.

Several observations predict that expression of the HD-ZIPIII genes is responsive to auxin. HD-ZIPIII expression is induced at sites marked by auxin maxima, such as in incipient leaf primordia and at sites where vascular strands will develop (chapter 8; Baima *et al.*, 1995; Bayer *et al.*, 2009; Heisler *et al.*, 2005; Ilegems *et al.*, 2010; Reinhardt *et al.*, 2003). In one case at least, HD-ZIPIII expression is under direct positive control of an auxin response factor (ARF), transcription factors that mediate auxin-induced gene expression responses. MONOPTEROS (MP, aka ARF5) binds an auxin response element in the promoter of the HD-ZIPIII gene *AtHB8* and activates its expression in the initiating vasculature (Donner *et al.*, 2009). However, transcript level analysis following the application of exogenous auxin shows that, at least in rice, most of the family members involved in leaf polarity have only a slow, slight positive response to auxin (Itoh *et al.*, 2008). This suggests that auxin-mediated regulation of some HD-ZIPIII family members may be indirect. Auxin could regulate HD-ZIPIII activity via the AP2-domain transcription factor DORNROSCHEN (DRN), which is a target of MP and interacts with a C-terminal domain of PHB and PHV (Chandler *et al.*, 2007; Cole *et al.*, 2009).

In addition to the homeodomain and leucine zipper motifs, members of the HD-ZIPIII family share a START domain (Ponting and Aravind, 1999; Schrick *et al.*, 2004). START domains are well known in animal systems where they are activated through high-affinity binding to lipophilic ligands. The presence of a START domain in the HD-ZIPIII proteins suggests the possibility that HD-ZIPIII activity is further regulated through interaction with a ligand. This concept is supported by the existence of a loss-of-function *rev* allele containing an amino acid substitution in the START domain (Otsuga *et al.*, 2001). Part of the communication between the meristem and adaxial domain could thus include the regulation of HD-ZIPIII function via a mobile lipophilic signal. However, the HD-ZIPIII proteins are unlikely to be receptors for the meristem-derived "Sussex signal" as miRNA-resistant HD-ZIPIII mutants have polarity defects, indicating that these proteins can function in the abaxial zone where this adaxializing signal is not predicted to be active.

7. Determinants of Abaxial Fate–The KAN Genes

Members of the KAN gene family, which encode transcription factors containing a Myb-like GARP DNA-binding domain, promote abaxial cell fate. KAN family members are expressed in a pattern complementary to that of the HD-ZIPIII genes, on the abaxial side of organ primordia and in the developing phloem of the vasculature (Candela *et al.*, 2008; Emery *et al.*, 2003; Henderson *et al.*, 2006; Kerstetter *et al.*, 2001; Zhang *et al.*, 2009). In addition, KAN expression is found in the internode regions below developing organs. Analogous expression in the hypocotyl of developing *Arabidopsis* embryos is critical for the correct positioning of cotyledons (Eshed *et al.*, 2004; Izhaki and Bowman, 2007; Gillmor *et al.*, 2010); KAN proteins repress expression of the auxin efflux carrier PIN-FORMED1 (PIN1) and thereby delineate the location of auxin maxima that presage organ initiation. *Arabidopsis kan1 kan2 kan4* triple mutants develop additional cotyledon-like outgrowths (Izhaki and Bowman, 2007). Double or triple *kan* mutant seedlings also develop radially symmetric adaxialized organs or narrow leaves with ectopic blade outgrowths on their abaxial leaf surface, confirming a requirement for KAN function in abaxial cell fate and control of outgrowth (Eshed *et al.*, 2001, 2004; Kerstetter *et al.*, 2001).

KAN function is conserved among angiosperms. Loss-of-function mutations in the maize KAN gene *milkweed pod1* (*mwp1*) develop ectopic outgrowths on the abaxial surface of the leaf's proximal sheath region, consistent with localized loss of abaxial cell fate. Expression analysis indicates that *mwp1* acts redundantly with other family members in the leaf blade and early in leaf development (Candela *et al.*, 2008). Mutations in the rice ortholog of *KAN1* and *mwp1, SHALLOT-LIKE1* (*SLL1*), also affect organ polarity, altering cell identity of a subset of abaxial cell types (Zhang *et al.*, 2009).

In both *Arabidopsis kan1 kan2* and maize *mwp1* mutant leaves, the ectopic blade outgrowths are associated with sectors of adaxial tissue and HD-ZIPIII misexpression (Candela *et al.*, 2008; Eshed *et al.*, 2001, 2004). HD-ZIPIII transcripts also accumulate ectopically throughout radialized *kan1 kan2* organs. Conversely, ectopic expression of *KAN1* or *KAN2* throughout the developing leaf leads to fully abaxialized organs, with a concomitant loss of HD-ZIPIII expression (Eshed *et al.*, 2004; Kerstetter *et al.*, 2001). These findings implicate the KAN proteins as negative regulators of the HD-ZIPIII genes (Fig. 5.3). Whereas reduced KAN expression in HD-ZIPIII gain-of-function mutants points to a reciprocal role for the HD-ZIPIII genes in negatively regulating KAN expression. There is no evidence that the antagonistic relationship between these polarity determinants is direct. This mutual inhibition may be due to a complex cell signaling and inhibition pathway, which prevents cells expressing both adaxial and abaxial cell fates simultaneously. Considering that KAN proteins negatively regulate PIN1 expression, a role for auxin in this inhibitory pathway may need to be considered.

A further antagonistic interaction exists between KAN proteins and the AS pathway (Fig. 5.3). KAN proteins directly downregulate *AS2* expression (Wu *et al.*, 2008). Loss of AS2 in the abaxial domain restricts the AS1/AS2 interaction to the adaxial side of the leaf where it promotes adaxial cell fate. This, in part, via repression of *KAN* genes, as expression of these abaxial determinants is reduced in plants that ectopically express *AS2* on the abaxial side (Wu *et al.*, 2008). The repressive nature of the AS pathway (Guo *et al.*,

2008) makes it plausible that this latter regulatory interaction is also direct, but this remains to be demonstrated. The antagonistic interactions of the KAN proteins with AS2 and the HD-ZIPIII genes prevent cells expressing both adaxial and abaxial cell fates simultaneously. Such mutually exclusive and opposing cell fates, are key to generating and maintaining the boundary between adaxial and abaxial domains in the developing leaf (see below).

8. Determinants of Abaxial Fate—The AUXIN RESPONSE FACTORS

The ARF proteins ARF3 (aka ETTIN (ETT)) and ARF4 represent a fourth class of highly conserved transcription factors required to polarize developing organs. Loss of either *ARF3* or *ARF4* activity has little effect on adaxial–abaxial patterning, but double mutants develop adaxialized leaves that resemble *kan1 kan2* leaves (Pekker *et al.*, 2005). Mutations in *ARF3* suppress the phenotypes resulting from ectopic *KAN1* expression, positioning *ARF3* as a possible downstream target of KAN activity. However, neither the *ARF3* nor the *ARF4* expression domains are altered in a *kan1 kan2* background (Pekker *et al.*, 2005). Perhaps, KAN proteins regulate the activity of ARF3 or these transcription factors interact to form a heteromeric protein complex. Alternatively, *ARF* transcription may be regulated redundantly also through a KAN-independent pathway. Consistent with the latter possibility, expression of *ARF3* and *ARF4* is induced by auxin (Pekker *et al.*, 2005; Yoon *et al.*, 2010).

Auxin also regulates the activity of ARF4 and other ARF family members at the posttranslational level. Without auxin, ARF proteins dimerize with members of the Aux/IAA family of proteins that repress their transcriptional activity. Accumulation of auxin triggers the proteosome-mediated degradation of Aux/IAA proteins and thereby releases ARF proteins from inhibition, allowing them to alter expression of target genes (see Guilfoyle and Hagen, 2007). ARF3, however, lacks the domains required for interaction with Aux/IAA proteins, such that its activity is likely not responsive to auxin levels (Tiwari *et al.*, 2003; Ulmasov *et al.*, 1999).

9. Regulation of *ARF3* and *ARF4* by the *TAS3* ta-siRNA Pathway

Just like the adaxializing HD-ZIPIII genes, *ARF3* and *ARF4* are targets of small RNA-mediated gene regulation (Fig. 5.3). Expression of *ARF3* and *ARF4* is regulated by the tasiR-ARF class of trans-acting short-interfering RNAs (ta-siRNAs) (Allen *et al.*, 2005; Chitwood *et al.*, 2009; Fahlgren *et al.*, 2006; Hunter *et al.*, 2006). These small RNAs are processed from noncoding *TAS3* precursor transcripts through a specialized RNAi pathway that relies uniquely on the activity of ARGONAUTE7/ZIPPY (AGO7/ZIP) and its selective binding of miR390 (Montgomery *et al.*, 2008). The miR390–AGO7 complex directs the cleavage of *TAS3* transcripts at a 3' target site but interacts in a noncleavage mode also at a 5' site (Montgomery *et al.*, 2008). As a result, *TAS3* cleavage fragments are stabilized and converted into long double-stranded RNAs (dsRNAs) through the activities of SUPPRESSOR OF GENE SILENCING3 (SGS3) and RNA-DEPENDENT RNA POLYMERASE6 (RDR6), respectively. DICER-LIKE4 (DCL4) then processes these dsRNAs into 21-nt ta-siRNAs, a subset of which posttranscriptionally represses the expression of target genes (see Chapman and Carrington, 2007).

The *TAS3* ta-siRNA pathway and its regulation of ARFs are conserved throughout land plant evolution. Nonetheless, the contribution of this small RNA pathway to adaxial–abaxial patterning differs greatly between plant species. *Arabidopsis* mutants that block the biogenesis of tasiR-ARFs have phenotypes consistent with an accelerated transition from juvenile to adult leaf development, but reveal no obvious organ polarity defects (Hunter *et al.*, 2006; Peragine *et al.*, 2004). In contrast, mutations in genes required for tasiR-ARF biogenesis in maize, rice, and the legume, *Lotus japonicus*, result in the formation of centric abaxialized organs (Nagasaki *et al.*, 2007; Nogueira *et al.*, 2007; Yan *et al.*, 2010). Exactly why the loss of tasiR-ARF activity in *Arabidopsis* confers no obvious leaf polarity phenotypes has yet to be resolved. Functional overlap exists between the *TAS3* ta-siRNA and AS pathways, as tasiR-ARF biogenesis mutants enhance the *as1* and *as2* phenotypes and the expression levels of abaxial determinants, such as miR165/166, are elevated specifically in mutants compromised for both pathways (Garcia *et al.*, 2006; Li *et al.*, 2005; Xu *et al.*, 2006). However, as the leaves of such double mutants largely retain correct adaxial–abaxial polarity, additional redundancies or differences in the patterning of the *TAS3* ta-siRNA pathway and downstream targets must exist in *Arabidopsis* as compared to maize, rice, and *Lotus*.

Localization of *TAS3* ta-siRNA pathway components reveals that tasiR-ARF biogenesis in *Arabidopsis* is restricted to the two adaxial most cell layers of developing leaf primordia. Although miR390 accumulates broadly throughout the vegetative apex, tasiR-ARF biogenesis is restricted to the adaxial side of leaf primordia by the localized expression of *AGO7* and *TAS3* (Chitwood *et al.*, 2009). Interestingly, tasiR-ARFs accumulate outside this defined region of biogenesis. Intercellular movement of tasiR-ARFs creates a graded pattern of accumulation across the leaf, strongest near their adaxial source of biogenesis and dissipating toward the abaxial side.

This gradient of tasiR-ARF accumulation does not simply translate into an inverse expression gradient of its targets. Instead, it is interpreted into discrete regions of high and low tasiR-ARF activity that creates a sharply defined domain of ARF3 and ARF4 expression on the abaxial side of leaf primordia (Chitwood *et al.*, 2009; Pekker *et al.*, 2005).

How tasiR-ARF activity becomes patterned in *Arabidopsis* leaves is not currently understood. Considering that PINHEAD/ZWILLE/AGO10 (AGO10) localizes specifically to the adaxial side of leaf primordia (Lynn *et al.*, 1999), this may involve regulation by small RNA effector complexes. Alternatively, conversion of the tasiR-ARF gradient into discrete regions of ARF3 and ARF4 accumulation may reflect a concentration-dependent, morphogen-like small RNA read-out (Chitwood *et al.*, 2009). Mathematical modeling of the interaction between tasiR-ARFs and ARF3 supports the theoretical feasibility of a morphogen-like small RNA-mediated patterning mechanism (Levine *et al.*, 2007). Moreover, increasing the levels of nontargeted *ARF3* transcripts within the leaf results in increasingly severe defects (Fahlgren *et al.*, 2006; Hunter *et al.*, 2006). This indicates that leaf development is sensitive to the levels of *ARF3* expression and supports a concentration-dependent effect of the *TAS3* ta-siRNA pathway on leaf development.

Key differences exist in the spatio-temporal regulation of the *TAS3* ta-siRNA biogenesis between *Arabidopsis* and maize that might further explain the distinct contributions of this pathway to adaxial–abaxial patterning in the two species. In *Arabidopsis*, tasiR-ARFs are found uniformly throughout the SAM and their contribution to adaxial–abaxial patterning may not be realized until later in organ development (Chitwood *et al.*, 2009). In contrast, tasiR-ARF activity in maize is polarized in incipient primordia and is required to generate the abaxial gradient of miR166 accumulation in the incipient leaf (Nogueira *et al.*, 2007). Mutations in *leafbladeless1* (*lbl1*), which encodes the functional ortholog of SGS3, lead to the uniform accumulation of miR166 throughout incipient and developing leaf primordia. As a result, *lbl1* leaves frequently develop as centric abaxialized organs in which HD-ZIPIII expression is lost or dramatically reduced (Juarez *et al.*, 2004b; Timmermans *et al.*, 1998).

A further distinction is that in maize, miR390 is the restrictive factor that precisely positions *TAS3* ta-siRNA biogenesis. miR390 localizes to the adaxial side of incipient and young leaf primordia, whereas other components of the *TAS3* ta-siRNA biogenesis pathway are expressed more broadly (Noguiera *et al.*, 2009). Importantly, miR390 constitutes the most upstream component in the network that specifies adaxial–abaxial polarity in the incipient maize leaf. Expression of miR390 remains restricted to the upper side of *lbl1* primordia that are morphologically and molecularly completely abaxialized. The polar accumulation of miR390 is thus regulated independently of the ta-siRNA pathway and downstream polarity determinants, such as miR166 and the HD-ZIPIII genes, and appears instead to be regulated by positional information inherent to the SAM. This positional information may act to pattern small RNA-related factors. Expression of the miR390 precursors in the SAM is not limited to the incipient primordium, indicating that the discrete accumulation of mature miR390 in just the incipient leaf is regulated at the level of small RNA processing and/or stability (Husbands *et al.*, 2009; Nogueira *et al.*, 2009).

Adaxial–abaxial patterning of the maize leaf thus involves a novel mechanism in which miR390 is positioned at the top of a cascade of small RNA activities. miR390 localizes to the adaxial side of the incipient primordium where it triggers the biogenesis of *TAS3* ta-siRNAs that then restrict miR166 abaxially (Nogueira *et al.*, 2007, 2009). Such a small RNA cascade might also contribute to organ polarity in *Arabidopsis*, albeit within older leaf primordia, as tasiR-ARFs, additively with the AS pathway, repress the levels of miR165/166 accumulation (Li *et al.*, 2005; Xu *et al.*, 2006).

10. How Are These Polarized Gene Expression Domains Established?

Expression analyses indicate that many of the adaxial and abaxial determinants exhibit a polarized expression pattern in the incipient primordium (Heisler *et al.*, 2005; Itoh *et al.*, 2008; Juarez *et al.*, 2004a, b; Nogueira *et al.*, 2009). How these expression domains are first established remains a pressing question in the field. Leaf primordia form at sites of auxin maxima in the peripheral zone (PZ) of the SAM, which is positioned between the apical central zone (CZ) and basal internode. Like earlier botanists, we are tempted to speculate that adaxial–abaxial polarity in the incipient leaf is a manifestation of apical–basal patterning in the meristem (Sussex, 1951, 1954; Wardlaw, 1949). The observation that miR390 remains adaxially localized in molecularly "abaxialized" *lbl1* primordia supports this idea and suggests that the incipient organ may be prepatterned into adaxial and abaxial domains (Husbands *et al.*, 2009; Nogueira *et al.*, 2009).

Polarization of the PZ along the apical–basal meristem axis could form the basis for such a prepattern (Fig. 5.4A). If so, what is the nature of the positional information that patterns the PZ? HD-ZIPIII genes are expressed in the CZ, and the genetic interaction between the KAN proteins and the auxin efflux carrier PIN1 indicates that these abaxial determinants function in the internode (Heisler *et al.*, 2005; Itoh *et al.*, 2008; Izhaki and Bowman, 2007; Juarez *et al.*, 2004b; McConnell *et al.*, 2001) (Fig. 5.2). In addition, the maize KAN gene *ZmKAN2* was shown to be expressed in this basal domain of the apex (Henderson *et al.*, 2006). The possibility that the HD-ZIPIII and KAN expression domains

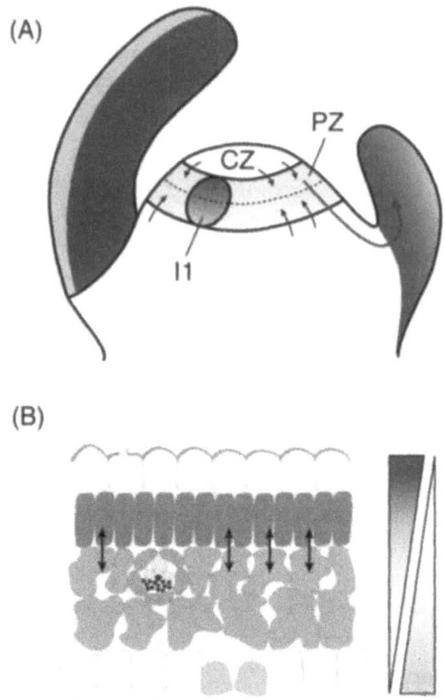

Figure 5.4 Mobile signals in the establishment and maintenance of leaf polarity. (A) The peripheral zone (PZ) is envisioned to be patterned into apical/centric and basal/outer regions based on positional signals derived from the HD-ZIPIII expressing CZ and the internode region below the PZ, which is marked by KAN and miR166 expression. The incipient leaf (I1) forms at an auxin maximum in the PZ and is divided in adaxial and abaxial domains by this prepattern in the PZ. As the leaf develops, the meristem grows away from it, but a signal through the L1 (long red arrow) maintains this initial polarity until the P2 stage of organ development when maintenance mechanisms within the organ are in place. (B) Mutually antagonistic interactions between cell-autonomous polarity determinants contributes to the maintenance of organ polarity, but is unlikely sufficient to define a precise boundary between adaxial and abaxial organ domains. Superimposed intercellular signals derived from the adaxial and abaxial sides can provide positional inputs to place and refine the adaxial–abaxial boundary. Gradients of adaxially derived mobile tasiR-ARFs (purple) and abaxially-derived miR166 (yellow) are candidates for such signals. Additionally, cell–cell communication that allows cells to acquire information about the identity of their neighbors and to coordinate cell division in the different cell layers (small arrows) is required to coordinate blade outgrowth at the adaxial–abaxial boundary. (See Color Insert.)

prepattern the PZ is intriguing considering the positive relationship that meristems have with adaxial fate (McConnell and Barton, 1998; Waites *et al.*, 1998). However, neither the HD-ZIPIII nor KAN proteins, nor any of the other polarity transcription factors, are known to move between cells. Patterning of the PZ and the incipient leaf therefore must depend on intercellularly mobile signals that act downstream of the HD-ZIPIII and

KAN proteins or other factors that mark the apical CZ and basal internode (Fig. 5.4A).

The pattern of miR166 accumulation in the maize SAM (strongest immediately below the incipient leaf and dissipating into the abaxial domain of this I1 primordium) along with the recent finding that miR166 acts non-cell autonomously suggest that miR166 might act as a mobile positional signal from the SAM that polarizes the new leaf (Carlsbecker *et al.*, 2010; Juarez *et al.*, 2004a). Most other signals that have been proposed (auxin, the HD-ZIPIII steroid ligand, and tasiR-ARF) unlikely fulfill such a role. Auxin, like the "Sussex" signal, is transported primarily through the L1 of the meristem (Bayer *et al.*, 2009; Heisler *et al.*, 2005; Reinhardt *et al.*, 2003). Nonetheless, auxin is unlikely to function as a positional signal in the establishment of organ polarity. The pattern of PIN1 polarization in the meristem would result in a nonpolar accumulation of auxin in the incipient primordium. Instead, the auxin maximum at I1 may serve to induce expression of polarity determinants (e.g., the HD-ZIPIII and ARF genes) in the initiating organ.

TAS3 ta-siRNAs can traffic between cells and, in maize, function to polarize the incipient leaf (Chitwood *et al.*, 2009; Nogueira *et al.*, 2007). However, unlike miR166, tasiR-ARFs seem to act downstream of the factors that prepattern the PZ. These unknown patterning factors are probably required to correctly position miR390, whose adaxial specific accumulation in I1 is regulated independently of known polarity determinants (Husbands *et al.*, 2009; Nogueira *et al.*, 2009). However, as miR390 and tasiR-ARF accumulate throughout the *Arabidopsis* SAM (Chitwood *et al.*, 2009), the PZ prepattern may not affect dorsiventral polarity via regulation of small RNA accumulation in all plant species.

11. Maintenance of Organ Polarity—Cross-Talk between Polarity Determinants

Prepatterning of the PZ could be sufficient to set up leaf polarity but surgical experiments show that carry-over of positional information from the zones of the meristem is not sufficient to produce a polarized mature leaf. During the P1 stage of leaf development, when the primordium grows outward and its cells lose contact with the meristem, the "Sussex" signal transferred through the L1 is critical to maintain organ polarity (Reinhardt *et al.*, 2005; Sussex, 1951; Fig. 5.1C–E). By the P2 stage, mechanisms within the organ are in place that maintain polarity independent of this adaxializing signal (Reinhardt *et al.*, 2005). Antagonistic interactions between cell-autonomous adaxial and abaxial determinants, such as those between the HD-ZIPIII and KAN proteins or between the AS pathway and KAN genes, would generate mutually exclusive and opposing cell fates (Fig. 5.3). However, to stably maintain the separation of adaxial and abaxial domains throughout primordium growth, this network must also include positive interactions that reinforce the identity within each domain, as well as signaling between the domains to provide positional inputs that place and refine the boundary (Fig. 5.4B).

tasiR-ARFs represent the first identified signaling component in the network of interactions that maintains the separation of the adaxial and abaxial domains throughout leaf development (Chitwood *et al.*, 2009). Movement of this low abundant small RNA from its defined source of biogenesis on the adaxial side generates an accumulation gradient across the leaf. Possibly via a threshold effect, the tasiR-ARF gradient is converted into discrete domains of ARF3 and ARF4 expression. The recent finding that miR166 is mobile within the *Arabidopsis* root (Carlsbecker *et al.*, 2010) identifies this small RNA as a plausible signal to mediate such interdomain communication from the abaxial side. This presents the interesting scenario that adaxial–abaxial polarity in the leaf employs a novel patterning mechanism, relying on opposing small RNA gradients: (1) the tasiR-ARF gradient that defines the adaxial side of leaf primordia by spatially restricting the domain of ARF3 and ARF4 accumulation and (2) the miR166 gradient, which delineates the abaxial side by restricting expression of the adaxializing HD-ZIPIII genes (Fig. 5.4B). These gradients are at least partially interconnected, as the *TAS3* ta-siRNA pathway creates the abaxial gradient of miR166. A similar feedback of the miR165/166–HD-ZIPIII pathway on tasiR-ARF production would create a self-stabilizing network.

12. Downstream Genes Regulating Medio-Lateral Blade Outgrowth

Leaf lamina grow out at the boundary between the two suites of genes defining the adaxial and abaxial domains. This boundary is required not only for growth in the lateral direction but also for the leaf to increase in length by proximo-distal growth (Johnston *et al.*, 2010). Exactly how an adaxial–abaxial boundary coordinates blade outgrowth is still unclear. There must be a degree of cell–cell communication to allow cells to acquire information about the identity of their neighbors and to coordinate cell division in the different cell layers (Fig. 5.4B). The precise cell signaling processes required for blade outgrowth remain to be discovered, but several transcription factor families that act downstream of the polarity determinants to regulate medio-lateral blade growth have been described.

Lamina outgrowth requires the activity of YABBY genes, a seed plant-specific group of transcription factors (Eshed *et al.*, 2004). Loss of YABBY function in *Arabidopsis* and *Antirrhinum* results in narrow lateral organs (Eshed *et al.*, 2004; Goltz *et al.*, 2004). Clonal analysis in *graminifolia* (*gram*) mutants shows that this is due to loss of marginal cell divisions in the L2. This is a non-cell-autonomous effect as expression of GRAM in the abaxial L1 is sufficient to rescue the phenotype (Goltz *et al.*, 2004). Narrow leaves alone would not be enough to support a role in lamina outgrowth but YABBY activity is also required for the ectopic outgrowths seen in *kan1 kan2* mutants (Eshed *et al.*, 2004).

The YABBY gene family has roles beyond lamina outgrowth. One clade of the gene family, the CRC group, is required for nectary, carpel, and outer integument development (Alvarez and Smyth, 1999; Bowman and Smyth, 1999; Villanueva *et al.*, 1999). In moncots, this clade has evolved an additional role in midrib development and vascular polarity (Fourquin *et al.*, 2005; Ishikawa *et al.*, 2009; Liu *et al.*, 2007; Wang *et al.*, 2009; Yamaguchi *et al.*, 2004). YABBY genes also have a broader role in regulation of proliferation. Changes at a YABBY locus are responsible for the increase in carpel number, which occurred during the domestication of tomato (Cong *et al.*, 2008), and YABBY genes are required to maintain repression of KNOX genes in the *Arabidopsis* leaf (Kumaran *et al.*, 2002).

YABBY expression is polar in lateral organs and mutants were first characterized based on defects in abaxial cell fate, in accordance with the abaxial expression in early *Arabidopsis* and *Antirrhinum* leaf development (Goltz *et al.*, 2004; Sawa *et al.*, 1999; Siegfried *et al.*, 1999). However, in basal angiosperms and monocots, expression of non-CRC YABBY genes is adaxial (Fourquin *et al.*, 2005; Juarez *et al.*, 2004b; Yamada *et al.*, 2003). During later leaf development, in both monocots and eudicots, expression becomes limited to the presumptive boundary between the adaxial and abaxial domains, at the leaf margins. This expression at the margins is limited to regions where lamina outgrowth occurs (Eshed *et al.*, 2004; Gleissberg *et al.*, 2005; Juarez *et al.*, 2004b). Regulation of the YABBY expression domain in developing leaf primordia is likely to be a key step in the control of lamina outgrowth.

In both monocots and eudicots, YABBY gene expression is downstream of polarity pathway genes, such as the KAN, HD-ZIPIII, *TAS3* ta-siRNA, and AS pathways (Eshed *et al.*, 2004; Garcia *et al.*, 2006; Juarez *et al.*, 2004b; Li et al., 2005; Siegfried *et al.*, 1999). YABBY gene expression is also regulated by members of the WOX family of homeodomain transcription factors. This family includes *narrow sheath1* (*ns1*) and *ns2* of maize, *PRESSED FLOWER* (*PRS*) of *Arabidopsis*, and *MAEWEST* (*MAW*) of *Petunia*. In all three species, these play a role in lateral growth (Nardman *et al.*, 2004; Scanlon *et al.*, 1996; Vandenbussche *et al.*, 2009). In maize, the *ns* genes are expressed at the margins, and in twin foci in the PZ of the meristem that drive the lateral recruitment of founder cells from the SAM into the incipient leaf primordium (Nardman *et al.*, 2004). In *Arabidopsis*, lateral recruitment during leaf development is limited to the cells that will produce the stipules, and these are missing in *prs* mutants (Nardman *et al.*, 2004). *maw* mutants have severely reduced lateral outgrowth in all organs but dorsiventral polarity defects are limited to the margins.

YABBY expression is reduced in the *ns* mutant (Zhang *et al.*, 2007); however, the interaction between the WOX and YABBY genes is likely to be complex, as overexpression of the rice *ns* ortholog, Os*WOX3*, also leads to reduced *YAB3* expression (Dai *et al.*, 2007). There also appears to have been some shuffling of function between WOX family members. *MAW* belongs to a different clade of the family than *ns1, ns2, PRS*, and rice *WOX3*; it is more closely related to *WOX1* of *Arabidopsis*. *wox1* mutants in *Arabidopsis* have no developmental phenotype but when combined with *prs*, a strong lateral outgrowth phenotype is revealed (Vandenbussche *et al.*, 2009). Although the NS/PRS clade and the MAW clade have different expression patterns in the leaf they overlap at the margins, where the clades could have overlapping roles in regulating YABBY-mediated outgrowth.

Juxtaposition of adaxial and abaxial cell fates does not always lead to outgrowth. In most plant species, the petiole is polarized but blade outgrowth here is repressed. This repression in *Arabidopsis* is regulated by the redundant *BLADE-ON-PETIOLE* (*BOP*) genes—*BOP1* and *BOP2*. These genes encode BTB/POZ domain proteins with ankyrin repeats that are thought to regulate higher-order chromatin structure during transcription (Jun *et al.*, 2010; Norberg *et al.*, 2005). BOP genes are expressed in the proximal regions of leaf primordia where they downregulate JAGGED, a C2HC zinc finger transcription factor involved in regulation of cell division at leaf margins chapter 7; (Dinneny *et al.*, 2004; Ohno *et al.*, 2004; Norberg *et al.*, 2005). Additionally, recent results show that the BOP genes directly activate AS2, which sets up polarity in the petiole region. BOP genes therefore set up polarity as well as regulate the growth response to it (Ha *et al.*, 2007; Jun *et al.*, 2010).

13. Why So Complicated?

One of the surprising aspects of leaf polarity is that many of the genes involved have different mutant phenotypes in different species (Kidner and Timmermans, 2007). This suggests that the genetic pathways underling leaf development are labile within angiosperms. The key families involved in leaf polarity evolved before the origin of leaves (Floyd and Bowman, 2007). Genes may have retained ancestral functions or other derived functions. The AS pathway is required to silence KNOX genes in the leaf, and AS1 has additional roles in disease resistance through regulation of jasmonic acid synthesis (Nurmberg *et al.*, 2007). HD-ZIPIII genes have multiple roles in meristem function and vascular patterning, which may be related to their roles in leaf polarity (Prigge *et al.*, 2005). As well as their roles in leaf development, KAN function is required in the embryo to exclude the auxin transporter PIN1 from the hypocotyl (Izhaki and Bowman, 2007). Conflicting selective pressure on the different roles that the leaf polarity genes play could lead to adaptive conflict. Gene duplication followed by sub- or neo-functionalization of the paralogs allows escape from this conflict (Des Marais and Rausher, 2008; Hughes, 1994). Several of the polarity genes exist as gene families and there is evidence for variable subfunctionalization within them (e.g., Candela *et al.*, 2008; Floyd and Bowman, 2007; Prigge *et al.*, 2005; Nogueira *et al.*, 2009). The presence of semiredundant pathways could also result in escape from adaptive conflict by varying reliance on different pathways in different species. This could result in some species being more dependent on the ta-siRNA pathway for leaf polarity, others on the AS pathway.

The varying reliance on the different polarity pathways among angiosperms raises the question whether these pathways also operate in basal lineages. The adaxializing L1 signal must be seed plant-specific, as only the seed plants have a layered meristem (Harrison *et al.*, 2007; White and Turner, 1995). The carry-over of pattern from the meristem may also be limited to seed plants as the organization of non-seed plant meristems is so different. HD-ZIPIII genes are found in all land plants and they are expressed in the meristem of the lycophyte *Selaginella kraussiana*, although there is no evidence that they specify a distinct zone there. Any meristem patterning in *Selaginella* is unlikely to transfer into patterning of the leaf, as the *Selaginella* leaf derives from two adjacent epidermal cells, presumably both in the same zone of the meristem and the expression pattern of HD-ZIPIII on the adaxial side of leaves is not conserved in *Selaginella* (Floyd and Bowman, 2006; Harrison *et al.*, 2007).

The HD-ZIPIII and KAN gene families are distinctive in that both are necessary and sufficient for adaxial or abaxial cell fate, respectively, in angiosperm leaves, and that both have additional roles in vascular patterning and meristem function. This points to ancient, conserved roles in vascular plant development that may underlie and constrain the roles they play in leaf polarity determination. Their mutual inhibitory interaction, which prevents both gene families being expressed in the same cell types, may be a key selective feature in recruiting the HD-ZIPIII–KAN module for several different aspects of plant development. Together with cell-signaling pathways that allow cells to detect and respond to the cell fate of their neighbors, this mutual antagonistic interaction can create and stabilize two distinctly defined domains. Identifying the mobile signals and signaling pathways functioning with the HD-ZIPIII–KAN module will be a key advance in the polarity field and plant development in general. It can reveal whether or not the same signaling pathways function with the HD-ZIPIII–KAN module in meristem, vasculature, and leaf polarity. If so, all three might simply be aspects of the plant's apical–basal/internal–external polarity; a polarity that is fundamental to all land plants and predates divergence of the many pathways by which leaves have evolved.

ACKNOWLEDGMENTS

The authors thank current and former members of the Kidner and Timmermans labs for helpful discussions, and Cris Kuhlemeier for the contribution of images. Research on leaf polarity in the laboratory of M.T. is supported by grants from the USDA (06-03420) and the NSF (IOS-0615752 and IOS-0820610). Work on the evolution of leaf form in the laboratory of C.A.K. is funded by grants from the BBSRC and the M. L. MacIntyre Begonia Research Scholarship Trust.

REFERENCES

Allen, E., Xie, Z., Gustafson, A., and Carrington, J. (2005). microRNA-directed phasing during trans-acting siRNA biogenesis in plants. *Cell* **121**, 207–221.

Alvarez, J., Pekker, I., Goldshmidt, A., Blum, E., Amsellem, Z., and Eshed, Y. (2006). Endogenous and synthetic microRNAs stimulate simultaneous, efficient, and localized regulation of multiple targets in diverse species. *Plant Cell* **18**, 1134–1151.

Alvarez, J., and Smyth, D. (1999). *CRABS CLAW* and *SPATULA*, two *Arabidopsis* genes that control carpel development in parallel with *AGAMOUS*. *Development* **126**, 2377–2386.

Baima, S., Nobili, F., Sessa, G., Lucchetti, S., Ruberti, I., and Morelli, G. (1995). The expression of the *Athb-8* homeobox gene is restricted to provascular cells in *Arabidopsis thaliana*. *Development* **121**, 4171–4182.

Bayer, E., Smith, R., Mandel, T., Nakayama, N., Sauer, M., Prusinkiewicz, P., and Kuhlemeier, C. (2009). Integration of transport-based models for phyllotaxis and midvein formation. *Genes Dev.* **23**, 373–384.

Beerling, D. J., Osborne, C. P., and Chaloner, W. G. (2001). Evolution of leaf-form in land plants linked to atmospheric CO2 decline in the Late Palaeozoic era. *Nature* **410**, 352–354.

Bowman, J., and Smyth, D. (1999). *CRABS CLAW*, a gene that regulates carpel and nectary development in *Arabidopsis*, encodes a novel protein with zinc finger and helix-loop-helix domains. *Development* **126**, 2387–2396.

Boyce, C. (2010). The evolution of plant development in a paleontological context. *Curr. Opin. Plant Biol.* **13**, 102–107.

Byrne, M., Barley, R., Curtis, M., Arroyo, J., Dunham, M., Hudson, A., and Martienssen, R. (2000). *Asymmetric leaves1* mediates leaf patterning and stem cell function in *Arabidopsis*. *Nature* **408**, 967–971.

Candela, H., Johnston, R., Gerhold, A., Foster, T., and Hake, S. (2008). The *milkweed pod1* gene encodes a KANADI protein that is required for abaxial/adaxial patterning in maize leaves. *Plant Cell* **20**, 2073–2087.

Carlsbecker, A., Lee, J. Y., Roberts, C. J., Dettmer, J., Lehesranta, S., Zhou, J., Lindgren, O., Moreno-Risueno, M. A., Vatén, A., Thitamadee, S., Campilho, A., Sebastian, J., et al. (2010). Cell signalling by microRNA165/6 directs gene dose-dependent root cell fate. *Nature* **465**, 316–321.

Chandler, J., Cole, M., Flier, A., Grewe, B., and Werr, W. (2007). The AP2 transcription factors DORNROSCHEN and DORNROSCHEN-LIKE redundantly control *Arabidopsis* embryo patterning via interaction with PHAVOLUTA. *Development* **134**, 1653–1662.

Chapman, E., and Carrington, J. (2007). Specialization and evolution of endogenous small RNA pathways. *Nat. Rev. Genet.* **8**, 884–896.

Chitwood, D., Nogueira, F., Howell, M., Montgomery, T., Carrington, J., and Timmermans, M. (2009). Pattern formation via small RNA mobility. *Genes Dev.* **23**, 549–554.

Cole, M., Chandler, J., Weijers, D., Jacobs, B., Comelli, P., and Werr, W. (2009). DORNROSCHEN is a direct target of the auxin response factor MONOPTEROS in the Arabidopsis embryo. *Development* **136**, 1643–1651.

Cong, B., Barrero, L., and Tanksley, S. D. (2008). Regulatory change in YABBY-like transcription factor led to evolution of extreme fruit size during tomato domestication. *Nat. Genet.* **40**, 800–804.

Dai, M., Hu, Y., Zhao, Y., Liu, H., and Zhou, D. X. (2007). A *WUSCHEL-LIKE HOMEOBOX* gene represses a *YABBY* gene expression required for rice leaf development. *Plant Physiol.* **144**, 380–390.

Des Marais, D., and Rausher, M. (2008). Escape from adaptive conflict after duplication in an anthocyanin pathway gene. *Nature* **454**, 762–765.

Dinneny, J., Yadegari, R., Fischer, R., Yanofsky, M., and Weigel, D. (2004). The role of JAGGED in shaping lateral organs. *Development* **131**, 1101–1110.

Donner, T., Sherr, I., and Scarpella, E. (2009). Regulation of preprocambial cell state acquisition by auxin signaling in *Arabidopsis* leaves. *Development* **136**, 3235–3246.

Dunoyer, P., Schott, G., Himber, C., Meyer, D., Takeda, A., Carrington, J. C., and Voinnet, O. (2010). Small RNA duplexes function as mobile silencing signals between plant cells. *Science* **328**, 912–916.

Emery, J., Floyd, S., Alvarez, J., Eshed, Y., Hawker, N., Izhaki, A., Baum, S., and Bowman, J. (2003). Radial patterning of *Arabidopsis* shoots by class III HD-ZIP and KANADI genes. *Curr. Biol.* **13**, 1768–1774.

Eshed, Y., Baum, S., Perea, J., and Bowman, J. (2001). Establishment of polarity in lateral organs of plants. *Curr. Biol.* **11**, 1251–1260.

Eshed, Y., Izhaki, A., Baum, S., Floyd, S., and Bowman, J. (2004). Asymmetric leaf development and blade expansion in *Arabidopsis* are mediated by KANADI and YABBY activities. *Development* **131**, 2997–3006.

Evans, M. (2007). The indeterminate gametophyte1 gene of maize encodes a LOB domain protein required for embryo sac and leaf development. *Plant Cell* **19**, 46–62.

Fahlgren, N., Montgomery, T., Howell, M., Allen, E., Dvorak, S., Alexander, A., and Carrington, J. (2006). Regulation of *AUXIN RESPONSE FACTOR3* by *TAS3* ta-siRNA affects developmental timing and patterning in *Arabidopsis*. *Curr. Biol.* **16**, 939–944.

Floyd, S. K., and Bowman, J. L. (2007). The ancestoral develomental tool kit of land plants. *Int. J. Plant Sci.* **168**, 1–35.

Floyd, S. K., and Bowman, J. L. (2006). Distinct developmental mechanisms reflect the independent origins of leaves in vascular plants. *Curr. Biol.* **16**, 1911–1917.

Fourquin, C., Vinauger-Douard, M., Fogliani, B., Dumas, C., and Scutt, C. (2005). Evidence that CRABS CLAW and TOUSLED have conserved their roles in carpel development since the ancestor of the extant angiosperms. *Proc. Natl. Acad. Sci. U.S.A* **102**, 4649–4654.

Garcia, D., Collier, S., Byrne, M., and Martienssen, R. (2006). Specification of leaf polarity in *Arabidopsis* via the trans-acting siRNA pathway. *Curr. Biol.* **16**, 933–938.

Gillmor, C., Park, M., Smith, M., Pepitone, R., Kerstetter, R., and Poethig, R. (2010). The MED12-MED13 module of Mediator regulates the timing of embryo patterning in *Arabidopsis*. *Development* **137**, 113–122.

Gleissberg, S., Groot, E. P., Schmalz, M., Eichert, M., Kölsch, A., and Hutter, S. (2005). Developmental events leading to peltate leaf structure in *Tropaeolum majus* (Tropaeolaceae) are associated with expression domain changes of a *YABBY* gene. *Dev. Genes Evol.* **215**, 313–319.

Golz, J., Roccaro, M., Kuzoff, R., and Hudson, A. (2004). GRAMINIFOLIA promotes growth and polarity of *Antirrhinum* leaves. *Development* **131**, 3661–3670.

Guilfoyle, T., and Hagen, G. (2007). Auxin response factors. *Curr. Opin. Plant Biol.* **10**, 453–460.

Guo, M., Thomas, J., Collins, G., and Timmermans, M. (2008). Direct repression of *KNOX* loci by the ASYMMETRIC LEAVES1 complex of Arabidopsis. *Plant Cell* **19**, 48–58.

Ha, C. M., Jun, J. H., Nam, H. G., and Fletcher, J. (2007). *BLADE-ON-PETIOLE* 1 and 2 control *Arabidopsis* lateral organ fate through regulation of LOB domain and adaxial-abaxial polarity genes. *Plant Cell* **19**, 1809–1825.

Harrison, C., Corley, S., Moylan, E., Alexander, D., Scotland, R., and Langdale, J. (2005). Independent recruitment of a conserved developmental mechanism during leaf evolution. *Nature* **434**, 509–514.

Harrison, C. J., Rezvani, M., and Langdale, J. A. (2007). Growth from two transient apical initials in the meristem of *Selaginella kraussiana*. *Development* **134**, 881–889.

Heisler, M., Ohno, C., Das, P., Sieber, P., Reddy, G., Long, J., and Meyerowitz, E. (2005). Patterns of auxin transport and gene expression during primordium development revealed by live imaging of the *Arabidopsis* inflorescence meristem. *Curr. Biol.* **15**, 1899–1911.

Henderson, D., Zhang, X., Brooks, L., III, and Scanlon, M. (2006). RAGGED SEEDLING2 is required for expression of *KANADI2* and *REVOLUTA* homologues in the maize shoot apex. *Genesis* **44**, 372–382.

Hughes, A. (1994). The evolution of functionally novel proteins after gene duplication. *Proc. R. Soc. Lond. B* **256**, 119–124.

Hunter, C., Willmann, M., Wu, G., Yoshikawa, M., dela Luz Guierrez-Nava, M., and Poethig, S. (2006). Trans-acting siRNA-mediated repression of *ETTIN* and *ARF4* regulates heteroblasty in *Arabidopsis*. *Development* **133**, 2973–2981.

Husbands, A., Chitwood, D., Plavskin, Y., and Timmermans, M. (2009). Signals and prepatterns: new insights into organ polarity in plants. *Genes Dev.* **23**, 1986–1997.

Ilegems, M., Douet, V., Meylan-Bettex, M., Uyttewaal, M., Brand, L., Bowman, J. L., and Stieger, P. A. (2010). Interplay of auxin, KANADI and Class III HD-ZIP transcription factors in vascular tissue formation. *Development* **137**, 975–984.

Ishikawa, M., Ohmori, Y., Tanaka, W., Hirabayashi, C., Murai, K., Ogihara, Y., Yamaguchi, T., and Hirano, H. Y. (2009). The spatial expression patterns of *DROOPING LEAF* orthologs suggest a conserved function in grasses. *Genes Genet. Syst.* **84**, 137–146.

Itoh, J. -I., Hibara, K. -I., Sato, Y., and Nagato, Y. (2008). Developmental role and auxin responsiveness of Class III homeodomain leucine zipper gene family members in rice. *Plant Physiol.* **147**, 1960–1975.

Iwakawa, H., Iwasaki, M., Kojima, S., Ueno, Y., Soma, T., Tanaka, H., Semiarti, E., Machida, Y., and Machida, C. (2007). Expression of the *ASYMMETRIC LEAVES2* gene in the adaxial domain of *Arabidopsis* leaves represses cell proliferation in this domain and is critical for the development of properly expanded leaves. *Plant J.* **51**, 173–184.

Izhaki, A., and Bowman, J. (2007). KANADI and Class III HD-ZIP gene families regulate embryo patterning and modulate auxin flow during embryogenesis in *Arabidopsis*. *Plant Cell* **19**, 495–508.

Johnston, R., Candela, H., Hake, S., and Foster, T. (2010). The maize *milkweed pod1* mutant reveals a mechanism to modify organ morphology. *Genesis* [Epub ahead of print].

Juarez, M., Kui, J., Thomas, J., Heller, B., and Timmermans, M. (2004a). microRNA-mediated repression of *rolled leaf1* specifies maize leaf polarity. *Nature* **428**, 84–88.

Juarez, M., Twigg, R., and Timmermans, M. (2004b). Specification of adaxial cell fate during maize leaf development. *Development* **131**, 4533–4544.

Jun, J., Ha, C., and Fletcher, J. (2010). *BLADE-ON-PETIOLE1* coordinates organ determinacy and axial polarity in *Arabidopsis* by directly activating *ASYMMETRIC LEAVES2*. *Plant Cell* **22**, 62–76.

Kerstetter, R., Bollman, K., Taylor, R., Bomblies, K., and Poethig, R. (2001). *KANADI* regulates organ polarity in *Arabidopsis*. *Nature* **411**, 706–709.

Kidner, C., and Martienssen, R. (2004). Spatially restricted microRNA directs leaf polarity through ARGONAUTE1. *Nature* **428**, 81–84.

Kidner, C., and Timmermans, M. (2007). Mixing and matching pathways in leaf polarity. *Curr. Opin. Plant Biol.* **10**, 13–20.

Kim, M., McCormick, S., Timmermans, M., and Sinha, N. (2003). The expression domain of *PHANTASTICA* determines leaflet placement in compound leaves. *Nature* **424**, 438–443.

Kim, Y. -S., Kim, S. -G., Lee, M., Lee, I., Park, H. -Y., Seo, P., Jung, J. -H., Kwon, E. -J., Suh, S., Paek, K. -H., et al. (2008). HD-ZIP III activity is modulated by competitive inhibitors via a feedback loop in *Arabidopsis* shoot apical meristem development. *Plant Cell* **20**, 920–933.

Kumaran, M. K., Bowman, J. L., and Sundaresan, V. (2002). *YABBY* polarity genes mediate the repression of *KNOX* homeobox genes in *Arabidopsis*. *Plant Cell* **14**, 2761–2770.

Levine, E., McHale, P., and Levine, H. (2007). Small regulatory RNAs may sharpen spatial expression patterns. *PLoS Comp. Biol.* **3**, e233.

Li, H., Xu, L., Wang, H., Yuan, Z., Cao, X., Yang, Z., Zhang, D., Xu, Y., and Huang, H. (2005). The putative RNA-dependent RNA polymerase *RDR6* acts synergistically with *ASYMMETRIC LEAVES1* and *2* to repress *BREVIPEDICELLUS* and micro-RNA165/166 in *Arabidopsis* leaf development. *Plant Cell* **17**, 2157–2171.

Lin, W. -C., Shuai, B., and Springer, P. (2003). The *Arabidopsis LATERAL ORGAN BOUNDARIES*-domain gene *ASYMMETRIC LEAVES2* functions in the repression of *KNOX* gene expression and in adaxial-abaxial patterning. *Plant Cell* **15**, 2241–2252.

Liu, H. L., Xu, Y. Y., Xu, Z. H., and Chong, K. (2007). A rice *YABBY* gene, *OsYABBY4*, preferentially expresses in developing vascular tissue. *Dev. Genes Evol.* **217**, 629–637.

Luo, J. H., Yan, J., Lin, W., Jun, Y., Zhong, Z., Cao, J., Hu, C., Xiao, H., and Luo, D. (2005). Different expression patterns of duplicated *PHANTASTICA*-like genes in *Lotus japonicus* suggest their divergent functions during compound leaf development. *Cell Res.* **15**, 665–677.

Lynn, K., Fernandez, A., Aida, M., Sedbrook, J., Tasaka, M., Masson, P., and Barton, M. (1999). The *PINHEAD/ZWILLE* gene acts pleiotropically in *Arabidopsis* development and has overlapping functions with the *ARGONAUTE1* gene. *Development* **126**, 469–481.

McConnell, J., and Barton, M. (1998). Leaf polarity and meristem formation in *Arabidopsis*. *Development* **125**, 2935–2942.

McConnell, J., Emery, J., Eshed, Y., Bao, N., Bowman, J., and Barton, M. (2001). Role of *PHABULOSA* and *PHAVOLUTA* in determining radial patterning in shoots. *Nature* **411**, 709–713.

McHale, N., and Koning, R. (2004a). *PHANTASTICA* regulates development of the adaxial mesophyll in *Nicotiana* leaves. *Plant Cell* **16**, 1251–1262.

McHale, N., and Koning, R. (2004b). MicroRNA-directed cleavage of *Nicotiana sylvestris PHAVOLUTA* mRNA regulates the vascular cambium and structure of apical meristems. *Plant Cell* **16**, 1730–1740.

Montgomery, T., Howell, M., Cuperus, J., Li, D., Hansen, J., Alexander, A., Chapman, E., Fahlgren, N., Allen, E., and Carrington, J. (2008). Specificity of ARGONAUTE7-miR390 interaction and dual functionality in *TAS3* trans-acting siRNA formation. *Cell* **133**, 128–141.

Nagasaki, H., Itoh, J., Hayashi, K., Hibara, K., Satoh-Nagasawa, N., Nosaka, M., Mukouhata, M., Ashikari, M., Kitano, H., Matsuoka, M., Nagato, Y., and Sato, Y. (2007). The small interfering RNA production pathway is required for shoot meristem initiation in rice. *Proc. Natl. Acad. Sci. U.S.A* **104**, 14867–14871.

Nardmann, J., Ji, J., Werr, W., and Scanlon, M. (2004). The maize duplicate genes *narrow sheath1* and *narrow sheath2* encode a conserved homeobox gene function in a lateral domain of shoot apical meristems. *Development* **131**, 2827–2839.

Nogueira, F., Madi, S., Chitwood, D., Juarez, M., and Timmermans, M. (2007). Two small regulatory RNAs establish opposing fates of a developmental axis. *Genes Dev.* **21**, 750–755.

Nogueira, F., Chitwood, D., Madi, S., Ohtsu, K., Schnable, P., Scanlon, M., and Timmermans, M. (2009). Regulation of small RNA accumulation in the maize shoot apex. *PLoS Genet.* **5**, e1000320.

Norberg, M., Holmlund, M., and Nilsson, O. (2005). The *BLADE ON PETIOLE* genes act redundantly to control the growth and development of lateral organs. *Development* **132**, 2203–2213.

Nurnberg, P., Knox, K., Yun, B., Morris, P., Shafiei, R., Hudson, A., and Loake, G. (2007). The developmental selector *AS1* is an evolutionarily conserved regulator of the plant immune response. *Proc. Natl. Acad. Sci. U.S.A* **104**, 18795–18800.

Ohno, C., Reddy, G., Heisler, M., and Meyerowitz, E. (2004). . The Arabidopsis *JAGGED* gene encodes a zinc finger protein that promotes leaf tissue development. *Development* **131**, 1111–1122.

Otsuga, D., DeGuzman, B., Prigge, M., Drews, G., and Clark, S. (2001). *REVOLUTA* regulates meristem initiation at lateral positions. *Plant J.* **25**, 223–236.

Pekker, I., Alvarez, J. P., and Eshed, Y. (2005). Auxin response factors mediate *Arabidopsis* organ asymmetry via modulation of *KANADI* activity. *Plant Cell* **17**, 2899–2910.

Peragine, A., Yoshikawa, M., Wu, G., Albrecht, H., and Poethig, R. (2004). *SGS3* and *SGS2/SDE1/RDR6* are required for juvenile development and the production of trans-acting siRNAs in *Arabidopsis*. *Genes Dev.* **18**, 2368–2379.

Phelps-Durr, T., Thomas, J., Vahab, P., and Timmermans, M. (2005). Maize *rough sheath2* and its *Arabidopsis* orthologue *ASYMMETRIC LEAVES1* interact with HIRA, a predicted histone chaperone, to maintain *knox* gene silencing and determinacy during organogenesis. *Plant Cell* **17**, 2886–2898.

Pinon, V., Etchells, J., Rossignol, P., Collier, S., Arroyo, J., Martienssen, R., and Byrne, M. E. (2008). Three PIGGYBACK genes that specifically influence leaf patterning encode ribosomal proteins. *Development* **135**, 1315–1324.

Ponting, C., and Aravind, L. (1999). START: a lipid-binding domain in StAR, HD-ZIP and signalling proteins. *Trends Biochem. Sci.* **24**, 130–132.

Prigge, M., Otsuga, D., Alonso, J., Ecker, J., Drews, G., and Clark, S. (2005). Class III homeodomain-leucine zipper gene family members have overlapping, antagonistic, and distinct roles in *Arabidopsis* development. *Plant Cell* **17**, 61–76.

Reinhardt, D., Frenz, M., Mandel, T., and Kuhlemeier, C. (2005). Microsurgical and laser ablation analysis of leaf positioning and dorsoventral patterning in tomato. *Development* **132**, 15–26.

Reinhardt, D., Pesce, E., Stieger, P., Mandel, T., Baltensperger, K., Bennett, M., Traas, J., Friml, J., and Kuhlemeier, C. (2003). Regulation of phyllotaxis by polar auxin transport. *Nature* **426**, 255–260.

Sawa, S., Watanabe, K., Goto, K., Liu, Y., Shibata, D., Kanaya, E., Morita, E., and Okada, K. (1999). *FILAMENTOUS FLOWER*, a meristem and organ identity gene of

Arabidopsis, encodes a protein with a zinc finger and HMG-related domains. *Genes Dev.* **13**, 1079–1088.

Scanlon, M., Schneeberger, R., and Freeling, M. (1996). The maize mutant narrow sheath fails to establish leaf margin identity in a meristematic domain. *Development* **122**, 1683–1691.

Schneeberger, R., Tsiantis, M., Freeling, M., and Langdale, J. (1998). The *rough sheath2* gene negatively regulates homeobox gene expression during maize leaf development. *Development* **125**, 2857–2865.

Schrick, K., Nguyen, D., Karlowski, W. M., and Mayer, K. (2004). START lipid/sterol-binding domains are amplified in plants and are predominantly associated with homeodomain transcription factors. *Genome Biol.* **5**, R41.

Siegfried, K., Eshed, Y., Baum, S., Otsuga, D., Drews, G., and Bowman, J. (1999). Members of the *YABBY* gene family specify abaxial cell fate in *Arabidopsis*. *Development* **126**, 4117–4128.

Steemans, P., Hérissé, A., Melvin, J., Miller, M., Paris, F., Verniers., J., and Wellman., C. (2009). Origin and radiation of the earliest vascular land plants. *Science* **2009**(324), 353.

Sussex, I. (1951). Experiments on the cause of dorsiventrality in leaves. *Nature* **167**, 651–652.

Sussex, I. (1954). Morphogenesis in *Solanum tuberosum* L.: experimental investigation of leaf dorsiventrality and orientation in the juvenile shoot. *Phytomorphology* **5**, 286–300.

Tang, G., Reinhart, B., Bartel, D., and Zamore, P. (2003). A biochemical framework for RNA silencing in plants. *Genes Dev.* **17**, 49–63.

Tattersall, A., Turner, L., Knox, M., Ambrose, M., Ellis, T., and Hofer, J. (2005). The mutant *crispa* reveals multiple roles for PHANTASTICA in pea compound leaf development. *Plant Cell* **17**, 1046–1060.

Timmermans, M., Schultes, N., Jankovsky, J., and Nelson, T. (1998). *Leafbladeless1* is required for dorsoventrality of lateral organs in maize. *Development* **125**, 2813–2823.

Timmermans, M., Hudson, A., Becraft, P., and Nelson, T. (1999). ROUGH SHEATH2: A Myb protein that represses *knox* homeobox genes in maize lateral organ primordia. *Science* **284**, 151–153.

Tiwari, S. B., Hagen, G., and Guilfoyle, T. (2003). The roles of auxin response factor domains in auxin-responsive transcription. *Plant Cell* **15**, 533–543.

Tsiantis, M., Schneeberger, R., Golz, J., Freeling, M., and Langdale, J. (1999). The maize *rough sheath2* gene and leaf development programs in monocot and dicot plants. *Science* **284**, 154–156.

Ueno, Y., Ishikawa, T., Watanabe, K., Terakura, S., Iwakawa, H., Okada, K., Machida, C., and Machida, Y. (2007). Histone deacetylases and *ASYMMETRIC LEAVES2* are involved in the establishment of polarity in leaves of *Arabidopsis*. *Plant Cell* **19**, 445–457.

Ulmasov, T., Hagen, G., and Guilfoyle, T. (1999). Activation and repression of transcription by auxin-response factors. *Proc. Natl. Acad. Sci. U.S.A* **96**, 5844–5849.

Vandenbussche, M., Horstman, A., Zethof, J., Koes, R., Rijpkema, A., and Gerats, T. (2009). Differential recruitment of WOX transcription factors for lateral development and organ fusion in *Petunia* and *Arabidopsis*. *Plant Cell* **21**, 2269–2283.

Villanueva, J., Broadhvest, J., Hauser, B., Meister, R., Schneitz, K., and Gasser, C. (1999). *INNER NO OUTER* regulates abaxial- adaxial patterning in *Arabidopsis* ovules. *Genes Dev.* **13**, 3160–3169.

Waites, R., and Hudson, A. (1995). *phantastica*: a gene required for dorsoventrality of leaves in *Antirrhinum majus*. *Development* **121**, 2143–2154.

Waites, R., Selvadurai, H., Oliver, I., and Hudson, A. (1998). The *PHANTASTICA* gene encodes a MYB transcription factor involved in growth and dorsoventrality of lateral organs in *Antirrhinum*. *Cell* **93**, 779–789.

Wang, A., Tang, J., Li, D., Chen, C., Zhao, X., and Zhu, L. (2009). Isolation and functional analysis of *LiYAB1*, a *YABBY* family gene, from lily (*Lilium longiflorum*). *J. Plant. Physiol.* **166**, 988–995.

Wardlaw, C. (1949). Experiments on organogenesis in ferns. *Growth (suppl.).* **13**, 93–131.
Wenkel, S., Emery, J., Hou, B. -H., Evans, M., and Barton, M. (2007). A feedback regulatory module formed by *LITTLE ZIPPER* and *HD-ZIPIII* genes. *Plant Cell* **19**, 3379–3390.
White, R., and Turner, M. (1995). Anatomy and development of the fern sporophyte. *Bot. Rev.* **61**, 281–305.
Wu, G., Lin, W.-c., Huang, T., Poethig, R., Springer, P., and Kerstetter, R. (2008). *KANADI1* regulates adaxial-abaxial polarity in *Arabidopsis* by directly repressing the transcription of *ASYMMETRIC LEAVES2. Proc. Natl. Acad. Sci. U.S.A* **105**, 16392–16397.
Xu, L., Yang, L., Pi, L., Liu, Q., Ling, Q., Wang, H., Poethig, R., and Huang, H. (2006). Genetic interaction between the AS1-AS2 and RDR6-SGS3-AGO7 pathways for leaf morphogenesis. *Plant Cell Physiol.* **47**, 853–863.
Yamada, T., Ito, M., and Kato, M. (2003). Expression pattern of INNER NO OUTER homologue in *Nymphaea* (water lily family, Nymphaeaceae). *Dev. Genes Evol.* **213**, 510–513.
Yamaguchi, T., Nagasawa, N., Kawasaki, S., Matsuoka, M., Nagato, Y., and Hirano, H. -Y. (2004). The *YABBY* gene *DROOPING LEAF* regulates carpel specification and midrib development in *Oryza sativa. Plant Cell* **16**, 500–509.
Yan, J., Cai, X., Luo, J., Sato, S., Jiang, Q., Yang, J., Cao, X., Hu, X., Tabata, S., Gresshoff, P., and Luo, D. (2010). The *REDUCED LEAFLET* genes encode key components of the trans-acting small interfering RNA pathway and regulate compound leaf and flower development in *Lotus japonicus. Plant Physiol.* **152**, 797–807.
Yoon, E., Yang, J., Lim, J., Kim, S., Kim, S., and Lee, W. (2010). Auxin regulation of the microRNA390-dependent transacting small interfering RNA pathway in *Arabidopsis* lateral root development. *Nucleic Acids Res.* **2010**(38), 1382–1391.
Zhang, X., Madi, S., Borsuk, L., Nettleton, D., Elshire, R., Buckner, B., Janick-Buckner, D., Beck, J., Timmermans, M., Schnable, P., and Scanlon, M. (2007). Laser microdissection of narrow sheath mutant maize uncovers novel gene expression in the shoot apical meristem. *PLoS Genet.* **3**, e101.
Zhang, G. -H., Xu, Q., Zhu, X. -D., Qian, Q., and Xue, H. -W. (2009). SHALLOT-LIKE1 is a KANADI transcription factor that modulates rice leaf rolling by regulating leaf abaxial cell development. *Plant Cell* **21**, 719–735.

CHAPTER SIX

Evolution Of Leaf Shape: A Pattern Emerges

Daniel Koenig *and* Neelima Sinha

Contents

Abstract

Leaf shape is a highly variable trait. Ancestrally, all leaves are proposed to have derived from modifications of branched shoot systems. The formation of blade, smooth margins or serrations on the blade, or distinct leaflets that are the characteristic features of some leaves provides an opportunity to study the generation of morphogenesis in organs that are evolutionarily homologous and yet developmentally distinct in patterning. Intense research in several model species with distinct leaf morphologies has revealed a complex network of genes that interact to pattern the leaf. Several parallels between leaf patterning and shoot patterning exist. The plant growth hormone auxin is emerging as a key player in the specification of both shoot and leaf patterning. The outcome of this underlying auxin pattern may be determined by variation in the two opposing developmental forces of differentiation and indeterminancy. The expanded suite of genetic and physiological factors regulating leaf shape has provided interesting insight into the mechanisms by which morphological innovation is accomplished.

Department of Plant Biology, University of California, Davis, California, USA

Current Topics in Developmental Biology, Volume 91
ISSN 0070-2153, DOI 10.1016/S0070-2153(10)91006-5

1. Introduction

Leaf shape is one of the most diverse characters in all of biology, despite the highly conserved function of leaves in photosynthetic light capture and gas exchange. A major difference in leaf shape arises from the fact that while some leaf blades are dissected into leaflets others are composed of a single unfractionated blade. The degree to which the blade is lobed, in leaves or leaflets, represents another aspect of shape variation. Leaf dissection in dicot species is a highly variable trait, with many documented transitions between simple and compound states over evolutionary time. The proposed ancestral state of seed plants is a compound leaf, derived from limited blade outgrowth of a modified shoot system, while the proposed ancestral state of angiosperms is the simple leaf (Beerling, 2005; Beerling and Fleming, 2007). The molecular basis for shape variation in leaves has been an area of intense study in the last decade, revealing a complex network of signals that contribute to the development of leaves. In this review we discuss the roles of several molecular and physiological inputs regulating compound leaf development and speculate on how these inputs might be varied to generate compound and simple leaf forms. Particular emphasis is given to the implications of recent work describing the role of the plant phytohormone auxin in leaf shape. We attempt to place the patterning framework as directed by auxin into context with previously described components of leaf shape regulation. Finally, we compare leaf shape regulation in simple- and compound-leafed model species and speculate on the homology of these two structures.

2. Determinacy and Leaf Dissection

KNOX genes were the first molecular regulators of leaf dissection identified. In simple-leafed species such as *Arabidopsis thaliana* and maize, *KNOX* genes are involved in the specification and maintenance of indeterminate identity in the shoot apical meristem (SAM) (chapter 4; Hake *et al.*, 2004). Mutants and transgenic plants overexpressing members of this gene family produce leaves with characteristics indicating ectopic acquisition of indeterminate identity such as the production of ectopic meristems. In simple-leafed dicot species, such as tobacco and *A. thaliana*, *KNOX* overexpression consistently results in the production of lobed leaves (Chuck *et al.*, 1996; Sinha *et al.*, 1993). Functional experiments in the model compound-leafed species tomato and in *Cardamine hirsuta* have suggested that *KNOX* activity in developing leaf primordia is necessary for compound leaf development (Hay and Tsiantis, 2006; Shani *et al.*, 2009), but overexpression experiments and evolutionary surveys of *KNOX* expression pattern have clearly shown that it *KNOX* activity is not sufficient to produce compound leaves (Bharathan *et al.*, 2002; Chen *et al.*, 1997; Hareven *et al.*, 1996; Janssen *et al.*, 1998; Parnis *et al.*, 1997). Further examination of *KNOX* gene function and expression in many compound-leafed species firmly established members of the *KNOX* gene family as important regulators of leaf morphogenesis throughout the dicot lineage, with an important exception amongst a derived clade (containing pea) in the legumes (Bharathan *et al.*, 1999; Champagne *et al.*, 2007; Hofer *et al.*, 1997). Together, these results have lead to the conclusion that the role of *KNOX* genes in compound leaf development is to promote an indeterminate environment competent to initiate leaflets. Interestingly, though truncation or removal of *KNOX*-promoted indeterminancy results in leaf simplification, so does high levels of constitutive *KNOX* expression during leaf development. Thus, *KNOX* expression driven in leaves by the highly active *YABBY* and FILAMENTOUS FLOWER (FIL) promoters causes leaf simplification in tomato (Shani *et al.*, 2009), while regulated *KNOX* overexpression through fusion of KNOX proteins to the glucocorticoid receptor or by using a promoter with more defined activity causes elaboration of the leaf margin (Hay *et al.*, 2006). This indicates that exit from the indeterminate stage of compound leaf development is also necessary for appropriate lateral growth during compound leaf development (Fig. 6.1).

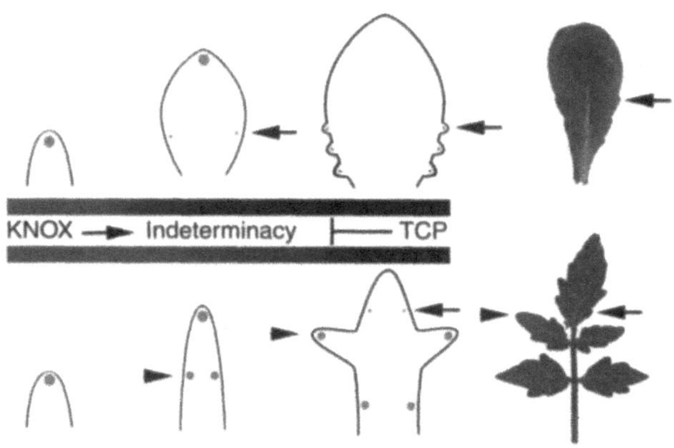

Figure 6.1 The genetic and physiological control of leaf shape. Diagrammatic representation of *A. thaliana* (upper) and tomato (lower) leaf development with real images of final leaf morphologies. The progression from indeterminate to determinate cell fates is indicated by the red to black gradient shown in the two diagrams, and the roles of KNOX and TCP transcription factors are indicated with respect to indeterminacy/determinacy. The tomato leaf is shown to maintain indeterminacy for a more prolonged developmental window when compared to *A. thaliana*. The transition to determinate cell fates is also indicated by the initiation of blade outgrowth in the *A. thaliana* leaf. Auxin maxima are indicated by green spots, and these maxima form in *A. thaliana* after blade outgrowth is initiated. Arrowheads indicate leaflet primordia or leaflets, and arrows indicate lobes/serration primordia or mature lobes/serrations.

The dual role of *KNOX* genes in shoot and compound leaf development suggests homology between these two structures, but some inconsistency exists. In the SAM, downregulation of *KNOX* expression is consistently observed during the earliest stages of leaf primordia initiation (Hake *et al.*, 2004). *KNOX* downregulation during leaf initiation is considered an important molecular event differentiating indeterminacy in the meristems from determinant incipient leaf primordia. The most well studied dissected leaf species, tomato, does not show downregulation of KNOX expression immediately subsequent to leaflet initiation (Chen *et al.*, 1997; Janssen *et al.*, 1998). Contrastingly, *KNOX* downregulation has been observed subsequent to the initiation of *C. hirsuta* leaflets (Hay and Tsiantis, 2006). One possible explanation for this inconsistency is that the leaves of *C. hirsuta* are dissected along the primary axis only, while tomato leaves continue to produce leaflets along the secondary axis. These results are consistent with the hypothesis that maintenance of KNOX expression during leaf development provides a prolonged competence to initiate leaflets in the leaf margin.

The simple-leafed, dominant tomato mutant *Lanceolate* (*La*) provides further insight into the role of indeterminancy in leaf shape. *LA* encodes a *TCP* transcription factor that promotes differentiation during leaf development of tomato (Ori *et al.*, 2007). The *La* mutant is caused by increased expression of the *LA* transcript due to mutation of a miR319 binding site found in the mature mRNA. As a result of increased *LA* expression, mutant leaves prematurely differentiate and possess a simple leaf morphology. Conversely, overexpression of miR319 promotes leaf dissection in tomato by prolonging leaf maturation through downregulation of *LA*. These results show parallels with the development of serrations in *A. thaliana*, in which overexpression of the miR319 gene *JAW* produces serrated leaves (Palatnik *et al.*, 2003). *La* is epistatic to *KNOX* overexpression phenotypes, and *KNOX* expression is not downregulated in *La* simple leaves (Ori *et al.*, 2007). Thus, the effect of *KNOX* expression on leaf development may be dependent on the timing of *TCP* expression. It is possible that the combination of *KNOX* expression in leaves and a delay or reduction in *TCP* expression prolongs indeterminate growth and promotes the transition from simple to compound leaf development (Fig. 6.1).

3. The Patterning of Growth During Leaf Development

The preceding observations provide significant insight into the importance of timing of leaf maturation on morphogenesis but do not reveal a clear mechanism by which vastly different compound and simple leaf growth patterns are achieved. Dicot simple leaves can be divided into two morphological domains along the proximo-distal axis. The distal regions of the leaf initiate lateral outgrowth of blade tissues resulting in a flattened lamina, while in the proximal region lateral outgrowth is suppressed resulting in a near bladeless petiole. For many simple-leafed species, this pattern can be observed in young primordia shortly after initiation from the shoot around the P3 or P4 stages of leaf development. Compound leaves, and some simple leaves, diverge from this two-domain pattern of lateral growth. During early leaf development, compound leaves initiate marginal outgrowth in a punctuated fashion, giving rise to discrete leaflet primordia. The lack of understanding of how lateral growth along the proximo-distal axis of leaf primordia is patterned has led several groups to investigate the regulation of this process.

In *A. thaliana*, the redundant genes *BLADE ON PETIOLE* (*BOP1* and *BOP2*) and *JAGGED* (*JAG*) act antagonistically to specify the two developing domains of the simple leaf (Dinneny *et al.*, 2004, 2006; Ha *et al.*, 2004; Hepworth *et al.*, 2005; Ohno *et al.*, 2004). *JAG* promotes cell division and lateral outgrowth in the distal portion of the developing leaf, while the *BOP* genes repress *JAG* in the proximal petiole (chapter 7). Examination of the function of the tomato ortholog of *JAG*, *LYRATE* (*LYR*), reveals a substantial role in leaf dissection (David-Schwartz *et al.*, 2009). Unlike *JAG*, *LYR* is expressed in punctuated discrete domains corresponding to initiating leaflets (Fig. 6.2). Disruption of *LYR* function results in reduced blade and leaflet

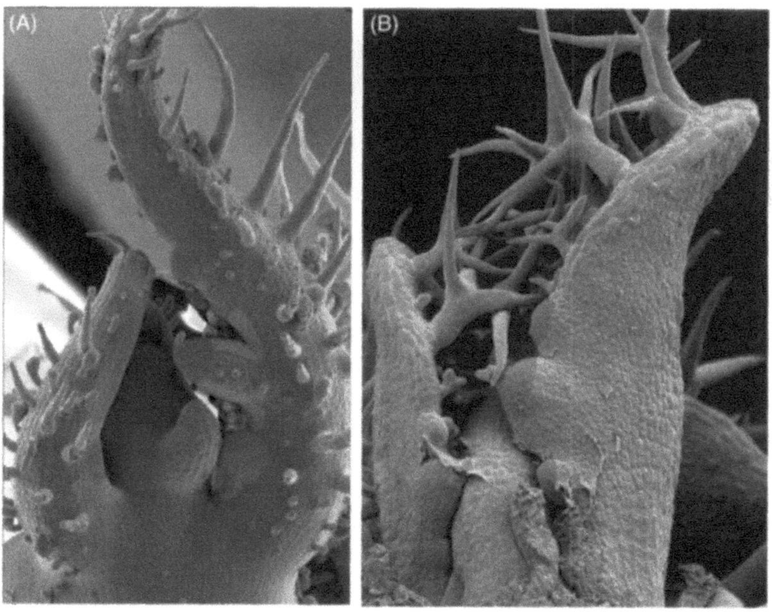

Figure 6.2 Comparison of marginal dissection in compound and simple leaves. (A) Scanning electron micrograph of a tomato shoot apex. (B) Scanning electron micrograph of a developing *A. thaliana* leaf. Purple indicates the activity of growth-promoting factors such as auxin and *LYR*. Orange indicates the activity of repressing factors like *E* and *NAM/CUC* genes. As leaflets elongate, *NAM/CUC* expression remains restricted (dark orange), while *E* is likely to be expanded to all areas with low auxin expression. (See Color Insert.)

initiation, and ubiquitous expression of *LYR* results in ectopic production of blade and leaflets along the rachis. The distribution of *LYR* expression along the proximo-distal axis of leaf primordia in *A. thaliana* and tomato is drastically different. Thus, it is possible that transitions between compound and simple leaf states are mediated by the patterning of this key regulator that directly promotes blade outgrowth. Interestingly, *lyr* plants are reduced in expression of genes positively regulated by the plant hormone auxin (David-Schwartz *et al.*, 2009). This result suggests that the auxin and *LYR* genetic pathways might interact during the specification of leaflet and blade outgrowth.

4. The Role of Auxin in Leaf Patterning

Promotion of lateral growth by *LYR* is a key component of compound leaf development, but how is the pattern initially established? An excellent candidate for this function is the plant hormone auxin. Auxin has been implicated in almost every angiosperm developmental process (Vanneste and Friml, 2009). The described functions of auxin are highly diverse and tissue dependent, suggesting that context plays a major role in determining auxin response. In meristematic tissues, significant evidence has accumulated to implicate auxin maxima as important signals delineating organ initiation (Benkova *et al.*, 2003; Reinhardt *et al.*, 1998, 2000). These events are associated with two phenomena: the promotion of rapid cell division and a shift from anticlinal to periclinal cell division culminating in the rapid and discrete outgrowth of lateral organs.

Accumulation of auxin in initiating organs is driven by the activity of the *PIN* family of auxin efflux carriers. *PIN* proteins drive directional auxin flow through their polar localization within the cell (Fig. 6.3), by transporting auxin in the direction of their localization (Wisniewska *et al.*, 2006). In the SAM, the most well studied PIN protein regulating auxin flow is PIN1. PIN1 forms convergence points in the epidermis of the meristem that predict the site of organ initiation, and disruption of PIN1 activity results in defects in or loss of lateral organ initiation (Reinhardt *et al.*, 2000, 2003). The events associated with leaf initiation, such as activation of cell division and the reorientation of cell division plane, show clear parallels with events observed during the initiation of leaflets in compound leaves, making auxin and *PIN* obvious candidates as regulators of leaf morphogenesis.

Recent studies have revealed a role for auxin transport in leaf morphogenesis post initiation. The earliest described molecular marker specifying leaflet development is the formation of PIN1 convergence points in the epidermis of the margin (Fig. 6.3), and the initiation of leaflets is dependent on PIN activity in many compound leaf species (Barkoulas *et al.*, 2008; DeMason and Chawla, 2004; Koenig *et al.*, 2009).

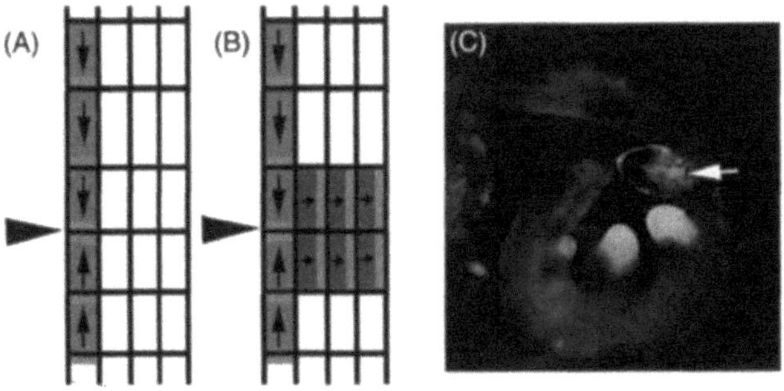

Figure 6.3 Specification of leaflet or lobe initiation by auxin. (A) Formation of auxin peaks (arrowhead) by epidermal PIN1 convergence. The flow of auxin is indicated by the arrows, and the green lines indicate PIN1 localization. Red colored cells are the epidermal cell file. The leaflet or lobe primordium will initiate at this convergence point. (B) Subsequent to epidermal convergence, internal cells below the convergence adopt internal PIN1 localization and specify the cell files that will eventually differentiate into vascular tissue (purple cells). Thus, the two processes are developmental linked. (C) Fluorescence imaging of *PIN1:PIN1::GFP* expression in tomato showing upregulation in leaflets and lobes. The provascular expression of GFP is indicated by the arrow. (See Color Insert.)

Auxin response is activated in initiating leaflets (Figs. 6.2 and 6.3), and exogenous auxin application is sufficient to generate both leaflet and blade from tomato leaf margins (Koenig *et al.*, 2009). After leaflet initiation, PIN1 localization transitions to apical epidermal localization and basal internal localization (Barkoulas *et al.*, 2008; Koenig *et al.*, 2009). This sequence of PIN1 localization and auxin accumulation bears close resemblance to leaf initiation processes and implies that a shoot-like patterning mechanism is active during the development of leaflets. Recent studies have interpreted these observations as evidence for homology of the shoot and the compound leaf as described by Agnes Arbor over half a century ago (Barkoulas *et al.*, 2008).

5. What Distinguishes Lobes, Leaflets, and Blade?

Despite the resemblance between leaflet initiation and leaf initiation, several distinctions exist between organ initiation in the shoot and lateral growth in the leaf (Figs. 6.1 and 6.2). First, auxin application to developing leaves can promote both leaflet and blade outgrowth (Koenig *et al.*, 2009). Second, the initiation of lobes in tomato compound leaves was nearly identical to that of leaflets with respect to auxin transport and response. Finally, the formation of marginal auxin maxima is not unique to compound leaves, but occurs in simple leaves as a guide for vascular patterning and serration formation (chapter 8; Hay *et al.*, 2006; Scarpella *et al.*, 2006). These observations demonstrate that the existence of auxin maxima in the margins of developing leaves is not sufficient to confer a compound leaf phenotype. Instead the pattern appears to be inherent in leaf development and is recruited into morphological patterning of compound, and perhaps lobed, leaves. The overlap in the patterning mechanism of vascular tissue and marginal growth resolves the classical argument regarding the evolutionary linkage between the two processes (Dengler and Kang, 2001), but does not address the question of what actually differentiates compound and simple leaf development. It may be more accurate to draw parallels between all angiosperm leaves and shoots regardless of homology. The leaf is widely accepted to be ancestrally related to the shoot, and it may be that the role of PIN1 and auxin in developing leaf margins to pattern both vascular and morphological development is a remnant of this ancestry.

If the basic auxin pattern is a general phenomenon amongst dicot leaves, then why are the outcomes so varied? There are many possible answers to this question, and it is likely that the answer will differ by case, but several non-mutually exclusive hypotheses might be proposed.

5.1. Context of the auxin maximum: timing of differentiation

Leaves initiate as nearly radial, peg-like structures. As the leaf elongates along the proximo-distal axis, medio-lateral growth is initiated from a band of cells at the junction between adaxial ("top") and abaxial ("bottom") cell identities. In simple leaves, this band of cells initiates the earliest stages of outgrowth of blade, but is not maintained during subsequent development that produces the entire lamina (Poethig, 1987). Instead, after initiation of marginal growth, the lamina is produced through cell divisions throughout the intervening regions between the margin and the midrib. Thus, if these marginal cells can be described as "meristem-like," it is only at the earliest stages of leaf development, and the transition to lateral growth might be seen as their differentiation. It may be that only in the window of time in which the marginal tissues remain undifferentiated are they capable of perceiving the auxin signal to produce leaflets, while the specification of vasculature is unaffected by the initiation of marginal outgrowth. Under this hypothesis, the respective timing of marginal differentiation and auxin maxima-driven pattern formation might separate compound from simple leaf development (Fig. 6.1). For example, if the timing of auxin pattern formation remains static but the differentiation program is advanced, then lateral outgrowth would occur prior to the formation of maxima and would be continuous (perhaps organized by the distal auxin maximum driving leaf outgrowth).

Alternatively, if the differentiation program were delayed without changing the timing of maxima formation, then punctuated growth could be achieved. Similar transitions might occur if the timing of the auxin pattern formation were changed without altering the timing of differentiation. These simple scenarios postulate independence of the timing of two possibly overlapping processes, but even subtle relative variation might produce different shapes. From a molecular prospective, the *LA* and *KNOX* pathways are intriguing candidates for modulators of marginal blastozone/meristem indeterminancy in the context of this model.

5.2. Refinement of auxin maxima: both physically and in perception

Another possibility is that the physical characteristics of the auxin maximum itself might influence the resulting outgrowth. For example, a broader auxin peak might produce a lobe rather then a leaflet. How this might actually be accomplished over evolutionary time is not clear, but it remains a possibility. Perhaps alteration of *PIN* transcriptional regulation might influence the efficiency of auxin transport into developing maxima. Alternatively, the concentration of factors that restrict lateral growth (such as *ENTIRE*, discussed below) could also affect the refinement of the response peak. Low expression of such factors might result in increased sensitivity to auxin, therefore causing growth over a broader region.

5.3. Duration of auxin maximum formation

Somewhat related to the refinement of auxin peaks, the maintenance of the auxin peak might also influence the development of distinct leaflets. Once the auxin sink is lost at the tip of the developing leaflet, the concentration of auxin in adjacent tissues increases, possibly resulting in the activation of growth. Therefore, the longer the maximum is maintained, the more discrete the leaflet will be. As with refinement of maxima, the prolonged maintenance of auxin maxima might be influenced by the level of *PIN* protein present. Alternatively, the rate of development of the vasculature, or at least the timing of provascular *PIN1* expression, might influence auxin maxima maintenance and refinement by providing a conduit for auxin exit from the leaflet tip.

Any of these described hypotheses might contribute to natural variation in leaf phenotype, and certainly none of them is independent. For example, the timing of differentiation might influence the expression of PIN, thus influencing the maintenance of the auxin sink. It will be fascinating to determine how this patterning system is modified to produce varied leaf shape.

6. The Perception of Auxin Maxima and Specification of Differential Growth

The development of leaflets requires not just promotion of growth by *LYR* and auxin, but also repression of growth in adjacent tissues. In tomato this is accomplished by the AUX/IAA gene *ENTIRE (E)* (Dengler, 1984; Koenig *et al.*, 2009; Wang *et al.*, 2005; Zhang *et al.*, 2007). The AUX/IAA gene family is directly responsive to auxin, inhibiting auxin response in the absence of the hormone, and becoming degraded in its presence (Gray *et al.*, 2001). *e* mutants initiate blade outgrowth between leaflets resulting in the formation of a continuous blade reminiscent of ectopic auxin application to tomato leaves (Dengler, 1984; Koenig *et al.*, 2009). *E* RNA is localized throughout the developing leaf primordia, and *E* protein is presumably degraded at the site of leaflet initiation, leaving protein in the adjacent tissues where auxin concentration is low (Koenig *et al.*, 2009). Ubiquitous expression of auxin resistant versions of *E* results in inhibited lateral growth in leaves, consistent with this hypothesis. Auxin response maxima are less distinct in *e* leaves when compared to wild type in accordance with *E* function as a repressor of auxin response between leaflets. Combined with the above positive regulators of lateral growth, a model begins to emerge describing the patterning of tomato leaf shape. First, auxin is transported to the site of leaflet specification by PIN1. The accumulation of auxin results in the degradation of E resulting in the promotion of auxin response, cell cycle activity, and ultimately primordium outgrowth. *LYR* is also expressed in the initiating primordium and likely facilitates the activation of auxin response and cell cycle activity. Growth is inhibited in tissues adjacent to the primordium by the depletion of auxin and the concomitant failure to degrade E. Much remains to be understood even amongst these factors. For example, genetic analysis has not resolved the relationship between *E* and *LYR*. It is therefore unclear whether *LYR* expression is regulated by auxin. Integrating these individual components into a network is an important step in our understanding of leaf morphological patterning.

In addition to *E*, repression of growth between serrations in simple leaves and leaflets in compound leaves requires the activity of NAC domain transcription factors. The *A. thaliana CUP-SHAPED COTYLEDON2* (*CUC2*) gene is reported to play a role in the generation of serrations via repression of growth in serration sinuses (Nikovics *et al.*, 2006). Similarly, loss-of-function mutation of the tomato *NAC* gene *GOBLET* (*GOB*) results in a reduction in leaf complexity, including the fusion of leaflets through ectopic blade outgrowth (Berger *et al.*, 2009; Brand *et al.*, 2007). The role of *NAM/CUC* genes in compound leaf development has been extended across several species with independently derived compound leaves suggesting a conserved role in boundary specification during leaflet initiation (Blein *et al.*, 2008). In tomato, the role of *GOB* and *E* and the phenotypes of both loss-of-function mutants bear a significant degree of resemblance (Berger *et al.*, 2009; Brand *et al.*, 2007). Double-mutant analysis does not indicate a linear relationship between these two factors (Berger *et al.*, 2009), and it is therefore difficult to connect auxin accumulation in the margin with the downstream events specifying punctuated growth such as expression of GOB and E between leaflets. The molecular identity of *E* suggests that it may act early in auxin perception, but the relationship between *GOB* and auxin remains unresolved. Regardless, the roles of auxin and *NAM/CUC* genes appear to be conserved amongst compound leaf species, even in species (like pea) that do not utilize *KNOX* genes during the specification of leaf indeterminacy. Furthermore, the conserved utilization of both auxin maxima and boundary genes, like *NAM/CUC3*, in the morphogenesis of shoots and compound leaves supports the evolutionary derivation of leaves from branched, shoot-like structures (Blein *et al.*, 2008).

7. Homologies in Leaf Development

The molecular regulators of compound leaf development described in this review fall into two categories: those that control indeterminate identity in the leaf (*KNOX* and *TCP*) and those that specify lateral outgrowths (auxin, *E*, *GOB*, and *LYR*). The role played by indeterminacy promoting factors in leaf dissection indicates a parallel with shoot development. *KNOX* activity plays a critical role in the maintenance of SAM indeterminacy, and seems to play a similar role in the developing leaf margin to generate outgrowths that will become leaflets. In addition, the patterning of a compound leaf primordium by auxin, *PIN1*, *LYR*, and *NAC* genes also has strong parallels with processes in the SAM during organ initiation. These combined data suggest that, at least in tomato and *C. hirsuta* (and to a limited extent in pea as validated by studies on *NAM/CUC* genes), compound leaf development involves the recruitment of shoot developmental pathways. One possible explanation for this observation is that most of these factors had an ancestral role in patterning branching in the leafless ancestor of land plants. Under this hypothesis, compound leaf development is a partial reversion to this ancestral state. Why is the patterning framework maintained over large evolutionary distances in simple-leafed lineages? One possibility is that some aspect of the patterning framework (e.g., the PIN1-driven epidermal auxin maxima) underlies both vascular development and leaf shape specification (Barkoulas *et al.*, 2008; Koenig *et al.*, 2009; Scarpella *et al.*, 2006). As a result, the maintenance of vascular patterning provides the "skeletal" framework for morphological patterning and facilitates transition from simple to compound leaf development by repeated co-option into the leaf developmental program. The commonality of the patterning mechanism for these two developmental processes may explain the long-debated linkage between vascular patterning and leaf shape (Dengler *et al.*, 2001).

Despite the many parallels between shoot and compound leaf development, important differences exist. The discontinuous meristematic character of the marginal blastozone, the bilateral symmetry of the leaf, and the determinate nature of the leaf are amongst the critical divergences between the two structures. In addition, leaflets in most examined compound leaves are not separated by regions lacking lateral outgrowth, as observed between leaves at the apex, but instead by a rachis on which a vestigial blade is present (Kim *et al.*, 2003b). Therefore, the specification of lateral growth in compound leaves is not completely discrete; initiation of growth occurs throughout the margin, but subsequent differential maintenance of growth determines the ultimate shape of the leaf. It is important to note that of the major molecular and physiological factors regulating compound leaf development described above, all except the *KNOX* genes have been described to hay development play a role in *A. thaliana* serration development (Dinneny *et al.*, 2004; Hay *et al.*, 2006; Koenig *et al.*, 2009; Nikovics *et al.*, 2006). This observation suggests that marginal dissection across simple and compound leaf morphologies might utilize a similar molecular network.

Though it is clear that exceptions exist with respect to the role of any of these factors during compound leaf development, and it is likely that more will be revealed, it is interesting that so much of the network appears to be conserved across a wide evolutionary distance over which many transitions in leaf morphology have occurred. We propose that response to this network (or individual factors in the network) is modified by the maturation status of the developing leaf primordium, as regulated by factors such as the *KNOX* and *TCP* genes. Although this hypothesis might partially explain the many apparent homologies in leaf shape and the superficial continuity between simple entire, simple lobed, and compound leaves, it remains to be seen whether it is broadly applicable.

Some unresolved issues that need to be considered deal with the details of how leaflets are initiated on the leaf primordium. Unlike the initiation of leaves from the shoot, leaflet initiation in compound leaves occurs in both acropetal and basipetal patterns. It is unclear how the same auxin patterning mechanism might specify these two processes unless placed in the framework of discontinuous differentiation across such leaves. In addition, compound leaf morphology can be further divided into pinnate and palmate forms. In palmate leaves, leaflets are attached to the rachis at a single point rather than at intervals. This transition is poorly understood, though a limited number of palmate morphologies are correlated with alteration of dorsiventral (adaxial-abaxial) polarity. Palmate leaves are divided into two classes: those which initiate leaflets in around the entire circumference of the leaf tip (peltate) and those which do not initiate leaflets on the adaxial face of the junction between the petiole and the rachis (pedate). In peltate leaves the PHANTASTICA (PHAN) transcription factor was shown to be a key regulator of the extent and placement of the adaxial domain in the developing leaf primordium (Kim *et al.*, 2003a). In such instances, the petiole is abaxialized and radially symmetric, without vestigial blade outgrowths. The initiation of such leaves at the SAM has not been molecularly investigated. Furthermore, far more extreme modifications in morphology are observed amongst lineages such as the *Sarraceniaceae*, where blade outgrowth is followed by growth of a radial tendril-like structure that terminates in a cup-shaped pitcher. No patterning mechanism has yet been proposed for development of these unique leaf morphologies. The existence of such differing patterns of blade outgrowth, leaflet initiation, and positioning across species highlights the necessity to investigate the role of indeterminacy factors, genetic regulators of blade outgrowth and fractionation, and auxin distribution in many lineages. The investigation of the abundance of remaining diversity in leaf shape, in addition to the further dissection of leaf shape in models such as tomato and *C. hirsuta*, is likely to continue expanding our knowledge of plant morphological patterning for years to come.

ACKNOWLEDGMENTS

We would like to thank Dr. Julie Kang and Dr. Dan Chitwood for their critical review of this manuscript. Funding was provided by the National Science Foundation Grants IOS 0641696 and the Elsie Taylor Stocking Graduate Fellowship (to DK).

REFERENCES

Barkoulas, M., Hay, A., Kougioumoutzi, E., and Tsiantis, M. (2008). A developmental framework for dissected leaf formation in the Arabidopsis relative Cardamine hirsuta. *Nat. Genet.* **40**, 1136–1141.

Beerling, D. J. (2005). Leaf evolution: gases, genes and geochemistry. *Ann. Bot.* **96**, 345–352.

Beerling, D. J., and Fleming, A. J. (2007). Zimmermann's telome theory of megaphyll leaf evolution: a molecular and cellular critique. *Curr. Opin. Plant Biol.* **10**, 4–12.

Benkova, E., Michniewicz, M., Sauer, M., Teichmann, T., Seifertova, D., Jurgens, G., and Friml, J. (2003). Local, efflux-dependent auxin gradients as a common module for plant organ formation. *Cell* **115**, 591–602.

Berger, Y., Harpaz-Saad, S., Brand, A., Melnik, H., Sirding, N., Alvarez, J. P., Zinder, M., Samach, A., Eshed, Y., and Ori, N. (2009). The NAC-domain transcription factor GOBLET specifies leaflet boundaries in compound tomato leaves. *Development* **136**, 823–832.

Bharathan, G., Goliber, T. E., Moore, C., Kessler, S., Pham, T., and Sinha, N. R. (2002). Homologies in leaf form inferred from KNOXI gene expression during development. *Science* **296**, 1858–1860.

Bharathan, G., Janssen, B. J., Kellogg, E. A., and Sinha, N. (1999). Phylogenetic relationships and evolution of the KNOTTED class of plant homeodomain proteins. *Mol. Biol. Evol.* **16**, 553–563.

Blein, T., Pulido, A., Vialette-Guiraud, A., Nikovics, K., Morin, H., Hay, A., Johansen, I. E., Tsiantis, M., and Laufs, P. (2008). A conserved molecular framework for compound leaf development. *Science* **322**, 1835–1839.

Brand, A., Shirding, N., Shleizer, S., and Ori, N. (2007). Meristem maintenance and compound-leaf patterning utilize common genetic mechanisms in tomato. *Planta* **226**, 941–951.

Champagne, C. E. M., Goliber, T. E., Wojciechowski, M. F., Mei, R. W., Townsley, B. T., Wang, K., Paz, M. M., Geeta, R., and Sinhaa, N. R. (2007). Compound leaf development and evolution in the legumes. *Plant Cell* **19**, 3369–3378.

Chen, J. J., Janssen, B. J., Williams, A., and Sinha, N., 1997. A gene fusion at a homeobox locus: alterations in leaf shape and implications for morphological evolution. *Plant Cell* **9**, 1289–1304.

Chuck, G., Lincoln, C., and Hake, S., 1996. KNAT1 induces lobed leaves with ectopic meristems when overexpressed in Arabidopsis. *Plant Cell* **8**, 1277–1289.

David-Schwartz, R., Koenig, D., and Sinha, N. R. (2009). LYRATE is a key regulator of leaflet initiation and lamina outgrowth in tomato. *Plant Cell* **21**, 3093–3104.

DeMason, D. A., and Chawla, R. (2004). Roles for auxin during morphogenesis of the compound leaves of pea (Pisum sativum). *Planta* **218**, 894–894.

Dengler, N. G. (1984). Comparison of leaf development in normal (+/+), entire (E/E), and lanceolate (La/+) plants of tomato, Lycopersicon-esculentum Ailsa Craig. *Bot. Gaz.* **145**, 66–77.

Dengler, N., and Kang, J. (2001). Vascular patterning and leaf shape. *Curr. Opin. Plant Biol.* **4**, 50–6.

Dinneny, J. R., Weigel, D., and Yanofsky, M. F. (2006). NUBBIN and JAGGED define stamen and carpel shape in Arabidopsis. *Development* **133**, 2285–2285.

Dinneny, J. R., Yadegari, R., Fischer, R. L., Yanofsky, M. F., and Weigel, D. (2004). The role of JAGGED in shaping lateral organs. *Development* **131**, 1101–1110.

Gray, W. M., Kepinski, S., Rouse, D., Leyser, O., and Estelle, M. (2001). Auxin regulates SCFTIR1-dependent degradation of AUX/IAA proteins. *Nature* **414**, 271–276.

Ha, C. M., Jun, J. H., Nam, H. G., and Fletcher, J. C. (2004). BLADE-ON-PETIOLE1 encodes a BTB/POZ domain protein required for leaf morphogenesis in Arabidopsis thaliana. *Plant Cell Physiol.* **45**, 1361–1370.

Hake, S., Smith, H. M. S., Holtan, H., Magnani, E., Mele, G., and Ramirez, J. (2004). The role of knox genes in plant development. *Annu. Rev. Cell Dev. Biol.* **20**, 125–151.

Hareven, D., Gutfinger, T., Parnis, A., Eshed, Y., and Lifschitz, E. (1996). The making of a compound leaf: genetic manipulation of leaf architecture in tomato. *Cell* **84**, 735–744.

Hay, A., Barkoulas, M., and Tsiantis, M. (2006). ASYMMETRIC LEAVES1 and auxin activities converge to repress BREVIPEDICELLUS expression and promote leaf development in Arabidopsis. *Development* **133**, 3955–3961.

Hay, A., and Tsiantis, M. (2006). The genetic basis for differences in leaf form between Arabidopsis thaliana and its wild relative Cardamine hirsuta. *Nat. Genet.* **38**, 942–947.

Hepworth, S. R., Zhang, Y. L., McKim, S., Li, X., and Haughn, G. (2005). BLADE-ON-PETIOLE-dependent signaling controls leaf and floral patterning in Arabidopsis. *Plant Cell* **17**, 1434–1448.

Hofer, J., Turner, L., Hellens, R., Ambrose, M., Matthews, P., Michael, A., and Ellis, N. (1997). UNIFOLIATA regulates leaf and flower morphogenesis in pea. *Curr. Biol.* **7**, 581–587.

Janssen, B. J., Williams, A., Chen, J. J., Mathern, J., Hake, S., and Sinha, N. (1998). Isolation and characterization of two knotted-like homeobox genes from tomato. *Plant Mol. Biol.* **36**, 417–425.

Kim, M., McCormick, S., Timmermans, M., and Sinha, N. (2003a). The expression domain of PHANTASTICA determines leaflet placement in compound leaves. *Nature* **424**, 438–443.

Kim, M., McCormick, S., Timmermans, M., and Sinha, N. (2003b). The expression domain of PHANTASTICA determines leaflet placement in compound leaves. *Nature* **425**, 102.

Koenig, D., Bayer, E., Kang, J., Kuhlemeier, C., and Sinha, N. (2009). Auxin patterns Solanum lycopersicum leaf morphogenesis. *Development* **136**, 2997–3006.

Nikovics, K., Blein, T., Peaucelle, A., Ishida, T., Morin, H., Aida, M., and Laufs, P. (2006). The balance between the MIR164A and CUC2 genes controls leaf margin serration in Arabidopsis. *Plant Cell* **18**, 2929–2945.

Ohno, C. K., Reddy, G. V., Heisler, M. G. B., and Meyerowitz, E. M. (2004). The Arabidopsis JAGGED gene encodes a zinc finger protein that promotes leaf tissue development. *Development* **131**, 1111–1122.

Ori, N., Cohen, A. R., Etzioni, A., Brand, A., Yanai, O., Shleizer, S., Menda, N., Amsellem, Z., Efroni, I., Pekker, I., Alvarez, J. P., Blum, E., *et al.* (2007). Regulation of LANCEOLATE by miR319 is required for compound-leaf development in tomato. *Nat. Genet.* **39**, 787–791.

Palatnik, J. F., Allen, E., Wu, X. L., Schommer, C., Schwab, R., Carrington, J. C., and Weigel, D. (2003). Control of leaf morphogenesis by microRNAs. *Nature* **425**, 257–263.

Parnis, A., Cohen, O., Gutfinger, T., Hareven, D., Zamir, D., and Lifschitz, E. (1997). The dominant developmental mutants of tomato, Mouse-ear and Curl, are associated with distinct modes of abnormal transcriptional regulation of a knotted gene. *Plant Cell* **9**, 2143–2158.

Poethig, R. S. (1987). Clonal analysis of cell lineage patterns in plant development. *Am. J. Bot.* **74**, 581–594.

Reinhardt, D., Mandel, T., and Kuhlemeier, C. (2000). Auxin regulates the initiation and radial position of plant lateral organs. *Plant Cell* **12**, 507–518.

Reinhardt, D., Pesce, E. R., Stieger, P., Mandel, T., Baltensperger, K., Bennett, M., Traas, J., Friml, J., and Kuhlemeier, C. (2003). Regulation of phyllotaxis by polar auxin transport. *Nature* **426**, 255–260.

Reinhardt, D., Wittwer, F., Mandel, T., and Kuhlemeier, C. (1998). Localized upregulation of a new expansin gene predicts the site of leaf formation in the tomato meristem. *Plant Cell* **10**, 1427–1437.

Scarpella, E., Marcos, D., Friml, J., and Berleth, T. (2006). Control of leaf vascular patterning by polar auxin transport. *Genes Dev.* **20**, 1015–1027.

Shani, E., Burko, Y., Ben-Yaakov, L., Berger, Y., Amsellem, Z., Goldshmidt, A., Sharon, E., and Ori, N. (2009). Stage-specific regulation of solanum lycopersicum leaf maturation by class 1 KNOTTED1-LIKE HOMEOBOX proteins. *Plant Cell* **21**, 3078–3092.

Sinha, N. R., Williams, R. E., and Hake, S. (1993). Overexpression of the maize homeo box gene, KNOTTED-1, causes a switch from determinate to indeterminate cell fates. *Genes Dev.* **7**, 787–795.

Vanneste, S., and Friml, J. (2009). Auxin: a trigger for change in plant development. *Cell* **136**, 1005–1016.

Wang, H., Jones, B., Li, Z. G., Frasse, P., Delalande, C., Regad, F., Chaabouni, S., Latche, A., Pech, J. C., and Bouzayen, M. (2005). The tomato Aux/IAA transcription factor IAA9 is involved in fruit development and leaf morphogenesis. *Plant Cell* **17**, 2676–2692.

Wisniewska, J., Xu, J., Seifertova, D., Brewer, P. B., Ruzicka, K., Blilou, I., Rouquie, D., Scheres, B., and Friml, J. (2006). Polar PIN localization directs auxin flow in plants. *Science* **312**, 883–883.

Zhang, J. H., Chen, R. G., Xiao, J. H., Qian, C. J., Wang, T. T., Li, H. X., Bo, O. Y., and Ye, Z. B. (2007). A single-base deletion mutation in SlIAA9 gene causes tomato (Solanum lycopersicum) entire mutant. *J. Plant Res.* **120**, 671–678.

CHAPTER SEVEN

Control of Tissue and Organ Growth in Plants

Holger Breuninger *and* Michael Lenhard

Contents

Abstract

Plant organs grow to characteristic, species-specific sizes and shapes. At the cellular level, organ growth is initially characterized by cell proliferation, which gives way to cell expansion at later stages. Using mainly *Arabidopsis thaliana* as a model species, a number of factors have been isolated in recent years that promote or restrict organ growth, with the altered organ size being associated with changes in cell number, in cell size, or in both. However, cells in an organ

Department of Cell & Developmental Biology, John Innes Centre, Norwich, UK

Current Topics in Developmental Biology, Volume 91
ISSN 0070-2153, DOI 10.1016/S0070-2153(10)91007-7

do not appear to follow a strictly autonomous program of proliferation and expansion, and their behavior is coordinated in at least three different respects: normally sized organs can be formed consisting of altered numbers of cells with compensatory changes in the size of the individual cells, suggesting that cellular behavior is subject to organ-wide control; the growth of cells derived from more than one clonal origin is coordinated within a plant lateral organ with its different histological layers; and growth of cells in different regions of an organ is coordinated to generate a reasonably flat leaf or floral organ. Organ growth is strongly modulated by environmental factors, and the molecular basis for this regulation is beginning to be understood. Given the complexity of organ growth as a dynamic four-dimensional process, precise quantification of growth parameters and mathematical modeling are increasingly used to understand this fascinating problem of plant biology.

1. Introduction

Much of the morphological variation we see around us in nature is due to differences in the shapes and sizes of organs. For example, plants within one taxonomic family generally have a very similar floral architecture, yet size and shape of the floral organs can vary very widely (Stebbins, 1974). These differences in size and shape are highly heritable, indicating that the species-specific genetic program tightly controls the growth of organs to their final form (Fig. 7.1). Understanding the genetic and molecular mechanisms that regulate organ growth is not just a fascinating problem of basic biology; given our dependence on plants for the production of food, feed, and fuel,

Figure 7.1 Daffodil (*Narcissus*) field in Norfolk, UK. Individual plants of the same genotype develop organs of similar size, indicating that size is heritable and thus genetically controlled. (Image courtesy of Andrew Davis).

knowledge about the endogenous control of growth is likely to have important implications for the rational manipulation of plant yields.

In this review, we will discuss our current understanding of the genetic control of plant organ growth, focusing on lateral shoot organs and drawing mainly on results obtained with the dicotyledonous model *Arabidopsis thaliana* (called *Arabidopsis* in the following). After describing the control of the two basic cellular processes by which plant organs grow, proliferation and cell expansion, we will discuss organ-wide processes involved in growth control and modulation of growth by the environment, before outlining recent modeling approaches to enable an understanding of the complexities of growth that exceed simple intuition.

2. Description of Organ Growth

Lateral shoot organs are initiated by an iterative process at the shoot apical meristem, a group of long-term stem cells and their immediate daughters at the growing tip of the plant (Steeves and Sussex, 1989). Before the floral transition, the meristem produces leaves, while after the onset of flowering, the shoot meristem gives rise to floral meristems, which then initiate floral organs. From sector analysis it has been estimated that the early leaf primordium consists of about 100–150 cells in tobacco (Poethig and Sussex, 1985) and about 100 cells in cotton (Dolan and Poethig, 1998), while in *Arabidopsis* the number of founder cells for the first true leaves appears to be considerably smaller (Irish and Sussex, 1992; Schnittger *et al.*, 1996). In the *Arabidopsis* flower, sepals are formed by two rows of eight cells each in the epidermis and subepidermal layer, while petals arise from two cells in each layer (Bossinger and Smyth, 1996). After initiation, the organ grows by cell proliferation, i.e., the coupled accumulation of cytoplasmic mass, increase in cell size, and subsequent cell division to yield two daughter cells of comparable size to the original cell. While proliferation occurs initially throughout the entire primordium, at later stages it becomes gradually restricted to defined regions, while the remaining parts of the organ continue growing by cell expansion (Figs. 7.2 and 7.3). In the lamina of *Arabidopsis* leaves, the arrest of proliferation occurs first at the distal tip of the primordium, after which cells in progressively more proximal parts of the leaf stop proliferating (Donnelly *et al.*, 1999). Although this process has been described as an arrest front that travels from the tip to the base of the leaf, it is currently not clear whether the absolute length of the region in which cells proliferate shrinks, or whether the distal regions merely grow out of a field with proliferative competence that has a relatively constant size and is anchored at the leaf base. Eventually, proliferation ceases throughout all of the primordium and further growth occurs by cell expansion until the final size and shape is reached (Fig. 7.3).

Figure 7.2 *Arabidopsis* rosette expressing the *pCYCB1;1::CDB-GUS* reporter that marks mitotic cells. Only a small fraction of the plant is actively dividing. During organ growth, proliferation ceases along an arrest front from distal to proximal, as indicated by the loss of reporter expression at the tip of the outgrowing leaf. (See Color Insert.)

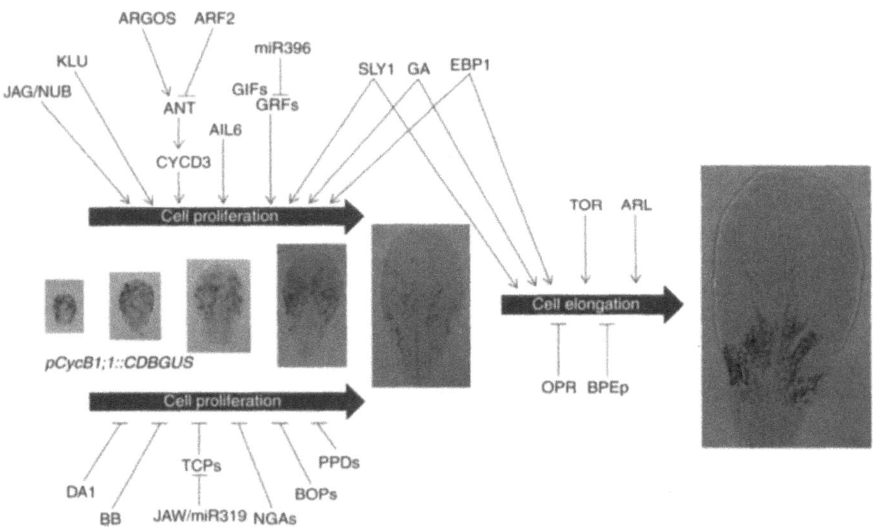

Figure 7.3 Pathways controlling cell proliferation and cell elongation. Central pictures show *pCycB1;1:GUS* expression in progressively older developing petals as a model organ. During development, cell division is restricted to the central part of the growing petal and ceases at later stages, after which the organ grows merely by cell elongation. Genetic pathways promoting cell proliferation or elongation are shown in the upper part; growth-restricting pathways are shown in the lower part of the figure. See also Table 7.1. (*pCycB1;1* expression pictures are taken from Disch *et al.*, 2006, copyright Current Biology). (See Color Insert.)

Growth of *Arabidopsis* leaves has been described using kinematic analysis (Beemster *et al.*, 2005; Cookson *et al.*, 2005; Wiese *et al.*, 2007). The maximal relative growth rate (RGR) is found very early during leaf growth, when most of the cells in the primordium are proliferating. RGR then drops, as proliferation starts to cease, and continues falling during the phase when most of the leaf grows by cell expansion. Spatially, the distribution of RGR mirrors the basipetal gradient in cell proliferation activity, remaining higher at the leaf base than toward the tip (Wiese *et al.*, 2007).

Based on this description of organ growth, changes in final organ size could be due to a change in the number of organ founder cells or due to an altered rate or duration of growth. The number of cells recruited into organ primordia appears to be an important determinant of leaf size in grasses (Scanlon and Freeling, 1997). However, to our knowledge no clear cases of changes in organ size resulting from a different number of founder cells have been documented in dicotyledonous plants, placing the emphasis on the control of post-initiation growth of the organ.

There has been a long-standing debate about what drives growth in plants and the relationship between cellular processes and overall organ growth. Organismal theory views growth and its control as a tissue property, with cell division and cell expansion merely as two different modes of "filling in" the increasing volume of tissue (Kaplan, 2001; Kaplan and Hagemann, 1991). By contrast, cell theory stresses the importance of the cell-based processes of proliferation and expansion in determining overall organ growth (Tsukaya, 2003). Given the examples of growth compensation to be discussed below, "neo-cell" theory has been proposed to reconcile the two extreme views. It acknowledges the importance of cell division and cell expansion in providing the building blocks for growing organs and the effect that purely cellular changes can have on final organ size, but emphasizes at the same time regulatory feedback from organ-wide control systems on these cellular processes. What this discussion underlines is our profound ignorance about how growing organs determine for how long to grow and in which directions. Against this background, it is important to bear in mind that for many of the factors discussed below it is not clear how exactly they act to promote or restrict growth. Their classification under "factors controlling growth by proliferation" and "factors controlling growth by expansion" is based on the cellular changes associated with the different final organ sizes; these could, however, arise from effects primarily on cellular processes or on a supracellular control system. Also, regarding growth by proliferation, it is important to emphasize that cell division in itself would not lead to an increase in size of a tissue or organ, but that it is only the combination of cytoplasmic growth (i.e., the accumulation of proteins, of membrane and of cell wall growth) and of cell division that enables the increase of a tissue composed of cells of a relatively constant size (Mitchison, 2003).

3. Proliferative Growth Control

After specification of the founder cell pool, organs initially grow mainly by cell proliferation. In the following we will discuss the function of several key regulators of proliferative growth (Fig. 7.3, Table 7.1).

3.1. AINTEGUMENTA and related AP2-domain transcription factors as regulators of proliferative growth

Transcription factors of the AINTEGUMENTA (ANT) subfamily within the larger APETALA2 (AP2)-domain family stimulate growth by proliferation (Horiguchi *et al.*, 2009; Krizek, 1999, 2009; Mizukami and Fischer, 2000). While *ant* mutants have small organs with fewer cells, plants overexpressing *ANT* form larger organs composed of more cells due to an extended period of proliferative growth. Recent studies addressing the expression patterns of members of the *ANT* subfamily, termed *AINTEGUMENTA-LIKE* (*AIL*) genes, and their function in plant development suggest that potentially all *AIL* genes are involved in the control of proliferation. Several members are expressed in young proliferating tissues, and loss- and gain-of-function analyses for *AIL6* showed similar effects on growth by proliferation as observed for *ANT* itself (Krizek, 2009; Nole-Wilson *et al.*, 2005). Thus, the family of *ANT* and *AIL* genes represents a set of central growth regulators.

ANT acts as a transcriptional activator, positively regulating the expression of cell cycle genes such as *CYCLIN D3;1* (*CYCD3;1*) (Mizukami and Fischer, 2000). Increased *CYCD3;1* levels are sufficient to increase the proportion of cells in the G2-phase of the cell cycle by stimulating the G1–S phase transition (Dewitte *et al.*, 2003; Menges *et al.*, 2006). However, increased *CYCD3;1* activity does not *per se* lead to larger organs. In fact, plants overexpressing *CYCD3;1* are smaller, and organs consist of more, but smaller cells that fail to differentiate and expand normally (Dewitte *et al.*, 2003). This suggests either that *ANT* is able to stimulate *CYCD3;1* activity in a finely balanced and patterned manner, so as not to interfere with cell differentiation and expansion, or that *ANT* stimulates cellular growth by an independent pathway and the upregulated *CYCD3;1* activity mainly helps to divide the more strongly growing cells in a timely manner. Loss-of-function studies with *cycd3* mutants are consistent with the latter possibility, as they suggest that *CYCD3* activity is important for a normal number of cell divisions, but not for organ growth (Dewitte *et al.*, 2007). Cell numbers and *CYCB1;1* expression levels are reduced in the organs of triple mutants lacking all three *CYCD3* family members, yet the plants develop normally, and organ size is not reduced, as the lower cell number is offset by an increase in cell size (discussed below).

Table 7.1 Arabidopsis genes involved in organ size/growth control

Gene name	Abbreviation	Locus identifier	Protein type	Cell proliferation	Cell elongation	Organ growth	Redundant with	Affects expression of	Reference*
12-Oxophytodienoic acid reductase	OPR3	AT2G06050	OPDA-Reductase	0	–	–		BPEp	Brioudes (2009)
AINTEGUMENTA	ANT	AT4G37750	AP2-domain TF	+	–	+		CYCD3;1, CYCB1;1	Krizek (2009), Krizek (1999), Mizukami (2000), Horiguchi (2009)
AINTEGUMENTA LIKE 5	AIL5	AT5G57390	AP2-domain TF	NA	NA	NA			Nole-Wilson (2005)
AINTEGUMENTA LIKE 6	AIL6	AT5G10510	AP2-domain TF	+	–	+	ANT		Nole-Wilson (2005), Krizek (2009)
AINTEGUMENTA LIKE 7	AIL7	AT5G65510	AP2-domain TF	0	0	0			Nole-Wilson (2005), Krizek (2009)
ARGOS-LIKE	ARL	AT2G44080	novel plant-specific	0	+	+			Hu (2006)
AtRPT2a	AtRPT2a	AT4G29040	19S proteasome subunit	0	–[1]	–		CYCD3;1, CDC6b, CDT1a, CDT1b, HISH4, CYCA3;1	Sonoda (2009)
AUXIN RESPONSE FACTOR 2	ARF2	AT5G62000	ARF TF	–	–	–		ANT, CYCD3;1	Ellis (2005), Okushima (2005), Schruff (2006)
AUXIN-REGULATED GENE INVOLVED IN ORGAN SIZE	ARGOS	AT3G59900	novel plant-specific	+	0	+		ANT, CYCD3;1, CYCB1;1	Hu (2003)

(*Continued*)

Table 7.1 (*Continued*)

Gene name	Abbreviation	Locus identifier	Protein type	Cell proliferation	Cell elongation	Organ growth	Redundant with	Affects expression of	Reference*
BIG BROTHER	BB	AT3G63530	E3-ubiquitin ligase	-	0	-			Disch (2006)
BIGPETAL	BPE	AT1G59640	bHLH TF	0	-	-			Szecsi (2006)
BLADE ON PETIOLE 1	BOP 1	AT3G57130	nuclear BTB/POZ domain protein with ankyrin repeats.	-	0	-	BOP2	JAG, NUB	Ha (2007), Ha (2004), Ha (2003), Norberg (2005)
BLADE ON PETIOLE 2	BOP 2	AT2G41370	nuclear BTB/POZ domain protein with ankyrin repeats.	-	0	-	BOP1		Ha (2007), Ha (2004), Ha (2003), Norberg (2005)
BRASSINOSTEROID-INSENSITIVE4	BIN4	AT5G24630	Component of Topoisomerase VI complex	0	+[1]	+		CYCB1;1	Breuer (2007)
CYCLIN D3;1	CYCD3;1	AT4G34160	CYCLIN	+	-	0	CYCD3;2, CYCD3;3		Mizukami (2000), Dewitte (2003), Dewitte (2007)
CYCLIN D3;2	CYCD3;2	AT5G67260	CYCLIN	+	-	0	CYCD3;1, CYCD3;3		Dewitte (2007)
CYCLIN D3;3	CYCD3;3	AT3G50070	CYCLIN	+	-	0	CYCD3;1, CYCD3;2		Dewitte (2007)
CYP78A7	CYP78A7	AT5G09970	Cytochrom P450	+	0	+	KLU/ CYP78A5		Wang (2008)
DA1	DA1	AT1G19270	ubiquitin interaction domain containing protein	-	0	-			Li (2008)
ErbB-3 epidermal growth factor receptor binding protein	EBP1	AT3G51800	part of a pre-ribosomal nucleoprotein complex	+	+	+		CDKB1;1, RNR2, CYCD3;1	Horvath (2006)

FIZZY-RELATED 2	FZR2	AT4G22910	Fizzy-related	0	+[1]	+		Larson-Rabin (2009)
GA1	GA1	AT4G02780	Ent-coplayl Diphosphate Synthetase	+	+	+		Ubeda-Thomas (2009)
GRF-INTERACTING FACTOR 1/ ANGUSTIFOLIA3	GIF1/ AN3	AT5G28640	SYT-like transcriptional coactivator	+	0	+	GIF2, GIF3	Horiguchi (2005), Kim (2004), Lee (2009)
GRF-INTERACTING FACTOR 2	GIF2	AT1G01160	SYT-like transcriptional coactivator	+	0	+	GIF1, GIF3	Horiguchi (2005), Kim (2004), Lee (2009)
GRF-INTERACTING FACTOR 3	GIF3	AT4G00850	SYT-like transcriptional coactivator	+	0	+	GIF1, GIF2	Horiguchi (2005), Kim (2004), Lee (2009)
GROWTH-REGULATING FACTOR 1	GRF 1	AT2G22840	QLQ/WRC-domain containing TF	+	0	+	GRF2, GRF3	Kim (2003)
GROWTH-REGULATING FACTOR 2	GRF 2	AT4G37740	QLQ/WRC-domain containing TF	+	0	+	GRF1, GRF3	Kim (2003)
GROWTH-REGULATING FACTOR 3	GRF 3	AT2G36400	QLQ/WRC-domain containing TF	+	0	+	GRF1, GRF2	Kim (2003)
GROWTH-REGULATING FACTOR 4	GRF 4	AT3G52910	QLQ/WRC-domain containing TF	NA	NA	NA		Kim (2003)
GROWTH-REGULATING FACTOR 5	GRF 5	AT3G13960	QLQ/WRC-domain containing TF	+	0	+		Kim (2003), Horiguchi (2005)
GROWTH-REGULATING FACTOR 6	GRF 6	AT2G06200	QLQ/WRC-domain containing TF	NA	NA	NA		Kim (2003)
GROWTH-REGULATING FACTOR 7	GRF 7	AT5G53660	QLQ/WRC-domain containing TF	NA	NA	NA		Kim (2003)

(*Continued*)

Table 7.1 (*Continued*)

Gene name	Abbreviation	Locus identifier	Protein type	Cell proliferation	Cell elongation	Organ growth	Redundant with	Affects expression of	Reference*
GROWTH-REGULATING FACTOR 8	GRF 8	AT4G24150	QLQ/WRC-domain containing TF	NA	NA	NA			Kim (2003)
GROWTH-REGULATING FACTOR 9	GRF 9	AT2G45480	QLQ/WRC-domain containing TF	+	0	+			Kim (2003), Horiguchi (2005)
JAGGED	JAG	AT1G68480	Zn finger TF	+	0	+	NUB		Dinneny (2006), Dinneny (2004), Ohno (2004)
JAW/miR319a	jaw	AT4G23713	- (miRNA)	+	0	+		TCP2, 3, 4, 10, 24	Palatnik (2003), Nag (2009)
KLUH/CYP78A5	KLU	AT1G13710	Cytochrom P450	+	0	+			Anastasiou (2007), Wang (2008)
miR319b	miR319b	AT5G41663	- (miRNA)	+	0	+		TCP2, 3, 4, 10, 24	Nag (2009)
miR319c	miR319c	AT2G40805	- (miRNA)	+	0	+		TCP2, 3, 4, 10, 24	Nag (2009)
miR396a	miR396a	At2g10606	- (miRNA)	-	0	-		GRF1, 2, 3, 4, 7, 9, CYCB1;1	Rodriguez (2010)
miR396b	miR396b	At5g35407	- (miRNA)	-	0	-		GRF1, 2, 3, 4, 7, 9, CYCB1;1	Rodriguez (2010)
NGATHA1	NGA1	At2g46870	B3 TF	-	0	-	NGA2, NGA3, NGA4		Alvarez (2009), Trigueros (2009), Kwon (2009)

NGATHA2	NGA2	At3g61970	B3 TF	-	0	-	NGA1, NGA3, NGA4	Alvarez (2009), Trigueros (2009), Kwon (2009)
NGATHA3	NGA3	At1g01030	B3 TF	-	0	-	NGA1, NGA2, NGA4	Alvarez (2009), Trigueros (2009), Kwon (2009)
NGATHA4	NGA4	At4g01500	B3 TF	-	0	-	NGA1, NGA2, NGA3	Alvarez (2009), Trigueros (2009), Kwon (2009)
NUBBIN	NUB	AT1G13400	Zn finger TF	+	0	+	JAG	Dinneny (2006)
PEAPOD1	PPD1	At4g14713	nuclear localized DNA binding protein	-	0	-	PPD2	White (2006)
PEAPOD2	PPD2	At4g14720	nuclear localized DNA binding protein	-	0	-	PPD1	White (2006)
ROOT HAIRLESS 1	RHL1	AT1G48380	Component of Topoisomerase VI complex	0	+[1]	+		Sugimoto-Shirasu (2005)
ROOT HAIRLESS 2	RHL2	AT5G02820	Component of Topoisomerase VI complex	+	+[1]	+		Sugimoto-Shirasu (2002)
ROOT HAIRLESS 3/ HYPOCOTYL6	RHL3/ HYP6	AT3G20780	Component of Topoisomerase VI complex	+	+[1]	+		Sugimoto-Shirasu (2002)
SLEEPY 1	SLY1	AT4G24210	F-box protein	+	+	+		Achard (2009)

(Continued)

Table 7.1 (*Continued*)

Gene name	Abbreviation	Locus identifier	Protein type	Cell proliferation	Cell elongation	Organ growth	Redundant with	Affects expression of	Reference*
TARGET OF RAPAMYCIN	TOR	AT1G50030	Protein kinase	0	+	+			Menand (2002), Deprost (2007)
TCP protein 2	TCP2	At4g18390	TCP TF	-	0	-			Palatnik (2003), Nag (2009)
TCP protein 3	TCP3	At1g53230	TCP TF	-	0	-			Palatnik (2003), Nag (2009)
TCP protein 4	TCP4	At3g15030	TCP TF	-	0	-		miR396	Palatnik (2003), Nag (2009), Rodriguez (2010)
TCP protein 10	TCP10	At2g31070	TCP TF	-	0	-			Palatnik (2003), Nag (2009)
TCP protein 24	TCP24	AT1G30210	TCP TF	-	0	-			Palatnik (2003), Nag (2009)

*, for space constrains only first authors are listed; TF, transcription factor; 0, not affected; +, positively affected; -, negatively affected, 1, involved in endoreduplication; NA, no data available.

Given their important role in controlling growth by proliferation, how is the expression of *ANT* and *AIL* genes regulated? Several hormonal pathways appear to influence *ANT* expression. One of these is represented by the auxin-inducible gene *ARGOS* (for A*UXIN*-R*EGULATED GENE INVOLVED IN ORGAN SIZE*) (Hu *et al.*, 2003). Loss of *ARGOS* function reduces organ size, whereas overexpression leads to enlarged organs with more cells, suggesting that ARGOS mediates auxin-stimulated cell proliferation. Changes in *ARGOS* activity lead to similar changes in the expression of *ANT, CYCB1;1*, and *CYCD3;1*, and loss of *ANT* function abolishes the growth-promoting activity of *ARGOS* overexpression. Thus, *ARGOS* appears to promote organ growth by stimulating *ANT* expression, which in turn leads to upregulation of *CYCD3;1* and *CYCB1;1. ARGOS* encodes a novel, plant-specific protein, and its high transcriptional induction by auxin suggests that *ARGOS* is downstream of the transcription machinery directly controlled by auxin, represented by members of the *AUXIN RESPONSE FACTOR* (*ARF*) family. The closest homolog of *ARGOS*, called *ARGOS-LIKE* (*ARL*), also promotes growth. Interestingly, however, *ARL* does so by acting exclusively on cell expansion (discussed below), and its expression is stimulated by brassinosteroid (BR), but not auxin (Hu *et al.*, 2006). Thus, two closely related members of the same gene family stimulate growth by acting exclusively on either of the two cellular processes underlying overall organ growth. Further research into the function of this gene pair promises to refine our understanding of the relationship between these two processes.

ARF family members influence multiple aspects of plant development, with *ARF2* playing a prominent role in growth regulation. Loss of *ARF2* function enlarges leaves and floral organs, indicating that *ARF2* limits growth (Ellis *et al.*, 2005; Okushima *et al.*, 2005; Schruff *et al.*, 2006). Cell number and cell size in *arf2* mutants are increased, suggesting that *ARF2* controls organ growth by restricting the extent of cell proliferation and also cell elongation. In contrast to *ARGOS, ARF2* represses *ANT* and *CYCD3;1* expression, suggesting that it antagonizes *ARGOS* in the control of *ANT*-mediated growth stimulation. ARF2 may act as a point of integration for the synergistic action of brassinosteroids (BRs) and auxin (Vert *et al.*, 2008). The BR-regulated kinase BIN2 directly interacts with ARF2 upon induction by BR, phosphorylating ARF2, and thus modulating its DNA-binding activity. While ARF2 alone has high affinity to DNA and is thought to act as a repressor for auxin-mediated transcription, phosphorylation by BIN2 leads to a highly reduced DNA-binding ability. Thus, the BR-induced release of ARF2 from its binding sites may enable other activating ARFs to bind those sites and upregulate previously repressed genes (Vert *et al.*, 2008). Thus, ARF2 negatively regulates *ANT* expression and cell proliferation, and this negative function appears to be repressed by BR-mediated phosphorylation.

3.2. GRFs and GIF, a growth control duet

The *GROWTH-REGULATING FACTOR* (*GRF*) gene family represents another plant-specific group of transcription factors involved in organ size control (van der Knaap *et al.*, 2000). GRF proteins are characterized by highly conserved QLQ and WRC domains, and the *Arabidopsis* genome contains nine genes with similar features termed *AtGRF1* to *AtGRF9* (Kim *et al.*, 2003). All *Arabidopsis AtGRF* genes show elevated expression levels in growing organs. Triple mutants of *grf1 grf2 grf3* form smaller organs, whereas overexpression of *AtGRF1* and *AtGRF2* increases leaf and cotyledon area. These changes in organ size are due to changes mainly in cell number, suggesting that GRFs mainly promote growth by proliferation (Kim and Kende, 2004; Kim *et al.*, 2003). Expression of seven of the nine *GRFs* is controlled by the microRNA miR396. miR396 expression is induced during leaf maturation, and overexpression leads to down-regulation of several *GRFs* and *CYCB1;1* suggesting that cell proliferation during organ growth is controlled by the antagonistic function of *GRFs* and miR396 (Rodriguez *et al.*, 2010). The GRF proteins physically interact with a family of putative transcriptional coactivators termed GRF-INTERACTING FACTORs (GIFs) (Horiguchi *et al.*, 2005; Kim and Kende, 2004). *Arabidopsis* contains three *GIF* genes, one of which is the *ANGUSTIFOLIA3* (*AN3*) gene, which mutates to a characteristic phenotype with narrow leaves (Horiguchi *et al.*, 2005). Like their interaction partners, *GIF* genes act redundantly in organ size control. Mutations in *GIF* genes cause retarded organ outgrowth and smaller final size, with the strongest effects seen in *gif1 gif2 gif3* triple mutants (Horiguchi *et al.*, 2005; Kim and Kende, 2004; Lee *et al.*, 2009). In contrast to many other factors, such as *BIG BROTHER (BB), DA1*, and *KLUH*, that appear to act exclusively on the duration of the cell proliferation phase, multiple *gif* mutants show not only a shortened period of cell proliferation but also a strong reduction in maximal proliferation rates, highlighting their central role in promoting growth by proliferation (Lee *et al.*, 2009).

3.3. Control of cell proliferation via JAGGED, NUBBIN, and NGATHA: a link to patterning?

The Zn-finger-containing putative transcription factors NUBBIN (NUB) and JAGGED (JAG) are a pair of homologous genes that control plant organ growth by proliferation (Dinneny *et al.*, 2004, 2006; Ohno *et al.*, 2004). Mutations in *JAG* lead to reduced growth of floral organs, with petals most severely affected, because of a premature arrest of cell proliferation. These defects are strongly enhanced in *jag nub* double mutants. While *nub* single mutants exhibit only subtle phenotypes, leaves in the double mutants are reduced in size, and floral organs exhibit retarded growth leading to severe malformations. As in *jag* single mutants, the phase of cell proliferation is terminated early in development, leading to the loss of distal parts of the organ, as seen most strongly in petals. By contrast, overexpression of *JAG* leads to ectopic outgrowth of lamina tissue from the leaf petiole and to the ectopic formation of bracts (leaves subtending flowers), while ectopic *NUB* activity promotes an increase in the number of cell layers in organs, indicating that both genes are sufficient to promote growth and proliferation. The striking observations that loss of *jag* and *nub* function ablates growth in the distal regions of organs and that overexpression leads to the formation of ectopic organs suggest that both genes mediate pattern-specific cell proliferation, rather than controlling general proliferative ability.

This view is supported by the characterization of the novel developmental regulators *BLADE ON PETIOLE1* (*BOP1*) and *BOP2*, encoding proteins with BTB/POZ domains and ankyrin repeats. Loss-of-function mutants for *bop1* show formation of ectopic lamina tissue on the petiole, together with increased expression of class I *KNOTTED*-like homeobox genes and *LATERAL ORGAN BOUNDARIES* genes (Ha *et al.*, 2003, 2004, 2007). In *bop1 bop2* double mutants, the leaf overgrowth including the ectopic outgrowth of lamina tissue along the petiole is enhanced (Norberg *et al.*, 2005). Also, *bop1 bop2* double mutants form ectopic bracts. The similarities in the phenotype of *bop1 bop2* mutants and *JAG*-overexpressing plants are also seen at the molecular level. *bop1 bop2* plants show strong ectopic expression of *JAG* and *NUB*, suggesting that *BOP* genes restrict tissue growth in a region-specific manner by repressing *JAG* and *NUB* expression (Norberg *et al.*, 2005). Overexpression of *BOP* genes strongly reduces organ size and causes phenotypes that resemble double mutants of the class I *KNOX* gene *BREVIPEDICELLUS* and the BELL-like homeobox gene *BELLRINGER* (Ha *et al.*, 2007). Thus, *BOP* genes appear to regulate leaf growth by repressing *JAG*, *NUB*, and *KNOX* gene expression.

The recently identified RAV-B3 type transcription factors NGATHA (NGA) present another likely example for a link between organ patterning and cell proliferation control. *Arabidopsis* contains four *NGA* genes, which redundantly are required for style formation and gynoecia development (Alvarez *et al.*, 2009; Trigueros *et al.*, 2009). The style of *nga* quadruple mutants is replaced by valve-like projections, suggesting that *NGA* genes are involved in apical-basal patterning of the gynoecia. Furthermore, *nga* mutants exhibit increased organ growth mainly caused by increased cell proliferation, indicating that at least part of *NGA* function is to suppress cell proliferation (Alvarez *et al.*, 2009; Kwon *et al.*, 2009; Trigueros *et al.*, 2009).

Most of the factors described above appear to act as transcriptional regulators, yet our understanding of their critical targets in controlling tissue and organ growth remains incomplete. Thus, unraveling how these transcription factors are connected to the growth machinery of the cell will be an important goal for future studies in the field.

3.4. Beyond mRNA: protein synthesis, protein turnover and size

Ribosomes are the cellular factories underlying cytoplasmic growth. For cells to maintain a constant size while proliferating, cytoplasmic growth, i.e., the synthesis of new proteins and other macromolecules, and cell division have to be precisely coordinated. Indeed, in yeast and animal systems it is well established that ribosome biogenesis and translational control play an important role in regulating growth (reviewed in Cook and Tyers, 2007). Biogenesis of ribosomes occurs mainly in the nucleolus. In animals the ErbB-3 epidermal growth factor receptor-binding protein EBP1 is located in the nucleolus as part of a pre-ribosomal nucleoprotein complex and has been suggested to provide a potential link between ribosome biogenesis and proliferation control (Squatrito *et al.*, 2004). In addition, EBP1 also associates with mature ribosomes to sustain protein translation during stressful conditions. EBP1 is conserved in plants and influences growth, suggesting that also in plants ribosome biogenesis and translational control regulate growth. Decreased levels of *EBP1* expression reduce organ growth by limiting cell number as well as cell size, whereas increased levels of *EBP1* activity lead to larger organs (Horvath *et al.* 2006). These enlarged organs are mainly due to increased cell numbers, suggesting that *EBP1* functions primarily to promote growth by proliferation, possibly via stimulation of ribosome biogenesis and protein translation. Another possible link that could coordinate the fundamental processes of cellular growth and cell division is suggested by the analysis of the class I TCP protein TCP20 (see below for a more detailed discussion of TCP proteins). TCP20 binds to the GCCCR element in the promoters of genes encoding ribosomal proteins and the mitotic *CYCB1;1* gene to activate their expression (Li *et al.*, 2005). Thus, *TCP20* activity could coordinately regulate the expression of critical factors for cytoplasmic growth and cell division. However, as will be discussed below, ribosome biogenesis appears not just to influence cell proliferation. Some of the conserved pathways between animals, yeast, and plants are also linked to cell expansion (discussed below).

The stability of a protein is an important factor influencing its *in vivo* function. One key mechanism that controls protein stability is the ubiquitin-mediated control of protein degradation. Proteins marked with polyubiquitin chains by ubiquitin ligases are designated for degradation by the 26S proteasome (Vierstra, 2009). The transfer of ubiquitin to target proteins is catalyzed by an enzymatic cascade of three activities, termed E1, E2, and E3, with specificity for the targets generally being determined by the E3 ubiquitin ligases. The ubiquitin–proteasome pathway is central in regulating many aspects of plant development (Vierstra, 2009), including the control of organ size. This is demonstrated, for example, by the E3 ubiquitin ligase BIG BROTHER (BB), which acts as a central negative growth regulator (Disch *et al.*, 2006). Plants lacking *BB* activity show a dramatic increase in final organ size, mainly in flowers and the stem, whereas overexpression leads to a severe reduction in growth. While cell sizes are relatively unaffected by *BB* mRNA levels, the duration of the proliferation phase and thus final cell number within an organ depend strictly on the level of *BB* expression. The accumulation of BB protein closely follows cell proliferation activity in a growing organ, suggesting that *BB* acts as an intrinsic growth brake to prevent overproliferation by targeting positive growth stimulators for protein degradation.

The recently described *DA1* gene also limits organ growth in *Arabidopsis* (Li *et al.*, 2008). The DA1 protein contains two putative ubiquitin interaction motifs, typical for ubiquitin receptors. *da1* mutations strongly enhance *bb* mutant phenotypes, suggesting a synergistic function of both factors. Thus, it is conceivable that DA1, like *BB*, acts in the ubiquitin–proteasome pathway. Similar to *BB*, *DA1* activity limits the duration of the cell proliferation phase, yet by itself overexpression of *DA1* does not seem sufficient to reduce growth below wild-type levels. Taken together, BB and DA1 are two examples of proteins with putative functions in controlling the stability of positive growth regulators, acting to determine the duration of proliferative growth.

3.5. Long-range growth control

Given their presumed molecular functions, it is plausible to assume that all of the pathways described above affect organ size in a cell-autonomous manner. However, for an organ to reach its appropriate final size and shape, the constituent cells must stop growing in a coordinated manner at the right place and time. Although in theory cell-autonomous timers could achieve this, such a strategy would seem prone to deviations from the target shape and size because of random fluctuations in the performance of the independent timers. An alternative would be to use a system that can coordinate cell behavior by cell–cell signaling. The cytochrome P450 KLUH (KLU)/CYP78A5 appears to be involved in such a coordinating system (Anastasiou *et al.*, 2007; Eriksson *et al.*, 2010; Wang *et al.*, 2008). Mutants for *klu* form smaller organs due to a premature arrest of cell proliferation, while increased *KLU* expression leads to the formation of larger organs with more cells, indicating that KLU is necessary and sufficient for cell proliferation. *KLU* shows asymmetric redundancy with another member of the *CYP78A* subfamily, *CYP78A7*; while *cyp78a7* single mutants do not show a phenotype, *klu cyp78a7* double mutants are severely dwarfed and often even embryo lethal (Wang *et al.*, 2008). Unlike other tested growth-regulating factors, the region of *KLU* expression and the domain of highest proliferation in outgrowing primordia are separated: while young primordia still show high levels of *KLU* expression throughout the organ, in later stages *KLU* expression becomes restricted to the margins, around the entire periphery in petals and mainly at the base in leaves. Cell proliferation, however, occurs in much larger domains of the organs, suggesting that *KLU* acts non-cell-autonomously to maintain cell proliferation. The transcriptional response to induced *KLU* activity and physiological experiments indicate that KLU is not involved in the synthesis or degradation of a classical plant hormone, suggesting that its non-cell-autonomous action is mediated by a novel mobile growth regulator (Anastasiou *et al.*, 2007). The range of action of this postulated growth signal has recently been analyzed using chimeric plants consisting of wild-type and *klu* mutant tissue (Eriksson *et al.*, 2010). Within an inflorescence, organ size was largely independent of the genotype of the individual organ, but correlated with the overall amount of *KLU* wild-type tissue in the inflorescence, suggesting that the level of the presumed signal is integrated across an inflorescence. Thus, rather than being used to measure the size of individual growing primordia, as suggested earlier (Anastasiou *et al.*, 2007), the developmental function of *KLU* appears to coordinate growth at a higher level, i.e., growth of the individual organs within a flower and potentially even of individual flowers within an inflorescence, in order to ensure uniformity of organ size (Eriksson *et al.*, 2010).

4. Growth by Cell Elongation

The arrest of proliferation is followed by a second growth phase in which the organ grows by cell expansion, driven by loosening of the cell walls and uptake of water into the vacuole. Compared to the relatively large set of factors controlling proliferation, less is known about pathways regulating cell elongation (Fig. 7.3, Table 7.1).

4.1. Cell biology of cell expansion—the role of expansins

Cell expansion requires the synthesis of new cell wall material and controlled loosening of the wall to allow it to stretch and increase in area. Cell-wall-associated proteins of the so-called expansin family are key components in this process. Expansins increase wall extensibility, presumably by breaking noncovalent bonds between wall polysaccharides and allowing them to slide relative to each other (Cosgrove, 2005). Expansin activity is required for normal cell expansion during organ growth, as shown by downregulation of expansin gene activity in different species (Cho and Cosgrove, 2000; Choi *et al.*, 2003; Zenoni *et al.*, 2004). Conversely, overexpression of expansins leads to increased organs consisting of larger cells, suggesting that expansin activity is limiting for growth by expansion in wild-type organs (Cho and Cosgrove, 2000; Choi *et al.*, 2003). However, the function of expansins appears not to be strictly limited to the phase of post-mitotic growth by expansion. When increased expansin expression was locally induced in early leaf primordia growing by cell proliferation, excess lamina outgrowth was observed at the sites of induction in mature leaves (Pien *et al.*, 2001; Wyrzykowska *et al.*, 2002). This excess tissue was reported to consist of normally sized cells, suggesting that increased wall extensibility in early stages of organ development induced increased growth by cell proliferation. Expansin activity appears to be tightly controlled in wild-type organs, as suggested by experiments using a single pulse of induced expansin overexpression (Sloan *et al.*, 2009). Only leaves toward the end of the proliferative phase showed overgrowth, while younger or older leaves were not responsive; also, increased *in vitro* expansin activity was only extractable from leaves induced toward the end of the proliferative phase, suggesting that endogenous expansin activity may be controlled by a stage-specifically expressed modulator.

4.2. Linking elongation and identity: BIGPETAL

All plant organs grow to a specific size and shape determined by their identity. In fact, next to color and patterns of cell differentiation, shape and size are the major criteria for assigning organ identity. This raises the question how the systems that determine organ identity modulate local growth patterns. In general, our knowledge about this fundamental problem remains rudimentary, yet fascinating insights have recently been obtained concerning the control of cell expansion specifically in petals. The *BIGPETAL* (*BPE*) gene encodes a bHLH transcription factor with two different mRNAs generated by alternative splicing (Szecsi *et al.*, 2006). One transcript, termed *BPEub*, is ubiquitously expressed in all tested organs; the other, termed *BPEp*, is only found in petals. Decreasing the levels of *BPEp* mRNA by RNAi or T-DNA insertions leads to the enlargement of petals due to increased cell expansion. The alternative splicing event that generates the *BPEp* mRNA is stimulated by the class B homeotic proteins PISTILLATA and APETALA3. Thus, depending on the identity of the organ as a petal, a specific splice form of a growth repressor is generated that limits cell expansion.

The proteins encoded by the *BPEp* and *BPEub* mRNAs belong to subfamily XII of the large bHLH transcription factor family (Toledo-Ortiz *et al.*, 2003). Members of this subfamily are involved in jasmonate (JA)-signaling, suggesting a link between JA and *BPEp* function (Brioudes *et al.*, 2009). Indeed, plants impaired in JA synthesis due to a mutation in 12-oxophytodienoic acid reductase (*OPR3*) show a specific downregulation of the splice event that generates the *BPEp* mRNA, which can be rescued by JA application. Like *bpe* mutants, *opr3* plants show a significant increase in petal size due to larger cells; however, overexpression of *BPEp* in an *opr3* mutant background can not rescue the phenotype. This suggests that besides *BPEp* other components act in parallel to limit cell elongation in petals downstream of JA. It will be interesting to determine in more detail the relationship between floral patterning mediated by organ-identity factors, growth control, and JA signaling.

4.3. Ribosomes and cell expansion

Research in yeast has shown that cell growth is strongly controlled by ribosome biogenesis (reviewed by Cook and Tyers, 2007; discussed above). One important regulator of ribosome biogenesis in yeast is the TARGET OF RAPAMYCIN (TOR) kinase. At least part of the TOR signaling pathway is conserved and plays an important role in plants; a *TOR* ortholog is found in *Arabidopsis*, whose mutation leads to embryo lethality (Menand *et al.*, 2002). More limited downregulation of *TOR* function using RNAi and overexpression of *TOR* demonstrated that *TOR* activity promotes cell expansion. The changes in *TOR* activity correlated with altered levels of translationally active polyribosomes and of *EBP1* mRNA expression, suggesting that *TOR* promotes cell growth by stimulating translation and/or ribosome biogenesis (Deprost *et al.*, 2007). Whether the effects on cell expansion indicate that ongoing ribosome biogenesis is required after the arrest of cell proliferation for cells to expand to their normal size or whether a smaller amount of ribosomes and possibly cytoplasm at the time of proliferation arrest limits the future potential for expansion will be an important question for future studies.

4.4. Endoreduplication

Bypassing mitosis in the cell cycle leads to an increase in DNA content in the cell (Sugimoto-Shirasu and Roberts, 2003). This process is called endoreduplication and a widespread phenomenon in animals and plants. Cell size shows a tight positive correlation with DNA content. Although this basic correlation has been known for decades, its mechanistic basis remains largely unknown.

Imbalanced cell cycle gene expression can induce endoreduplication in plants. The 19S proteasome subunit AtRPT2a seems to be involved in the regulation of cell cycle genes (Sonoda *et al.*, 2009). Cells in *atrpt2a* mutants expand to larger sizes than wild-type cells, leading to plants with larger organs. The excess cell expansion is correlated with higher endoreduplication levels, and several G1- and S-phase specific genes are overexpressed in *atrpt2a* mutants. The *Arabidopsis FIZZY-RELATED* (*FZR*) genes have opposite effects on endoreduplication (Larson-Rabin *et al.*, 2009). *Arabidopsis* has three *FZR* genes and loss-of-function mutations of *fzr2* lead to a strong reduction in endoreduplication, while *FZR2* overexpression causes increased endoreduplication. *Drosophila* Fizzy protein triggers degradation of A- and B-type cyclins (Sigrist and Lehner, 1997; Sigrist *et al.*, 1995;), suggesting by analogy that imbalanced activity of cell cycle regulators causes the effects on endoreduplication from changes in *FZR2* activity. Notably, overall organ sizes were normal in *fzr2* mutants and weak overexpression lines, with the changes in cell size offset by changes in cell number (discussed below).

Plants impaired in the topoisomerase VI complex have strongly reduced levels of endoreduplication and exhibit a dwarf phenotype with strongly reduced cell sizes (Breuer *et al.*, 2007; Sugimoto-Shirasu *et al.*, 2002, 2005). Ploidy levels in any mutants of the topoisomerase VI complex never exceed 8C, suggesting that higher rounds of endoreduplication require decatenation of DNA by this enzymatic activity. Interestingly, callus growth is not impaired by loss of topoisomerase VI function, suggesting that structural DNA integrity during cell proliferation is controlled by different topoisomerase(s).

5. Compensation

Although we have discussed the two processes of proliferation and expansion separately above, this is a major simplification. Within the same leaf primordium, cells in one region may expand, while they are still proliferating in another region. More importantly, cell expansion does not appear to be governed by a strictly cell-autonomous program (John and Qi, 2008), and the extent of cell expansion is in many cases negatively correlated with cell number in the primordium, a phenomenon called compensation (Tsukaya, 2006; Fig. 7.4). In numerous mutants or transgenic lines with reduced cell numbers, cell sizes are strongly increased, thus partially making up for the shortfall in cell number at the level of overall organ size (e.g., Bemis and Torii, 2007; Hemerly *et al.*, 1995; Horiguchi *et al.*, 2005; Kim and Kende, 2004; Mizukami and Fischer, 2000; Wang *et al.*, 2000). Similarly, increased expression of the cyclin *CYCD2;1* leads to an accelerated G1-/S-phase transition, resulting in the formation of more smaller cells per unit tissue in young leaf primordia; however, overall leaf size is not increased, as the cells expand to a smaller final size than in wild type (Qi and John, 2007). Also, as mentioned above, reduced or increased endoreduplication resulting from changes in *FZR2* activity causes corresponding changes in final cell size, yet not in overall leaf size because of compensatory changes in cell number (Larson-Rabin *et al.*, 2009).

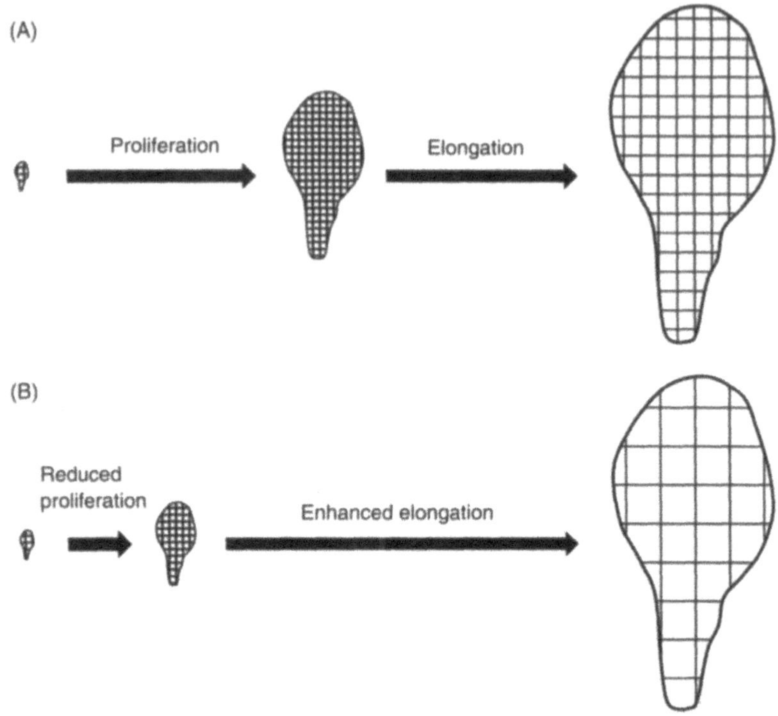

Figure 7.4 Schematic representation of the compensation syndrome. (A) In normal plant development organs grow initially mainly by cell proliferation. This phase is followed by elongation growth until the final size of the organ is reached. (B) Mutants impaired in cell proliferation often show growth compensation. Here the reduced phase of cell proliferation is compensated for by an enhanced cell elongation, leading to organs of similar size with fewer, but larger cells.

Despite the many examples documenting the phenomenon, our mechanistic understanding of compensation remains limited. In particular, the question of whether it involves active sensing of cell numbers and/or primordium size, coupled to signaling in order to modify cell expansion, is not fully answered. An attractive explanation that would not involve active sensing and signaling could be that growth of the organ as a whole is regulated according to a parameter that is at least partly independent of cell numbers, such as total cytoplasmic mass or total tissue volume (Potter and Xu, 2001). In this view, if the primordium grew to a final size determined by the target value of this cell-number-independent parameter, a change in the timing of when cells stop proliferating and begin to expand would automatically cause an opposite change in final cell sizes. This notion would explain the described examples of "perfect compensation" (*CYCD2;1*-overexpressing plants; *fzr2* mutants and *FZR2*-overexpressing plants and others), and it would imply that factors that can promote growth of organs to sizes larger than wild type ultimately increase the target value of the cell-number-independent parameter (although it would not explain why in some cases the larger organs contain more, and in other cases larger cells). At the same time, the cases of "imperfect compensation" where leaves stay smaller because of a strong reduction in cell numbers despite compensated cell enlargement would suggest that a minimal cell number has to be attained for the organ to reach its normal final size, possibly because cell expansion can only be increased up to a certain limit.

Despite these uncertainties, progress has recently been made in identifying factors that are required for compensated cell enlargement. *Arabidopsis* mutants of the *extra-small sisters* (*xs*) class reduce final size of leaf cells without affecting cell numbers (Fujikura *et al.*, 2007). When combined with a mutation that on its own causes compensated cell enlargement, some of the *xs* mutants were able to suppress the excess cell expansion, suggesting that pathways required for normal cell expansion in leaves are hyperactivated during compensation. Measurements of ploidy levels indicate that compensated cell enlargement can occur without additional endoreduplication, suggesting that it is not merely the result of "skipped" cell divisions (Ferjani *et al.*, 2007; Fujikura *et al.*, 2007). Contradicting the hypothesis described above, it has been argued that cell number in a leaf needs to fall below a threshold value for compensation to occur (Fujikura *et al.*, 2009). While individual mutations in ribosomal subunit proteins only moderately reduce leaf cell number, but do not affect cell size, double mutants with more strongly reduced cell numbers show compensated cell enlargement, and similar results were obtained with different levels of reduction in *AN3* function (Fujikura *et al.*, 2009). A further genetic and molecular dissection of compensation will be required to fully understand this fascinating, but still enigmatic aspect of plant growth control.

6. Coordination of Growth Across and Within Tissue Layers

Organ growth is a multicellular process that as such critically relies on the coordinated behavior of the constituent cells. In plants, this coordination acts on two different levels, *between* tissue layers made up of largely clonally distinct cells and *within* the individual layers.

All the cells in aboveground organs of plants are ultimately derived from a small number of long-term stem cells located in the central zone of the shoot apical meristem (chapter 4; Rieu and Laux, 2009). In a typical dicotyledonous meristem, these stem cells are arranged in three tiers; the so-called L1 and L2 layers form the epidermis and the subepidermal palisade parenchyma of leaves, and an internal corpus of tissue, called L3, gives rise to the pith of the shoot and the spongy mesophyll mainly in the center of the leaf (Marcotrigiano, 2001). Although these layers remain largely clonally distinct, the behavior of the cells is not governed by their lineage, but rather by their position; cells that become displaced, for example, from the L1 into the L2 will behave fully in accordance with their new position (Marcotrigiano, 2001). Also, the contribution of L2- and L3-derived tissue to the final leaf is highly variable within and across leaves (Marcotrigiano, 2001).

There has been considerable debate about the relative roles of the different tissue layers in controlling the growth of lateral organs (reviewed in Marcotrigiano, 2001; Savaldi-Goldstein and Chory, 2008). On theoretical grounds, a role of the epidermis as a physical restraint on tissue growth appears attractive; epidermal cells are under strongly polarized outward pressure due to the turgor pressure of the internal tissue, suggesting that controlled yielding of the epidermal cells to this pressure could regulate overall tissue growth. Such a biophysical model has been championed for the control of organ initiation at the shoot meristem (Green, 1999); the organ outgrowth induced by local cell wall loosening has lent experimental support (Fleming *et al.*, 1997; Pien *et al.*, 2001), although the situation is clearly far more complex, involving polar auxin transport and patterned gene expression (Kuhlemeier, 2007). In leaves, the question of layer contribution to growth control and cell autonomy of growth-related mutations has been mainly addressed by layer-specific expression of transgenes and generation of periclinal chimeras, i.e., plants in which the genotype of one of the three layers differs from that of the others (Marcotrigiano, 2001). Studies that specifically limited or enhanced growth in the L1 support the concept of the epidermis as a physical barrier to growth (Bemis and Torii, 2007; Savaldi-Goldstein *et al.*, 2007). To what extent these effects are due to mere physical limitation by an epidermis that cannot grow as much as the internal tissue or whether they also involve active inter-layer signaling is an important problem for future research. However, there are also a number of cases where growth of the epidermis appears to follow a pattern dictated by the internal tissue (Bai *et al.*, 2009; McHale and Marcotrigiano, 1998), indicating that the L1 does not merely act as a physical barrier to growth. Rather, inter-layer communication seems likely to be involved, possibly through dedicated signaling pathways. In this view, the different results obtained with the different chimeras would reflect where in the presumed process of communication the individual genetic alterations act. However, clearly more work is required to untangle this problem of inter-layer coordination of growth.

Growth coordination is also required *within* tissue layers in at least two respects, i.e., between different regions of an organ and between different cell types in an organ. Achieving a flat organ requires that growth proceeds and arrests evenly across interconnected regions; otherwise buckling and distortions will occur. A case in point are mutants affecting transcription factors of class II of the TCP family, named after the founding factors *T*EOSINTE BRANCHED 1, *C*YCLOIDEA, and *P*CF (Cubas *et al.*, 1999). *Antirrhinum* plants defective in the TCP transcription factor CINCINNATA (CIN) form curly and wavy leaves, with an overall increase in leaf area (Nath *et al.*, 2003). Developmentally, this phenotype is due to a belated arrest of cell proliferation along the margin of the leaf relative to the central region. The resulting overgrowth of the marginal region leads to a relative excess of tissue there compared to the center, which is accommodated by buckling of the tissue out of the plane of the leaf. The same phenotype is seen in *Arabidopsis* plants overexpressing the *JAW* microRNA (miRNA319a), which downregulates several members of the *CIN*-family of *TCP* genes (*CIN-TCPs*), and even more extremely upon overexpression of both *JAW* and an artificial miRNA (amiRNA) targeting additional *CIN-TCPs* (Efroni *et al.*, 2008; Palatnik *et al.*, 2003). In contrast to the dramatic effects of downregulation of *CIN-TCP* genes throughout leaf development, more subtle changes in their activity levels only during the very early proliferative phase lead to strongly enlarged, but flat leaves. By contrast, constitutive upregulation of *TCP* genes in a loss-of-function mutant for miR319a leads to reduced floral organ growth (Nag *et al.*, 2009). Together with the results of detailed molecular and phenotypic analysis of *cin-tcp* mutant leaves, these data indicate that the primary target of *CIN-TCPs* is the promotion of cellular differentiation. This supports the notion that these transcription factors control an endogenous maturation program in leaves (Efroni *et al.*, 2008). How could this be achieved on a molecular level? Plants carrying a miRNA resistant *TCP4* gene show elevated levels of miR396 expression and downregulation of several *GRF* genes. As mentioned before, *GRFs* are important stimulators of cell proliferation. Thus, the *TCP4*-induced downregulation of *GRF* activity represents a convincing mechanism by which plants control the extent of cell proliferation during organ maturation (Rodriguez *et al.*, 2010).

The role of class II TCP family members in temporally controlling the proliferative capacity and differentiation status of the marginal region in leaves appears to be widely conserved, also in species with compound leaves (Ori *et al.*, 2007). In tomato, gain-of-function mutations in the *LANCEOLATE* gene that disrupt the miRNA regulation of its mRNA and lead to overexpression of the encoded TCP protein cause a lack of lateral leaflets, thus converting the normally compound tomato leaf into a simple one (Ori *et al.*, 2007). This change is due to a premature loss of proliferative and morphogenetic potential at the margins of the leaf (see chapter 6). Conversely, as in *Arabidopsis* and *Antirrhinum*, downregulation of *TCP* genes by *miR319a* overexpression leads to larger leaves with overgrown, curvy margins. Besides their role during early leaf growth, miR319-regulated *TCP* genes have recently also been shown to promote JA-biosynthesis and leaf senescence during late stages of leaf development (Schommer *et al.*, 2008).

The need for coordinating the growth behavior of different cell types within an organ is underlined by the *peopod* (*ppd*) mutant of *Arabidopsis* (White, 2006). The *ppd* locus comprises two tandemly duplicated, redundant genes encoding nuclear DNA-binding proteins. Loss of *ppd* function leads to larger, bell-shaped leaves, due to overgrowth of the central leaf lamina relative to the margins. This overgrowth appears to be due to a delayed proliferation arrest specifically in cell types that also in the wild type proliferate for longer than most of the mesophyll and epidermal cells in the leaf lamina, such as meristemoids in the stomatal lineage or procambial cells in the developing vasculature. Thus, it has been suggested that there are two successive waves of proliferation arrest in leaves: a primary arrest front acting on general cell proliferation in the primordium, whose shape is regulated by the class II TCPs discussed above, and a secondary arrest front acting on meristemoids and procambial cells, which requires *PPD* function for its timely progression (White, 2006).

Considering the presumably quite different biophysical properties of vascular and mesophyll cells, it is likely that growth of the leaf vasculature and the intervening lamina tissue also has to be tightly coordinated to ensure a flat surface. In early stages of leaf development, when the vascular elements are being patterned, the successive, polar auxin transport-dependent establishment of new veins in unvascularized regions of the lamina provides a way of coupling the overall increase in primordium size to the formation of more vascular strands (chapter 8; Scarpella *et al.*, 2006). However, how the length of established veins is matched to the size of the lamina during later stages of leaf growth is currently not well understood. The importance of this aspect of growth coordination is shown by plants with reduced expansin activity in the developing midvein (Cho and Cosgrove, 2000). The resulting leaves are curled and twisted to accommodate the relative excess of lamina tissue compared to the shortened midvein.

7. Modulation of Growth by Environmental Signals

Plants are sessile organisms and adapt their growth patterns according to environmental cues. Several examples of how the environment modifies organ growth have been intensively studied, including salt stress, photomorphogenesis, and shade avoidance (Achard *et al.*, 2006; Franklin, 2008; Nemhauser, 2008). Here, we focus on one pathway that appears to integrate various environmental stimuli to modify plant growth, represented by the DELLA transcription factors of the GRAS-domain superfamily (reviewed by Schwechheimer and Willige, 2009). DELLA proteins repress growth; the phytohormone gibberellin (GA) prevents this repression by inducing the degradation of DELLA proteins (reviewed by Schwechheimer and Willige, 2009). The stability of DELLA proteins is also influenced by ethylene, acting antagonistically to GA to stabilize DELLA proteins (Achard *et al.*, 2003). Auxin, working in concert with GA, destabilizes them, highlighting the integrative function of DELLA proteins in hormone responses (Fu and Harberd, 2003). Although best studied for their role in controlling cell expansion, DELLA proteins and GA have recently also been found to influence cell proliferation. GA-insensitive mutants have fewer cells in leaf primordia and in roots that divide for a longer period and do not elongate, leading to a smaller final organ size, whereas plants mutated in four of the five DELLA proteins, i.e., mimicking constitutive GA activation, have more cells that divide for a shorter period, leading overall to larger organs (Achard *et al.*, 2009; Ubeda-Tomas *et al.*, 2009). GA-deficient plants show altered levels of cyclin-dependent kinase inhibitors, suggesting that GA signaling directly regulates the cell cycle, which is further supported by the phenotypic rescue of the GA-deficient mutant *ga1-3* by the overexpression of *CYCD3;1* (Achard *et al.*, 2009). Thus, DELLA regulation and the GA-signaling pathway represent an elegant example of how proliferation and cell elongation can be regulated through a complex interaction of different growth control pathways on a common molecular target.

How DELLA proteins act molecularly has recently been elucidated. DELLA proteins interact with and inhibit transcriptional regulation by members of the PHYTOCHROM INTERACTING FACTOR (PIF) subfamily of basic helix–loop–helix (bHLH) transcription factors (chapter 2; de Lucas *et al.*, 2008; Feng *et al.*, 2008). This provides an explanation for how DELLA-mediated transcriptional control is linked to environmental input via light signaling and shows how GA-induced cell elongation is effected by the release of PIFs. With high GA levels and in the absence of light, PIFs are able to bind to DNA and activate transcription, leading to elongated hypocotyls due to increased cell expansion. This GA-mediated cell elongation is overridden by light signaling due to the rapid degradation of PIFs after phytochrome B activation by light. This example highlights the complex interplay between phytohormone and light signaling in controlling elongation growth. It also shows how plants efficiently integrate independent environmental and developmental clues on one central regulatory pathway in order to assess the most favorable growth pattern.

8. Modeling Plant Growth

The growth of organs to their final sizes and shapes is a dynamic four-dimensional process. Different regions within an organ can grow at different rates that change over time, and the resulting complexity quickly exceeds what can be grasped by intuition alone. In particular, the connectedness between regions growing with different rates or different main directions tends to produce effects on shape that are very difficult to predict without computational modeling of the growing structure. Basic examples include the different forms of buckling that result when regions of varying size in the center or around the periphery of a structure grow at a different rate from the remainder (discussed above). By the same token, it can be difficult to predict from observing the final shape or size of a mutant organ how exactly the affected gene influences growth rates, growth directions, and/or the duration of growth in different regions of an organ. Thus, a full understanding of how a given gene exerts its effect on organ size or shape ideally requires detailed, spatially and temporally resolved measurements of growth and assessment of the resulting hypotheses using mathematical modeling. A framework for how this can be achieved has recently been described with particular reference to lateral organs (Coen *et al.*, 2004) or the shoot apical meristem (Kwiatkowska, 2004).

Growth of an organ can be described quantitatively by determining four basic parameters for each of its component regions: growth rate, anisotropy, the main direction of growth, and rotation relative to other regions (Coen *et al.*, 2004). These parameters can be measured most directly by determining the displacement of landmarks over time, using either natural features, such as trichomes or veins, or marks applied externally, e.g., ink dots (Avery, 1933; Erickson, 1976; Granier and Tardieu, 1998; Wiese *et al.*, 2007). Methods for tracking have largely been restricted to relatively late stages of organ growth, when organs have already attained a minimum size that makes them accessible for manipulation and imaging. Especially in cases like *Arabidopsis*, this largely restricts the approach to following growth by cell expansion. Also, direct tracking is not readily applicable to organs that are buried within larger structures, such as petals. Extending current methods in three-dimensional imaging at different spatial scales (Lee *et al.*, 2006; Truernit *et al.*, 2008) to the dynamic observation of live organs should help to overcome these limitations. An alternative to tracking is provided by clonal analysis, in which individual cells are stably marked and the location of their progeny is determined at later stages of growth (Rolland-Lagan *et al.*, 2003; Szymkowiak and Sussex, 1996). While the size and shape of the resulting marked clones give information about the growth rate and anisotropy, the direction of growth with respect to the main axes of the organ and the rotation of this direction cannot be directly inferred from clones resulting from only one time point of marking. These two parameters are, however, accessible by marking cells at several successive time points, and then using the *difference* between the clones resulting from two successive marking events to estimate growth parameters in the intervening period between when the cells were marked (Rolland-Lagan *et al.*, 2003). Once growth parameters and their variation in different genetic backgrounds have been determined by either of the strategies described above, they can be used in computational models of a growing organ to determine whether the observed differences in growth parameters can adequately explain the changes in final size and shape. As a result, gene activities can be incorporated in the models as locally modifying growth parameters. The different strategies for modeling plant growth have been reviewed elsewhere (Coen *et al.*, 2004; Erickson, 1976; Grieneisen and Scheres, 2009; Prusinkiewicz and Lindenmayer, 1996).

The general strategy outlined above has been applied to different aspects of plant growth, e.g., organ initiation at the shoot apical meristem and growth of floral organs. Studies of the shoot meristem have mainly used tracking of cells by either confocal microscopy or resin casts (Grandjean *et al.*, 2004; Kwiatkowska and Dumais, 2003; Reddy *et al.*, 2004) to determine rates of growth and cell division in different regions of the meristem. Information about different growth rates has been incorporated into mathematical models of auxin transport-driven organ initiation (Smith *et al.*, 2006). The question of how dorsal petals in *Antirrhinum* flowers acquire their characteristic asymmetric shape has been addressed using marked clones (Rolland-Lagan *et al.*, 2003). Clones were induced at different time points to estimate growth parameters during the periods between inductions in a spatially resolved manner. The estimated parameter values for different regions of the growing petal were then incorporated into a spring-based growth model. Surprisingly, the main factor governing the final asymmetric shape of the dorsal petal lobe was the rotation of the lobe relative to the basal petal tube through roughly 45°. This rotation of the lobe as a whole is accompanied by a complementary rotation of the main direction of anisotropic growth in the different regions of the lobe. Differences in local growth rates or the extent of anisotropy did not seem to contribute significantly to the establishment of the final shape. Thus, it was proposed that a long-range polarizing signal aligns local growth directions with the proximo-distal axis of the petal as a whole (Rolland-Lagan *et al.*, 2003).

More progress in this area can be expected in the future, as the number of characterized factors, the sophistication of imaging and quantifying growth, and the power of mathematical models continue to increase.

9. Concluding Remarks

Our knowledge about the genetic regulation of plant organ growth has come a long way in the last years. However, despite this impressive progress many basic questions remain unanswered. How are the activities of the identified factors orchestrated, and is there a unifying "size control network" or merely a number of independent gene activities that all impinge on growth as one of the most integrating traits? What is the functional relationship between the cellular activities of proliferation and expansion and any regulatory systems above the level of individual cells? How do organ primordia keep track of their size? How does organ identity modify local growth patterns to determine the appropriate final size and shape? Answering these questions will require not only the identification of additional factors acting together with the ones already described, but also the development of more sophisticated approaches for manipulating gene activities only in parts of an organ and for quantifying the resulting effects on growth; more powerful imaging and image processing technology; and the integration of quantitative modeling as an essential aspect of studying growth. Although these are no small challenges, answering some of the major questions about growth control promises to be highly rewarding both from an intellectual and an applied point of view.

REFERENCES

Achard, P., Cheng, H., De Grauwe, L., Decat, J., Schoutteten, H., Moritz, T., Van Der Straeten, D., Peng, J., and Harberd, N. P. (2006). Integration of plant responses to environmentally activated phytohormonal signals. *Science* **311**, 91–94.

Achard, P., Gusti, A., Cheminant, S., Alioua, M., Dhondt, S., Coppens, F., Beemster, G. T., and Genschik, P. (2009). Gibberellin signaling controls cell proliferation rate in Arabidopsis. *Curr Biol.* **19**, 1188–1193.

Achard, P., Vriezen, W. H., Van Der Straeten, D., and Harberd, N. P. (2003). Ethylene regulates Arabidopsis development via the modulation of DELLA protein growth repressor function. *Plant Cell.* **15**, 2816–2825.

Alvarez, J. P., Goldshmidt, A., Efroni, I., Bowman, J. L., and Eshed, Y. (2009). The *NGATHA* distal organ development genes are essential for style specification in Arabidopsis. *Plant Cell.* **21**, 1373–1393.

Anastasiou, E., Kenz, S., Gerstung, M., MacLean, D., Timmer, J., Fleck, C., and Lenhard, M. (2007). Control of plant organ size by KLUH/CYP78A5-dependent intercellular signalling. *Dev Cell.* **13**, 843–856.

Avery, G. S. (1933). Structure and Development of the Tobacco Leaf. *Am J Bot.* **20**, 565–592.

Bai, Y., Falk, S., Schnittger, A., Jakoby, M. J., and Hulskamp, M. (2009). Tissue layer specific regulation of leaf length and width in Arabidopsis as revealed by the cell autonomous action of ANGUSTIFOLIA. *Plant J.* **61**, 191–199.

Beemster, G. T., De Veylder, L., Vercruysse, S., West, G., Rombaut, D., Van Hummelen, P., Galichet, A., Gruissem, W., Inze, D., and Vuylsteke, M. (2005). Genome-wide analysis of gene expression profiles associated with cell cycle transitions in growing organs of Arabidopsis. *Plant Physiol.* **138**, 734–743.

Bemis, S. M., and Torii, K. U. (2007). Autonomy of cell proliferation and developmental programs during Arabidopsis aboveground organ morphogenesis. *Dev Biol.* **304**, 367–381.

Bossinger, G., and Smyth, D. R. (1996). Initiation patterns of flower and floral organ development in *Arabidopsis thaliana*. *Development* **122**, 1093–1102.

Breuer, C., Stacey, N. J., West, C. E., Zhao, Y., Chory, J., Tsukaya, H., Azumi, Y., Maxwell, A., Roberts, K., and Sugimoto-Shirasu, K. (2007). BIN4, a novel component of the plant DNA topoisomerase VI complex, is required for endoreduplication in Arabidopsis. *Plant Cell.* **19**, 3655–3668.

Brioudes, F., Joly, C., Szecsi, J., Varaud, E., Leroux, J., Bellvert, F., Bertrand, C., and Bendahmane, M. (2009). Jasmonate controls late development stages of petal growth in *Arabidopsis thaliana*. *Plant J.* **60**, 1070–1080.

Cho, H. T., and Cosgrove, D. J. (2000). Altered expression of expansin modulates leaf growth and pedicel abscission in *Arabidopsis thaliana*. *Proc Natl Acad Sci U S A.* **97**, 9783–9788.

Choi, D., Lee, Y., Cho, H. T., and Kende, H. (2003). Regulation of expansin gene expression affects growth and development in transgenic rice plants. *Plant Cell.* **15**, 1386–1398.

Coen, E., Rolland-Lagan, A. G., Matthews, M., Bangham, J. A., and Prusinkiewicz, P. (2004). The genetics of geometry. *Proc Natl Acad Sci U S A.* **101**, 4728–4735.

Cook, M., and Tyers, M. (2007). Size control goes global. *Curr Opin Biotechnol.* **18**, 341–350.

Cookson, S. J., Van Lijsebettens, M., and Granier, C. (2005). Correlation between leaf growth variables suggest intrinsic and early controls of leaf size in *Arabidopsis thaliana*. *Plant Cell Environ.* **28**, 1355–1366.

Cosgrove, D. J. (2005). Growth of the plant cell wall. *Nat Rev Mol Cell Biol.* **6**, 850–861.

Cubas, P., Lauter, N., Doebley, J., and Coen, E. (1999). The TCP domain: a motif found in proteins regulating plant growth and development. *Plant J.* **18**, 215–222.

de Lucas, M., Daviere, J. M., Rodriguez-Falcon, M., Pontin, M., Iglesias-Pedraz, J. M., Lorrain, S., Fankhauser, C., Blazquez, M. A., Titarenko, E., and Prat, S. (2008). A molecular framework for light and gibberellin control of cell elongation. *Nature* **451**, 480–484.

Deprost, D., Yao, L., Sormani, R., Moreau, M., Leterreux, G., Nicolai, M., Bedu, M., Robaglia, C., and Meyer, C. (2007). The Arabidopsis TOR kinase links plant growth, yield, stress resistance and mRNA translation. *EMBO Rep.* **8**, 864–870.

Dewitte, W., Riou-Khamlichi, C., Scofield, S., Healy, J. M., Jacqmard, A., Kilby, N. J., and Murray, J. A. (2003). Altered cell cycle distribution, hyperplasia, and inhibited differentiation in Arabidopsis caused by the D-type cyclin CYCD3. *Plant Cell.* **15**, 79–92.

Dewitte, W., Scofield, S., Alcasabas, A. A., Maughan, S. C., Menges, M., Braun, N., Collins, C., Nieuwland, J., Prinsen, E., Sundaresan, V., and Murray, J. A. (2007). Arabidopsis CYCD3 D-type cyclins link cell proliferation and endocycles and are rate-limiting for cytokinin responses. *Proc Natl Acad Sci U S A.* **104**, 14537–14542.

Dinneny, J. R., Weigel, D., and Yanofsky, M. F. (2006). NUBBIN and JAGGED define stamen and carpel shape in Arabidopsis. *Development* **133**, 1645–1655.

Dinneny, J. R., Yadegari, R., Fischer, R. L., Yanofsky, M. F., and Weigel, D. (2004). The role of JAGGED in shaping lateral organs. *Development* **131**, 1101–1110.

Disch, S., Anastasiou, E., Sharma, V. K., Laux, T., Fletcher, J. C., and Lenhard, M. (2006). The E3 ubiquitin ligase BIG BROTHER controls Arabidopsis organ size in a dosage-dependent manner. *Curr Biol.* **16**, 272–279.

Dolan, L., and Poethig, R. S. (1998). Clonal analysis of leaf development in cotton. *Am J Bot.* **85**, 315–321.

Donnelly, P. M., Bonetta, D., Tsukaya, H., Dengler, R. E., and Dengler, N. G. (1999). Cell cycling and cell enlargement in developing leaves of Arabidopsis. *Dev Biol.* **215**, 407–419.

Efroni, I., Blum, E., Goldshmidt, A., and Eshed, Y. (2008). A protracted and dynamic maturation schedule underlies Arabidopsis leaf development. *Plant Cell.* **20**, 2293–2306.

Ellis, C. M., Nagpal, P., Young, J. C., Hagen, G., Guilfoyle, T. J., and Reed, J. W. (2005). AUXIN RESPONSE FACTOR1 and AUXIN RESPONSE FACTOR2 regulate senescence and floral organ abscission in *Arabidopsis thaliana*. *Development* **132**, 4563–4574.

Erickson, R. O. (1976). Modeling of Plant Growth. *Annual Review of Plant Physiology and Plant Molecular Biology.* **27**, 407–434.

Eriksson, S., Stransfeld, L., Adamski, N. M., Breuninger, H., and Lenhard, M. (2010). KLUH/CYP78A5-dependent growth signaling coordinates floral organ growth in Arabidopsis. *Curr Biol.* **20**, 527–532.

Feng, S., Martinez, C., Gusmaroli, G., Wang, Y., Zhou, J., Wang, F., Chen, L., Yu, L., Iglesias-Pedraz, J. M., Kircher, S., Schafer, E., Fu, X., *et al.* (2008). Coordinated regulation of *Arabidopsis thaliana* development by light and gibberellins. *Nature* **451**, 475–479.

Ferjani, A., Horiguchi, G., Yano, S., and Tsukaya, H. (2007). Analysis of leaf development in *fugu* mutants of Arabidopsis reveals three compensation modes that modulate cell expansion in determinate organs. *Plant Physiol.* **144**, 988–999.

Fleming, A. J., McQueen-Mason, S., Mandel, T., and Kuhlemeier, C. (1997). Induction of leaf primordia by the cell wall protein expansin. *Science* **276**, 1415–1418.

Franklin, K. A. (2008). Shade avoidance. *New Phytol.* **179**, 930–944.

Fu, X., and Harberd, N. P. (2003). Auxin promotes Arabidopsis root growth by modulating gibberellin response. *Nature* **421**, 740–743.

Fujikura, U., Horiguchi, G., Ponce, M. R., Micol, J. L., and Tsukaya, H. (2009). Coordination of cell proliferation and cell expansion mediated by ribosome-related processes in the leaves of *Arabidopsis thaliana*. *Plant J.* **59**, 499–508.

Fujikura, U., Horiguchi, G., and Tsukaya, H. (2007). Dissection of enhanced cell expansion processes in leaves triggered by a defect in cell proliferation, with reference to roles of endoreduplication. *Plant Cell Physiol.* **48**, 278–286.

Grandjean, O., Vernoux, T., Laufs, P., Belcram, K., Mizukami, Y., and Traas, J. (2004). In vivo analysis of cell division, cell growth, and differentiation at the shoot apical meristem in Arabidopsis. *Plant Cell.* **16**, 74–87.

Granier, C., and Tardieu, F. (1998). Spatial and temporal analyses of expansion and cell cycle in sunflower leaves. A common pattern of development for all zones of a leaf and different leaves of a plant. *Plant Physiol.* **116**, 991–1001.

Green, P. B. (1999). Expression of pattern in plants: combining molecular and calculus-based biophysical paradigms. *Am J Bot.* **86**, 1059–1076.

Grieneisen, V. A., and Scheres, B. (2009). Back to the future: evolution of computational models in plant morphogenesis. *Curr Opin Plant Biol.* **12**, 606–614.

Ha, C. M., Jun, J. H., Nam, H. G., and Fletcher, J. C. (2004). *BLADE-ON-PETIOLE1* encodes a BTB/POZ domain protein required for leaf morphogenesis in *Arabidopsis thaliana*. *Plant Cell Physiol.* **45**, 1361–1370.

Ha, C. M., Jun, J. H., Nam, H. G., and Fletcher, J. C. (2007). BLADE-ON-PETIOLE 1 and 2 control Arabidopsis lateral organ fate through regulation of LOB domain and adaxial-abaxial polarity genes. *Plant Cell.* **19**, 1809–1825.

Ha, C. M., Kim, G. T., Kim, B. C., Jun, J. H., Soh, M. S., Ueno, Y., Machida, Y., Tsukaya, H., and Nam, H. G. (2003). The *BLADE-ON-PETIOLE 1* gene controls leaf pattern formation through the modulation of meristematic activity in Arabidopsis. *Development* **130**, 161–172.

Hemerly, A., Engler Jde, A., Bergounioux, C., Van Montagu, M., Engler, G., Inze, D., and Ferreira, P. (1995). Dominant negative mutants of the Cdc2 kinase uncouple cell division from iterative plant development. *Embo J.* **14**, 3925–3936.

Horiguchi, G., Gonzalez, N., Beemster, G. T., Inze, D., and Tsukaya, H. (2009). Impact of segmental chromosomal duplications on leaf size in the *grandifolia-D* mutants of *Arabidopsis thaliana*. *Plant J.* **60**, 122–133.

Horiguchi, G., Kim, G. T., and Tsukaya, H. (2005). The transcription factor AtGRF5 and the transcription coactivator AN3 regulate cell proliferation in leaf primordia of *Arabidopsis thaliana*. *Plant J.* **43**, 68–78.

Horvath, B. M., Magyar, Z., Zhang, Y., Hamburger, A. W., Bako, L., Visser, R. G., Bachem, C. W., and Bogre, L. (2006). EBP1 regulates organ size through cell growth and proliferation in plants. *EMBO J.* **25**, 4909–4920.

Hu, Y., Poh, H. M., and Chua, N. H. (2006). The Arabidopsis *ARGOS-LIKE* gene regulates cell expansion during organ growth. *Plant J.* **47**, 1–9.

Hu, Y., Xie, Q., and Chua, N. H. (2003). The Arabidopsis auxin-Inducible gene *ARGOS* controls lateral organ size. *Plant Cell.* **15**, 1951–61.

Irish, V. F., and Sussex, I. M. (1992). A fate map of the *Arabidopsis* embryonic shoot apical meristem. *Development* **117**, 745–753.

John, P. C., and Qi, R. (2008). Cell division and endoreduplication: doubtful engines of vegetative growth. *Trends Plant Sci.* **13**, 121–127.

Kaplan, D. R. (2001). Fundamental concepts of leaf morphology and morphogenesis: a contribution to the interpretation of molecular genetic mutants. *Int J Plant Sci.* **162**, 465–474.

Kaplan, D. R., and Hagemann, W. (1991). The relationship of cell and organism in vascular plants. *BioScience* **41**, 693–703.

Kim, J. H., Choi, D., and Kende, H. (2003). The AtGRF family of putative transcription factors is involved in leaf and cotyledon growth in Arabidopsis. *Plant J.* **36**, 94–104.

Kim, J. H., and Kende, H. (2004). A transcriptional coactivator, AtGIF1, is involved in regulating leaf growth and morphology in Arabidopsis. *Proc Natl Acad Sci U S A.* **101**, 13374–13379.

Krizek, B. (2009). AINTEGUMENTA and AINTEGUMENTA-LIKE6 act redundantly to regulate Arabidopsis floral growth and patterning. *Plant Physiol.* **150**, 1916–1929.

Krizek, B. A. (1999). Ectopic expression of *AINTEGUMENTA* in Arabidopsis plants results in increased growth of floral organs. *Dev Genet.* **25**, 224–236.

Kuhlemeier, C. (2007). Phyllotaxis. *Trends Plant Sci.* **12**, 143–150.

Kwiatkowska, D. (2004). Structural integration at the shoot apical meristem: Models, measurements, and experiments. *Am J Bot.* **91**, 1277–1293.

Kwiatkowska, D., and Dumais, J. (2003). Growth and morphogenesis at the vegetative shoot apex of *Anagallis arvensis L. J Exp Bot.* **54**, 1585–1595.

Kwon, S. H., Lee, B. H., Kim, E. Y., Seo, Y. S., Lee, S., Kim, W. T., Song, J. T., and Kim, J. H. (2009). Overexpression of a *Brassica rapa NGATHA* gene in *Arabidopsis thaliana* negatively affects cell proliferation during lateral organ and root growth. *Plant Cell Physiol.* **50**, 2162–2173.

Larson-Rabin, Z., Li, Z., Masson, P. H., and Day, C. D. (2009). *FZR2/CCS52A1* expression is a determinant of endoreduplication and cell expansion in Arabidopsis. *Plant Physiol.* **149**, 874–884.

Lee, B. H., Ko, J. H., Lee, S., Lee, Y., Pak, J. H., and Kim, J. H. (2009). The Arabidopsis GRF-INTERACTING FACTOR gene family performs an overlapping function in determining organ size as well as multiple developmental properties. *Plant Physiol.* **151**, 655–668.

Lee, K., Avondo, J., Morrison, H., Blot, L., Stark, M., Sharpe, J., Bangham, A., and Coen, E. (2006). Visualizing plant development and gene expression in three dimensions using optical projection tomography. *Plant Cell.* **18**, 2145–2156.

Li, C., Potuschak, T., Colon-Carmona, A., Gutierrez, R. A., and Doerner, P. (2005). Arabidopsis TCP20 links regulation of growth and cell division control pathways. *Proc Natl Acad Sci U S A.* **102**, 12978–12983.

Li, Y., Zheng, L., Corke, F., Smith, C., and Bevan, M. W. (2008). Control of final seed and organ size by the *DA1* gene family in *Arabidopsis thaliana. Genes Dev.* **22**, 1331–1336.

Marcotrigiano, M. (2001). Genetic mosaics and the analysis of leaf development. *Int J Plant Sci.* **162**, 513–525.

McHale, N. A., and Marcotrigiano, M. (1998). LAM1 is required for dorsoventrality and lateral growth of the leaf blade in *Nicotiana. Development* **125**, 4235–4243.

Menand, B., Desnos, T., Nussaume, L., Berger, F., Bouchez, D., Meyer, C., and Robaglia, C. (2002). Expression and disruption of the Arabidopsis TOR (target of rapamycin) gene. *Proc Natl Acad Sci U S A.* **99**, 6422–6427.

Menges, M., Samland, A. K., Planchais, S., and Murray, J. A. (2006). The D-type cyclin CYCD3;1 is limiting for the G1-to-S-phase transition in Arabidopsis. *Plant Cell.* **18**, 893–906.

Mitchison, J. M. (2003). Growth during the cell cycle. *Int Rev Cytol.* **226**, 165–258.

Mizukami, Y., and Fischer, R. L. (2000). Plant organ size control: AINTEGUMENTA regulates growth and cell numbers during organogenesis. *Proc Natl Acad Sci U S A.* **97**, 942–947.

Nag, A., King, S., and Jack, T. (2009). miR319a targeting of *TCP4* is critical for petal growth and development in Arabidopsis. *Proc Natl Acad Sci U S A.* **106**, 22534–22539.

Nath, U., Crawford, B. C., Carpenter, R., and Coen, E. (2003). Genetic control of surface curvature. *Science* **299**, 1404–1407.

Nemhauser, J. L. (2008). Dawning of a new era: photomorphogenesis as an integrated molecular network. *Curr Opin Plant Biol.* **11**, 4–8.

Nole-Wilson, S., Tranby, T. L., and Krizek, B. A. (2005). *AINTEGUMENTA-like* (*AIL*) genes are expressed in young tissues and may specify meristematic or division-competent states. *Plant Mol Biol.* **57**, 613–628.

Norberg, M., Holmlund, M., and Nilsson, O. (2005). The *BLADE ON PETIOLE* genes act redundantly to control the growth and development of lateral organs. *Development* **132**, 2203–2213.

Ohno, C. K., Reddy, G. V., Heisler, M. G., and Meyerowitz, E. M. (2004). The Arabidopsis *JAGGED* gene encodes a zinc finger protein that promotes leaf tissue development. *Development* **131**, 1111–1122.

Okushima, Y., Mitina, I., Quach, H. L., and Theologis, A. (2005). AUXIN RESPONSE FACTOR 2 (ARF2): a pleiotropic developmental regulator. *Plant J.* **43**, 29–46.

Ori, N., Cohen, A. R., Etzioni, A., Brand, A., Yanai, O., Shleizer, S., Menda, N., Amsellem, Z., Efroni, I., Pekker, I., Alvarez, J. P., Blum, E., *et al.* (2007). Regulation of *LANCEOLATE* by miR319 is required for compound-leaf development in tomato. *Nat Genet.* **39**, 787–791.

Palatnik, J. F., Allen, E., Wu, X., Schommer, C., Schwab, R., Carrington, J. C., and Weigel, D. (2003). Control of leaf morphogenesis by microRNAs. *Nature* **425**, 257–263.

Pien, S., Wyrzykowska, J., McQueen-Mason, S., Smart, C., and Fleming, A. (2001). Local expression of expansin induces the entire process of leaf development and modifies leaf shape. *Proc Natl Acad Sci U S A.* **98**, 11812–11817.

Poethig, R. S., and Sussex, I. M. (1985). The developmental morphology and growth dynamics of the tobacco leaf. *Planta* **165**, 158–169.

Potter, C. J., and Xu, T. (2001). Mechanisms of size control. *Curr Opin Genet Dev.* **11**, 279–286.

Prusinkiewicz, P., and Lindenmayer, A., 1996. *The Algorithmic Beauty of Plants.* Springer, New York.

Qi, R., and John, P. C. (2007). Expression of genomic *AtCYCD2;1* in Arabidopsis induces cell division at smaller cell sizes: implications for the control of plant growth. *Plant Physiol.* **144**, 1587–1597.

Reddy, G. V., Heisler, M. G., Ehrhardt, D. W., and Meyerowitz, E. M. (2004). Real-time lineage analysis reveals oriented cell divisions associated with morphogenesis at the shoot apex of *Arabidopsis thaliana. Development* **131**, 4225–4237.

Rieu, I., and Laux, T. (2009). Signaling pathways maintaining stem cells at the plant shoot apex. *Semin Cell Dev Biol.* **20**, 1083–1088.

Rodriguez, R. E., Mecchia, M. A., Debernardi, J. M., Schommer, C., Weigel, D., and Palatnik, J. F. (2010). Control of cell proliferation in *Arabidopsis thaliana* by microRNA miR396. *Development* **137**, 103–112.

Rolland-Lagan, A. G., Bangham, J. A., and Coen, E. (2003). Growth dynamics underlying petal shape and asymmetry. *Nature* **422**, 161–163.

Savaldi-Goldstein, S., and Chory, J. (2008). Growth coordination and the shoot epidermis. *Curr Opin Plant Biol.* **11**, 42–48.

Savaldi-Goldstein, S., Peto, C., and Chory, J. (2007). The epidermis both drives and restricts plant shoot growth. *Nature* **446**, 199–202.

Scanlon, M. J., and Freeling, M. (1997). Clonal sectors reveal that a specific meristematic domain is not utilized in the maize mutant narrow sheath. *Dev Biol.* **182**, 52–66.

Scarpella, E., Marcos, D., Friml, J., and Berleth, T. (2006). Control of leaf vascular patterning by polar auxin transport. *Genes Dev.* **20**, 1015–1027.

Schnittger, A., Grini, P. E., Folkers, U., and Hulskamp, M. (1996). Epidermal fate map of the Arabidopsis shoot meristem. *Dev Biol.* **175**, 248–255.

Schommer, C., Palatnik, J. F., Aggarwal, P., Chetelat, A., Cubas, P., Farmer, E. E., Nath, U., and Weigel, D. (2008). Control of jasmonate biosynthesis and senescence by miR319 targets. *PLoS Biol.* **6**, e230.

Schruff, M. C., Spielman, M., Tiwari, S., Adams, S., Fenby, N., and Scott, R. J. (2006). The AUXIN RESPONSE FACTOR 2 gene of Arabidopsis links auxin signalling, cell division, and the size of seeds and other organs. *Development* **133**, 251–261.

Schwechheimer, C., and Willige, B. C. (2009). Shedding light on gibberellic acid signalling. *Curr Opin Plant Biol.* **12**, 57–62.

Sigrist, S., Jacobs, H., Stratmann, R., and Lehner, C. F. (1995). Exit from mitosis is regulated by *Drosophila fizzy* and the sequential destruction of cyclins A, B and B3. *EMBO J.* **14**, 4827–4838.

Sigrist, S. J., and Lehner, C. F. (1997). *Drosophila fizzy-related* down-regulates mitotic cyclins and is required for cell proliferation arrest and entry into endocycles. *Cell* **90**, 671–681.

Sloan, J., Backhaus, A., Malinowski, R., McQueen-Mason, S., and Fleming, A. J. (2009). Phased control of expansin activity during leaf development identifies a sensitivity window for expansin-mediated induction of leaf growth. *Plant Physiol.* **151**, 1844–1854.

Smith, R. S., Guyomarc'h, S., Mandel, T., Reinhardt, D., Kuhlemeier, C., and Prusinkiewicz, P. (2006). A plausible model of phyllotaxis. *Proc Natl Acad Sci U S A.* **103**, 1301–1306.

Sonoda, Y., Sako, K., Maki, Y., Yamazaki, N., Yamamoto, H., Ikeda, A., and Yamaguchi, J. (2009). Regulation of leaf organ size by the Arabidopsis RPT2a 19S proteasome subunit. *Plant J.* **60**, 68–78.

Squatrito, M., Mancino, M., Donzelli, M., Areces, L. B., and Draetta, G. F. (2004). EBP1 is a nucleolar growth-regulating protein that is part of pre-ribosomal ribonucleoprotein complexes. *Oncogene* **23**, 4454–4465.

Stebbins, G. L., 1974. *Flowering Plants: Evolution Above the Species Level.* Belknap Press, Cambridge, Mass.

Steeves, T. A., and Sussex, I. M., 1989. *Patterns in Plant Development.* Cambride University Press, Cambridge.

Sugimoto-Shirasu, K., Roberts, G. R., Stacey, N. J., McCann, M. C., Maxwell, A., and Roberts, K. (2005). RHL1 is an essential component of the plant DNA topoisomerase VI complex and is required for ploidy-dependent cell growth. *Proc Natl Acad Sci U S A.* **102**, 18736–18741.

Sugimoto-Shirasu, K., and Roberts, K. (2003). "Big it up": endoreduplication and cell-size control in plants. *Curr Opin Plant Biol.* **6**, 544–553.

Sugimoto-Shirasu, K., Stacey, N. J., Corsar, J., Roberts, K., and McCann, M. C. (2002). DNA Topoisomerase VI Is Essential for Endoreduplication in Arabidopsis. *Curr Biol.* **12**, 1782–1786.

Szecsi, J., Joly, C., Bordji, K., Varaud, E., Cock, J. M., Dumas, C., and Bendahmane, M. (2006). BIGPETALp, a bHLH transcription factor is involved in the control of Arabidopsis petal size. *Embo J.* **25**, 3912–3920.

Szymkowiak, E. J., and Sussex, I. M. (1996). What chimeras can tell us about plant development. *Annu Rev Plant Physiol Plant Mol Biol.* **47**, 351–376.

Toledo-Ortiz, G., Huq, E., and Quail, P. H. (2003). The Arabidopsis basic/helix-loop-helix transcription factor family. *Plant Cell.* **15**, 1749–1770.

Trigueros, M., Navarrete-Gomez, M., Sato, S., Christensen, S. K., Pelaz, S., Weigel, D., Yanofsky, M. F., and Ferrandiz, C. (2009). The *NGATHA* genes direct style development in the Arabidopsis gynoecium. *Plant Cell.* **21**, 1394–1409.

Truernit, E., Bauby, H., Dubreucq, B., Grandjean, O., Runions, J., Barthelemy, J., and Palauqui, J. C. (2008). High-resolution whole-mount imaging of three-dimensional tissue organization and gene expression enables the study of Phloem development and structure in Arabidopsis. *Plant Cell.* **20**, 1494–1503.

Tsukaya, H. (2003). Organ shape and size: a lesson from studies of leaf morphogenesis. *Curr Opin Plant Biol.* **6**, 57–62.

Tsukaya, H. (2006). Mechanism of leaf-shape determination. *Annu Rev Plant Biol.* **57**, 477–496.

Ubeda-Tomas, S., Federici, F., Casimiro, I., Beemster, G. T., Bhalerao, R., Swarup, R., Doerner, P., Haseloff, J., and Bennett, M. J. (2009). Gibberellin signaling in the endodermis controls Arabidopsis root meristem size. *Curr Biol.* **19**, 1194–1199.

van der Knaap, E., Kim, J. H., and Kende, H. (2000). A novel gibberellin-induced gene from rice and its potential regulatory role in stem growth. *Plant Physiol.* **122**, 695–704.

Vert, G., Walcher, C. L., Chory, J., and Nemhauser, J. L. (2008). Integration of auxin and brassinosteroid pathways by Auxin Response Factor 2. *Proc Natl Acad Sci U S A.* **105**, 9829–9834.

Vierstra, R. D. (2009). The ubiquitin-26S proteasome system at the nexus of plant biology. *Nat Rev Mol Cell Biol.* **10**, 385–397.

Wang, H., Zhou, Y., Gilmer, S., Whitwill, S., and Fowke, L. C. (2000). Expression of the plant cyclin-dependent kinase inhibitor ICK1 affects cell division, plant growth and morphology. *Plant J.* **24**, 613–623.

Wang, J. W., Schwab, R., Czech, B., Mica, E., and Weigel, D. (2008). Dual effects of miR156-targeted *SPL* genes and *CYP78A5/KLUH* on plastochron length and organ size in *Arabidopsis thaliana*. *Plant Cell.* **20**, 1231–1243.

White, D. W. (2006). PEAPOD regulates lamina size and curvature in Arabidopsis. *Proc Natl Acad Sci U S A.* **103**, 13238–13243.

Wiese, A., Christ, M. M., Virnich, O., Schurr, U., and Walter, A. (2007). Spatio-temporal leaf growth patterns of *Arabidopsis thaliana* and evidence for sugar control of the diel leaf growth cycle. *New Phytol.* **174**, 752–761.

Wyrzykowska, J., Pien, S., Shen, W. H., and Fleming, A. J. (2002). Manipulation of leaf shape by modulation of cell division. *Development* **129**, 957–964.

Zenoni, S., Reale, L., Tornielli, G. B., Lanfaloni, L., Porceddu, A., Ferrarini, A., Moretti, C., Zamboni, A., Speghini, A., Ferranti, F., and Pezzotti, M. (2004). Downregulation of the *Petunia hybrida* alpha-expansin gene *PhEXP1* reduces the amount of crystalline cellulose in cell walls and leads to phenotypic changes in petal limbs. *Plant Cell.* **16**, 295–308.

CHAPTER EIGHT

Vascular Pattern Formation in Plants

Enrico Scarpella* *and* Ykä Helariutta†

Contents

Abstract

Reticulate tissue systems exist in most multicellular organisms, and the principles underlying the formation of cellular networks have fascinated philosophers, mathematicians, and biologists for centuries. In particular, the beautiful and varied arrangements of vascular tissues in plants have intrigued mankind since antiquity, yet the organizing signals have remained elusive. Plant vascular tissues form systems of interconnected cell files throughout the plant body. Vascular cells are aligned with one another along continuous lines, and vascular tissues differentiate at reproducible positions within organ environments. However, neither the precise path of vascular differentiation nor the exact geometry of vascular networks is fixed or immutable. Several recent advances converge to reconcile the seemingly conflicting predictability

* Department of Biological Sciences, University of Alberta, Edmonton, Canada
† Department of Biological and Environmental Sciences, Institute of Biotechnology, University of Helsinki, Helsinki, Finland

Current Topics in Developmental Biology, Volume 91
ISSN 0070-2153, DOI 10.1016/S0070-2153(10)91008-9

and plasticity of vascular tissue patterns. A control mechanism in which an apical–basal flow of signal establishes a basic coordinate system for body axis formation and vascular strand differentiation, and in which a superimposed level of radial organizing cues elaborates cell patterns, would generate a reproducible tissue configuration in the context of an underlying robust, self-organizing structure, and account for the simultaneous regularity and flexibility of vascular tissue patterns.

1. Introduction

One of the most significant series of evolutionary events in the history of life on Earth was the colonization of land environments by water-based complex life forms. These events began with the migration of aquatic photosynthetic organisms onto land some 475 million years ago, and resulted in an explosion of evolutionary innovation and consequent diversification of terrestrial ecosystems. Prior to the input of energy into terrestrial ecosystems by land plants, terrestrial environments were occupied by bacteria, fungi, and simple algae (Gray, 1993; Gray and Shear, 1992; Kenrick and Crane, 1997; Knoll, 1994). However, within 75 million years of the origin of land plants, terrestrial photosynthetic organisms had undergone an amazing evolutionary radiation, and established conditions for the colonization and subsequent diversification of various multicellular lineages on land.

Photosynthesis in an aerial environment requires high levels of gas exchange and is facilitated in land plants by the formation of openings in the outer surface of the plant. These pores, however, also result in considerable losses of gas-phase water from the internal tissues of the plant body to the external environment. Therefore, to remain hydrated, terrestrial plants larger than a few centimeters require specialized conducting tissues to transport water from the plant–soil interface to the aerial portions of the plant body (Raven, 1993). Biophysical models suggest that any significant increase in the stature of land plants was dependent on the evolution of highly specialized hollow, dead water-conducting cells (Raven, 1984; 1993), and the fossil record documents that land plants remained extremely small and structurally simple until the appearance, ca. 425 million years ago, of developmentally complex water-conducting cells defined by the presence of secondary cell walls, lignification, and programmed cell death (Gensel and Andrews, 1987; Gray and Shear, 1992; Kenrick and Crane, 1997; Knoll and Niklas, 1987; Knoll and Rothwell, 1981; Raven, 1993).

Unlike water transport, evolution of an intercellular pathway for translocation of photoassimilates is not thought to be associated with invasion of the Earth's surface, but rather with the increasing size of the plant body and its functional differentiation into physiologically distinct organs (Cook and Graham, 1999; Graham, 1993; Niklas, 1997; Schulz, 1998; Taylor, 1990; van Bel, 1999). In fact, long-distance photosynthate transport through specialized cell arrays already occurs in algae (Schmitz, 1990); however, in land plants, elements for the transport of photoassimilates and water-conducting elements differentiate from the same precursor cells and remain in intimate spatial association with one another throughout the plant's life. Therefore, the solitary presence of the evolutionary more ancient system for photosynthate translocation in algae suggests that the appearance of a similar transport path in land plants may have evolved independently, rather than being a phylogenic continuation of the same conducting tissue.

The patterned formation of the tissue system responsible for long-distance transport of water and nutrients in plants and its integration within the local organ context, with emphasis on plant growth by lengthening, is the subject of this chapter; further aspects of plant vascular development, including differentiation during plant growth by radial thickening, have been thoughtfully reviewed elsewhere (e.g., Aloni, 2001; Baucher *et al.*, 2007; Berleth *et al.*, 2000; Berleth and Sachs, 2001; Dengler, 2001; 2006; Dengler and Kang, 2001; Fukuda, 2004; Fukuda *et al.*, 2007; Jung and Park, 2007; Nelson and Dengler, 1997; Rolland-Lagan, 2008; Samuels and Deyholos, 2006; Sieburth *et al.*, 2006; Spicer and Groover, 2010; Turner *et al.*, 2007; Turner and Sieburth, 2003; van Bel *et al.*, 2002; Ye, 2002).

2. The Plant Vascular System

The vascular system of extant seed plants is a network of continuous files of cells organized in strands (or bundles) that extend through each organ and throughout the entire plant, functionally connecting every part of the shoot with the root system (Esau, 1965; Fig. 8.1A). At maturity, vascular strands typically represent the organization of two vascular tissues, the phloem and the xylem. The phloem represents the route for translocation of dissolved carbohydrates, which are transported from tissues that are net producers of photoassimilates to tissues that are net users, while the xylem is the main conduit for transport of water and minerals from the root to the sites of evapo-transpiration in the shoot. Furthermore, phloem and xylem are responsible for the distribution of a variety of regulatory molecules involved in plant development, response to the environment, and defense against pathogens (Cambridge and Morris, 1996; Citovsky and Zambryski, 2000; Hirose *et al.*, 2008; Jiang and Hartung, 2008; Lough and Lucas, 2006; Oparka and Cruz, 2000; van Bel *et al.*, 2002). Finally, both vascular tissues typically comprise a number of specialized cell types, including conducting elements, and cells that fulfill nourishing and supporting functions.

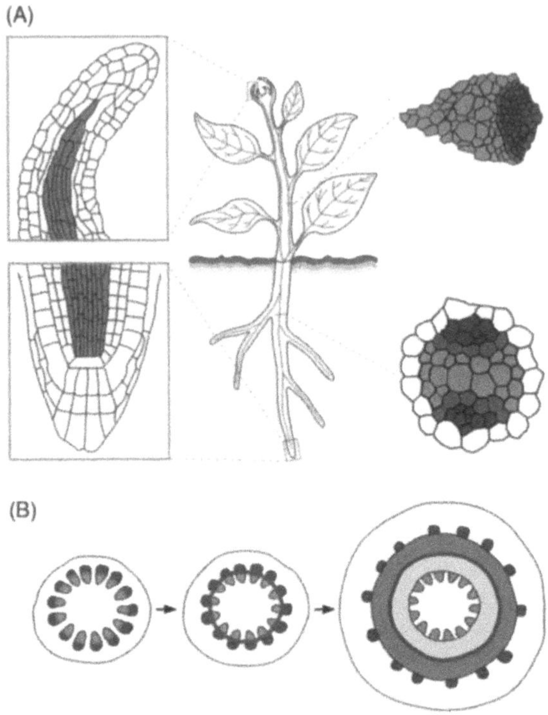

Figure 8.1 The vascular system of plants and its origin. (A) The vascular system is a network of continuous vascular strands (green) extending throughout the entire plant body. At maturity, vascular strands represent the organization of two vascular tissues, the phloem (dark blue) and the xylem (light blue). In the stem vascular strands of most plant species, the phloem pole is located at the outer side and the xylem pole at the inner side of the strand. In root vascular strands, the xylem forms a solid central core with projections extending toward the periphery of the vascular strand, and the phloem is organized in separate strands alternating with the xylem ridges. Both vascular tissues differentiate in immature parts of plant organs from a common precursor tissue, the procambium (brown), recognizable as uninterrupted files of terminally aligned, elongated cells. In many species, a layer of undifferentiated procambial cells separates phloem and xylem in mature strands. Redrawn from: Fig. 8, Girolami, G. (1954), Leaf histogenesis in *Linum usitatissimum*, *Am. J. Bot.* 41, 264–273, with kind permission of the Botanical Society of America; Fig. 2J,K, Scarpella, E., and Meijer, A. H. (2004), Pattern formation in the vascular system of monocot and dicot plant species, *New Phytologist* 164, 209–242, with kind permission of John Wiley and Sons; Figs. 1C, 2A, B; van den Berg, C., Willemsen, V., Hage, W., Weisbeek, P., and Scheres, B. (1995), Cell fate in the Arabidopsis root meristem determined by directional signalling. *Nature* 378, 62–65, with kind permission of Nature Publishing Group and Macmillan Publishers. (B) During stem growth by radial thickening, the residual procambium within vascular strands and parenchyma cells between vascular strands resume division activity and originate a hollow cylinder—a circle in the transverse section shown here—of cambium tissue, from which secondary phloem (dark gray) and xylem (light gray) differentiate. Arrows temporally connect successive stages of secondary vascular tissue formation. A formally similar process occurs during root secondary growth. (See Color Insert.)

3. Origin of Vascular Tissues

During growth of the plant by lengthening (primary growth), all vascular cell types differentiate from procambial cells: cytoplasmically dense, elongated cells typically aligned with one another to form continuous files (Esau, 1943; Fig. 8.1A). Procambial cells located further from the shoot and root tip tend to be, however, much longer and more vacuolated. Furthermore, procambial cells in lateral organs (e.g., leaves and flowers) may never attain the length of procambial cells situated along the main axis of the plant. Because cytoplasm density and cell length are relative criteria, transgenic lines in which initiation of reporter gene expression is tightly associated with appearance of procambial cells represent invaluable tools to assist in the identification of the first anatomically recognizable procambial cells and of the dynamics of procambium formation in organ development (e.g., *DOF2.4::GUS*, Konishi and Yanagisawa, 2007; *Q0990::mGFP5er*, Sawchuk *et al.*, 2007; *ET1135::GUS*, Scarpella *et al.*, 2004; *GA1:GUS*, Silverstone *et al.*, 1997).

In plants that undergo radial thickening (secondary growth), secondary vascular tissues differentiate from the (vascular) cambium, the initials of which originate in part from the residual procambium within vascular strands (intrafascicular cambium), and in part from the parenchyma between vascular strands (interfascicular cambium) (Esau, 1965; Fig. 8.1B).

Finally, vascular elements can differentiate from nonvascular parenchyma cells in order to regenerate vascular strand continuity when interrupted as a consequence of wounding or grafting (Sachs, 1981).

4. Levels of Vascular Organization

Two elements of spatial organization are reflected in the structure of the vascular system (Dengler and Kang, 2001; Esau, 1965): the longitudinal pattern, which is the three-dimensional configuration of vascular strands within an organism; and the radial pattern, which is the arrangement of phloem and xylem within each vascular bundle. Both levels of organization are organ specific, as reflected, for example, in the different positions of vascular and nonvascular tissues in different organs such as stems, leaves, and roots (Fig. 8.2); however, within a given organ, species-specific cues further shape vascular organizations, as shown, for example, by the successful use of species-specific leaf vascular patterns as taxonomic diagnostic features (e.g., Klucking, 1995; Fig. 8.2).

While organ- and species-specific vascular organizations are highly reproducible, the precise course and arrangement of vascular strands is neither completely predictable nor immutable. For example, the configuration of the network of vascular strands in a leaf is variable, although the degree of variability differs among plant species. Further, the vascular system, at least in dicot species, retains a high level of plasticity: vascular bridges can be formed to circumvent a wound in mature organs and reconnect severed vascular strands (Sachs, 1981).

The apparent contradiction of coexistence of variability and reproducibility in vascular organizations suggests the presence of two types of patterning cues: directional signals in combination with self-organizing capabilities of vascular tissues, which generate continuous files of vascular cells that are aligned along unpredictable routes and organized into variable networks; and genetic cues, which constrain the variety of possible arrangements and internal compositions of vascular strands, and ensure the reproducible differentiation of nonvascular cells in fixed spatial relations to vascular tissues. These two types of control are the object of the next two sections.

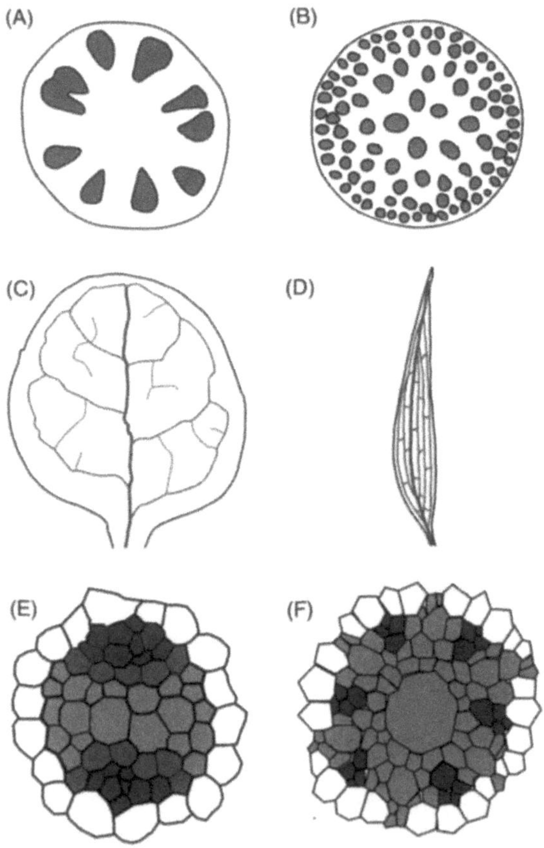

Figure 8.2 *(Continued)*

5. Control of Longitudinal Vascular Patterning and Integration with Organ Formation

5.1. Vascular strand regeneration in mature organs

Evidence for the existence of mechanisms that integrate vascular strand formation with organismal signals initially derived from experiments of local application of compounds of the auxin class of indole-related molecules to plant tissues (Sachs, 1981; Fig. 8.3A). It is unlikely, of course, that auxin acts alone, and a role of other factors must be considered. Concrete evidence is available that a variety of other substances promote vascular differentiation, among them adenine-type compounds of the cytokinin family, brassinosteroids, ethylene, fatty acid-derived jasmonates, peptides,

Figure 8.2 Levels of vascular organization. Vascular patterns in stems (A,B), leaves (C,D), and roots (E,F) of dicots (A,C,E) and monocots (B,C,F) as examples of organ- and species-specific vascular organizations. Two levels of spatial organization are evident in the structure of the vascular system: the longitudinal pattern, which is the three-dimensional array of vascular strands within an organism; and the radial pattern, which is the arrangement of phloem and xylem within each vascular strand. Both levels of organization are organ specific, as reflected in the different positions of vascular and nonvascular tissues in stems, leaves, and roots of the same species (compare A with C and E, and B with D and F). However, within a given organ, species-specific cues further limit the variety of possible vascular organizations. For example, in most dicot stems (A), vascular strands (green) are arranged in a circle at the periphery of the organ, while, in most monocots (B), they are scattered throughout the organ. Further, in dicot leaves (C), veins (green) are connected in a reticulate pattern, while, in monocots (D), they are aligned in a striate pattern. Finally, in the root, vascular tissues are organized in a central vascular cylinder, in which the xylem (dark blue) forms a solid central core with extensions toward the periphery of the vascular cylinder, while the phloem (light blue) is organized in separate strands alternating with the xylem arms. The number of xylem ridges varies in different species; typically, dicots have up to four and monocots have more than four. Furthermore, procambial cells (brown) separate xylem and phloem in dicot, but not in monocot, roots. Note, however, that while vascular differentiation must be continuous, while the orientations of the cells within each strand must be precisely integrated to ensure conductivity, and while the arrangements and compositions of vascular strands must be constrained to ensure functionality in the local organ context, the precise route and arrangement of vascular strands cannot be predicted. Redrawn from: Fig. 1A, Donner, T. J., and Scarpella, E. (2009), Auxin-transport-dependent leaf vein formation, Botany-Botanique 87, 678–684, with kind permission of National Research Council Canada Research Press; Fig. 2D,H, Scarpella, E., and Meijer, A. H. (2004), Pattern formation in the vascular system of monocot and dicot plant species, *New Phytol.* 164, 209–242, with kind permission of John Wiley and Sons. (See Color Insert.)

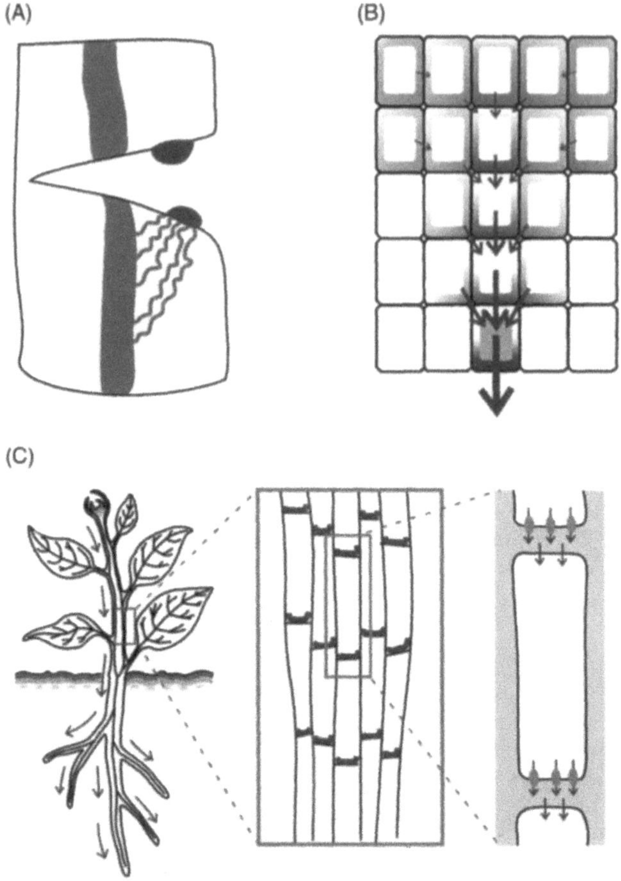

Figure 8.3 Vascular strand regeneration and polar auxin transport. (A) Vascular strand formation in response to local auxin application. A source of auxin (brown blob) locally applied to suitable plant tissues triggers the formation of vascular strands (brown wavy lines) that connect the auxin source to an existing vascular strand (green). While the precise path of vascular strand formation is not reproducible, the differentiation process proceeds toward the basal pole of the plant (note absence of vascular strand formation in correspondence with inverse auxin application) is restricted to narrow bundles of uniterrupted cell files, leads to continuity with existing vasculature, and is strictly dependent on polar auxin transport. (B) Auxin canalization hypothesis. In this simplified view, it is assumed that intensity and direction of auxin flow are solely controlled through the quantity and distribution of auxin efflux carriers (blue) in the plasma membranes of cells (rectangles). Arrows of different strength indicate the intensity of auxin flow. The hypothesis proposes that auxin flow and cell polarization are connected in a positive feedback loop, illustrated here by restriction of auxin efflux to the basal side of each cell as an expression of cell polarization. Under the influence of a localized sink such as the preexisting vasculature (here basal to the cell grid), cells mutually integrate apical–basal polarity and restrict auxin flow to preferred routes, which subsequently differentiate into vascular tissues (light brown). (C) Organismal auxin transport routes. Auxin is primarily produced in immature, apical parts of the plant and transported to the root tips mainly through vascular strands. The apical–basal polarity of auxin transport results at the cellular level from the asymmetric distribution of proteins members of the PIN family of auxin transporters (blue) at the basal plasma membrane of auxin transport-competent cells.

polyamines, and terpenoids of the gibberellin class (e.g., Aloni, 1987; Brioudes *et al.*, 2009; Cano-Delgado *et al.*, 2004; Fukuda, 2004; Fukuda *et al.*, 2007; Ibañes *et al.*, 2009; Vera-Sirera *et al.*, 2010). Auxin is unique, however, among known substances. In fact, not only does auxin trigger vascular cell differentiation, but application of auxin to plant tissues evokes formation of new vascular strands that connect the auxin source with preexisting vasculature (Sachs, 1981; Fig. 8.3A). This observation goes beyond the mere identification of a vascular differentiation-promoting substance: it establishes a spatial relation between an auxin source and a patterned cellular response that is characterized by distinctive and reproducible properties (Berleth *et al.*, 2000; Sachs, 1981; Fig. 8.3A). First, the auxin-induced vascular differentiation response is local, as it is initiated at the specific site of auxin application; second, it is polar, as it is oriented from the auxin source toward the basal side of the plant; third, it is continuous, as it generates uninterrupted files of vascular cells; fourth, it is constrained in the planes perpendicular to the main axis of the vascular differentiation response, as it originates slender bundles of vascular cell files; fifth, it is dependent on auxin flow, as it requires efficiently transported auxins and is obstructed in the presence of auxin transport inhibitors. These observations and considerations form the basis of the "auxin canalization hypothesis," which postulates that a positive feedback between auxin transport through a cell and the cell's auxin conductivity would progressively restrict an initially dispersed auxin flow to narrow preferred canals, which would finally differentiate into vascular strands (Sachs, 1981; Fig. 8.3B). Evidence for similar directional effects has not been reported for substances other than auxin. It is possible, therefore, that the unique effect of auxin integrates various influences on cell orientation, which would be readily disrupted if it were determined by various independent factors. Vascular strand formation in response to auxin application does not readily occur in all species (e.g., Aloni and Plotkin, 1985), suggesting that additional factors that act in concert with auxin are probably required. Nevertheless, the capability of a simple signal to trigger a complex and oriented cellular response such as that of vascular strand formation suggests that the underlying mechanism recruits directional cues already present in the organism. Because auxin is synthesized predominantly in young regions of the plant, such as leaf primordia (Normanly, 2010), and transported basally (Michniewicz *et al.*, 2007; Fig. 8.3C), the source of asymmetric information co-opted by the auxin-induced vascular differentiation response probably coincides with the polar flow of auxin itself. The apical–basal polarity of auxin transport results from the fact that, as a weak acid, auxin is selectively charged at intracellular pH. Therefore, its export from the cell is dependent upon specific proteins, which are assumed to be concentrated at the basal end of each cell (Raven, 1975; Rubery and Sheldrake, 1974; Fig. 8.3B). Among the gene families implicated in the cellular efflux of auxin, five Arabidopsis genes encoding integral plasma membrane proteins of the PIN-FORMED (PIN) family exert highly redundant functions, and their subcellular localizations mark the presumed auxin-efflux side of cells in highly dynamic, yet robust patterns (Petrasek and Friml, 2009; Fig. 8.3B). PIN proteins are instrumental in defining the polarity of auxin flow, and therefore its directions can be inferred from the asymmetric localization of PIN proteins during development. Consistent with predictions of the auxin canalization hypothesis, local application of auxin to pea tissues activates PIN1 expression in broad domains that connect the auxin source to the preexisting vascular strands (Sauer *et al.*, 2006). Over time, wide PIN1 expression domains narrow to individual cell files in which PIN1 polarity indicates auxin transport from the site of auxin application to the preexisting vasculature, and which define the sites of future vascular strand formation. Studies in the past few years, however, have not only added molecular details to long-standing models of auxin-induced vascular differentiation in fully polarized organs but have also accumulated evidence for a crucial role of auxin transport in defining and integrating paths of vascular strand formation in embryo and organ development.

5.2. Vascular strand formation in embryo development

The pattern of a dicot seedling can be formalized as sequences of apically–basally and centrally–peripherally arranged elements (Capron *et al.*, 2009; Fig. 8.4A). The main elements along the apical–basal axis are the shoot apical meristem, the cotyledons, the hypocotyl, the root, and the root apical meristem. Most of the seedling body takes the form of a cylinder with concentric layers of vascular, ground, and epidermal tissues arranged in a central-peripheral pattern. The hypocotyl and root sections along this cylinder have distinguishable anatomical and physiological properties, and the radial pattern of the root is perpetuated by the activity of the root apical meristem. Both the longitudinal and transverse patterns of the seedling are generated during embryogenesis, and all the structural elements of these patterns are apparent at the heart stage (chapter 1; Fig. 8.4B).

Available experimental and genetic evidence suggests that acquisition of apical–basal polarity in the embryo and formation of the procambial strand in its center depend on auxin flow. Application of auxin transport inhibitors to dicot embryos dramatically interferes with apical–basal axis formation, and occasionally results in entirely apolar, ball-shaped embryos and seedlings (Friml *et al.*, 2003; Hadfi *et al.*, 1998; Liu *et al.*, 1993; Fig. 8.4C). Likewise, pertubation of polar auxin transport in embryos through simultaneous mutation of multiple *PIN* genes leads to aberrant auxin distribution and pronounced defects in apical–basal polarity; in the strongest cases, no apical–basal axis is established (Benkova *et al.*, 2003; Friml

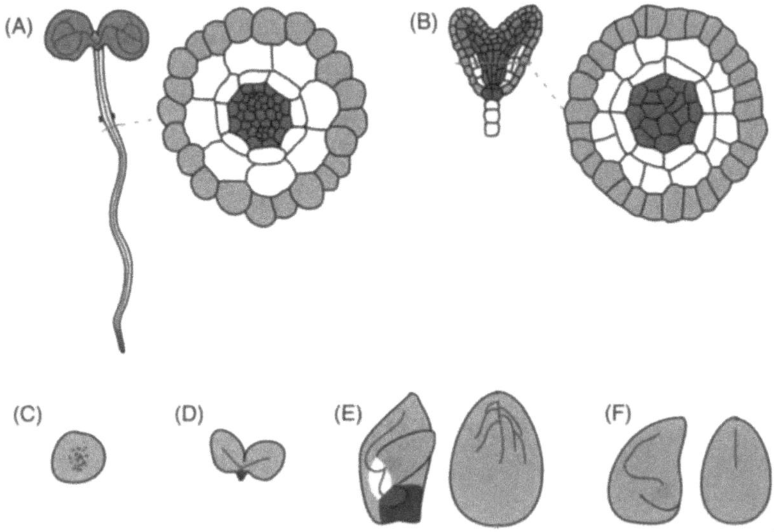

Figure 8.4 Vascular strand formation in embryos. (A, B) Embryonic origin of seedling structures. The reproducibility of Arabidopsis embryo development enables tracing the origin of seedling organs and tissues (A) to progenitor cells in the early embryo. Although cells in most tissues will only complete differentiation after germination, the complexity of the tissue pattern in the heart-stage embryo (B) basically equals that of the seedling. Colors identify corresponding regions in seedling and embryo. Longitudinal subdivision: shoot apical meristem (light gray), cotyledons (light blue), embryo axis (white), root apical meristem (dark gray), vascular strands (brown). Radial subdivision: epidermis (light brown), ground tissue (white), vascular strand (brown). (C) Auxin transport-defective Arabidopsis seedling. Pharmacological or genetic interference with auxin transport during embryogenesis results in seedlings completely lacking apical–basal polarity and with randomly oriented vascular cells. (D) Auxin response or biosynthesis Arabidopsis mutants. The vascular system is strongly reduced; hypocotyl and root are missing and a conical structure devoid of vascular tissues (dark blue) forms instead. (E) Mature wild-type rice embryo. Left, median longitudinal section; right, adaxial view. A procambial strand is positioned at the center of the C-shaped embryo axis (white). The shoot and root apical meristem (light and dark gray, respectively) are surrounded by the coleoptile (light brown) and coleorhiza (dark blue), respectively, which are not present in the dicot embryo. The vascular system of the scutellum, which represents the single cotyledon, is composed of a primary procambial strand that branches at the apex into a series of secondary strands. Occasionally, tertiary strands branch from secondary strands. (F) Mature embryo of the auxin-insensitive *ral1* mutant of rice. Left, median longitudinal section; right, adaxial view. The embryo axis with its central procambial strand is absent; the vascular system of the scutellum is limited to a short primary procambial strand and higher-order strands are completely missing. Redrawn from: Fig. 3B,C,D, Berleth, T., and Scarpella, E. (2004), Polar signals in vascular development, In "Polarity in Plants" (K. Lindsey, Ed.), pp. 264–287, Blackwell Publishing Ltd, Oxford with kind permission of Wiley-Blackwell; Fig. 2E, Dolan, L., Janmaat, K., Willemsen, V., Linstead, P., Poethig, S., Roberts, K., and Scheres, B. (1993), Cellular-Organization of the Arabidopsis-Thaliana Root. *Development* 119, 71–84, Fig. 3G, Mayer, U., Buttner, G., and Jürgens, G. (1993), Apical–Basal Pattern-Formation in the Arabidopsis Embryo—Studies on the Role of the Gnom Gene, *Development* 117, 149–162, Fig. 1J, Scheres, B., Dilaurenzio, L., Willemsen, V., Hauser, M. T., Janmaat, K., Weisbeek, P., and Benfey, P. N. (1995), Mutations Affecting the Radial Organization of the Arabidopsis Root Display Specific Defects Throughout the Embryonic Axis, *Development* 121, 53–62, with kind permission of the Company of Biologists. (See Color Insert.)

et al., 2003; Fig. 8.4C). However, mutations in a single Arabidopsis gene, *EMBRYO DEFECTIVE30/GNOM* (*EMB30/GN*), are sufficient to result in embryonic defects comparable to those reported for embryos treated with high doses of auxin transport inhibitors or mutant for multiple *PIN* genes (Mayer *et al.*, 1993; Fig. 8.4C). Embryos and seedlings mutant for *EMB30/GN* can be entirely ball-shaped with no detectable apical–basal polarity, and the vascular system of these mutants consists of randomly oriented cells. Therefore, mutations in *EMB30/GN* seem to identify a specific cellular mechanism underlying alignment of vascular differentiation with the apical–basal axis of the embryo. The EMB30/GN protein is a guanine nucleotide exchange factor (GEF) implicated in the control of ADP ribosylation factor (ADP-RF) activity (Shevell *et al.*, 1994; Steinmann *et al.*, 1999). ADP-RFs are small GTP-binding proteins that function as molecular switches to regulate vesicle trafficking (Chavrier and Goud, 1999). GDP-bound ADP-RFs are inactive in the cytoplasm, and spontaneous GDP release from ADP-RFs (a prerequisite for GTP binding) is very slow and must be catalyzed by GEFs. The GDP to GTP transition induces a conformational change, which allows ADP-RFs to interact with the specific membrane where the GEF is localized. Therefore, the location of the GEF is a determinant for the definition of the appropriate membrane sites of ADP-RF activation. Membrane-bound ADP-RF–GTP triggers the recruitment of coat protein complexes to the donor compartment, and assembly of coat subunits at the membrane deforms the donor membrane and captures cargo molecules in the forming vesicle. A vesicle-associated GTPase activating protein (GAP) stimulates the hydrolysis of bound GTP to release ADP-RF–GDP and to trigger coat dissociation prior to vesicle fusion. By promoting the switch from ADP-RF–GDP to ADP-RF–GTP, ADP-RF GEFs like EMB30/GN are positive regulators of ADP-RF function, and mutations in *EMB30/GN* perturb the positioning of PIN proteins at the basal plasma membrane (Geldner *et al.*, 2003; Kleine-Vehn *et al.*, 2008; Steinmann *et al.*, 1999). Continuous EMB30/GN-dependent vesicle transport may be necessary to maintain auxin transport because auxin efflux carrier components are not statically anchored to the plasma membrane but subject to constant reshuffling from intracellular locations (Friml *et al.*, 2002; Geldner *et al.*, 2003; 2001; Kleine-Vehn *et al.*, 2008), which could explain the central importance of EMB30/GN for auxin transport and embryo polarity. The strict dependence of procambial strand formation and apical–basal polarity acquisition on auxin transport in the embryo suggests that the apical–basal axis and vascular strand formation may have a common origin in the orientation and distribution of auxin flow.

In addition to the polar localization of membrane proteins ensuring the directionality of auxin transport, proper auxin perception is required to promote vascular differentiation. Mutations in three genes,

MONOPTEROS (*MP*), *BODENLOS* (*BDL*), and *AUXIN RESISTANT6* (*AXR6*), interfere to varying degrees with formation of the central procambial strand in the embryo and with formation of the embryo axis (Berleth and Jurgens, 1993; Hamann *et al.*, 1999; Hobbie *et al.*, 2000; Fig. 8.4D). Mutations in all three genes have been associated with defective auxin responses (Hamann *et al.*, 1999; Hobbie *et al.*, 2000; Mattsson *et al.*, 2003), and the products of all three genes are directly linked to auxin signal transduction. *MP* encodes a transcriptional regulator of the AUXIN RESPONSE FACTOR (ARF) family (Hardtke and Berleth, 1998). This family of transcription factors can confer auxin responsiveness to downstream genes by binding to conserved promoter elements, and their activity is repressed by interaction with nuclear proteins of the AUXIN/INDOLE-3-ACETIC ACID (AUX/IAA) family (Guilfoyle and Hagen, 2007). The semi-dominant *bdl* mutation confers enhanced stability and elevated abundance to the negative ARF regulator IAA12 (Hamann *et al.*, 2002; Weijers *et al.*, 2005), and the embryonic phenotype of *bdl* mutants results from inhibition of MP activity (Weijers *et al.*, 2006). Finally, the two semi-dominant *axr6* mutations affect a crucial residue in the CULLIN1 protein, which affects its ability to assemble into stable SKP1/CULLIN/F-BOX complexes (Hellmann *et al.*, 2003). These complexes, in turn, are required for efficient degradation of negative regulators of auxin response of the AUX/IAA family (Mockaitis and Estelle, 2008). Severe embryonic patterning defects have only been found associated with single mutations in *MP, BDL*, and *AXR6*, but contributions of additional genes implicated in auxin biosynthesis and signaling have been obscured by widespread functional redundancy (e.g., Cheng *et al.*, 2007; Dharmasiri *et al.*, 2003; 2005; 2007; Hardtke *et al.*, 2004; Stepanova *et al.*, 2008; Thomas *et al.*, 2009).

What cell differentiation events are precisely dependent on the function of these genes is not entirely clear. They could be involved in relaying a signal that has a general role in oriented cell differentiation in embryos. Consistent with a role in cellular events beyond those that form the procambial strand in the center of the embryo axis, *MP* is initially expressed in broad domains that correlate with sites of aligned cell differentiation (Hardtke and Berleth, 1998); only after the embryo has acquired an anatomically distinct apical–basal axis does the *MP* transcript become restricted to the central procambial strand. Failure to confine *MP* expression to the core of the embryo axis might then be expected to mostly affect procambial strand, rather than embryo axis, formation. Indeed, simultaneous mutation of the protein phosphatase-encoding genes *POLTERGEIST* (*POL*) and *POLTERGEIST-LIKE1* (*PLL1*) of Arabidopsis results in wide *MP* expression domains that are not laterally constrained during embryogenesis (Gagne *et al.*, 2008), and *pol pll1* embryos eventually differentiate a cylindrical embryo axis devoid of the central procambial core (Song *et al.*,

2008). Alternatively, auxin signal transduction may exclusively control formation of the embryonic procambial cylinder, which, in turn, could then provide a scaffold system to align a variety of morphological features, and positional effects of procambial strands on patterning of surrounding tissues have been reported (Aida *et al.*, 2004; Helariutta *et al.*, 2000; Sabatini *et al.*, 1999).

Mature monocot embryos are the result of morphogenetic processes that generate a series of embryonic organs not present in dicot embryos (Itoh *et al.*, 2005; Fig. 8.4E). Further, shoot development proceeds to a more advanced stage in monocot embryos, such that a number of leaf primordia are initiated by the shoot apical meristem before embryogenesis is completed. The vast divergence in the appearance of monocot and dicot embryos therefore justifies the question of whether any general conclusion can be drawn from studies in dicot species. As in dicots, however, application of auxin transport inhibitors to wheat embryos disturbs apical–basal axis formation (Fischer and Neuhaus, 1996). Further, despite their divergent appearance, there may still be landmarks to be assessed for similarities between dicot and monocot embryos, for example, the presence of a procambial strand positioned in the center of the C-shaped embryo axis that joins the shoot apical meristem with the root apical meristem. Mutation in the *RADICLELESS1* (*RAL1*) gene of rice is characterized by defects related to those of *mp, bdl*, and *axr6* mutants of Arabidopsis (Scarpella *et al.*, 2003; Fig. 8.8.F). Mutant embryos fail to establish an embryo axis, recognizable by the absence of a central procambial strand and of a root apical meristem, and *ral1* mutants are auxin insensitive. These observations suggest that related auxin-dependent patterning mechanisms control apical–basal polarity acquisition, embryo axis formation, and vascular differentiation in dicots and monocots.

5.3. Vascular strand formation in leaf development

In dicot leaves, vascular strands are arranged in a ramified pattern that largely reflects the shape of the leaf (Candela *et al.*, 1999; Dengler and Kang, 2001; Nelson and Dengler, 1997). Lateral veins branch from a conspicuous central vein (midvein) that is continuous with the stem vasculature (Fig. 8.5A). In many species, lateral veins extend along the leaf edge to form marginal veins, which connect to distal veins to form prominent closed loops. Finally, a series of higher-order veins branch from midvein and loops, and can either terminate in the lamina (free-ending veins) or join two veins (connected veins). Definition of the major elements of leaf shape temporally coincides with formation of major veins (i.e., the midvein and lateral veins) during leaf development (Candela *et al.*, 1999; Dengler and Kang, 2001; Nelson and Dengler, 1997; Scarpella *et al.*, 2004), and a suite of mutants exists in which leaf shape and vascular pattern defects are coupled (e.g., Bohmert *et al.*, 1998; Calderon-Villalobos *et al.*, 2005; Ceserani *et al.*, 2009; Clay and Nelson, 2005; Cnops *et al.*, 2006; Deyholos *et al.*, 2003; Kang *et al.*, 2007; Li *et al.*, 2005; Navarro *et al.*, 2004; Petricka *et al.*, 2008; Petricka and Nelson, 2007; Robles *et al.*, 2010; Semiarti *et al.*, 2001; Shirakawa *et al.*, 2009; Sun *et al.*, 2002; Yu *et al.*, 2004; Zgurski *et al.*, 2005). These observations suggest that either the same factors regulate leaf morphogenesis and vein patterning or that the two processes influence one another. While the interdependency between leaf form acquisition and vein network configuration remains largely unexplored, intertwined pathways of auxin transport and signal transduction have been implicated in the control of both leaf primordium initiation and vascular strand formation.

Asymmetric distribution of PIN1 at the plasma membrane of epidermal cells at the shoot apex suggests that auxin transport routes converge toward sites of leaf primordium formation (chapter 4; Bayer *et al.*, 2009; Benkova *et al.*, 2003; Heisler *et al.*, 2005; Reinhardt *et al.*, 2003; Scarpella *et al.*, 2006; Fig. 8.5B). Epidermal "convergence points" of PIN1 polarity identifying locations of primordium initiation become associated with subsequently emerging, broad subepidermal domains of PIN1 expression that will narrow to define the position of midvein formation. Similarly, lateral outgrowth of the leaf primordium and positioning of lateral vein-forming PIN1 subepidermal domains seem to be anchored to one another through epidermal PIN1 convergence points at the primordium margin (Hay *et al.*, 2006; Scarpella *et al.*, 2006; Wenzel *et al.*, 2007; Fig. 8.5C). Excess auxin resulting from local auxin application or auxin transport inhibition reduces spacing of epidermal convergence points in the leaf and leads to the surplus formation of closely spaced loops, similar to vein arrangements observed in *pin1* mutant leaves (Mattsson *et al.*, 1999; Scarpella *et al.*, 2006; Sieburth, 1999; Fig. 8.5D). Because leaf expansion is inhibited by superabundant auxin (Keller *et al.*, 2004; Ljung *et al.*, 2001; Sachs, 1989), efficient auxin drainage through incipient veins may allow leaf growth; the consequent absence of auxin buildup may act as a signal that the need for tissue vascularization has been met, therefore contributing to the tight and robust relation between leaf development and vein formation. Consistent with a functional similarity between midvein and lateral vein positioning in the leaf, on the one hand, and primordium formation at the shoot apex, on the other, auxin overload (through direct auxin application or auxin transport inhibition) also reduces separation between successively initiated leaf primordia (Bennett *et al.*, 1995; Okada *et al.*, 1991; Reinhardt *et al.*, 2000). These observations suggest that positioning of convergence points is strictly dependent on auxin supply and transport.

Mechanisms integrating aligned cell differentiation during vascular strand formation were probably not reinvented in the evolution of leaves, but more likely recruited and revised by leaf-specific controls. Epidermal convergence of auxin flow in the formation of lateral shoot organs and of major veins within them is consistent with the prevailing hypothesis of how leaves formed during the evolution of land plants. Fossil evidence suggests that leaves with broad laminae evolved from simple, leafless branching systems (Beerling and Fleming, 2007; Zimmermann, 1930; 1952). However, the leaf venation is more than a series of branching veins: it comprises closed networks, where most veins are connected to others at both ends with no obvious polarity (Sachs, 1975; 1989). Visualization of early auxin transport routes during lamina formation provided a possible explanation that could reconcile the polarity of auxin flow with a closed vein network. During normal and experimentally altered leaf vein formation, PIN1 expression in subepidermal cells precedes and converges toward sites of procambial strand formation and, at the PIN1 expression level, all veins appear to be generated through two basic ontogenies (Scarpella *et al.*, 2006; Wenzel *et al.*, 2007; Fig. 8.5B, C). While the midvein and lateral veins

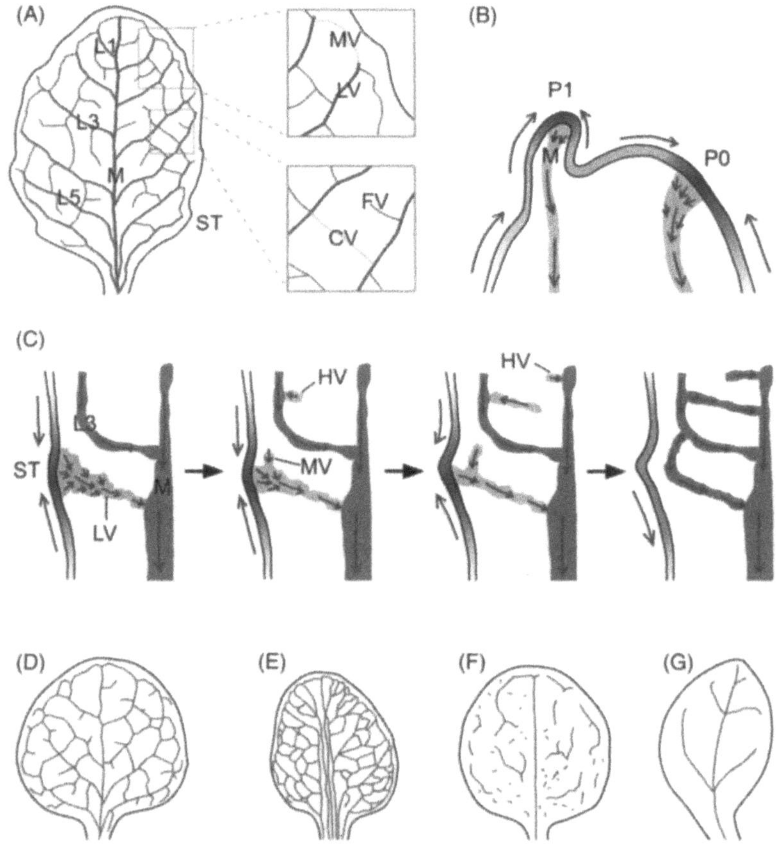

Figure 8.5 *(Continued)*

Figure 8.5 Vein formation in leaf development. (A) Simplified mature dicot leaf illustrating midvein (M), first, third, and fifth loops (L1, L3, and L5, respectively)—each derived from corresponding lateral (LV) and marginal (MV) veins—free-ending (FV) and connected (CV) higher-order veins, and serration tips (ST). (B, C) Routes of auxin transport (brown arrows) and epidermal convergence points of auxin flow (dark blue) in leaf primordium initiation and midvein formation (B), and in marginal leaf outgrowth and lateral vein formation (C) as inferred from subcellular localization of PIN1. (B) Localization of PIN1 suggests that auxin flows in a base-to-tip direction in epidermal cells of the stem, basal to the site of primordium formation (P0). Auxin also flows in a tip-to-base direction in all epidermal cells of the shoot apical meristem apical to the site of primordium formation, with resulting formation of a point of converging auxin flows; there, auxin is transported in internal tissues where it gradually induces formation of a vascular strand (light blue), which transports auxin in a tip-to-base direction. Epidermal auxin flows also converge at the tip of older primordia (P1); there, auxin is drained through the center of the primordium, marking the position of midvein formation. Older primordia additionally deplete the surrounding area of auxin through its transport in epidermal cells toward the site of new primordium formation. (C) Positions of marginal expansion zones in leaf development and sites of lateral vein formation are associated with epidermal convergence of auxin flow. Marginal veins emerge in continuity with lateral veins from PIN1 domains initiated within the growing lamina. Open-ended marginal vein precursors form the upper part of each vein loop and display uniform auxin transport polarity toward preexisting lateral veins, but they switch to bipolarity as they become connected at both ends to give rise to closed vein loops. Similar to marginal veins, higher-order veins (HV) appear in continuity with preexisting vasculature from PIN1 domains initiated within the expanding blade, and can end freely in the lamina or become connected on proximity to other PIN1 domains. Arrows temporally connect successive stages of vein formation. (D) Vascular pattern in Arabidopsis wild-type mature leaf. (E) Vein pattern of auxin transport-inhibited or mutant leaves. Auxin accumulates to inductive levels closer to the marginal sources. More cells are exposed to high levels of auxin for longer periods and therefore vascular differentiation occurs in broader zones and at more locations. Vascular strands are broader and increased in number, and marginal veins become relatively more pronounced. (F) Vein continuity mutant pattern. Segments of vascular strands differentiate discontinuously along paths defined by initially continuous routes of auxin flow. (G) Auxin-reponse or biosynthesis mutant vein pattern. The vascular pattern is reduced to the midvein and a few lateral veins; neither marginal nor higher-order veins differentiate. Impaired sensitivity to or reduced levels of vascular-inducing auxin signals restricts vascular differentiation to central and basal positions, the sites of convergence of auxin transport routes from across the lamina. Redrawn from: Fig. 7, Alonso-Peral, M. M., Candela, H., del Pozo, J. C., Martinez-Laborda, A., Ponce, M. R., and Micol, J. L. (2006), The *HVE/CAND1* gene is required for the early patterning of leaf venation in Arabidopsis, *Development* 133, 3755–3766, by kind permission of the Company of Biologists; Fig. 2C, Berleth, T., Scarpella, E., and Prusinkiewicz, P. (2007), Towards the systems biology of auxin-transport-mediated patterning, *Trends in Plant Science* 12, 151–159, by kind permission of Elsevier; Fig. 4E, Przemeck, G. K., Mattsson, J., Hardtke, C. S., Sung, Z. R., and Berleth, T. (1996), Studies on the role of the Arabidopsis gene *MONOPTEROS* in vascular development and plant cell axialization, *Planta* 200, 229–237, with kind permission of Springer Science and Business Media; Fig. 1A, Sawchuk, M. G., Donner, T. J., and Scarpella, E. (2008), Auxin transport-dependent, stage-specific dynamics of leaf vein formation, *Plant Signaling & Behavior* 3, 286–289, with kind permission of Landes Bioscience; Fig. 3B,G,H,I, 6R, Scarpella, E., Marcos, D., Friml, J., and Berleth, T. (2006), Control of leaf vascular patterning by polar auxin transport, *Genes Dev.* 20, 1015–1027, with kind permission of Cold Spring Harbor Laboratory Press.

originate from subepidermal PIN1 domains associated with convergence points of PIN1 polarity in the epidermis, higher-order veins emerge from PIN1 domains initiated within the expanding lamina. These internal domains are initially free-ending but can become connected upon proximity to other PIN1 domains. Interestingly, individual loops are composed of a "lateral" PIN1 domain, which is initiated at an epidermal convergence point, and a "marginal" domain, which is ontogenically equivalent to a connected higher-order PIN1 domain (Fig. 8.5B). In mature leaves, the composite origin of the first and second loop pairs is obscured by the smooth amalgamation of lateral and marginal veins, but this origin can be recognized in the third and subsequent loop pairs (Kang and Dengler, 2004; Scarpella *et al.*, 2006; Fig. 8.5A). Heightened auxin levels, occurring naturally, in association with leaf margin projections, or experimentally, as a consequence of either direct auxin application or auxin transport inhibition, lead to the expansion of PIN1 expression domains (Aloni *et al.*, 2003; Hay *et al.*, 2006; Mattsson *et al.*, 2003; Scarpella *et al.*, 2006; Wenzel *et al.*, 2007). Broad PIN1 domains eventually taper to a few cell files that predict vein position, and this narrowing process is dependent on auxin transport. Subcellular PIN1 localization indicates that PIN1 polarity may not be uniformly directed toward preexisting veins across wide PIN1 domains, but it is usually so along each domain's midline. Within narrow PIN1 domains, subcellular localization indicates auxin transport toward preexisting veins: in free-ending domains, a single polarity exists, while in connected domains, two opposite polarities are bridged by a bipolar cell (Scarpella *et al.*, 2006; Wenzel *et al.*, 2007). Therefore, closed vein networks originate from open-ended precursors with uniform auxin-transport polarity toward preexisting veins that switches to bipolarity as they become connected at both ends (Fig. 8.5B).

These observations are consistent with the notion that the vasculature is formed along core areas of gradually restricted domains of elevated auxin transport (Mitchison, 1980; Rolland-Lagan and Prusinkiewicz, 2005; Sachs, 1981). Traditional models of auxin flow, which rely on high levels of auxin-exporter expression in cells specialized for auxin transport, predict auxin depletion from developing veins. Computational simulations suggest that cells that become incorporated into a preferred auxin transport route need only contain relatively low levels of auxin at the beginning of the process, when they are acting as sinks (Feugier *et al.*, 2005; Fujita and Mochizuki, 2006a, b; Mitchison, 1981); at later stages, auxin could still flow into transporting cells even if these have accumulated relatively high concentrations of auxin. These predictions are difficult to reconcile with experimental evidence suggesting that maximum auxin levels characterize early stages of vein formation (Aloni *et al.*, 2003; Mattsson *et al.*, 2003; Scarpella *et al.*, 2004; 2006). However, polarization of peripherally expressed PIN1 toward the midline of broad PIN1 expression domains is predicted to be sufficient to generate high auxin concentrations in central cells, which show basal PIN1 localization and will differentiate into vascular cells (Bayer *et al.*, 2009). Similar results are obtained by supplementing auxin export-based models with additional carrier-mediated auxin uptake in vascular cells (Kramer, 2004; 2009), and this hypothesis can now be experimentally tested as multiple mutant combinations of members of the AUXIN RESISTANT1 (AUX1)/LIKE AUX1 family of auxin importers in Arabidopsis have recently become available (Bainbridge *et al.*, 2008; Ugartechea-Chirino *et al.*, 2010). Nevertheless, in its most basic form, an auxin transport model generates open ramified structures (Feugier *et al.*, 2005; Fujita and Mochizuki, 2006b; Mitchison, 1980), while most, if not all, leaves are characterized by networks of connected veins, the most conspicuous examples of which are the prominent vascular loops visible in the leaves of many species. Vein connection could originate from repeated shifts in the centers of auxin synthesis (Aloni, 2003; Rolland-Lagan and Prusinkiewicz, 2005; Runions *et al.*, 2005; Sachs, 1975; 1989) or through auxin-draining vein segments extending within an initially homogeneous auxin concentration field and meeting at the point of highest auxin concentration (Dimitrov and Zucker, 2006). Both hypotheses are compatible with the existence of bipolar cells integrating opposite polarities along the length of connected veins, and further predict maximum auxin levels in these cells; however, whether local auxin concentration peaks exist at the locations of bipolar cells remains to be experimentally tested.

It is far less intuitive how PIN1 polarization toward auxin maxima at convergence points may be reconciled with an auxin transport model. The simplest explanation is that different tissues adopt different strategies for orienting PIN1: epidermal cells would polarize PIN1 against an auxin concentration gradient, while inner cells would align PIN1 polarity with the direction of auxin flow. Evidence of the existence of tissue-specific determinants of PIN1 polarity exists (Vieten *et al.*, 2005; Wisniewska *et al.*, 2006; Xu *et al.*, 2006), and this model seems to be able to simulate the formation of convergence points and of their associated midvein and lateral veins (Smith and Bayer, 2009). Alternatively, both strategies could simultaneously be operating in different tissues (Bayer *et al.*, 2009). Examination of early stages of convergence-point formation seems to indicate that all cells, whether epidermal or subepidermal, polarize PIN1 toward one or very few epidermal cells. The switch of PIN1 localization in the epidermal cells at the center of the convergence point and in the subepidermal cells immediately below it toward basal polarity would demarcate the transition between a convergence point-patterning mechanism that functions against an auxin gradient at low auxin levels and a flow-directed polarizing model that generates vascular strands at high concentrations of auxin. By contrast, a single strategy for PIN1 polarization could account for both patterns. In models that hypothesize PIN1 polarization toward auxin concentration peaks, the unstable maximum could move toward inner tissues, leaving behind a file of prospective vascular cells with PIN1 polarized in the direction of the traveling peak (Merks *et al.*, 2007). Such a mechanism relies, however, on the presence of an unknown signal that communicates information about auxin concentration between the adjacent cells and directs PIN1 recruitment to regions of the plasma membrane facing cells with relatively high auxin concentration. In models that assume alignment of PIN1 polarity with orientation of auxin flow, drainage of epidermally accumulated auxin through incipient vascular strands could generate a sink toward which epidermal cells reorient PIN1 localization (Stoma *et al.*, 2008). In order for this to happen, however, PIN1 reallocation must respond differently to auxin transport in epidermal cells and internal tissues: broad, laminar PIN1 expression domains in the epidermis could emerge if PIN1 response to auxin flow is linear, while strands of PIN1 expression could derive from a nonlinear response (Feugier *et al.*, 2005; Mitchison, 1980; Stoma *et al.*, 2008). This mechanism would still require cells to be able to sense net auxin transport across the plasma membrane, although cells could use their internal auxin concentration gradient to sense flow (Kramer, 2009; Mitchison, 1981). While the product of computational simulations may be very similar to the final vascular pattern of a mature leaf, future attempts will have to be evaluated by their capability to recapitulate vein ontogeny in the context of a developing primordium, including the very small fields of cells within which the patterning mechanism has been shown to occur. On the other hand, available observations are still of insufficient resolution for discriminating tests of mathematical models, and further experimentation is required to visualize the intercellular signals and cellular mechanisms postulated by the formalized theoretical descriptions, including improvement of critical visualization tools, routine use of dynamic, quantitative phenotyping systems (e.g., Rolland-Lagan *et al.*, 2009), and molecular characterization of new, informative components (e.g., Steynen and Schultz, 2003).

Uninterrupted differentiation of vascular cells along continuous lines is obviously a stringent requirement for long-distance transport through vascular strands, and continuous cell-to-cell transport of the auxin signal should apparently be excluded as an organizing principle in all discontinuous venation patterns. Mutations in the *VASCULAR NETWORK3/SCARFACE* (*VAN3/SFC*) gene of Arabidopsis results in highly fragmented vein networks (Deyholos *et al.*, 2000; Koizumi *et al.*, 2000; 2005; Sieburth *et al.*, 2006; Fig. 8.5E); a phenotype that is accentuated by simultaneous mutation in the three Arabidopsis genes most related to *VAN3/SFC* (Sieburth *et al.*, 2006). Vein fragments in the mutant backgrounds seem to be largely formed along the same paths where they would normally differentiate continuously, suggesting that the isolated stretches of procambium are visible segments of a larger, invisible pattern. Consistent with this observation, subepidermal PIN1 expression domains are initially continuous in the *van3/sfc* leaf primordia (Naramoto *et al.*, 2009; Scarpella *et al.*, 2006). However, continuity of PIN1 expression domains is fragile in *van3/sfc* and leads to discontinuous PIN1 expression domains at later stages of vein ontogeny, suggesting that the primary defect in *van3/sfc* mutants lies in the maintenance rather than in the initial establishment of a continuous vascular pattern. This obviously raises the question of why do initially continuous PIN1 expression domains break down during *van3/sfc* leaf development. At early stages of both wild-type and *van3/sfc* vein ontogeny, continuous PIN1 expression domains connecting to preexisting domains may initially comprise very few cells but extend by intercalary cell division during leaf development. In wild type, any cell inserted into a continuous PIN1 expression domain by rapid division would be expected to very quickly adopt position-dependent polarity and auxin transport properties, thereby maintaining continuity. During cell division, PIN1 is localized at the nascent cell plate (Geldner *et al.*, 2001), suggesting that PIN1 must be relocated to the appropriate plasma membranes of daughter cells after division. *VAN3/SFC* encodes an ADP ribosylation factor GTPase activating protein (ADP-RF GAP) (Koizumi *et al.*, 2005; Sieburth *et al.*, 2006). Conversion of the activated ADP-RF–GTP form into ADP-RF–GDP, which is promoted by GAPs, is necessary for vesicle uncoating, a prelude for vesicle fusion with specific target membranes (Chavrier and Goud, 1999). Therefore, it is possible that VAN3/SFC activity may be required to deliver PIN1 to the appropriate plasma membrane of cells resulting from a division event and that failure to integrate polarity of newly formed cells with that of neighboring ones may result in fragmentation of PIN1 expression domains. Irrespective of the precise mechanism through which *VAN3/SFC* preserves continuity of PIN1 expression domains, localization of VAN3/SFC to the proper endomembrane compartment should be crucial for the protein's function. The activity of the COTYLEDON VASCULAR PATTERN2 (CVP2) and CVP2-LIKE1 (CVL1) inositol polyphosphate 5′ phosphatases (5-PTases) of Arabidopsis is critical for VAN3/SFC intracellular localization, and simultaneous mutation in *CVP2* and *CVL1* results in collapse of continuity of the PIN1 expression domains and in discontinuous vein patterns (Carland and Nelson, 2009; Naramoto *et al.*, 2009). CVP2 and CVL1, and possibly other 5-PTases, act redundantly to generate phosphatidylinositol 4-monophosphate, the phosphoinositides species that is bound by the VAN3/SFC ADP-RF GAP and presumably targets VAN3/SFC to the right endomembrane compartment (Carland and Nelson, 2009; Koizumi *et al.*, 2005; Lin *et al.*, 2005; Naramoto *et al.*, 2009; Sieburth *et al.*, 2006). Therefore, available evidence supports a decisive role for auxin flow along continuous paths in defining the course of vascular strand formation; however, permanence of such continuity seems to be inconsequential for later stages of vascular differentiation.

Similarly to *van3/sfc* and *cvp2*, mutations in the Arabidopsis gene *CVP1* lead to vein fragmentation (Carland *et al.*, 1999). *CVP1* encodes STEROL METHYLTRANSFERASE2 (Carland *et al.*, 2002), an enzyme that regulates the ratio of brassinosteroids to sterols. This ratio in wild type is shifted toward production of sterols, but is inverted in *cvp1* to favor the brassinosteroid pathway. While the phenotypic consequences of the *cvp1* mutation could be the result of enrichment in brassinosteroids or depletion of sterols, available evidence supports the latter possibility (Carland *et al.*, 2002). Sterols are synthesized at the endoplasmic reticulum and transported via the Golgi to the plasma membrane, where they mostly accumulate (Boutte and Grebe, 2009). In animals, membranes contain sterol-enriched domains that compartmentalize cellular processes and that are referred to as lipid rafts. Proteins enriched in such domains are characterized by their insolubility in non-ionic detergents. Plant membrane fractions insoluble under those conditions are classified as detergent-resistant membranes, but there is currently no *in vivo* evidence that sterols and specific proteins cluster together in membrane domains in plants. Nevertheless, because PIN1 abundance is slightly enriched in detergent-resistant membrane fractions (Titapiwatanakun *et al.*, 2009), and PIN1 polarized distribution at the plasma membrane is disrupted in the sterol-deficient Arabidopsis mutant *orc/smt1* (Willemsen *et al.*, 2003), it is possible that vein continuity defects in *cvp1* are the result of mislocalized PIN1 expression.

Paths of auxin flow seem to integrate sites of tissue expansion and vein formation in leaf development; however, proper auxin biosynthesis and perception should be required for triggering leaf primordium outgrowth and vascular differentiation. Consistent with this observation, defects in auxin synthesis or response are associated with reduced leaf size and simplified vein networks (Alonso-Peral *et al.*, 2006; Candela *et al.*, 2007; Cheng *et al.*, 2006; Przemeck *et al.*, 1996; Steynen and Schultz, 2003; Fig. 8.5F), and severely compromised auxin biosynthesis or signaling results in failure to promote leaf primordium initiation (Cheng *et al.*, 2007; Hardtke *et al.*, 2004; Schuetz *et al.*, 2008). Furthermore, defective auxin sensitivity in *mp* mutants (Mattsson *et al.*, 2003) is associated with initiation of fewer and weaker PIN1 subepidermal expression domains in the leaf (Wenzel *et al.*, 2007). Finally, elevated levels of local auxin response precede procambium formation in both untreated and auxin transport-inhibited leaves (Mattsson *et al.*, 2003), and MP directly controls correct initiation of leaf expression of the Class III *HOMEODOMAIN LEUCINE ZIPPER* (*HD-ZIP III*) gene *ARABIDOPSIS THALIANA HOMEOBOX8* (*ATHB8*), a regulator of procambium development (Baima *et al.*, 1995; 2001; Ilegems *et al.*, 2010; Donner *et al.*, 2009). As in embryogenesis, failure to refine broad domains of *MP* expression to incipient vascular strands during lateral organ development might be expected to primarily affect procambial strand formation, rather than organogenesis *per se*. Wide domains of *MP* expression are initiated in lateral organ primordia of Arabidopsis seedlings burdened by mutation in the *NO VEIN* (*NOV*) gene, which encodes a nuclear protein of unknown function (Tsugeki *et al.*, 2009); however, strong *nov* alleles fail to maintain *MP* expression and generate misshapen lateral organs that can be devoid of any vasculature. Acute defects in auxin response and absence of vascular tissue formation also characterize the most extreme leaf phenotypes of the *bypass1* (*bps1*) mutant of Arabidopsis (Van Norman *et al.*, 2004). *BPS1*, a gene of unknown function, is required to prevent constitutive production of a novel root-derived, mobile signaling compound that is sufficient to inhibit shoot apical meristem activity and leaf development. However, the specificity of the auxin signaling and vascular abnormalities in *bps1* remains to be confirmed.

In dicots, even low concentrations of auxin transport inhibitors, which do not significantly affect the shape of the leaf, result in entirely new types of venation patterns (Mattsson *et al.*, 1999; Sieburth, 1999). Therefore, in these species, vein patterns may be largely self-organized. The situation is less clear for monocot leaves, which are characterized by a striate pattern where major longitudinal veins lie parallel along the proximo-distal axis of the leaf and are connected transversally by minor vascular strands (Nelson and Dengler, 1997; Fig. 8.2D). Typically, the distribution and arrangement of veins in monocot leaves are highly reproducible and may therefore be rigidly specified. Alternatively, it is possible that treatments that in dicots uncouple venation patterns from other aspects of leaf development, such as auxin transport inhibition, could have similar effects on monocot leaves. Application of polar auxin transport inhibitors to rice leaves results in alterations of vein spacing without affecting the overall vein arrangement or leaf shape (Scarpella *et al.*, 2002; Sakaguchi *et al.*, 2010), and subtle vein pattern defects have been reported in mutants of rice with defective auxin transport or response (Qi *et al.*, 2008; Scarpella *et al.*, 2003; Sakaguchi *et al.*, 2010). These observations suggest that monocot vein patterns have a certain amount of plasticity but simultaneously show that this plasticity is far more constrained than in dicot leaves, presumably by genetic cues that regulate the development of the leaf.

In summary, there is a growing body of mutually supporting evidence for an instrumental role of auxin as an intercellular signal that mediates the integrated polarization of plant cells and the differentiation of vascular strands along routes of elevated auxin transport and concentration. This concept reconciles plasticity and continuity of vascular patterns in species with variable venation patterns and may direct the route of vascular strands outside of leaves even in other species. However, the role of auxin may be restricted to the delineation of procambial strands, and subsequent stages of vascular differentiation may be controlled by unrelated cues.

6. Regulation of Radial Vascular Patterning

6.1. Origin of the radial vascular pattern

While vascular strands reproducibly occupy internal domains within the plant, the exact organization of xylem and phloem within each vascular bundle (i.e., the radial vascular pattern) varies depending on the organ and species.

The ontogeny of vascular tissues has been investigated in detail during Arabidopsis embryo development (Scheres *et al.*, 1995). The vascular strand of the embryo axis is first recognizable at the early globular stage as the innermost domain of three concentric cell layers (Fig. 8.6A). At the

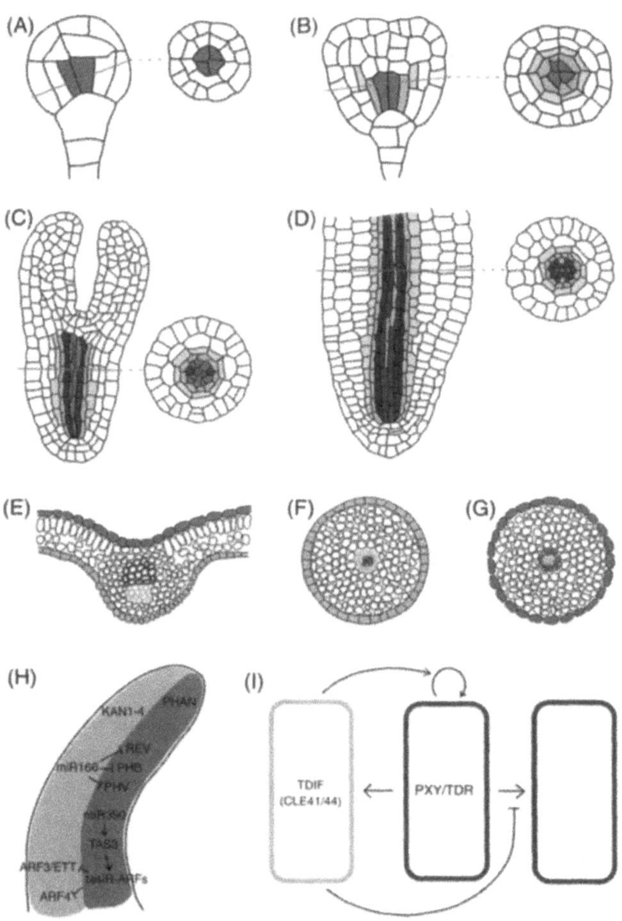

Figure 8.6 *(Continued)*

subsequent triangular stage, the pericycle becomes demarcated through asymmetric, periclinal cell divisions as the outermost cell layer of the vascular cylinder (Fig. 8.6B); this leaves a seemingly homogeneous core of procambial tissue at the center of the embryo axis. However, during subsequent stages of embryogenesis, highly regular, asymmetric divisions of procambial cells define the central xylem axis and the peripheral phloem domains (Bauby *et al.*, 2007; Mähönen *et al.*, 2000; Fig. 8.6C, D). Seemingly undifferentiated, intervening procambial cells are located between xylem and phloem, and the resulting arrangement represents the primary radial pattern of the root/hypocotyl vascular bundle, which is maintained postembryonically by asymmetric cell divisions in the root apical meristem and in the procambium proximal to it.

Anatomical details of radial vascular patterning during shoot postembryonic development are not available to as high a resolution as they are for embryo axis formation. However, the radial organization of tissues in the vascular bundle seems to reflect organ polarity. The formation of flat lateral organs such as leaves involves the specification of adaxial (dorsal) and abaxial (ventral) cell identities, and during leaf development of most species veins

---

Figure 8.6 Radial vascular patterning. (A–D) Schematic longitudinal (left) and transverse (right) representation of the radial vascular pattern of the embryo axis at selected stages of embryogenesis: early globular (A), triangular (B), torpedo (C), and mature embryo (D), respectively. Brown, apparently undifferentiated procambial cells; light brown, pericycle; dark blue, protoxylem; light blue, phloem; dark gray, metaxylem; light gray, endodermis. (E–G) Schematic representation of adaxial–abaxial pattern in wild-type (E), abaxialized (F), and adaxialized (G) leaf, respectively. Brown fill: adaxial epidermis; light brown fill: abaxial epidermis; dark blue line: xylem; light blue line: phloem; brown line: procambium. (H) Model of interactions among leaf adaxial (dark blue) and abaxial (light blue) identity promoting factors. Antagonistic interactions between polarity determinants generate mutually exclusive adaxial and abaxial domains that define organ polarity throughout leaf development. (I) Model of CLE receptor/ligand system maintaining procambial cell activity in a non-cell-autonomous manner. Phloem (light blue) secretes the TDIF/CLE41/44 peptide, which is perceived by the receptor PXY/TDR in procambial cells (brown); this signal promotes proliferation of procambium and inhibits its differentiation into xylem (dark blue). Redrawn from: Figs. 1A,C,E,F,N,O, S,T, Scheres, B., Dilaurenzio, L., Willemsen, V., Hauser, M. T., Janmaat, K., Weisbeek, P., and Benfey, P. N. (1995), Mutations Affecting the Radial Organization of the Arabidopsis Root Display Specific Defects Throughout the Embryonic Axis, *Development* 121, 53–62, Fig. 2K, McConnell, J. R., and Barton, M. K. (1998), Leaf polarity and meristem formation in Arabidopsis, *Development* 125, 2935–2942, Fig. 2A,F, Waites, R., and Hudson, A. (1995), *PHANTASTICA*—a Gene Required for Dorsoventrality of Leaves in *Antirrhinum majus*, *Development* 121, 2143–2154, by kind permission of the Company of Biologists; Fig. 4M, Hirakawa, Y., Shinohara, H., Kondo, Y., Inoue, A., Nakanomyo, I., Ogawa, M., Sawa, S., Ohashi-Ito, K., Matsubayashi, Y., and Fukuda, H. (2008), Non-cell-autonomous control of vascular stem cell fate by a CLE peptide/receptor system, Proc *Natl. Acad. Sci. U.S.A.* 105, 15208–15213, with kind permission of the National Academy of Science, U.S.A. (See Color Insert.)

are polarized according to the organ's adaxial–abaxial polarity (Esau, 1965; Fig. 8.6E): xylem is located at the upper (adaxial) side of the vein, whereas phloem occupies the lower (abaxial) part of it. Between them, a thin layer of seemingly undifferentiated procambium remains. A similar principle of arrangement of vascular tissues is observed during stem development in most species, with xylem differentiating toward the center and phloem toward the periphery of the organ (Fig. 8.1A).

6.2. Radial vascular patterning and organ polarity formation

An integrated regulation of vascular bundle and organ radial polarity was first suggested by the results of experiments in which developing leaf primordia had been surgically isolated from the shoot apical meristem (Sussex, 1951; 1955). The cultured primordia developed as cylindrical, radially symmetrical, abaxialized organs, i.e., they only contained cell types that would normally differentiate at the ventral side of the leaf (Fig. 8.6F). Accordingly, in their vascular bundles, the normally abaxially developing phloem completely surrounded the xylem. These observations suggested the existence of signals from the shoot apical meristem that would promote adaxial identity in the developing primordium. The nature of such signals is still elusive, but during the past decade significant understanding has been gained of the molecular mechanisms that specify the radial polarity of vascular bundles and of the organs in which they are embedded (see chapter 5). Many of these results have been obtained through genetic approaches in Arabidopsis and maize, and comparative analysis of the implicated loci in the two species has revealed slight differences in the genetic composition of the underlying regulatory pathways (Husbands *et al.*, 2009).

A function for the genes encoding the HD-ZIP III transcription factors REVOLUTA/INTERFASCICULAR FIBERLESS1 (REV/IFL1), PHABULOSA/ATHB14 (PHB), and PHAVOLUTA/ATHB9 (PHV) in the control of the radial organization of vascular bundles and the formation of transverse organ polarity was identified based on their gain-of-function phenotypes (Emery *et al.*, 2003; Juarez *et al.*, 2004a, b; McConnell *et al.*, 2001; Zhong and Ye, 2004). These mutants have adaxialized lateral organs, i.e., their leaves develop as radially symmetrical organs that lack discernible margin or lamina and contain cell types that would normally differentiate at the adaxial side of the leaf (Fig. 8.6G). The polarity shift in these leaves is reflected in the organization of vascular tissues in their single vascular bundle, in which the normally adaxially located xylem completely surrounds the phloem. *REV, PHB*, and *PHV* are normally expressed at the adaxial side of the developing primordium, including the xylem pole of the vascular bundles (Emery *et al.*, 2003; Juarez *et al.*, 2004a, b; McConnell *et al.*, 2001; Otsuga *et al.*, 2001; Prigge *et al.*, 2005; Zhong and Ye, 1999); however, the gain-of-function mutations result in ectopic expression of the corresponding mRNAs throughout the primordium. The opposite phenotype, i.e., abaxialized lateral organs with vascular bundles in which phloem surrounds xylem, is found in triple *rev phb phv* loss-of-function mutants (Emery *et al.*, 2003; Fig. 8.6F), suggesting that *HD-ZIP III* genes are key regulators of adaxial identity (Fig. 8.6H). However, both loss- and gain-of-function mutations in *REV* result in defects in vascular bundles of the stem in the absence of recognizable alterations in organ polarity (Emery *et al.*, 2003; Zhong *et al.*, 1997; 1999; Zhong and Ye, 1999; Zhong and Ye, 2004). These observations suggest that, at least in cylindrical organs, *HD-ZIP III* genes may have a role in vascular development that is separable from their function in organ polarity formation, a concept that seems to be further supported by root vascular phenotypes associated with loss-of-function mutations in *HD-ZIP III* genes (Carlsbecker *et al.*, 2010) (see below).

Gain-of-function mutations in the *HD-ZIP III* genes *REV*, *PHB*, and *PHV* map to a *microRNA165/166* (*miR165/166*) target sequence and lead to ectopic expression of the mutant transcripts, suggesting that *HD-ZIP III* genes are post-transcriptionally regulated by these microRNA species (Emery *et al.*, 2003; Itoh *et al.*, 2008; Juarez *et al.*, 2004a, b; Mallory *et al.*, 2004; McConnell *et al.*, 2001; Zhong and Ye, 2004). Mature *miR166* accumulates at the abaxial side of the leaf primordium (Juarez *et al.*, 2004a; Kidner and Martienssen, 2004), where it directs cleavage of the *HD-ZIP III* mRNAs (Tang *et al.*, 2003; Fig. 8.6H). How is the pattern of *miR166* expression established during organ polarity formation? An important insight into this question came with the identification of the molecular nature of the *leafbladeless1* (*lbl1*) mutation of maize (Nogueira *et al.*, 2007). In the *lbl1* mutant, expression of *HD-ZIP III* genes is strongly downregulated (Juarez *et al.*, 2004b; Timmermans *et al.*, 1998), a phenomenon that is accompanied by ectopic expression of *miR166* species throughout the leaf primordium (Nogueira *et al.*, 2007). *LBL1* encodes a functional ortholog of the Arabidopsis SUPPRESSOR OF GENE SILENCING3 (SGS3) (Nogueira *et al.*, 2007), which is an essential component of the *trans*-acting small interfering RNA (ta-siRNA) biogenesis pathway (Chapman and Carrington, 2007), suggesting that a ta-siRNA pathway is required for the spatial patterning of *miR166* species. In maize, *miR390* seems to represent a very early component of a regulatory cascade that triggers, at the adaxial domain of the primordium, the *LBL1*-dependent conversion of the *TAS3* ta-siRNA precursor into its mature forms (Nogueira *et al.*, 2007; 2009); *TAS3* ta-siRNAs restrict *miR166* expression to the abaxial side of the primordium, possibly through regulation of *ARF3* or other abaxial determinants (see below) (Fig. 8.6H), but it is unclear whether in maize the ta-siRNA pathway contributes to the vascular patterning aspect of adaxial–abaxial polarity formation (Juarez *et al.*, 2004a, b). The evidence in support of the involvement of a ta-siRNA pathway in the control of leaf polarity formation in Arabidopsis is even less clear. While adaxial-specific accumulation of ta-siRNAs occurs also in Arabidopsis (Chitwood *et al.*, 2009), mutations in components of the pathway, such as *SGS3*, do not appear to affect Arabidopsis leaf polarity (Peragine *et al.*, 2004; Vazquez *et al.*, 2004).

In cylindrical organs, patterns of *miR165/166* expression seem to be generated through a mechanism different from that which operates in flat organs such as leaves. In the vascular bundle of the Arabidopsis root, two phloem poles are separated by an intervening axis of xylem, which comprises peripheral, early differentiating protoxylem elements and central, late differentiating metaxylem elements (Bauby *et al.*, 2007; Mähönen *et al.*, 2000; Fig. 6C, D). *HD-ZIP III* genes act redundantly in Arabidopsis to promote xylem formation during root development, and their differential expression in xylem-forming procambial cells seems to control the specification of the mutually exclusive protoxylem and metaxylem fates (Carlsbecker *et al.*, 2010). The unequal expression of different *HD-ZIP III* genes in protoxylem-and metaxylem-precursor cells is achieved through post-transcriptional downregulation of *HD-ZIP III* expression by *miR165/166* at the protoxylem position. The specific *miRNA* loci involved are, however, transcribed in the endodermis, the innermost layer of ground tissue that surrounds the root vascular bundle (Fig. 6B–D), and act non-cell-autonomously in protoxylem-precursor cells. Endodermal expression of these microRNA species is, in turn, dependent on the activity of the SHORT-ROOT (SHR) transcription factor, which moves to the endodermis from the vascular bundle, where *SHR* is transcribed (Nakajima *et al.*, 2001). Therefore, patterning of the two xylem cell types in the Arabidopsis root is based on bidirectional signaling between the vascular bundle and the surrounding ground tissue, wherein SHR moves to the endodermal layer to activate expression of *miR165/166* transcripts that move to procambial cells to modulate levels of *HD-ZIP III* expression.

The adaxial identity-promoting locus *PHANTASTICA (PHAN)* was identified in *Antirrhinum* (Waites *et al.*, 1995); as in *rev phb phv* mutants of Arabidopsis, adaxial identity is lost in *phan* loss-of-function mutants and replaced by ectopic abaxial identity in the radially symmetrical leaves (Fig. 8.6F). The *PHAN* gene encodes a MYB domain transcription factor and is expressed throughout the leaf primordium (Waites *et al.*, 1998). Mutations in the Arabidopsis and maize orthologs of *PHAN* do not, however, result in phenotypic changes in the adaxial–abaxial polarity (Byrne *et al.*, 2000; Timmermans *et al.*, 1999; Tsiantis *et al.*, 1999), suggesting that the *PHAN* pathway in those species may be genetically redundant (Husbands *et al.*, 2009).

The *KANADI* genes (*KAN1-3*) encode GARP (from GOLDEN2 of maize, ARABIDOPSIS RESPONSE REGULATOR-B and PHOSPHATE

STARVATION RESPONSE1 of *Chlamydomonas*) transcriptional regulators that promote abaxial identity (Candela *et al.*, 2008; Eshed *et al.*, 2001; 2004; Kerstetter *et al.*, 2001; Zhang *et al.*, 2009). Double *kan1 kan2* and triple *kan1 kan2 kan3* loss-of-function mutants have adaxialized bladeless lateral organs, and ectopic expression of *KAN1* or *KAN2* result in fully abaxialized organs (Fig. 8.6F, G). *KAN* genes are expressed at the abaxial side of the primordia, and while *HD-ZIP III* transcripts accumulate ectopically in the *kan1 kan2 kan3* triple mutant, they are downregulated when *KAN* genes are ectopically expressed. Further, mutations in *KAN* genes can suppress the *rev phb phv* triple mutant. These observations suggest that *KAN* and *HD-ZIP III* genes interact antagonistically in adaxial–abaxial polarity formation (Fig. 8.6H). Simultaneous loss of function in the Arabidopsis genes encoding the ETTIN (ETT)/ARF3 and ARF4 auxin response transcription factors results in an adaxialized phenotype similar to that of *kan1 kan2* (Pekker *et al.*, 2005), and *ett/arf3* suppresses ectopic *KAN1* overexpression, suggesting that *ETT/ARF3* and *ARF4* may be interacting with *KAN* function. Finally, combinatorial loss-of-function mutants in the genes encoding the FILAMENTOUS FLOWER (FIL)/YABBY1, YABBY2 and YABBY3 transcription factors of Arabidopsis display weakly abaxialized phenotypes (Sawa *et al.*, 1999; Siegfried *et al.*, 1999); however, requirement for *YABBY* gene function appears to be specific for dicotyledonous species (Husbands *et al.*, 2009).

Ectopic expression of *KANADI* genes or of *FIL/YABBY1* results in abaxially radialized leaves that do not develop any vasculature (Eshed *et al.*, 2001; Kerstetter *et al.*, 2001; Sawa *et al.*, 1999); vascular bundles are also missing from most leaves of severely adaxialized *phb* gain-of-function mutants (McConnell and Barton, 1998). As both extreme phenotypes are incompatible with vascular tissue formation, it is possible that either intermediate positional values or adaxial–abaxial polarity *per se* may be essential for vascular strand formation. Adaxial–abaxial polarity may have derived from ancestral central–peripheral polarity in cylindrical organs (Tasaka, 2001), and this seems to be reflected in the organization of vascular bundles in stems and in the central–peripheral organization of hypocotyls, which in cases of extreme adaxial–abaxial shifts may be devoid of any vascular tissues (Ilegems *et al.*, 2010; Kerstetter *et al.*, 2001). Therefore, available evidence supports the integrated regulation of radial pattern formation within vascular strands and surrounding tissues in the majority of plant organs.

6.3. Vascular bundle-specific radial patterning mechanisms

While a concerted control of organ and vascular bundle polarity seems to be sufficient to account for the internal organization of vascular bundles and for their functional integration into the organ context, a number of mutations have been identified that specifically affect radial vascular patterning without noticeable alteration of organ polarity.

The recessive *woodenleg* (*wol*) mutation of Arabidopsis results in reduced cell proliferation and exclusive differentiation of the cell files of the root procambium as protoxylem (Mähönen *et al.*, 2006; Scheres *et al.*, 1995). The *wol* mutation maps to the *CYTOKININ RESPONSE1* (*CRE1*) locus, which encodes the CRE1/WOL/ARABIDOPSIS HISTIDINE KINASE4 (AHK4) cytokinin receptor (Inoue *et al.*, 2001; Mähönen *et al.*, 2000). Therefore, cytokinin signaling through the CRE1/WOL/AHK4 receptor is required to maintain undifferentiated, pluripotent cell identity during root procambium development. Consistent with this interpretation, *CRE1/WOL/AHK4* is first expressed in procambial cells of the embryo axis at the globular stage. The *ARABIDOPSIS HISTIDINE PHOSPHOTRANSFER PROTEIN6* (*AHP6*) locus was defined by the analysis of extragenic second-site suppressor mutations of *wol*, leading to a partial suppression of the all-protoxylem *wol* phenotype (Mähönen *et al.*, 2006), and the *ahp6* single mutants display sporadic protoxylem differentiation along the root vascular bundle. Protein members of the AHP family typically activate cytokinin signaling by shuttling to the nucleus after activation by the cytokinin receptors (Hwang and Sheen, 2001). However, the AHP6 protein lacks the conserved histidine residue critical for phosphorelay and has therefore been designated "pseudo" phosphotransfer protein (Mähönen *et al.*, 2006). Consequently, AHP6 acts as an inhibitor to cytokinin signaling and thereby promotes the default protoxylem identity in conditions of low cytokinin signaling. *AHP6* is expressed in protoxylem and adjacent pericycle cell lineages, and *AHP6* expression is already detectable at early stages of embryogenesis in both wild type and *wol* (Mähönen *et al.*, 2006). Therefore, mutant phenotype and expression pattern suggest that cytokinin signaling specifies the spatial domain of *AHP6* expression upstream of protoxylem differentiation.

The Arabidopsis root vascular bundle typically displays a diarch pattern with two xylem and two phloem poles (Dolan *et al.*, 1993; Fig. 8.2E). While reduction of procambial cell division in *wol* mutants is inconsequential for formation of the diarch vascular pattern of the Arbidopsis root, the diameter of the root vascular bundle in pea correlates with the number of xylem and phloem poles (Torrey, 1955; 1957), suggesting the presence of a cell proliferation-dependent patterning mechanism. The molecular nature of the underlying regulatory process is not understood, but in the recessive *lonesome highway* mutant of Arabidopsis a monoarch root vascular pattern is associated with severely reduced numbers of root vascular cells (Ohashi-Ito and Bergmann, 2007). The corresponding gene codes for a bHLH transcription factor that has a broad expression pattern in the root meristem, but the precise function of this gene in cell proliferation and patterning within the vascular bundle is yet to be resolved.

In the *pxy* (for *phloem intercalated with xylem*) mutant of Arabidopsis, the internal organization of vascular bundles in the stem is severely disrupted: cell divisions occur irregularly, and xylem and phloem are no longer properly separated (Fisher and Turner, 2007). The *PXY* locus encodes a leucine-rich receptor kinase homologous to *CLAVATA1 (CLV1)*. Together with CLV2, CLV1 constitutes an essential receptor component of a regulatory pathway that controls stem cell maintenance in the shoot apical meristem of Arabidopsis (chapter 4; Clark, 2001). The interaction between the CLV3 peptide and the CLV1 and CLV2 receptor complexes promotes activation of downstream components that negatively regulate the expression of the homeodomain transcription factor WUSCHEL (WUS); in turn, WUS acts non-cell-autonomously to maintain stem cell identity and promote CLV3 expression. Recently, a similar regulatory feedback has been uncovered that maintains the root apical meristem and involves the peptide CLAVATA3/EMBRYO-SURROUNDING REGION-RELATED40 (CLE40) and the WUS-RELATED HOMO-EOBOX5 (WOX5) transcription factor (chapter 3; Sarkar *et al.*, 2007; Stahl *et al.*, 2009). Could a similar pathway operate through PXY? The CLE dodecapeptide TRACHEARY ELEMENT DIFFERENTIATION INHIBITORY FACTOR (TDIF) was originally identified in a *Zinnia* cell culture system because it inhibited the differentiation of cultured cells into tracheary elements (Ito *et al.*, 2006). In Arabidopsis plants, exogenous TDIF application increases proliferation of procambial cells and suppresses xylem differentiation (Hirakawa *et al.*, 2008), and TDIF, encoded by the *CLE41* and *CLE44* loci in Arabidopsis, acts as ligand for PXY, also named TDR for putative TDIF RECEPTOR (Hirakawa *et al.*, 2008; Etchells and Turner, 2010). *CLE41/44*-encoded peptides are secreted by the phloem and bind their receptor in the procambium (Fig. 8.6I). Therefore, the CLE receptor/ligand type of regulon maintains procambial stem cell activity and controls its differentiation into xylem. Recently, *WOX4* has been shown to be expressed in procambial cells, and reduced *WOX4* levels interfere with xylem and phloem differentiation in Arabidopsis (Ji *et al.*, 2010). Furthermore, two *WUS*-related genes, *PttHB2* and *PttHB3*, have recently been found to be expressed in poplar cambium (Schrader *et al.*, 2004). Whether these or any other *WOX* genes are part of the PXY/TDIF feedback loop remains, at present, enigmatic.

7. Concluding Remarks

Deciphering the organizing mechanisms underlying the mesmerizing beauty of the diverse, intricate arrangements of plant vascular tissues has long challenged both theoretical and experimental biologists. The lessons learned from research in embryo, leaf, and root development converge to support earlier models and generate an emerging picture in which vascular tissues are patterned by co-opting apical–basal and central–peripheral/adaxial–abaxial polarity cues in the organism. Apical–basal transport of auxin could serve as a coordinating signal, adjusting local cell behavior to the overall growth requirements of the plant, and set up a coarse, basic pattern further refined by cascades of fine-tuning radial patterning gene activities. The body pattern of a plant is clearly more complex than the combination of two axes; however, three-dimensional organizing functions could use the axial architecture of the vascular system as a positional reference. It has long been suspected that numerous local cell-patterning events may be correlated with the positions of vascular tissues (e.g., Langdale *et al.*, 1987), and details of the molecular mechanisms through which centrally located vascular strands supply positional information in ground tissue patterning and root meristem formation have already become available. It is therefore possible that also in other organs vascular tissues represent an anchoring scaffold for adjacent cell-patterning events to ensure functional integration of vascular strands into surrounding tissue contexts (e.g., Carabelli *et al.*, 2007). This concept is consistent with the robust flexibility of vascular patterns, and accounts for the apparently paradoxical coexistence of variability in the paths of continuous vascular strand formation and reproducibility in the differentiation of nonvascular cell types in relation to vascular tissues. Obvious next steps will be identifying further components in the pathways controlling organismal polarities, determining how these pathways are integrated at the molecular and cellular levels to regulate vascular tissue organizations, and understanding how such integration ultimately generates the multitude of vascular arrangements apparent in nature. Conclusions presented here are based on work on selected model species. Although available evidence suggests that at least some components of the molecular pathways that control vascular patterning may be widely conserved, further work in appropriate species is required to evaluate the extent of this similarity and how it relates to divergent species-specific organizations, and may even reveal entirely new, unexpected levels of control. Work in suitable species will also be necessary to understand the mechanistic basis of vascular patterning during secondary development, in which cells are activated to generate a cambium, secondary xylem (i.e., wood), and phloem. This involves a specific control of cell proliferation and the formation of novel vascular cell types, processes for which molecular details are currently scarce, but that have very appealing economical applications. The preliminary nature of the interpretations presented should be emphasized, and stronger molecular groundwork is needed before these interpretations can be considered more than constructive working hypotheses. Nevertheless, the work reviewed here has originated a scaffold of hypotheses and derived predictions that can be experimentally tested in the near future.

ACKNOWLEDGMENTS

We would like to thank Edith Taylor for kindly providing relevant literature. We would further like to thank Andrew Hudson, Ben Scheres, Gerd Jürgens, Hiroo Fukuda, José Luis Micol, Thomas Berleth, the Botanical Society of America, Cold Spring Harbor Laboratory Press, the Company of Biologists, Elsevier, John Wiley and Sons, Landes Bioscience, Macmillan Publishers, the National Academy of Science, U.S.A., National Research Council Canada Research Press, Nature Publishing Group, Springer Science and Business Media, and Wiley-Blackwell for kindly granting permission to use published material as template for figures. Finally, we would like to thank Megan Sawchuk and Tyler Donner for critical comments to the manuscript. We apologize to colleagues whose results could not be included in the available space. This work was supported by a Discovery Grant of the Natural Sciences and Engineering Research Council of Canada (E.S.), the Canada Research Chairs Program (E.S.), the Academy of Finland (Y.H.), TEKES (Y.H.), and the European Science Foundation (Y.H.).

NOTE ADDED IN PROOF

After this manuscript had been accepted, Carland *et al.* (2010) reported that in the *cvp1* mutant brassinosteroid levels are unaffected but sterol composition is altered, and that *cvp1* roots are defective in auxin-inhibited endocytosis of the PIN2 auxin transporter.

REFERENCES

Aida, M., Beis, D., Heidstra, R., Willemsen, V., Blilou, I., Galinha, C., Nussaume, L., Noh, Y. S., Amasino, R., and Scheres, B. (2004). The PLETHORA genes mediate patterning of the Arabidopsis root stem cell niche. *Cell* **119**, 109–120.

Aloni, R. (1987). Differentiation of vascular tissues. *Annu. Rev. Plant Physiol. Plant Mol. Biol.* **38**, 179–204.

Aloni, R. (2001). Foliar and axial aspects of vascular differentiation: hypotheses and evidence. *J. Plant Growth Regul.* **20**, 22–34.

Aloni, R., and Plotkin, T. (1985). Wound-induced and naturally-occurring regenerative differentiation of xylem in Zea-Mays-L. *Planta* **163**, 126–132.

Aloni, R., Schwalm, K., Langhans, M., and Ullrich, C. I. (2003). Gradual shifts in sites of free-auxin production during leaf-primordium development and their role in vascular differentiation and leaf morphogenesis in Arabidopsis. *Planta* **216**, 841–853.

Alonso-Peral, M. M., Candela, H., del Pozo, J. C., Martinez-Laborda, A., Ponce, M. R., and Micol, J. L. (2006). The HVE/CAND1 gene is required for the early patterning of leaf venation in Arabidopsis. *Development* **133**, 3755–3766.

Baima, S., Nobili, F., Sessa, G., Lucchetti, S., Ruberti, I., and Morelli, G. (1995). The expression of the Athb-8 homeobox gene is restricted to provascular cells in Arabidopsis thaliana. *Development* **121**, 4171–4182.

Baima, S., Possenti, M., Matteucci, A., Wisman, E., Altamura, M. M., Ruberti, I., and Morelli, G. (2001). The Arabidopsis ATHB-8 HD-zip protein acts as a differentiation-promoting transcription factor of the vascular meristems. *Plant Physiol.* **126**, 643–655.

Bainbridge, K., Guyomarc'h, S., Bayer, E., Swarup, R., Bennett, M., Mandel, T., and Kuhlemeier, C. (2008). Auxin influx carriers stabilize phyllotactic patterning. *Genes Dev.* **22**, 810–823.

Bauby, H., Divol, F., Truernit, E., Grandjean, O., and Palauqui, J. C. (2007). Protophloem differentiation in early Arabidopsis thaliana development. *Plant Cell Physiol.* **48**, 97–109.

Baucher, M., El Jaziri, M., and Vandeputte, O. (2007). From primary to secondary growth: origin and development of the vascular system. *J. Exp. Bot.* **58**, 3485–3501.

Bayer, E. M., Smith, R. S., Mandel, T., Nakayama, N., Sauer, M., Prusinkiewicz, P., and Kuhlemeier, C. (2009). Integration of transport-based models for phyllotaxis and midvein formation. *Genes Dev.* **23**, 373–384.

Beerling, D. J., and Fleming, A. J. (2007). Zimmermann's telome theory of megaphyll leaf evolution: a molecular and cellular critique. *Curr. Opin. Plant Biol.* **10**, 4–12.

Benkova, E., Michniewicz, M., Sauer, M., Teichmann, T., Seifertova, D., Jurgens, G., and Friml, J. (2003). Local, efflux-dependent auxin gradients as a common module for plant organ formation. *Cell* **115**, 591–602.

Bennett, S.R.M., Alvarez, J., Bossinger, G., and Smyth, D. R. (1995). Morphogenesis in pinoid mutants of Arabidopsis thaliana. *Plant J.* **8**, 505–520.

Berleth, T., and Jurgens, G. (1993). The role of the monopteros gene in organizing the basal body region of the Arabidopsis embryo. *Development* **118**, 575–587.

Berleth, T., Mattsson, J., and Hardtke, C. S. (2000). Vascular continuity and auxin signals. *Trends Plant Sci.* **5**, 387–393.

Berleth, T., and Sachs, T. (2001). Plant morphogenesis: long-distance coordination and local patterning. *Curr. Opin. Plant Biol.* **4**, 57–62.

Berleth, T., and Scarpella, E. (2004). Polar signals in vascular development. In: Lindsey, K. (Ed.), *Polarity in Plants*. Blackwell Publishing Ltd, Oxford, pp. 264–287.

Berleth, T., Scarpella, E., and Prusinkiewicz, P. (2007). Towards the systems biology of auxin-transport-mediated patterning. *Trends Plant Sci.* **12**, 151–159.

Bohmert, K., Camus, I., Bellini, C., Bouchez, D., Caboche, M., and Benning, C. (1998). AGO1 defines a novel locus of Arabidopsis controlling leaf development. *EMBO J.* **17**, 170–180.

Boutte, Y., and Grebe, M. (2009). Cellular processes relying on sterol function in plants. *Curr. Opin. Plant Biol.* **12**, 705–713.

Brioudes, F., Joly, C., Szecsi, J., Varaud, E., Leroux, J., Bellvert, F., Bertrand, C., and Bendahmane, M. (2009). Jasmonate controls late development stages of petal growth in *Arabidopsis thaliana*. *Plant J.* **60**, 1070–1080.

Byrne, M., Barley, R., Curtis, M., Arroyo, J., Dunham, M., Hudson, A., and Martienssen, R. (2000). Asymmetric leaves1 mediates leaf patterning and stem cell function in Arabidopsis. *Nature* **408**, 967–971.

Calderon-Villalobos, L. I., Kuhnle, C., Dohmann, E. M., Li, H., Bevan, M., and Schwechheimer, C. (2005). The evolutionarily conserved TOUGH protein is required for proper development of *Arabidopsis thaliana*. *Plant Cell* **17**, 2473–2485.

Cambridge, A. P., and Morris, D. A. (1996). Transfer of exogenous auxin from the phloem to the polar auxin transport pathway in pea (*Pisum sativum* L). *Planta* **199**, 583–588.

Candela, H., Alonso-Peral, M. M., Ponce, M. R., and Micol, J. L. (2007). Role of HEMIVENATA and the ubiquitin pathway in venation pattern formation. *Plant Signal. Behav.* **2**, 258–259.

Candela, H., Johnston, R., Gerhold, A., Foster, T., and Hake, S. (2008). The milkweed pod1 gene encodes a KANADI protein that is required for abaxial/adaxial patterning in maize leaves. *Plant Cell* **20**, 2073–2087.

Candela, H., Martinez-Laborda, A., and Micol, J. L. (1999). Venation pattern formation in *Arabidopsis thaliana* vegetative leaves. *Dev. Biol.* **205**, 205–216.

Cano-Delgado, A., Yin, Y., Yu, C., Vafeados, D., Mora-Garcia, S., Cheng, J. C., Nam, K. H., Li, J., and Chory, J. (2004). BRL1 and BRL3 are novel brassinosteroid receptors that function in vascular differentiation in Arabidopsis. *Development* **131**, 5341–5351.

Capron, A., Chatfield, S., Provart, N., and Berleth, T. (2009). Embryogenesis: pattern formation from a single cell. In: Somerville, C. and Meyerowitz, E. (Eds.) *The Arabidopsis Book*. American Society of Plant Biologists, Rockville, MD.

Carabelli, M., Possenti, M., Sessa, G., Ciolfi, A., Sassi, M., Morelli, G., and Ruberti, I. (2007). Canopy shade causes a rapid and transient arrest in leaf development through auxin-induced cytokinin oxidase activity. *Genes Dev.* **21**, 1863–1868.

Carland, F.M., Berg, B.L, FitzGerald, J.N., Jinamornphongs, S., Nelson, T., Keith, B. (1999). Genetic regulation of vascular tissue patterning in Arabidopsis. *Plant Cell* **11**, 2123–2137.

Carland, F. M., Fujioka, S., Takatsuto, S., Yoshida, S., and Nelson, T. (2002). The identification of CVP1 reveals a role for sterols in vascular patterning. *Plant Cell* **14**, 2045–2058.

Carland, F., and Nelson, T. (2009). CVP2- and CVL1-mediated phosphoinositide signaling as a regulator of the ARF GAP SFC/VAN3 in establishment of foliar vein patterns. *Plant J.* **59**, 895–907.

Carland F, Fujioka S, Nelson T (2010). The sterol methyltransferases SMT1, SMT2, and SMT3 influence Arabidopsis development through nonbrassinosteroid products. *Plant Physiol* **153**, 741–56.

Carlsbecker, A., Lee, J.-Y., Roberts, C. J., Dettmer, J., Lehesranta, S., Zhou, J., Lindgren, O., Moreno-Risueno, M. A., Vatén, A., Thitamadee1, S., Campilho, A., *et al.* (2010). Cell signalling by microRNA165/6 directs gene dose-dependent root cell fate. *Nature* **465**, 316–321.

Ceserani, T., Trofka, A., Gandotra, N., and Nelson, T. (2009). VH1/BRL2 receptor-like kinase interacts with vascular-specific adaptor proteins VIT and VIK to influence leaf venation. *Plant J.* **57**, 1000–1014.

Chapman, E., and Carrington, J. (2007). Specialization and evolution of endogenous small RNA pathways. *Nat. Rev. Genet.* **8**, 884–896.

Chavrier, P., and Goud, B. (1999). The role of ARF and Rab GTPases in membrane transport. *Curr. Opin. Cell Biol.* **11**, 466–475.

Cheng, Y., Dai, X., and Zhao, Y. (2006). Auxin biosynthesis by the YUCCA flavin monooxygenases controls the formation of floral organs and vascular tissues in Arabidopsis. *Genes Dev.* **20**, 1790–1799.

Cheng, Y., Dai, X., and Zhao, Y. (2007). Auxin synthesized by the YUCCA flavin monooxygenases is essential for embryogenesis and leaf formation in Arabidopsis. *Plant Cell* **19**, 2430–2439.

Chitwood, D., Nogueira, F., Howell, M., Montgomery, T., Carrington, J., and Timmermans, M. (2009). Pattern formation via small RNA mobility. *Genes Dev.* **23**, 549–554.

Citovsky, V., and Zambryski, P. (2000). Systemic transport of RNA in plants. *Trends Plant Sci.* **5**, 52–54.

Clark, S. E. (2001). Cell signalling at the shoot meristem. *Nat. Rev. Mol. Cell Biol.* **2**, 276–284.

Clay, N. K., and Nelson, T. (2005). The recessive epigenetic swellmap mutation affects the expression of two step II splicing factors required for the transcription of the cell proliferation gene STRUWWELPETER and for the timing of cell cycle arrest in the Arabidopsis leaf. *Plant Cell* **17**, 1994–2008.

Cnops, G., Neyt, P., Raes, J., Petrarulo, M., Nelissen, H., Malenica, N., Luschnig, C., Tietz, O., Ditengou, F., Palme, K., Azmi, A., Prinsen, E., *et al.* (2006). The TORNADO1 and TORNADO2 genes function in several patterning processes during early leaf development in Arabidopsis thaliana. *Plant Cell* **18**, 852–866.

Cook, M. E., and Graham, L. E. (1999). Evolution of plasmodesmata. In: van Bel, W. J. P. v. K. A. J. E. (Ed.), *Plasmodesmata. Structure, Function, Role in Cell Communication.* Springer, Berlin, pp. 101–117.

Dello Ioio, R., Nakamura, K., Moubayidin, L., Perilli, S., Taniguchi, M., Morita, M. T., Aoyama, T., Costantino, P., and Sabatini, S. (2008). A genetic framework for the control of cell division and differentiation in the root meristem. *Science* **322**, 1380–1384.

Dengler, N. G. (2001). Regulation of vascular development. *J. Plant Growth Regul.* **20**, 1–13.

Dengler, N. G. (2006). The shoot apical meristem and development of vascular architecture. *Can. J. Bot.* **84**, 1660–1671.

Dengler, N., and Kang, J. (2001). Vascular patterning and leaf shape. *Curr. Opin. Plant Biol.* **4**, 50–56.

Deyholos, M. K., Cavaness, G. F., Hall, B., King, E., Punwani, J., Van Norman, J., and Sieburth, L. E. (2003). VARICOSE, a WD-domain protein, is required for leaf blade development. *Development* **130**, 6577–6588.

Deyholos, M. K., Cordner, G., Beebe, D., and Sieburth, L. E. (2000). The SCARFACE gene is required for cotyledon and leaf vein patterning. *Development* **127**, 3205–3213.

Dharmasiri, S., Dharmasiri, N., Hellmann, H., and Estelle, M. (2003). The RUB/Nedd8 conjugation pathway is required for early development in Arabidopsis. *EMBO J.* **22**, 1762–1770.

Dharmasiri, N., Dharmasiri, S., Weijers, D., Karunarathna, N., Jurgens, G., and Estelle, M. (2007). AXL and AXR1 have redundant functions in RUB conjugation and growth and development in Arabidopsis. *Plant J.* **52**, 114–123.

Dharmasiri, N., Dharmasiri, S., Weijers, D., Lechner, E., Yamada, M., Hobbie, L., Ehrismann, J. S., Jurgens, G., and Estelle, M. (2005). Plant development is regulated by a family of auxin receptor F box proteins. *Dev. Cell* **9**, 109–119.

Dimitrov, P., and Zucker, S. W. (2006). A constant production hypothesis guides leaf venation patterning. *Proc. Natl. Acad. Sci. U.S.A.* **103**, 9363–9368.

Dolan, L., Janmaat, K., Willemsen, V., Linstead, P., Poethig, S., Roberts, K., and Scheres, B. (1993). Cellular-organization of the *Arabidopsis thaliana* root. *Development* **119**, 71–84.

Donner, T. J., and Scarpella, E. (2009). Auxin-transport-dependent leaf vein formation. *Botany-Botanique* **87**, 678–684.

Donner, T. J., Sherr, I., and Scarpella, E. (2009). Regulation of preprocambial cell state acquisition by auxin signaling in Arabidopsis leaves. *Development* **136**, 3235–3246.

Emery, J., Floyd, S., Alvarez, J., Eshed, Y., Hawker, N., Izhaki, A., Baum, S., and Bowman, J. (2003). Radial patterning of Arabidopsis shoots by class III HD-ZIP and KANADI genes. *Curr. Biol.* **13**, 1768–1774.

Esau, K. (1943). Origin and development of primary vascular tissues in plants. *Botanical Rev.* **9**, 125–206.

Esau, K. (1965). *Plant Anatomy*. John Wiley, New York; London.

Eshed, Y., Baum, S., Perea, J., and Bowman, J. (2001). Establishment of polarity in lateral organs of plants. *Curr. Biol.* **11**, 1251–1260.

Eshed, Y., Izhaki, A., Baum, S., Floyd, S., and Bowman, J. (2004). Asymmetric leaf development and blade expansion in Arabidopsis are mediated by KANADI and YABBY activities. *Development* **131**, 2997–3006.

Etchells, J. P., and Turner, S. R. (2010). The PXY-CLE41 receptor ligand pair defines a multifunctional pathway that controls the rate and orientation of vascular cell division. *Development* **137**, 767–774.

Feugier, F. G., Mochizuki, A., and Iwasa, Y. (2005). Self-organization of the vascular system in plant leaves: inter-dependent dynamics of auxin flux and carrier proteins. *J. Theor. Biol.* **236**, 366–375.

Fischer, C., and Neuhaus, G. (1996). Influence of auxin on the establishment of bilateral symmetry in monocots. *Plant J.* **9**, 659–669.

Fisher, K., and Turner, S. (2007). PXY, a receptor-like kinase essential for maintaining polarity during plant vascular-tissue development. *Curr. Biol.* **17**, 1061–1066.

Friml, J., Vieten, A., Sauer, M., Weijers, D., Schwarz, H., Hamann, T., Offringa, R., and Jurgens, G. (2003). Efflux-dependent auxin gradients establish the apical–basal axis of Arabidopsis. *Nature* **426**, 147–153.

Friml, J., Wisniewska, J., Benkova, E., Mendgen, K., and Palme, K. (2002). Lateral relocation of auxin efflux regulator PIN3 mediates tropism in Arabidopsis. *Nature* **415**, 806–809.

Fujita, H., and Mochizuki, A. (2006a). The origin of the diversity of leaf venation pattern. *Dev. Dyn.* **235**, 2710–2721.

Fujita, H., and Mochizuki, A. (2006b). Pattern formation of leaf veins by the positive feedback regulation between auxin flow and auxin efflux carrier. *J. Theor. Biol.* **241**, 541–551.

Fukuda, H. (2004). Signals that control plant vascular cell differentiation. *Nat. Rev. Mol. Cell Biol.* **5**, 379–391.

Fukuda, H., Hirakawa, Y., and Sawa, S. (2007). Peptide signaling in vascular development. *Curr. Opin. Plant Biol.* **10**, 477–482.

Gagne, J. M., Song, S. K., and Clark, S. E. (2008). POLTERGEIST and PLL1 are required for stem cell function with potential roles in cell asymmetry and auxin signaling. *Commun. Integr. Biol.* **1**, 53–55.

Geldner, N., Anders, N., Wolters, H., Keicher, J., Kornberger, W., Muller, P., Delbarre, A., Ueda, T., Nakano, A., and Jurgens, G. (2003). The Arabidopsis GNOM ARF-GEF mediates endosomal recycling, auxin transport, and auxin-dependent plant growth. *Cell* **112**, 219–230.

Geldner, N., Friml, J., Stierhof, Y. D., Jurgens, G., and Palme, K. (2001). Auxin transport inhibitors block PIN1 cycling and vesicle trafficking. *Nature* **413**, 425–428.

Gensel, P. G., and Andrews, H. N. (1987). The evolution of early land plants. *Am. Sc.* **75**, 478–489.

Girolami, G. (1954). Leaf histogenesis in *Linum usitatissimum*. *Am. J. Bot.* **41**, 264–273.

Graham, L. E. (1993). *Origin of Land Plants*. Wiley, New York.

Gray, J. (1993). Major Palaeozoic land plant evolutionary bioevents. *Palaeogeogr. Palaeoclimatol. Palaeoecol.* **104**, 153–169.

Gray, J., and Shear, W. (1992). Early life on land. *Am. Sc.***80**, 444–456.

Guilfoyle, T. J., and Hagen, G. (2007). Auxin response factors. *Curr. Opin. Plant Biol.* **10**, 453–460.

Hadfi, K., Speth, V., and Neuhaus, G. (1998). Auxin-induced developmental patterns in *Brassica juncea* embryos. *Development* **125**, 879–887.

Hamann, T., Benkova, E., Baurle, I., Kientz, M., and Jurgens, G. (2002). The Arabidopsis BODENLOS gene encodes an auxin response protein inhibiting MONOPTEROS-mediated embryo patterning. *Genes Dev.* **16**, 1610–1615.

Hamann, T., Mayer, U., and Jurgens, G. (1999). The auxin-insensitive bodenlos mutation affects primary root formation and apical-basal patterning in the Arabidopsis embryo. *Development* **126**, 1387–1395.

Hanzawa, Y., Takahashi, T., Michael, A. J., Burtin, D., Long, D., Pineiro, M., Coupland, G., and Komeda, Y. (2000). ACAULIS5, an Arabidopsis gene required for stem elongation, encodes a spermine synthase. *EMBO J.* **19**, 4248–4256.

Hardtke, C. S., and Berleth, T. (1998). The Arabidopsis gene MONOPTEROS encodes a transcription factor mediating embryo axis formation and vascular development. *EMBO J.* **17**, 1405–1411.

Hardtke, C. S., Ckurshumova, W., Vidaurr, D. P., Singh, S. A., Stamatiou, G., Tiwari, S. B., Hagen, G., Guilfoyle, T. J., and Berleth, T. (2004). Overlapping and non-redundant functions of the Arabidopsis auxin response factors MONOPTEROS and NONPHOTOTROPIC HYPOCOTYL 4. *Development* **131**, 1089–1100.

Hay, A., Barkoulas, M., and Tsiantis, M. (2006). ASYMMETRIC LEAVES1 and auxin activities converge to repress BREVIPEDICELLUS expression and promote leaf development in Arabidopsis. *Development* **133**, 3955–3961.

Heisler, M. G., Ohno, C., Das, P., Sieber, P., Reddy, G. V., Long, J. A., and Meyerowitz, E. M. (2005). Patterns of auxin transport and gene expression during primordium development revealed by live Imaging of the Arabidopsis inflorescence meristem. *Curr. Biol.* **15**, 1899–1911.

Helariutta, Y., Fukaki, H., Wysocka-Diller, J., Nakajima, K., Jung, J., Sena, G., Hauser, M. T., and Benfey, P. N. (2000). The SHORT-ROOT gene controls radial patterning of the Arabidopsis root through radial signaling. *Cell* **101**, 555–567.

Hellmann, H., Hobbie, L., Chapman, A., Dharmasiri, S., Dharmasiri, N., del Pozo, C., Reinhardt, D., and Estelle, M. (2003). Arabidopsis AXR6 encodes CUL1 implicating SCF E3 ligases in auxin regulation of embryogenesis. *EMBO J.* **22**, 3314–3325.

Hirakawa, Y., Shinohara, H., Kondo, Y., Inoue, A., Nakanomyo, I., Ogawa, M., Sawa, S., Ohashi-Ito, K., Matsubayashi, Y., and Fukuda, H. (2008). Non-cell-autonomous control of vascular stem cell fate by a CLE peptide/receptor system. *Proc. Natl. Acad. Sci. U.S.A.* **105**, 15208–15213.

Hirose, N., Takei, K., Kuroha, T., Kamada-Nobusada, T., Hayashi, H., and Sakakibara, H. (2008). Regulation of cytokinin biosynthesis, compartmentalization and translocation. *J. Exp. Bot.* **59**, 75–83.

Hobbie, L., McGovern, M., Hurwitz, L. R., Pierro, A., Liu, N. Y., Bandyopadhyay, A., and Estelle, M. (2000). The axr6 mutants of *Arabidopsis thaliana* define a gene involved in auxin response and early development. *Development* **127**, 23–32.

Husbands, A. Y., Chitwood, D. H., Plavskin, Y., and Timmermans, M.C.P. (2009). Signals and prepatterns: new insights into organ polarity in plants. *Genes Dev.* **23**, 1986–1997.

Hwang, I., and Sheen, J. (2001). Two-component circuitry in Arabidopsis cytokinin signal transduction. *Nature* **413**, 383–389.

Ibanes, M., Fabregas, N., Chory, J., and Cano-Delgado, A. I. (2009). Brassinosteroid signaling and auxin transport are required to establish the periodic pattern of Arabidopsis shoot vascular bundles. *Proc. Natl. Acad. Sci. U.S.A.* **106**, 13630–13635.

Ilegems, M., Douet, V., Meylan-Bettex, M., Uyttewaal, M., Brand, L., Bowman, J.L., and Stieger, P.A. (2010) Interplay of auxin, KANADI and Class III HD-ZIP transcription factors in vascular tissue formation. *Development* **137**, 975–984.

Inoue, T., Higuchi, T., Hashimoto, Y., Seki, M., Kobayashi, M., Kato, T., Tabata, S., Shinozaki K., and Kakimoto, T. (2001). Identification of CRE1 as a cytokinin receptor from Arabidopsis. *Nature* **409**, 1060–1063.

Ito, Y., Nakanomyo, I., Motose, H., Iwamoto, K., Sawa, S., Dohmae, N., and Fukuda, H. (2006). Dodeca-CLE peptides as suppressors of plant stem cell differentiation. *Science* **313**, 842–845.

Itoh, J.-I., Hibara, K.-I., Sato, Y., and Nagato, Y. (2008). Developmental role and auxin responsiveness of Class III homeodomain leucine zipper gene family members in rice. *Plant Physiol.* **147**, 1960–1975.

Itoh, J., Nonomura, K., Ikeda, K., Yamaki, S., Inukai, Y., Yamagishi, H., Kitano, H., and Nagato, Y. (2005). Rice plant development: from zygote to spikelet. *Plant Cell Physiol.* **46**, 23–47.

Ji, J., Strable, J., Shimizu, R., Koenig, D., Sinha, N., and Scanlon, M. J. (2010). WOX4 promotes procambial development. *Plant Physiol.* **152**, 1346–1356.

Jiang, F., and Hartung, W. (2008). Long-distance signalling of abscisic acid (ABA): the factors regulating the intensity of the ABA signal. *J. Exp. Bot.* **59**, 37–43.

Juarez, M., Kui, J., Thomas, J., Heller, B., and Timmermans, M. (2004a). microRNA-mediated repression of rolled leaf1 specifies maize leaf polarity. *Nature* **428**, 84–88.

Juarez, M., Twigg, R., and Timmermans, M. (2004b). Specification of adaxial cell fate during maize leaf development. *Development* **131**, 453345–44.

Jung, J. H., and Park, C. M. (2007). Vascular development in plants: specification of xylem and phloem tissues. *J. Plant Biol.* **50**, 301–305.

Kang, J., and Dengler, N. (2004). Vein pattern development in adult leaves of *Arabidopsis thaliana*. *Int. J. Plant Sci.* **165**, 231–242.

Kang, J., Mizukami, Y., Wang, H., Fowke, L., and Dengler, N. G. (2007). Modification of cell proliferation patterns alters leaf vein architecture in *Arabidopsis thaliana*. *Planta* **226**, 1207–1218.

Keller, C. P., Stahlberg, R., Barkawi, L. S., and Cohen, J. D. (2004). Long-term inhibition by auxin of leaf blade expansion in bean and Arabidopsis. *Plant Physiol.* **134**, 1217–1226.

Kenrick, P., and Crane, P. R. (1997). The origin and early evolution of plants on land. *Nature* **389**, 33–39.

Kerstetter, R., Bollman, K., Taylor, R., Bomblies, K., and Poethig, R. (2001). KANADI regulates organ polarity in Arabidopsis. *Nature* **411**, 706–709.

Kidner, C., and Martienssen, R. (2004). Spatially restricted microRNA directs leaf polarity through ARGONAUTE1. *Nature* **428**, 81–84.

Kleine-Vehn, J., Dhonukshe, P., Sauer, M., Brewer, P. B., Wisniewska, J., Paciorek, T., Benkova, E., and Friml, J. (2008). ARF GEF-dependent transcytosis and polar delivery of PIN auxin carriers in Arabidopsis. *Curr. Biol.* **18**, 526–531.

Klucking, E. P. (1995). *Leaf Venation Patterns*. J. Cramer, Berlin.

Knoll, A. H. (1994). Proterozoic and early Cambrian protists—evidence for accelerating evolutionary tempo. *Proc. Natl. Acad. Sci. U.S.A.* **91**, 6743–6750.

Knoll, A. H., and Niklas, K. J. (1987). Adaptation, plant evolution, and the fossil record. *Rev. Paleobot. Palynol.* **50**, 127–149.

Knoll, A. H., and Rothwell, G. W. (1981). Palaeobotany: perspectives in 1980. *Paleobiology* **7**, 7–35.

Koizumi, K., Naramoto, S., Sawa, S., Yahara, N., Ueda, T., Nakano, A., Sugiyama, M., and Fukuda, H. (2005). VAN3 ARF-GAP-mediated vesicle transport is involved in leaf vascular network formation. *Development* **132**, 1699–1711.

Koizumi, K., Sugiyama, M., and Fukuda, H. (2000). A series of novel mutants of *Arabidopsis thaliana* that is defective in the formation of continuous vascular network: calling the auxin signal flow canalization hypothesis into question. *Development* **127**, 3197–3204.

Konishi, M., and Yanagisawa, S. (2007). Sequential activation of two Dof transcription factor gene promoters during vascular development in *Arabidopsis thaliana*. *Plant Physiol. Biochem.* **45**, 623–629.

Kramer, E. M. (2004). PIN and AUX/LAX proteins: their role in auxin accumulation. *Trends Plant Sci.* **9**, 578–582.

Kramer, E. M. (2009). Auxin-regulated cell polarity: an inside job? *Trends Plant Sci.* **14**, 242–247.

Langdale, J. A., Metzler, M. C., and Nelson, T. (1987). The argentia mutation delays normal development of photosynthetic cell-types in Zea-mays. *Dev. Biol.* **122**, 243–255.

Li, H., Xu, L., Wang, H., Yuan, Z., Cao, X., Yang, Z., Zhang, D., Xu, Y., and Huang, H. (2005). The putative RNA-dependent RNA polymerase RDR6 acts synergistically with ASYMMETRIC LEAVES1 and 2 to repress BREVIPEDICELLUS and MicroRNA165/166 in Arabidopsis leaf development. *Plant Cell* **17**, 2157–2171.

Lin, W. H., Wang, Y., Mueller-Roeber, B., Brearley, C. A., Xu, Z. H., and Xue, H. W. (2005). At5PTase13 modulates cotyledon vein development through regulating auxin homeostasis. *Plant Physiol.* **139**, 1677–1691.

Liu, C., Xu, Z., and Chua, N. H. (1993). Auxin polar transport is essential for the establishment of bilateral symmetry during early plant embryogenesis. *Plant Cell* **5**, 621–630.

Ljung, K., Bhalerao, R. P., and Sandberg, G. (2001). Sites and homeostatic control of auxin biosynthesis in Arabidopsis during vegetative growth. *Plant J.* **28**, 465–474.

Lough, T. J., and Lucas, W. J. (2006). Integrative plant biology: role of phloem long-distance macromolecular trafficking. *Annu. Rev. Plant Biol.* **57**, 203–232.

Mähönen, A. P., Bishopp, A., Higuchi, M., Nieminen, K. M., Kinoshita, K., Törmäkangas, K., Ikeda, Y., Oka, A., Kakimoto, T., and Helariutta, Y. (2006). Cytokinin signaling and its inhibitor AHP6 regulate cell fate during vascular development. *Science* **311**, 94–98.

Mähönen, A. P., Bonke, M., Kauppinen, L., Riikonen, M., Benfey, P. N., and Helariutta, Y. (2000). A novel two-component hybrid molecule regulates vascular morphogenesis of the Arabidopsis root. *Genes Dev.* **14**, 2938–2943.

Mallory, A., Reinhart, B., Jones-Rhoades, M., Tang, G., Zamore, P., Barton, M., and Bartel, D. (2004). MicroRNA control of PHABULOSA in leaf development: Importance of pairing to the microRNA 5′ region. *EMBO J.* **23**, 3356–3364.

Mattsson, J., Ckurshumova, W., and Berleth, T. (2003). Auxin signaling in Arabidopsis leaf vascular development. *Plant Physiol.* **131**, 1327–1339.

Mattsson, J., Sung, Z. R., and Berleth, T. (1999). Responses of plant vascular systems to auxin transport inhibition. *Development* **126**, 2979–2991.

Mayer, U., Buttner, G., and Jurgens, G. (1993). Apical-basal pattern-formation in the Arabidopsis embryo—studies on the role of the gnom gene. *Development* **117**, 149–162.

McConnell, J. R., and Barton, M. K. (1998). Leaf polarity and meristem formation in Arabidopsis. *Development* **125**, 2935–2942.

McConnell, J., Emery, J., Eshed, Y., Bao, N., Bowman, J., and Barton, M. (2001). Role of PHABULOSA and PHAVOLUTA in determining radial patterning in shoots. *Nature* **411**, 709–713.

Merks, R.M.H., Van de Peer, Y., Inze, D., and Beemster, G.T.S. (2007). Canalization without flux sensors: a traveling-wave hypothesis. *Trends Plant Sci.* **12**, 384–390.

Michniewicz, M., Brewer, P. B., and Friml, J. (2007). Polar auxin transport and asymmetric auxin distribution. In: Somerville, E.M.M.C.R. (Ed.), *The Arabidopsis Book*. American Society of Plant Biologists, Rockville, MD, doi/10.1199/tab.0108, http://www.aspb.org/publications/arabidopsis/.

Mitchison, G. J. (1980). A model for vein formation in higher plants. *Proc. R. Soc. Lond., B, Biol. Sci.* **207**, 79–109.

Mitchison, G. J. (1981). The polar transport of auxin and vein patterns in plants. *Philos. Trans. R. Soc. Lond., B, Biol. Sci.* **295**, 461–470.

Mockaitis, K., and Estelle, M. (2008). Auxin receptors and plant development: a new signaling paradigm. *Annu. Rev. Cell Dev. Biol.* **24**, 55–80.

Müller, B., and Sheen, J. (2008). Cytokinin and auxin interaction in root stem-cell specification during early embryogenesis. *Nature* **453**, 1094–1097.

Nakajima, K., Sena, G., Nawy, T., and Benfey, P. N. (2001). Intercellular movement of the putative transcription factor SHR in root patterning. *Nature* **413**, 307–311.

Naramoto, S., Sawa, S., Koizumi, K., Uemura, T., Ueda, T., Friml, J., Nakano, A., and Fukuda, H. (2009). Phosphoinositide-dependent regulation of VAN3 ARF-GAP localization and activity essential for vascular tissue continuity in plants. *Development* **136**, 1529–1538.

Navarro, C., Efremova, N., Golz, J. F., Rubiera, R., Kuckenberg, M., Castillo, R., Tietz, O., Saedler, H., and Schwarz-Sommer, Z. (2004). Molecular and genetic interactions between STYLOSA and GRAMINIFOLIA in the control of Antirrhinum vegetative and reproductive development. *Development* **131**, 3649–3659.

Nelson, T., and Dengler, N. (1997). Leaf vascular pattern formation. *Plant Cell* **9**, 1121–1135.

Niklas, K. J. (1997). *The Evolutionary Biology of Plants*. University of Chicago Press, Chicago, IL.

Nogueira, F., Chitwood, D., Madi, S., Ohtsu, K., Schnable, P., Scanlon, M., and Timmermans, M. (2009). Regulation of small RNA accumulation in the maize shoot apex. *PLoS Genet.* **5**, e1000320, doi:10.1371/journal.pgen.1000320.

Nogueira, F., Madi, S., Chitwood, D., Juarez, M., and Timmermans, M (2007). Two small regulatory RNAs establish opposing fates of a developmental axis. *Genes Dev.* **21**, 750–755.

Normanly, J. (2010). Approaching cellular and molecular resolution of auxin biosynthesis and metabolism. *Cold Spring Harb Perspect Biol.* **2**, a001594.

Ohashi-Ito, K., and Bergmann, D. C. (2007). Regulation of the Arabidopsis root vascular initial population by LONESOME HIGHWAY. Development **134**, 2959–2968.

Okada, K., Ueda, J., Komaki, M. K., Bell, C. J., and Shimura, Y. (1991). Requirement of the auxin polar transport system in early stages of Arabidopsis floral bud formation. *Plant Cell* **3**, 677–684.

Oparka, K. J., and Cruz, S. S. (2000). The great escape: phloem transport and unloading of macromolecules. *Annu. Rev. Plant Physiol. Plant Mol. Biol.* **51**, 323–347.

Otsuga, D., DeGuzman, B., Prigge, M. J., Drews, G. N., and Clark, S. E. (2001). REVOLUTA regulates meristem initiation at lateral positions. *Plant J.* **25**, 223–236.

Pekker, I., Alvarez, J., and Eshed, Y. (2005). Auxin response factors mediate Arabidopsis organ asymmetry via modulation of KANADI activity. *Plant Cell* **17**, 2899–2910.

Peragine, A., Yoshikawa, M., Wu, G., Albrecht, H., and Poethig R. (2004). SGS3 and SGS2/SDE1/RDR6 are required for juvenile development and the production of trans-acting siRNAs in Arabidopsis. *Genes Dev* **18**, 2368–2379.

Petrasek, J., and Friml, J. (2009). Auxin transport routes in plant development. *Development* **136**, 2675–2688.

Petricka, J. J., Clay, N. K., and Nelson, T. M. (2008). Vein patterning screens and the defectively organized tributaries mutants in *Arabidopsis thaliana*. *Plant J.* **56**, 251–263.

Petricka, J. J., and Nelson, T. M. (2007). Arabidopsis nucleolin affects plant development and patterning. *Plant Physiol.* **144**, 173–186.

Prigge, M. J., Otsuga, D., Alonso, J. M., Ecker, J. R., Drews, G. N., and Clark, S. E. (2005). Class III homeodomain-leucine zipper gene family members have overlapping, antagonistic, and distinct roles in Arabidopsis development. *Plant Cell* **17**, 61–76.

Przemeck, G. K., Mattsson, J., Hardtke, C. S., Sung, Z. R., and Berleth, T. (1996). Studies on the role of the Arabidopsis gene MONOPTEROS in vascular development and plant cell axialization. *Planta* **200**, 229–237.

Qi, J., Qian, Q., Bu, Q., Li, S., Chen, Q., Sun, J., Liang, W., Zhou, Y., Chu, C., Li, X., Ren, F., Palme, K., *et al.* (2008). Mutation of the rice Narrow leaf1 gene, which encodes a novel protein, affects vein patterning and polar auxin transport. *Plant Physiol.* **147**, 1947–1959.

Raven, J. A. (1975). Transport of indole acetic acid in plant cells in relation to pH and electrical potential gradients, and its significance for polar IAA transport. *New Phytol.* **74**, 163–172.

Raven, J. A. (1984). Physiological correlates of the morphology of early vascular plants. *Bot. J. Linn. Soc.* **88**, 105–126.

Raven, J. A. (1993). The evolution of vascular plants in relation to quantitative functioning of dead water-conducting cells and stomata. *Biol. Rev.* **68**, 337–363.

Reinhardt, D., Mandel, T., and Kuhlemeier, C. (2000). Auxin regulates the initiation and radial position of plant lateral organs. *Plant Cell* **12**, 507–518.

Reinhardt, D., Pesce, E. R., Stieger, P., Mandel, T., Baltensperger, K., Bennett, M., Traas, J., Friml, J., and Kuhlemeier, C. (2003). Regulation of phyllotaxis by polar auxin transport. *Nature* **426**, 255–260.

Robles, P., Fleury, D., Candela, H., Cnops, G., Alonso-Peral, M. M., Anami, S., Falcone, A., Caldana, C., Willmitzer, L., Ponce, M. R., Van Lijsebettens, M., and Micol, J. L. (2010). The RON1/FRY1/SAL1 gene is required for leaf morphogenesis and venation patterning in Arabidopsis. *Plant Physiol.* 152, 1357–1372.

Rolland-Lagan, A. G. (2008). Vein patterning in growing leaves: axes and polarities. *Curr. Opin. Genet. Dev.* **18**, 348–353.

Rolland-Lagan, A. G., Amin, M., and Pakulska, M. (2009). Quantifying leaf venation patterns: two-dimensional maps. *Plant J.* **57**, 195–205.

Rolland-Lagan, A. G., and Prusinkiewicz, P. (2005). Reviewing models of auxin canalization in the context of leaf vein pattern formation in Arabidopsis. *Plant J.* **44**, 854–865.

Rubery, P. H., and Sheldrake, A. R. (1974). Carrier-mediated auxin transport. *Planta* **118**, 101–121.

Runions, A., Fuhrer, M., Lane, B., Federl, P., Rolland-Lagan, A.-G., and Prusinkiewicz, P. (2005). Modeling and visualization of leaf venation patterns. *ACM Transactions on Graphics* **24**, 702–711.

Sabatini, S., Beis, D., Wolkenfelt, H., Murfett, J., Guilfoyle, T., Malamy, J., Benfey, P., Leyser, O., Bechtold, N., Weisbeek, P., and Scheres, B. (1999). An auxin-dependent distal organizer of pattern and polarity in the Arabidopsis root. *Cell* **99**, 463–472.

Sachs, T. (1975). Control of differentiation of vascular networks. *Ann. Bot.* **39**, 197–204.

Sachs, T. (1981). The control of the patterned differentiation of vascular tissues. *Adv. Bot. Res.* **9**, 151–262.

Sachs, T. (1989). The development of vascular networks during leaf development. *Curr. Top. Plant Biochem. Physiol.* **8**, 168–183.

Sakaguchi, J., Itoh, J.I., Ito, Y., Nakamura, A., Fukuda, H., Sawa, S. (2010). COE1, an LRR-RLK responsible for commissural vein pattern formation in rice. *Plant J.*, 2010 May 6. [Epub ahead of print] doi: 10.1111/j.1365-313X.2010.04250.x.

Samuels, A. L., Kaneda, M., and Rensing, K. H. (2006). The cell biology of wood formation: from cambial divisions to mature secondary xylem. *Can. J Botany* **84**, 631–639.

Sarkar, A. K., Luijten, M., Miyashima, S., Lenhard, M., Hashimoto, T., Nakajima, K., Scheres, B., Heidstra, R., and Laux, T. (2007). Conserved factors regulate signalling in *Arabidopsis thaliana* shoot and root stem cell organizers. *Nature* **446**, 811–814.

Sauer, M., Balla, J., Luschnig, C., Wisniewska, J., Reinohl, V., Friml, J., and Benkova, E. (2006). Canalization of auxin flow by Aux/IAA-ARF-dependent feedback regulation of PIN polarity. *Genes Dev.* **20**, 2902–2911.

Sawa, S., Watanabe, K., Goto, K., Liu, Y., Shibata, D., Kanaya, E., Morita, E., and Okada, K. (1999). FILAMENTOUS FLOWER, a meristem and organ identity gene of Arabidopsis, encodes a protein with a zinc finger and HMG-related domains. *Genes Dev.* **13**, 1079–1088.

Sawchuk, M. G., Donner, T. J., and Scarpella, E. (2008). Auxin transport-dependent, stage-specific dynamics of leaf vein formation. *Plant Signal. Behav.* **3**, 286–289.

Sawchuk, M. G., Head, P., Donner, T. J., and Scarpella, E. (2007). Time-lapse imaging of Arabidopsis leaf development shows dynamic patterns of procambium formation. *New Phytol.* **176**, 560–571.

Scarpella, E., Boot, K. J., Rueb, S., and Meijer, A. H. (2002). The procambium specification gene Oshox1 promotes polar auxin transport capacity and reduces its sensitivity toward inhibition. *Plant Physiol.* **130**, 1349–1360.

Scarpella, E., Francis, P., and Berleth, T. (2004). Stage-specific markers define early steps of procambium development in Arabidopsis leaves and correlate termination of vein formation with mesophyll differentiation. *Development* **131**, 3445–3455.

Scarpella, E., Marcos, D., Friml, J., and Berleth, T. (2006). Control of leaf vascular patterning by polar auxin transport. *Genes Dev.* **20**, 1015–1027.

Scarpella, E., and Meijer, A. H. (2004). Pattern formation in the vascular system of monocot and dicot plant species. *New Phytol.* **164**, 209–242.

Scarpella, E., Rueb, S., and Meijer, A. H. (2003). The RADICLELESS1 gene is required for vascular pattern formation in rice. *Development* **130**, 645–658.

Scheres, B., Dilaurenzio, L., Willemsen, V., Hauser, M. T., Janmaat, K., Weisbeek, P., and Benfey, P. N. (1995). Mutations affecting the radial organization of the Arabidopsis root display specific defects throughout the embryonic axis. *Development* **121**, 53–62.

Schmitz, K. (1990). Algae. In: Behnke, R.D.S.H.-D. (Ed.), *Sieve Elements. Comparative Structure, Induction and Development*. Springer, Berlin, pp. 1–18.

Schrader, J., Nilsson, J., Mellerowicz, E., Berglund, A., Nilsson, P., Hertzberg, M., and Sandberg, G. (2004). A high-resolution transcript profile across the wood-forming meristem of poplar identifies potential regulators of cambial stem cell identity. *Plant Cell* **16**, 2278–2292.

Schuetz, M., Berleth, T., and Mattsson, J. (2008). Multiple MONOPTEROS-dependent pathways are involved in leaf initiation. *Plant Physiol.* **148**, 870–880.

Schulz, A. (1998). Phloem. Structure related to function. *Prog. Bot.* **59**, 429–475.

Semiarti, E., Ueno, Y., Tsukaya, H., Iwakawa, H., Machida, C., and Machida, Y. (2001). The ASYMMETRIC LEAVES2 gene of *Arabidopsis thaliana* regulates formation of a symmetric lamina, establishment of venation and repression of meristem-related homeobox genes in leaves. *Development* **128**, 1771–1783.

Shevell, D. E., Leu, W. M., Gillmor, C. S., Xia, G., Feldmann, K. A., and Chua, N. H. (1994). EMB30 is essential for normal cell division, cell expansion, and cell adhesion in Arabidopsis and encodes a protein that has similarity to Sec7. *Cell* **77**, 1051–1062.

Shirakawa, M., Ueda, H., Shimada, T., Nishiyama, C., and Hara-Nishimura, I. (2009). Vacuolar SNAREs function in the formation of the leaf vascular network by regulating auxin distribution. *Plant Cell Physiol.* **50**, 1319–1328.

Sieburth, L. E. (1999). Auxin is required for leaf vein pattern in Arabidopsis. *Plant Physiol.* **121**, 1179–1190.

Sieburth, L. E., and Deyholos, M. K. (2006). Vascular development: the long and winding road. *Curr. Opin. Plant Biol.* **9**, 48–54.

Sieburth, L. E., Muday, G. K., King, E. J., Benton, G., Kim, S., Metcalf, K. E., Meyers, L., Seamen, E., and Van Norman, J. M. (2006). SCARFACE encodes an ARF-GAP that is required for normal auxin efflux and vein patterning in *Arabidopsis*. *Plant Cell* **18**, 1396–1411.

Siegfried, K., Eshed, Y., Baum, S., Otsuga, D., Drews, G., and Bowman, J. (1999). Members of the *YABBY* gene family specify abaxial cell fate in *Arabidopsis*. *Development* **126**, 4117–4128.

Silverstone, A. L., Chang, C. W., Krol, E., and Sun, T. P. (1997). Developmental regulation of the gibberellin biosynthetic gene GA1 in *Arabidopsis thaliana*. *Plant J.* **12**, 9–19.

Smith, R. S., and Bayer, E. M. (2009). Auxin transport–feedback models of patterning in plants. *Plant Cell Environ.* **32**, 1258–1271.

Song, S. K., Hofhuis, H., Lee, M. M., and Clark, S. E. (2008). Key divisions in the early Arabidopsis embryo require POL and PLL1 phosphatases to establish the root stem cell organizer and vascular axis. *Dev. Cell* **15**, 98–109.

Spicer, R., and Groover, A. (2010). Evolution of development of vascular cambia and secondary growth. *New Phytol.* **186**, 577–92.

Stahl, Y., Wink, R. H., Ingram, G. C., Simon, R. (2009). A signalling module controlling the stem cell niche in Arabidopsis root meristems. *Curr. Biol.* **19**, 909–914.

Steinmann, T., Geldner, N., Grebe, M., Mangold, S., Jackson, C. L., Paris, S., Galweiler, L., Palme, K., and Jurgens, G. (1999). Coordinated polar localization of auxin efflux carrier PIN1 by GNOM ARF GEF. *Science* **286**, 316–318.

Stepanova, A. N., Robertson-Hoyt, J., Yun, J., Benavente, L. M., Xie, D. Y., Dolezal, K., Schlereth, A., Jurgens, G., and Alonso, J. M. (2008). TAA1-mediated auxin biosynthesis is essential for hormone crosstalk and plant development. *Cell* **133**, 177–191.

Steynen, Q. J., and Schultz, E. A. (2003). The FORKED genes are essential for distal vein meeting in Arabidopsis. *Development* **130**, 4695–4708.

Stoma, S., Lucas, M., Chopard, J., Schaedel, M., Traas, J., and Godin, C. (2008). Flux-based transport enhancement asa plausible unifying mechanism for auxin transport in meristem development. *Plos Comput. Biol.* **4**, e1000207.

Sun, Y., Zhou, Q., Zhang, W., Fu, Y., and Huang, H. (2002). ASYMMETRIC LEAVES1, an Arabidopsis gene that is involved in the control of cell differentiation in leaves. *Planta* **214**, 694–702.

Sussex, I. (1951). Experiments on the cause of dorsiventrality in leaves. *Nature* ***167***, 651–652.

Sussex, I. (1955). Morphogenesis in *Solanum tuberosum* L.: Experimental investigation of leaf dorsiventrality and orientation in the juvenile shoot. *Phytomorphology* **5**, 286–300.

Tang, G., Reinhart, B., Bartel, D., and Zamore, P. (2003). A biochemical framework for RNA silencing in plants. *Genes Dev.* **23**, 49–63.

Tasaka, M. (2001). From central-peripheral to adaxial-abaxial. *Trends Plant Sci.* **6**, 548–550.

Taylor, E. L. (1990). Phloem evolution: an appraisal based on the fossil record. In: Behnke, R.D.S.H.-D. (Ed.), *Sieve Elements. Comparative Structure, Induction and Development.* Springer, Berlin, pp. 285–298.

Thomas, C. L., Schmidt, D., Bayer, E. M., Dreos, R., and Maule, A. J. (2009). Arabidopsis plant homeodomain finger proteins operate downstream of auxin accumulation in specifying the vasculature and primary root meristem. *Plant J.* **59**, 426–436.

Timmermans, M., Hudson, A., Becraft, P., and Nelson, T. (1999). ROUGH SHEATH2: A Myb protein that represses knox homeobox genes in maize lateral organ primordia. *Science* **284**, 151–153.

Timmermans, M., Schultes, N., Jankovsky, J., and Nelson, T. (1998). Leafbladeless1 is required for dorsoventrality of lateral organs in maize. *Development* **125**, 2813–2823.

Titapiwatanakun, B., Blakeslee, J. J., Bandyopadhyay, A., Yang, H., Mravec, J., Sauer, M., Cheng, Y., Adamec, J., Nagashima, A., Geisler, M., Sakai, T., Friml, J., *et al.* (2009). ABCB19/PGP19 stabilises PIN1 in membrane microdomains in Arabidopsis. *Plant Journal* **57**, 27–44.

Torrey, J. G. (1955). On the determination of vascular pattern formation during tissue differentiation in excised pea roots. *Am. J. Bot.* **42**, 183–198.

Torrey, J. G. (1957). Auxin control of vascular pattern formation in regenerating pea root meristems grown in vitro. *Am. J. Bot.* **44**, 859–870.

Tsiantis, M., Schneeberger, R., Golz, J., Freeling, M., and Langdale, J. (1999). The maize rough sheath2 gene and leaf development programs in monocot and dicot plants. *Science* **284**, 154–156.

Tsugeki, R., Ditengou, F. A., Sumi, Y., Teale, W., Palme, K., and Okada, K. (2009). NO VEIN mediates auxin-dependent specification and patterning in the Arabidopsis embryo, shoot, and root. *Plant Cell* **21**, 3133–3151.

Turner, S., Gallois, P., and Brown, D. (2007). Tracheary element differentiation. *Annu. Rev. Plant Biol.* **58**, 407–433.

Turner, S., and Sieburth, L. E. (2003). Vascular patterning. In Somerville, C. R. , and Meyerowitz, E. M. (Ed.), *The Arabidopsis Book.* American Society of Plant Biologists, Rockville, MD: American Society of Plant Biologists. doi: 10.1199/tab.0073, http://www.aspb.org/publications/arabidopsis/

Ugartechea-Chirino, Y., Swarup, R., Swarup, K., Peret, B., Whitworth, M., Bennett, M., and Bougourd, S. (2010). The AUX1 LAX family of auxin influx carriers is required for the establishment of embryonic root cell organization in *Arabidopsis thaliana. Ann. Bot.* **105**, 277–289.

van Bel, A.J.E. (1999). Evolution, polymorphology and multifunctionality of the phloem system. *Perspect. Plant Ecol., Evol. Syst* **2**, 163–184.

van Bel, A. J., Ehlers, K., and Knoblauch, M. (2002). Sieve elements caught in the act. *Trends Plant Sci.* **7**, 126–32.

van den Berg, C., Willemsen, V., Hage, W., Weisbeek, P., and Scheres, B. (1995). Cell fate in the Arabidopsis root meristem determined by directional signalling. *Nature* **378**, 62–65.

Van Norman, J. M., Frederick, R. L., and Sieburth, L. E. (2004). BYPASS1 negatively regulates a root-derived signal that controls plant architecture. *Curr. Biol.* **14**, 1739–1746.

Vazquez, F., Vaucheret, H., Rajagopalan, R., Lepers, C., Gasciolli, V., Mallory, A., Hilbert, J., Bartel, D., Crete, P. (2004). Endogenous trans-acting siRNAs regulate the accumulation of Arabidopsis mRNAs. *Mol Cell* **16**, 69–79.

Vera-Sirera, F., Minguet, E. G., Singh, S. K., Ljung, K., Tuominen, H., Blázquez, M. A., and Carbonell, J. (2010). Role of polyamines in plant vascular development. *Plant Physiol. Biochem.* 2010 Jan 22. [Epub ahead of print], doi: 10.1016/j.plaphy. 2010.01.011.

Vieten, A., Vanneste, S., Wisniewska, J., Benkova, E., Benjamins, R., Beeckman, T., Luschnig, C., and Friml, J. (2005). Functional redundancy of PIN proteins is accompanied by auxin-dependent cross-regulation of PIN expression. *Development* **132**, 4521–31.

Waites, R., and Hudson, A. (1995). Phantastica—a gene required for dorsoventrality of leaves in *Antirrhinum majus*. *Development* **121**, 2143–2154.

Waites, R., Selvadurai, H., Oliver, I., and Hudson, A. (1998). The PHANTASTICA gene encodes a MYB transcription factor involved in growth and dorsoventrality of lateral organs in Antirrhinum. *Cell* **93**, 779–789.

Waring, P. F., and Phillips, I.D.J. (1970). *The Control of Growth and Differentiation in Plants*. Pergamon Prss, Oxford.

Weijers, D., Benkova, E., Jager, K. E., Schlereth, A., Hamann, T., Kientz, M., Wilmoth, J. C., Reed, J. W., and Jurgens, G. (2005). Developmental specificity of auxin response by pairs of ARF and Aux/IAA transcriptional regulators. *EMBO J.* **24**, 1874–1885.

Weijers, D., Schlereth, A., Ehrismann, J. S., Schwank, G., Kientz, M., and Jurgens, G. (2006). Auxin triggers transient local signaling for cell specification in Arabidopsis embryogenesis. *Dev. Cell* **10**, 265–270.

Wenzel, C. L., Schuetz, M., Yu, Q., and Mattsson, J. (2007). Dynamics of MONOPTEROS and PIN-FORMED1 expression during leaf vein pattern formation in *Arabidopsis thaliana*. *Plant J.* **49**, 387–398.

Willemsen, V., Friml, J., Grebe, M., van den Toorn, A., Palme, K., and Scheres, B. (2003). Cell polarity and PIN protein positioning in Arabidopsis require STEROL METHYLTRANSFERASE1 function. *Plant Cell* **15**, 612–625.

Wisniewska, J., Xu, J., Seifertova, D., Brewer, P. B., Ruzicka, K., Blilou, I., Rouquie, D., Benkova, E., Scheres, B., and Friml, J. (2006). Polar PIN localization directs auxin flow in plants. *Science* **312**, 883.

Xu, J., Hofhuis, H., Heidstra, R., Sauer, M., Friml, J., and Scheres, B. (2006). A molecular framework for plant regeneration. *Science* **311**, 385–388.

Ye, Z. H. (2002). Vascular tissue differentiation and pattern formation in plants. *Annu. Rev. Plant Biol.* **53**, 183–202.

Yu, L., Yu, X., Shen, R., and He, Y. (2005). HYL1 gene maintains venation and polarity of leaves. *Planta* **221**, 231–242.

Zgurski, J. M., Sharma, R., Bolokoski, D. A., and Schultz, E. A. (2005). Asymmetric auxin response precedes asymmetric growth and differentiation of asymmetric leaf1 and asymmetric leaf2 Arabidopsis leaves. *Plant Cell* **17**, 77–91.

Zhang, G.-H., Xu, Q., Zhu X.-D., Qian, Q., and Xue, H.-W. (2009). SHALLOT-LIKE1 is a KANADI transcription factor that modulates rice leaf rolling by regulating leaf abaxial cell development. *Plant Cell* **21**, 719–735.

Zhong, R., Taylor, J. J., and Ye, Z. H. (1997). Disruption of interfascicular fiber differentiation in an Arabidopsis mutant. *Plant Cell* **9**, 2159–2170.

Zhong, R., Taylor, J. J., and Ye, Z. H. (1999). Transformation of the collateral vascular bundles into amphivasal vascular bundles in an Arabidopsis mutant. *Plant Physiol* **120**, 53–64.

Zhong, R., and Ye, Z. H. (1999). IFL1, a gene regulating interfascicular fiber differentiation in Arabidopsis, encodes a homeodomain-leucine zipper protein. *Plant Cell* **11**, 2139–2152.

Zhong, R., and Ye, Z. H. (2004). Amphivasal vascular bundle 1, a gain-of-function mutation of the IFL1/REV gene, is associated with alterations in the polarity of leaves, stems and carpels. *Plant Cell Physiol.* **45**, 369–385.

Zimmermann, W. (1930). *Die Phylogenie der Pflanzen*. Fischer, Jena.

Zimmermann, W. (1952). The main results of the telome theory. *Palaeobotanist* **1**, 456–470.

CHAPTER NINE

Stomatal Patterning and Development

Juan Dong *and* Dominique C. Bergmann

Contents

Abstract

Stomata are epidermal pores used for water and gas exchange between a plant and the atmosphere. Both the entry of carbon dioxide for photosynthesis and the evaporation of water that drives transpiration and temperature regulation are modulated by the activities of stomata. Each stomatal pore is surrounded by

Department of Biology, Stanford University, Stanford, California, USA

Current Topics in Developmental Biology, Volume 91
ISSN 0070-2153, DOI 10.1016/S0070-2153(10)91009-0

two highly specialized cells called guard cells (GCs), and may also be associated with neighboring subsidiary cells; this entire unit is referred to as the stomatal complex. Generation of GCs requires stereotyped asymmetric and symmetric cell divisions, and the pattern of stomatal complexes in the epidermis follows a "one-cell-spacing rule" (one complex almost never touches another one). Both stomatal formation and patterning are highly regulated by a number of genetic components identified in the last decade, including, but not limited to, secreted peptide ligands, plasma membrane receptors and receptor-like kinases, a MAP kinase module, and a series of transcription factors.

This review will elaborate on the current state of knowledge about components in signaling pathways required for cell fate and pattern, with emphasis on (1) a family of extracellular peptide ligands and their relationship to the TOO MANY MOUTHS receptor-like protein and/or members of the ERECTA receptor-like kinase family, (2) three tiers of a MAP kinase module and the kinases that confer novel regulatory effects in specific stomatal cell types, and (3) transcription factors that generate specific stomatal cell types and the regulatory mechanisms for modulating their activities. We will then consider two new proteins (BASL and PAN1, from *Arabidopsis* and maize, respectively) that regulate stomatal asymmetric divisions by establishing cell polarity.

1. Introduction

Stomata ("mouths" in Greek) are pores used for water and gas exchange between a plant and the atmosphere. Each stomatal pore is surrounded by two highly specialized cells called guard cells (GCs), and may also be associated with neighboring subsidiary cells (SCs); this entire unit is referred to as the stomatal complex. Stomata are found in different aerial organs and in species-specific patterns; nonetheless, in nearly all plants studied, the generation of stomata requires asymmetric and symmetric cell divisions, and the pattern of stomatal complexes in the epidermis follows a spacing rule whereby two stomatal complexes are always separated by at least one intervening epidermal cell.

Stomatal development in *Arabidopsis thaliana* requires a dedicated series of asymmetric cell divisions (ACDs) followed by a single symmetric cell division to form a pair of mature GCs (Fig. 9.1; Geisler *et al.*, 2003; Nadeau and Sack, 2003). In the developing epidermis of young leaves, the stomatal lineage is initiated from a subset of committed protodermal (Pr) cells called meristemoid mother cells (MMCs). ACD of an MMC generates two daughter cells: a small, triangular cell called a meristemoid and a larger cell called a stomatal lineage ground cell (SLGC). This type of division is also referred to as an *entry* division, because it creates stomatal lineage cells. Meristemoids may follow either of two paths (Fig. 9.1); they may transition into a round-shaped guard mother cell (GMC) and divide once to make

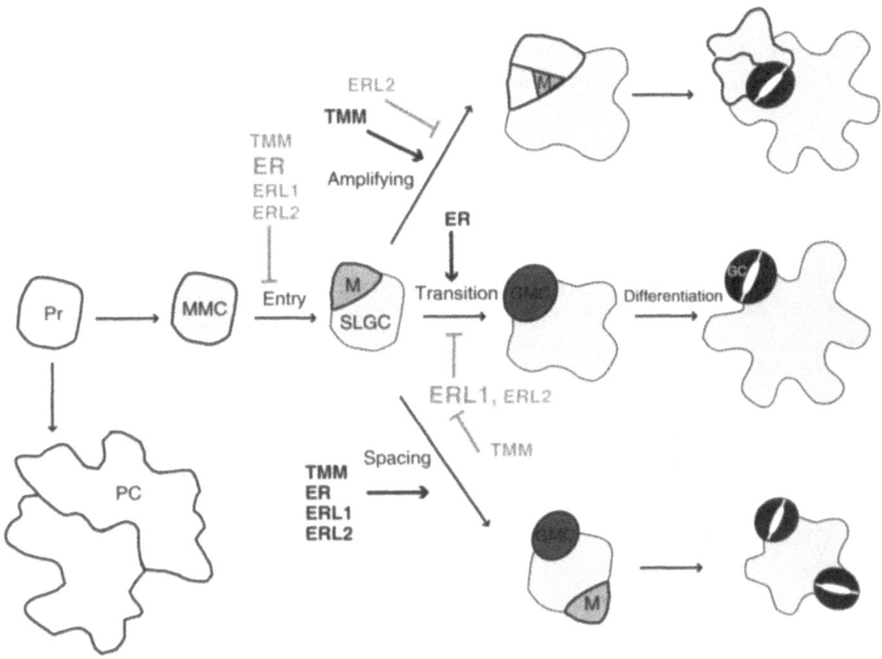

Figure 9.1 Schematic for stomatal development and regulation by TMM and members of the ERECTA family (ERf) receptors. Stomatal development initiates from meristemoid mother cells (MMCs), a set of committed protodermal (Pr) cells. In the stomatal *entry* division, asymmetric cell division of an MMC generates a small daughter cell, the meristemoid (M), and a larger daughter cell, the stomatal lineage ground cell (SLGC). A meristemoid may immediately *transition* into a guard mother cell (GMC) and further divide and *differentiate* into guard cells (GC), or it may divide again asymmetrically, up to three times, *amplifying* the number of cells in the stomatal lineage, before transitioning into a GMC. SLGCs may divide asymmetrically in a *spacing* division to generate secondary meristemoids, which are always spaced away from the existing stomatal precursor cell. Regulation of developmental progression by TMM ER, ERL1, and ERL2 is marked at each step where they function, with positive effects labeled with black arrows and negative effects labeled with gray T-bars. Font size used for TMM and ERf indicates the relative strength of each gene's effect, with larger fonts signifying stronger effects.

two GCs, or they may divide again asymmetrically. Continued asymmetric divisions of meristemoids (up to 3 times for each) are termed *amplifying divisions*. This second path serves to amplify the number of SLGCs, which are major contributors to the population of both stomatal and pavement cells in the leaf. SLGCs can directly expand and differentiate into pavement cells, or they may also undergo an ACD to produce a secondary meristemoid. When SLGCs produce secondary meristemoids, they always divide in such a way that this new cell is not next to existing meristemoids or stomata. This additional level of regulation using orientation of ACDs enforces the so-called "one-cell-spacing rule" that typifies stomatal pattern in *Arabidopsis* and nearly all other plants.

In monocot plants, such as maize and rice, stomata are also separated by at least one intervening nonstomatal cell, but the developmental events that create and pattern stomata are quite different (Hernandez *et al.*, 1999). For example, maize stomata are linearly arranged in specific cell files in the leaf epidermis. An individual maize stomatal complex is composed of a pair of GCs flanked by a pair of SCs. GC development in maize starts from an asymmetric division that generates a smaller daughter cell as the GMC. The GMC then signals to two lateral cells from the neighboring cell files to recruit them to become subsidiary mother cells (SMCs). SMC nuclei become polarized by migrating toward the GMC; subsequently, the SMCs divide asymmetrically to produce small SCs that flank the GMC, and will become accessory cells to help the stomatal pore open and close. The pore is formed in the next step when the GMC divides symmetrically and the daughters differentiate into a pair of dumbbell-shaped GCs.

The developmental processes and patterns that make up the stomatal systems in both *Arabidopsis* and maize are relatively simple, easily scored, and accessible for genetic and experimental manipulation. Also, stomatal development heavily relies on ACD, which is one of the most commonly used mechanisms for generating cellular diversity and overall morphology in multicellular organisms. Therefore, specific studies of the molecular mechanisms underlying stomatal development can facilitate our general understanding of many fundamental biological issues such as intracellular and intercellular communication, signal transduction cascades, cell fate determination, establishment of cell polarity, and the information for, and mechanics of, ACD.

2. The Signaling Cascade: From Extracellular Signals to Nuclear Factors

2.1. Receiving and sending signals

Plant growth and development requires a tremendous amount of cell-to-cell signaling and communication. Crosstalk mediated by extracellular peptide ligands and cell surface receptors is a key component, especially in maintaining the balance between stem cell populations and differentiated cells in the shoot and root apical meristems and vascular cambium (Byrne *et al.*, 2003; Gray *et al.*, 2008; Stahl and Simon, 2009). In plants, the term *stem cells* is typically used to refer to a small population of undifferentiated cells at the root and/or shoot apical meristem. By dividing asymmetrically, stem cells generate one daughter that will become one of the differentiated cells driving organ formation, but the other daughter will retain stem cell properties, thus maintaining a constant size to the stem cell pool. Peptide ligands, such as the ones containing the conserved CLAVATA/ESR (CLE) motif, are secreted from the meristematic cells and perceived by a set of receptors in the neighboring cells (Fiers *et al.*, 2007). The receptors are typically Leucine rich repeat (LRR) receptor-like proteins and/or kinases (LRR-RLPs and/or LRR-RLKs). Ligand binding is thought to activate the receptors and their downstream signaling cascade to promote differentiation and thereby prevent the neighbor cells from adopting stem cell fate and expanding the pool (chapters 3 and 4; Fletcher *et al.*, 1999; Ito *et al.*, 2006).

Two types of cells in the stomatal lineage population, MMCs and meristemoids, behave in a stem cell-like manner, such that these cells produces two daughter cells, one of which can maintain its MMC property of further division potential and the other which will generally differentiate into a pavement cell. In contrast to the stem cells in the apical meristems, however, stomatal MMCs and meristemoids are not in niches. Instead, they are dispersed on the leaf epidermis and their stem cell activity is transient. Therefore, it is not surprising that signaling components, including ligands and receptors, modulate stomatal development and patterning, but the exact ligands and receptors are distinct from those previously described in apical meristems. Newly discovered intercellular signals include the small secreted proteins EPIDERMAL PATTERNING FACTOR 1 (EPF1) (Hara *et al.*, 2007), EPF2 (Hara *et al.*, 2009; Hunt and Gray, 2009), CHALLAH (CHAL) (Abrash and Bergmann, 2010), and STOMAGEN (Kondo *et al.*, 2009; Sugano *et al.*, 2009), all of which belong to a small subfamily only distantly related to the CLE family.

2.2. Receptors at the cell surface

While the ligands are quite recent discoveries, the presumptive receptors were among the first components shown to regulate stomatal development and patterning (Yang and Sack, 1995). They include a receptor-like protein TOO MANY MOUTHS (TMM) (Nadeau and Sack, 2002) and three receptor-like kinases, ERECTA (ER), ERECTA-LIKE 1 (ERL1), and ERECTA-LIKE 2 (ERL2) (Shpak *et al.*, 2005), all containing an extracellular LRR domain. The three kinases are also collectively referred to as the ERECTA family (ERf). These four putative receptors are commonly described as negative regulators of stomatal development, mainly because loss of function of *TMM* and *ERf* genes results in overproliferation of stomata in the leaf epidermis.

The source of the excessive number of stomata in the *tmm* mutant was carefully analyzed and found to be a combination of three general defects including (1) more cells became MMCs and underwent entry divisions; (2) each meristemoid underwent fewer amplifying divisions, thereby becoming a GC more rapidly; and (3) divisions that create secondary meristemoids were misoriented, often placing a new meristemoid right next to a GC pair or a precursor (Nadeau and Sack, 2002). TMM expression is strong in the early stages of the stomatal lineage, with the highest level in the young MMCs and meristemoids and a lower, but detectable level in SLGCs. TMM is absent from mature GCs. These mutant phenotypes and expression patterns lead to the hypothesis that TMM in the MMCs and meristemoids functions in an autocrine loop to inhibit divisions in these cells, whereas TMM in the SLGCs perceives positional cues on the cell surface to control and orient the secondary meristemoid forming divisions (Nadeau and Sack, 2002).

The TMM protein contains extracellular and transmembrane domains, but lacks an intracellular kinase domain, suggesting that TMM might participate in signal transduction through interaction with other partners at the plasma membrane. The TMM-interacting kinase partners are likely to be the ERf LRR-RLKs, which were also characterized as a set of key regulators in stomatal production and patterning (Shpak *et al.*, 2005). In contrast to *tmm* whose apparent mutant phenotypes are restricted to stomatal development, the *erf* mutants have pleiotropic phenotypes, implicating them in the control of cell proliferation for patterning and organ development, in plant responses to environmental stresses, and even in phytohormone perception (reviewed in van Zanten *et al.*, 2009). In terms of stomatal development, the triple mutant *er;erl1;erl2* exhibits severe stomatal overproliferation and spacing defects (Shpak *et al.*, 2005). The three *ERf* genes largely function redundantly and synergistically in regulation of stomatal divisions, since these dramatic phenotypes are only revealed in the triple mutant. However, examination of the phenotypes of single and combinations of double mutants of *er, erl1*, and *erl2* in the leaf epidermis shows each family member plays subtly different roles (Fig. 9.1), for example, *ER* is predominantly involved in inhibition of MMC divisions, but may also promote meristemoid differentiation into GMCs (and eventually GCs); *ERL1* generally inhibits meristemoid differentiation and *ERL2* appears to promote continued asymmetric division of the SLGCs (Shpak *et al.*, 2005).

The functions of *TMM* and *ERf*, and, especially, their genetic interactions, are fairly complicated. One difficulty is that different genes play different roles depending on specific organs examined (Nadeau and Sack, 2002; Shpak *et al.*, 2005). For example, the *tmm* mutant has an elevated number of stomata on leaves (consistent with a negative role in stomatal development), yet *tmm* is devoid of stomata on stems (indicating a positive role, Yang and Sack, 1995). These roles are not completely opposite; in stems, the absence of stomata is due to transdifferentiation of meristemoids into pavement cells, not, as might be expected from the phenotype in leaves, because MMC divisions are absent (Bhave *et al.*, 2009). Combinations of the *tmm* mutation with mutations in the *ERf* members also result in unpredictable organ-specific phenotypes. For example, elimination of

ERL1 in a *tmm* mutant background restores stomata to the *tmm* stems, but elimination of *ER* in *tmm* siliques nearly eliminates stomatal production, although each single mutant tends to increase the number of stomata in this organ (Shpak *et al.*, 2005). A model derived from phenotypic analysis of various *TMM* and *ERf* mutant combinations in various organs was published by Shpak *et al.* (2005); we will revisit this model, incorporating new information from studies of ligands, in a later section of this review.

2.3. Extracellular ligands

The *Arabidopsis* genome contains more than 200 LRR - RLKs which mediate signaling through association of their extracellular LRR domains with a chemically heterogeneous collection of ligand molecules including steroid hormones, peptides, and small proteins (Torii, 2004). TMM and/or the ERf are likely engaged by four ligands from the same protein subfamily—EPF1, EPF2, CHAL, and STOMAGEN (Abrash and Bergmann, 2010; Hara *et al.*, 2007, 2009; Hunt and Gray, 2009; Kondo *et al.*, 2009; Sugano *et al.*, 2009). As we detail below, these peptide ligands enrich our understanding of the potential regulatory spectrum; they may signal from within the stomatal lineage or from underlying tissues, they may function as positive or negative regulators of stomata, and they display different requirements for the TMM and ERf receptors.

2.4. Extracellular ligands—negative regulators

Both EPF1 and EPF2 were discovered by their effects on stomatal development and patterning in a genome-wide overexpression screen of predicted small secreted proteins (Hara *et al.*, 2007, 2009) or by co-expression with known stomatal regulators (Hunt and Gray, 2009). Cauliflower mosaic virus 35S (35S) promoter-driven overexpression (OX) of EPF1 (EPF1-OX) or EPF2 (EPF2-OX) reduces stomata density in the leaf epidermis, but EPF2-OX inhibits the earliest asymmetric divisions, resulting in an epidermis composed entirely of pavement cells (Hara *et al.*, 2009; Hunt and Gray, 2009). EPF1-OX inhibits stomata formation at a later stage than EPF2-OX; although EPF1-OX leaves also lack mature stomata, asymmetric divisions can be observed (Hara *et al.*, 2007).

Consistent with their negative roles in the control of stomatal production, the loss-of-function mutants *epf1* and *epf2* displayed elevated numbers of stomata. Careful examination of the mutant phenotypes showed that *EPF1* acts mainly on orientation of the spacing division, so that without *EPF1*, the stomatal one-cell-spacing rule is disrupted and a slightly higher number of GC clusters are generated (Hara *et al.*, 2007). *EPF2* acts earlier to restrict Pr cells from acquiring stomatal lineage cell fates; *epf2* mutants display more entry divisions and mutant leaves feature clusters of small stomatal lineage cells that do not always become stomata (Hara *et al.*, 2009; Hunt and Gray, 2009). The cell-type-specific expression patterns of these ligands contributes to their specific cellular functions. *EPF2* is expressed at earlier stages (MMCs and meristemoids) than *EPF1* (meristemoids, GMCs, Fig. 9.2A) (Hara *et al.*, 2007, 2009; Hunt and Gray, 2009). Both *TMM* and *ERf* mutations are fully or partially epistatic to *epf1* and *epf2* and can block the overexpression effects of EPF1 and EPF2, supporting the idea that EPF1 and EPF2 function through these receptors.

EPF1 and EPF2 can explain how to regulate two lineage decisions in many organs. However, they do not solve the mystery of *tmm*'s opposite phenotypes in leaves versus stems. Identification and characterization of their paralog CHAL/EPFL6, however, may provide a reasonable explanation for these *tmm* mutant phenotypes. The *chal* mutant was identified as a *tmm* suppressor that restored stomata to *tmm* hypocotyls and stems. The phenotypes of *chal* are genotype specific, in that they are only revealed in a *tmm* background, and are largely organ specific, in that the phenotypes are observed in hypocotyls and stems, but not in leaves. Like EPF1 and EPF2, 35S-driven CHAL-OX reduces stomatal density in all organs. Also in parallel to EPF1 and EPF2, mutations in *ERf* are genetically epistatic to *CHAL*. Overexpression experiments in ERf family double mutants suggest that *ER, ERL1*, and *ERL2* are all capable of transducing the *CHAL* signal (Abrash and Bergmann, 2010).

Where CHAL differs from EPF1 and EPF2 is in its expression pattern and relationship with TMM. Together these two features help explain why no stomata are produced in *tmm* stems. CHAL is not expressed in stomatal lineage cells but is produced in internal tissues, especially in cells surrounding the vasculature (Abrash and Bergmann, 2010). This internal expression reaches a maximum in stems and in embryonic hypocotyls. CHAL does not appear to need *TMM* for its function; in fact TMM appears to inhibit CHAL signaling because the ability of CHAL-OX to repress stomatal formation is dramatically enhanced in a *tmm* background.

The current model is that CHAL is normally a ligand for the ERfs; in stems, the high level of CHAL would engage these receptors to inhibit stomatal production; however, TMM normally acts to "protect" the ERfs from CHAL, allowing stomata to be produced (Fig. 9.2A and B). When TMM is absent, CHAL overactivates its ERf receptors and stomatal production in stems is eliminated (Abrash and Bergmann, 2010).

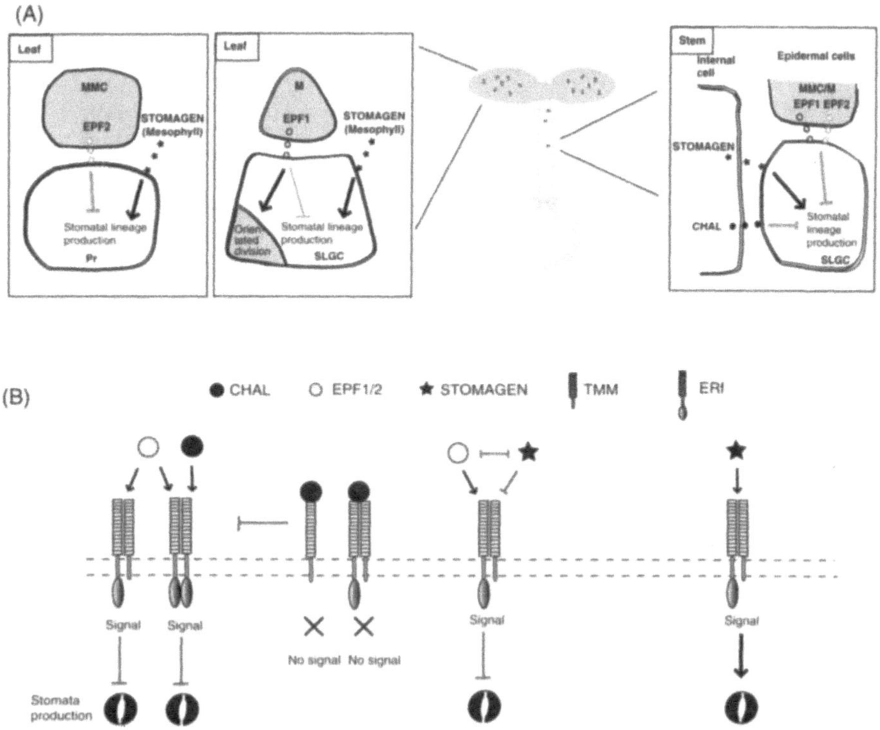

Figure 9.2 Speculative models for interactions among stomatal ligands and receptors. (A) In leaves and stems, ligands differentially regulate stomatal cell fates. Protodermal (Pr) cells or the stomatal lineage ground cell (SLGCs) in leaves receive EPF1 and EPF2 signals from the neighboring MMC or meristemoid, inhibiting the acquisition of stomatal lineage fate. These cells also receive the STOMAGEN signal from the underlying mesophyll that promotes stomatal lineage cell fate. In stems, SLGCs are additionally subject to inhibitory CHAL signals coming from internal tissues. (B) Possible regulatory interactions among ligands and receptors. If the receptors all act as negative regulators of stomatal development, then EPF1 and EPF2 could activate homo- or heterodimers of the ERf and TMM. TMM could dampen the CHAL–ERf inhibition of stomatal production activity in stems by sequestering CHAL in a nonfunctional complex, or by the CHAL–TMM complex antagonizing active CHAL–ERfs. STOMAGEN, as a positive regulator, might compete with negative regulators, EPF1 and EPF2, to bind receptor complexes. However, if TMM–ERf receptors also act as positive regulators, then STOMAGEN could stimulate stomatal production by associating directly with the TMM–ERf complexes.

2.5. Extracellular ligands—positive regulators

The negative regulators EPF1, EPF2, and CHAL can explain much about stomatal development and pattern. But are all the intercellular signals guiding this developmental pathway negative? The answer appears to be no, and surprisingly, the ligand responsible is part of the same EPF family. STOMAGEN/EPFL9 was identified in a co-expression analysis with TMM, STOMATAL DENSITY AND DISTRIBUTION-1 (SDD1), and EPF1 (Sugano *et al.*, 2009) and by its overexpression phenotypes (Kondo *et al.*, 2009). When ubiquitously overexpressed with the 35S promoter, STOMAGEN can generate extra stomata in the epidermis (Kondo *et al.*, 2009; Sugano *et al.*, 2009) and the loss-of-function phenotype of STOMAGEN obtained by RNA silencing is greatly reduced stomatal production, both consistent with a positive role of STOMAGEN in stomata formation (Sugano *et al.*, 2009).

Biochemical analysis and peptide sequencing of the mature form of STOMAGEN suggest that the protein is secreted and that it acts as a 45-amino-acid peptide. The STOMAGEN cDNA encodes a protein of 120 amino acids (aa) in length, including an N-terminal signal peptide and a second proteolytic cleavage site indicating the need for multiple processing events. A chemically synthesized 45-aa STOMAGEN peptide is sufficient to induce stomatal production when applied to plants (Kondo *et al.*, 2009; Sugano *et al.*, 2009), indicating that this ligand could work without any of the typical post-translational modifications found in other plant ligands (Kondo *et al.*, 2006). Interestingly, the sequences around the cleavage sites are highly conserved between STOMAGEN, EPF1, EPF2, and CHAL, suggesting that all of these proteins may be processed to yield an active ligand of 45–60 aa derived from the C-terminus.

The phenotypes of STOMAGEN-OX—higher density and clustered stomata—resemble that of *tmm* (and to a lesser extent *er;erl1;erl2* triple mutants). Increasing and decreasing expression levels of STOMAGEN in a *tmm* background no longer changes stomatal densities, indicating that *tmm* is epistatic to *STOMAGEN* and that STOMAGEN is likely to act through this receptor (Kondo *et al.*, 2009; Sugano *et al.*, 2009). Since the negative regulators EPF1 and EPF2 also require TMM, the genetic relationships between STOMAGEN and EPF1/EPF2 were also tested. STOMAGEN-OX in a *epf1;epf2* background does not further increase stomatal density (Kondo *et al.*, 2009), and reduction of *STOMAGEN* levels can reduce the number of stomata produced in an *epf1;epf2* background, suggesting that STOMAGEN could compete with EPF1 and EPF2 for signaling through TMM (Fig. 9.2B, Sugano *et al.*, 2009).

2.6. Models for stomatal peptide ligands and membrane receptors and future questions

In summary, the current count has stomatal development regulated at different stages and in different organs by four related secreted peptide ligands, EPF1, EPF2, CHAL, and STOMAGEN, and four LRR receptors or receptor-like kinases, TMM, ER, ERL1, and ERL2. How is communication among these players orchestrated to determine whether a given cell will ultimately become a stomatal GC?

There are similarities and important differences in the structure, behavior, and expression patterns of the ligands. Although clearly part of the same protein family, there are sequences unique to each of EPF1, EPf2, CHAL, and STOMAGEN, most notably in the region of the protein that forms a surface-exposed loop in STOMAGEN (Kondo *et al.*, 2009). These differences could lead to differential binding of receptors (CHAL vs. the TMM ligands) or different consequences of binding (STOMAGEN vs. the negative regulators). The ligands also show unique expression patterns: EPF1 and EPF2 are expressed in different stomatal lineage cell types but STOMAGEN and CHAL are not expressed in the stomatal lineage; instead, they are secreted from mesophyll cells of aerial organs and the internal tissue of stems, respectively, to regulate epidermal cell fate (Fig. 9.2A) (Kondo *et al.*, 2009; Sugano *et al.*, 2009; Abrash and Bergmann, 2010). Promoter swaps between EPF1 and EPF2 suggest that some, but not all of the differences in their function are due to expression pattern (Hara *et al.*, 2009).

Before any ligands were identified, the models for *TMM* and *ERf* functions were that ligand-bound ERf actively transduced a negative regulatory signal and that TMM could modulate ERf functions by either dimerizing with ERfs to form a malfunctional receptor complex or by binding to and titrating away a limiting pool of ERf ligands (Shpak *et al.*, 2005). Now that we know four ligands, we can revisit this model, modifying the possibilities based on new data. We must now consider that receptors mediate signaling that results in opposite outcomes, as with EPF1, EPF2, and STOMAGEN. How might this work? If we assume that the receptors always produce a stomatal repressive signal when activated, and that EPF1/2 are active ligands, then an "inactive" STOMAGEN could compete with them for association with the receptors, thereby preventing the receptors from sending a repressive signal (Fig. 9.2B). Alternatively, activated receptors might be able to produce both stomatal promoting and stomatal repressing signals depending on which receptors and which ligands are in a particular complex (Fig. 9.2B). The data on CHAL also make it clear that not all ligands will bear the same relationships to all receptors. If CHAL is a ligand for each of the ERf members, but does not act through TMM, then two reasonable regulatory scenarios are that CHAL binds to TMM, but this forms a nonfunctional association that serves to dampen the stimulatory ligand–receptor interaction between CHAL and ERf (Fig. 9.2B), or that ERf–ERf and ERf–TMM complexes compete and CHAL only participates in signaling through ERf–ERf complexes; the absence of TMM would shift the balance of receptor complexes toward more CHAL-responsive ones (Fig. 9.2B). Note that in these models we have not yet addressed organ-type specificity.

It is exciting that secreted peptide ligands from the same subfamily are stomatal regulators with a range of signaling properties. This greatly improves our understanding of how cell–cell communication determines cell fate. But there are still quite a few pieces missing from the puzzle, for example, the genetic interaction between STOMAGEN and the ERfs has not yet been investigated, nor have all combinations of receptors been tested with EPF1, EPF2, or CHAL. Other questions of future interest are:

- Do the characterized ligands indeed physically interact with TMM and/or ERf, as CLV3 does with CLV1? How many (if any) versions of receptor complexes are formed among TMM and the ERf? What are the preferred partners? Biochemical methods, such as *in vitro* assays for binding of CLV3 to CLV1 (Ogawa *et al.*, 2008), or cell-based assays such as FRET may be needed to rigorously identify and test the relevant physical associations.
- Are all of these peptide ligands post-translationally processed and modified? If so, how? The functional domains of some CLE proteins can be restricted to the short conserved CLE motif, itself hydroxylated at two proline residues (Ito *et al.*, 2006; Kondo *et al.*, 2006). The mature forms of CHAL, EPF1, and EPF2 have not been elucidated yet, but biochemical methods used on STOMAGEN would be ideal for characterizing these ligands.
- Do other EPFs or other ligand classes participate in stomatal development? SDD1, a subtilisin-like serine protease, has been known for many years to regulate stomatal development and pattern (Berger and Altmann, 2000; Von Groll *et al.*, 2002). This class of subtilases has been shown, in other cases, to process extracellular ligands. However, based on genetic analysis, EPF1, EPF2, STOMAGEN, and CHAL act independently of SDD1 (Abrash and Bergmann, 2010; Hara *et al.*, 2007, 2009; Kondo *et al.*, 2009). It would be intriguing to uncover the substrates of SDD1, some of which could represent new ligands acting through TMM and ERf, and may also provide some hints about ligand processing.
- Are TMM and EPf receptors sufficient to mediate all stomatal signaling events? Based on studies of signal transduction in the shoot apical meristem, another class of receptors, those possessing a kinase domain but no extracellular domains, and exemplified by CORYNE, form complexes with the TMM-like RLPs (Müller *et al.*, 2008). The possibility exists for the CORYNE type (or other unknown receptors) to also participate in stomatal development signals.

3. Signaling through a MAP Kinase Module

Regardless of the details of ligand and receptor interactions, the information originating at the cell surface must be relayed to targets inside. In *Arabidopsis*, some of the clearest and most complete signal transduction information comes from studies of plant innate immune responses. Here, a mitogen-activated protein kinase (MAPK) signaling cascade lies downstream of a LRR-RLK, FLS2, whose activity is triggered by binding flagellin peptide. Ultimately, the responses involve altered transcription, trafficking, and growth (Gomez-Gomez and Boller, 2002). Recent research progress reveals that the paradigm of MAPK signaling cascades functioning downstream of LRR receptors applies to stomatal development as well (Bergmann *et al.*, 2004; Lampard *et al.*, 2009; Wang *et al.*, 2007).

3.1. The YDA MAPK signaling module in stomatal development

MAPK cascades are evolutionary conserved across eukaryotes and play fundamental roles regulating numerous biological processes, including growth, development, and stress responses (reviewed in Colcombet and Hirt, 2008; Widmann *et al.*, 1999). The typical three-step phosphorylation relay proceeds from a MAPK kinase kinase (MAPKKK), to a MAPK kinase (MAPKK), and to a MAP kinase (MAPK) to regulate the activity of target proteins by phosphorylation. The *Arabidopsis* genome encodes a large number of components in the MAPK module, the majority of which are associated with plant responses to stresses (Colcombet and Hirt, 2008). Interestingly, MAPK modules consisting of one MAPKKK, YODA (YDA), four MAPKKs (MKK4/5/7/9), and two MAPKs (MPK3/6) appear to regulate stomatal development as well as contributing to stress responses (Colcombet and Hirt, 2008). The existence of this shared module may reflect one of the innate systems that enable a rapid and flexible modulation of plant developmental processes in response to external stresses.

The connection between individual MAPK signaling components and stomatal development came from several experimental sources. The MAPKKK YODA (YDA) was identified in a screen for stomatal pattern defects (Bergmann *et al.*, 2004). Soon after, MAPKK4/5 and MAPK3/6 were identified by a group interested in stress responses who noticed the stomatal phenotypes in double mutants (Wang *et al.*, 2007). Two additional MAPKKs (MAPKK7/9) were identified later in a stomatal-targeted constitutive activity screen (Lampard *et al.*, 2009). The phenotypes of *yda* mutants, the double mutant of *mpk3;mpk6*, or simultaneous downregulation of MKK4 and MKK5 by RNA interference (RNAi) are similar and resemble those of the *er;erl1;erl2* mutant. Each results in an excessive number of clustered stomata, indicating that MAPK signaling regulates both stomatal density and distribution. Consistent with this, expressing constitutively active versions of YDA (CA-YDA, Bergmann, 2004) or of NtMek2 (a MKK4/5 homolog, Wang *et al.*, 2007) completely represses stomatal production. Genetic analysis places *YDA* downstream of *TMM*, which is based upon the fact that one copy of CA-YDA (which has no phenotype on its own) suppresses *tmm* stomatal overproliferation and patterning defects (Bergmann, 2004). Based on similar genetic evidence, MKK4/5 and MPK3/6 act downstream of *YDA* (Wang *et al.*, 2007;

Lampard *et al.*, 2009), and all these factors function as negative players in stomatal development. It is interesting to note that only YDA has a noticeable effect on stomatal development as a single mutant. The redundant activities of MPK3 and MPK6 (as MAPKs) or MKK4 and MKK5 (as MAPKKs) were first documented in regard to regulation of plant responses to biotic and abiotic stresses, and it appears that this is also true for regulation of stomatal development and patterning.

3.2. Dissecting out the YDA module: negative or positive?

Components of the MAPK cascade are generally expressed at low levels throughout the plant. Some substantially impact plant growth and development at early stages; for example, *yda* mutants form malformed embryos and the plants are severely dwarfed and sterile (chapter 1; Lukowitz *et al.*, 2004) and the double mutants of *mpk3;mpk6* are embryo lethal (Wang *et al.*, 2007). The pleiotropic phenotypes generated by systemic manipulation of MAPK signaling (for example, via loss-of-function mutants, gene overexpression, or expression of constitutive-active versions) may hinder researchers from dissecting the roles of individual MAPK components in specific plant developmental contexts.

Recently, Lampard *et al.* (2009) assayed the impact of the MAPK cascade members on specific stomatal cell types by utilizing a set of stomatal lineage promoters. These promoters were based on a set of transcription factors (SPEECHLESS (SPCH), MUTE, and FAMA) whose specific functions will be discussed in detail below in section 4.1. Briefly, these promoters drive expression in (1) cells that undergo entry divisions (SPCHpro), (2) cells transitioning from meristemoid to GMC (MUTEpro), and (3) cells transitioning from GMC to GCs (FAMApro). By this strategy, the altered activity of MAP kinase components is limited to particular subset of stomatal cells, therefore substantially reducing phenotypes that are linked to other developmental events.

Lampard *et al.* first assayed two variants of YDA, including a dominant negative form and a constitutively active form, and constitutively active versions of six selected MAPKKs driven by cell-type specific promoters to examine their function in stomatal development at each aforementioned stage. As expected, when expressed at the earliest stage (SPCHpro), YDA, MKK4, and MKK5 behaved as negative regulators in the stomatal lineage, specifically inhibiting stomatal formation without introducing developmental defects outside of the epidermis (Lampard *et al.*, 2009). This study then expanded our understanding of MAPK cascade control of stomatal development by demonstrating that YDA negatively regulates stomatal formation at both entry and amplification stages, but then functions as a positive regulator at the final stage of development where the GMC differentiates into paired GCs. Activated MKK4 and MKK5 do not have any effects on stomatal development at this final stage, but the study revealed that MKK7 and MKK9 do (Lampard *et al.*, 2009). MKK7/9 signaling mirrors YDA at three discrete steps, negative for the first two transitions and positive at the last stage (Fig. 9.3 and Lampard *et al.*, 2009).

This positive function of YDA and MKK7/9 at the final stage of stomatal formation was unexpected however; perhaps it is not entirely surprising. It may reflect that a stomatal lineage cell prefers to form a mature GC once it has progressed to the GMC stage, rather than being arrested midway through this process. From observation of mutants, meristemoids and GMCs appear to have different commitments to becoming stomata.

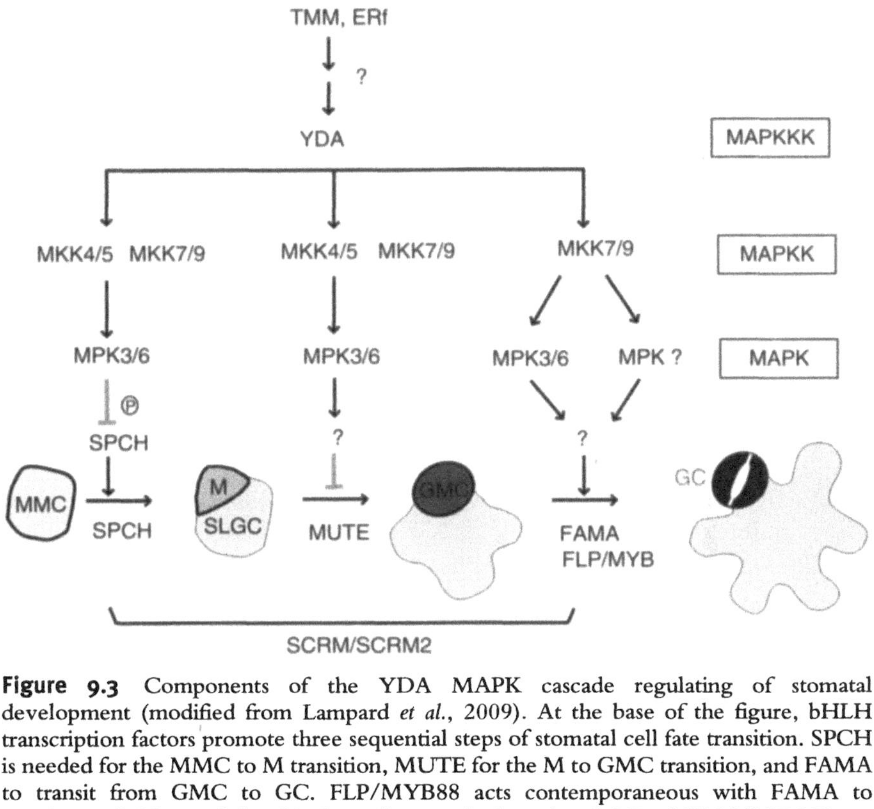

Figure 9.3 Components of the YDA MAPK cascade regulating of stomatal development (modified from Lampard *et al.*, 2009). At the base of the figure, bHLH transcription factors promote three sequential steps of stomatal cell fate transition. SPCH is needed for the MMC to M transition, MUTE for the M to GMC transition, and FAMA to transit from GMC to GC. FLP/MYB88 acts contemporaneous with FAMA to terminate divisions of the GMC to form a single pair of GCs. SCRM/SCRM2 are expressed throughout the stomatal lineage and could form heterodimers with SPCH, MUTE and FAMA, to facilitate progression through each step. Above the transcription factors, three tiers of the YDA MAPK cascade modulate the same three stomatal cell fate transitions. SPCH has been experimentally shown to be phosphorylated by MPK3/6. YDA and MKK4/5/7/9 negatively regulate the first and second fate transition. The third transition, however, is positively regulated by YDA and MKK7/9. MPKs are involved at each transition, as MPK3/6 were demonstrated for the first, with additional, currently unidentified, MPKs required especially at the third step.

Arrested GMCs are rarely seen, but meristemoids do not always complete the pathway; meristemoids can arrest, as seen in the leaves of the *er* mutant (Shpak *et al.*, 2005), or dedifferentiate into a SLGC, as observed on stems of the *tmm* mutant (Bhave *et al.*, 2009). Multiple control points during stomatal development might provide a way to naturally expand the flexibility of a plant cell confronting environmental challenges. Many environmental stresses induce an elevated number of stomata (Baena-Gonzalez *et al.*, 2007; Baena-Gonzalez and Sheen, 2008), and it has also been known for some time that many plant species respond to higher levels of CO_2 by producing altered numbers of stomata, suggesting both positive and negative regulators interpret the CO_2 effect for stomatal development (Woodward, 1987). It would be intriguing to know whether YDA and MKK4/5/7/9 are regulators that integrate CO_2 signals with developmental control.

Like with ligand–receptor interactions, new information raises new questions and both logical and technical challenges. In the next few years, we imagine several new directions for MAPK signaling. These include elucidating complete MAPK cascades in relevant cell types and monitoring their activities in real time. Some questions of particular interest are as follows:

- What is the connection between the highest level kinase (MAPKKK YDA) and the receptor-like kinases at the membrane? MAPKKKs are often activated by phosphorylation and require scaffold proteins for specificity. In the embryo, a cytoplasmic receptor-like kinase, SHORT SUSPSENSOR (SSP), acts upstream of YDA, but does not require its kinase activity to function (Bayer *et al.*, 2009). Could SSP or its homologs be a bridge between YDA and the ER family kinases?
- Which of the 20 MAPKs are required downstream of identified MKKs at the different stages of stomatal development? Are some required redundantly? In addition, MAPKs are also regulated by phosphatases, and again, genes encoding these proteins are numerous in plants. Using the information from many large-scale screens such as protein arrays and expression profiles may help reduce the candidates to a testable number.

4. Transcription Factors

How could the linear MAP kinase cascade, simply by phosphorylation relays, oppositely promote or repress stomatal cell fate at different transitions in the stomatal lineage? We have seen how different ligands can trigger different responses in different tissues. It is equally plausible that the YDA MAPK module phosphorylates distinct transcription factors, which are differentially expressed at specific cell types or stages, thereby switching on or off key developmental processes.

How might these transcriptional targets of MAPKs be identified? Recently, Popescu *et al.* (2009) utilized protein-microarray technology to reveal the potential substrates of a number of MAPKs, including MPK3 and MPK6 (activated by each of MKK4/5/7/9). The array is an incomplete proteome, but among 570 identified targets of MAPK phosphorylation, many transcription factors were identified (Popescu *et al.*, 2009). The array is by design an *in vitro* approach leaving open the question of how many of these targets would be relevant for stomatal development. However, comparing the transcription factors known to regulate stomatal development to the array data may be a very fruitful enterprise. Here we describe several related "master regulator" transcription factors in stomatal development; one has already been experimentally demonstrated to be phosphorylated by the downstream kinases in the YDA MAPK signaling pathway (Lampard *et al.*, 2008), and others may well incorporate phosphorylation as part of their regulatory strategies.

4.1. Genes that break silence

SPCH, MUTE, and FAMA are three key basic helix–loop–helix (bHLH) factors controlling three discrete and sequential cell fate transitions in stomatal development; each is absolutely essential for the plant to make stomata (Fig. 9.4; reviewed in Barton, 2007; Pillitteri and Torii, 2007). Loss-of-function of any one gives rise to a plant devoid of mature stomata, though they differ at which stages the stomatal lineage is blocked. Without SPCH, the epidermis consists only of pavement cells (MacAlister *et al.*, 2007), indicating that the primary function of SPCH is to initiate the stomatal lineage by promoting ACDs. The *mute* mutant, in contrast to *spch*, undergoes entry divisions to form meristemoids; however, the meristemoids are not able to transition from meristemoid to GMC and after excessive rounds of amplifying divisions, the cells arrest (Pillitteri and Torii, 2007). The key function of MUTE, therefore, is to terminate asymmetric divisions and allow the second phase transition in the stomatal lineage. The *fama* mutant produces caterpillar-like tumors composed of cells with GMC identity instead of forming recognizable stomata, implicating FAMA in the termination of the GMC divisions and in finalizing the differentiation of the GCs (Ohashi-Ito *et al.*, 2006). Combined with expression analysis and gain-of-function phenotypes, the picture that emerges is that each of these three bHLHs sequentially turn on and regulate a single different developmental transition in the stomatal lineage (Fig. 9.3).

MYB transcription factors often form complexes with bHLH proteins in plants (Payne *et al.*, 2000). Mutations in the MYB gene *FOUR LIPS (FLP)* or double mutants of *FLP* and its close paralog *MYB88* generate plants whose epidermis contains clusters of GCs and GMCs, suggesting that they redundantly function to terminate symmetric GMC divisions, thereby

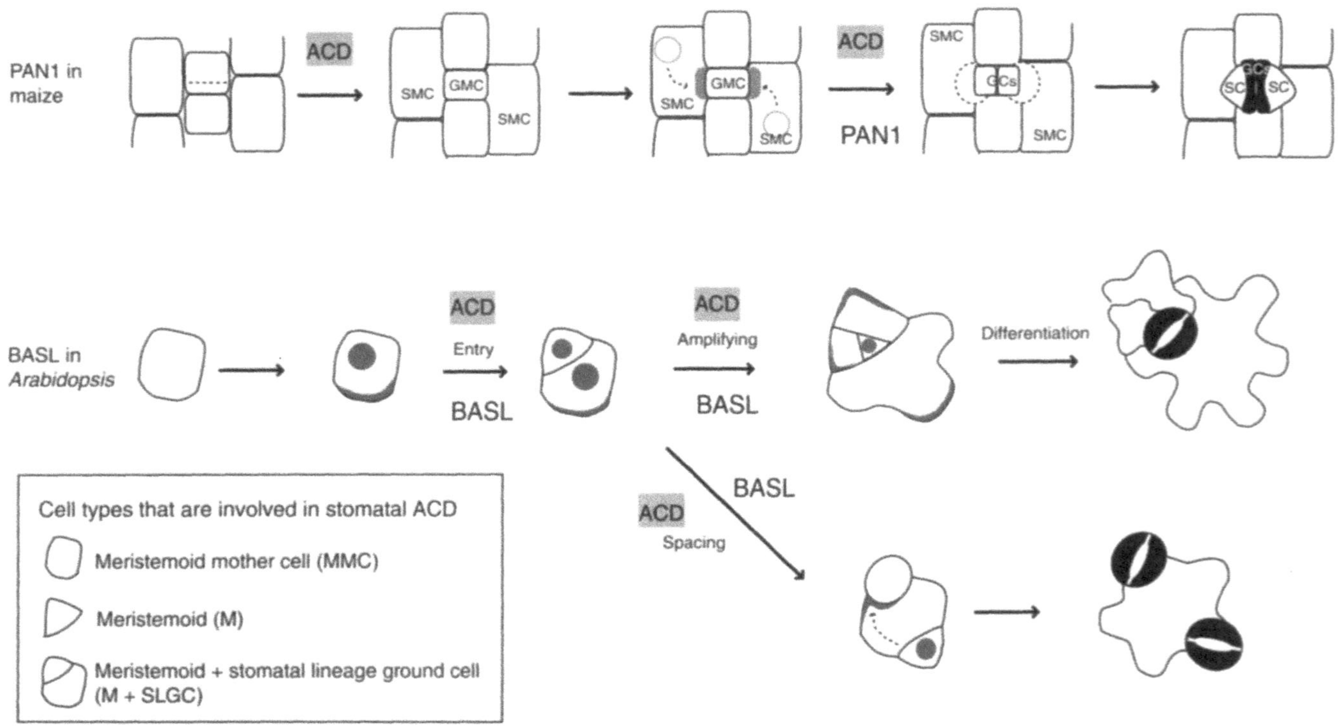

Figure 9.4 *(Continued)*

allowing cells to complete the final step in GC differentiation (Lai *et al.*, 2005). These stomatal arrest phenotypes are similar, but weaker than *fama*'s, and along with expression data indicating high expression in GMCs, it is tempting to consider FLP/MYB88 as partners for FAMA. However, genetic interaction results do not support a model of FLP/MYB88 working together with FAMA, nor do they physically interact with FAMA in split GFP assays (Ohashi-Ito *et al.*, 2006).

If not complexed with MYB or other proteins, bHLH transcription factors can also form homo- or heterodimers with bHLH proteins. Several interaction partners of SPCH, MUTE, and/or FAMA have been reported (Ohashi-Ito *et al.*, 2006; Kanaoka *et al.*, 2008); the most likely to be biologically relevant are the ICE/SCREAM (SCRM) proteins (Fig. 9.3, Kanaoka *et al.*, 2008). Originally characterized as a bHLH involved in response to cold stress, ICE1 (Chinnusamy *et al.*, 2003) was also recognized by its effects on stomatal development (Kanaoka *et al.*, 2008). A dominant mutation in *ICE1* (*scrm-D*) that replaces arginine 236 with histidine leads to an epidermis composed nearly completely of GCs. *ICE1/SCRM* encodes a bHLH-leucine zipper protein expressed during several stages of stomatal development. Its paralog SCRM2 is also expressed in the stomatal lineage, and together they appear to regulate all three key transition steps in stomatal development (Kanaoka *et al.*, 2008). This hypothesis is supported genetically by the phenotypes of *scrm*, *scrm;scrm2/+*, and *scrm; scrm2*, which display defects comparable to those in *fama*, *mute*, and *spch* mutants, respectively. Yeast-two-hybrid and split-GFP assays indicate that while SPCH, MUTE, and FAMA do not strongly homodimerize, each physically interacts with SCRM-D and (to a lesser extent) with SCRM (Kanaoka *et al.*, 2008). These data lead to the model that SCRM/SCRM2 proteins act in a dosage-dependent manner as partners with SPCH, MUTE, and FAMA (Kanaoka *et al.*, 2008 and Fig. 9.3).

Figure 9.4 Polarized proteins PAN1 and BASL in stomatal development. Asymmetric cell divisions (ACD) are labeled and localization of PAN1 and BASL proteins are indicated by gray shading. (Top) In maize, the guard mother cell (GMC) is generated from an ACD in a specific lineage (first panel, dashed line). GMCs provide a positional cue interpreted by the neighboring subsidiary mother cells (SMCs), inducing their nuclei to migrate toward the GMC in preparation for a highly asymmetric division. The smaller daughters of the SMC division become subsidiary cells (SCs) that, together with the GCs, comprise the stomatal complex. In SMCs, PAN1 appears in patches at the GMC contact sites. (Bottom) *Arabidopsis* BASL first appears in the nucleus of the MMC, and then as a crescent in the cell periphery. After the MMC divides, only the larger daughter cell inherits the peripheral BASL. The nuclear BASL initially present in both daughter cells may disappear causing the cell to exit an asymmetric division phase; loss of nuclear BASL from SLGCs is associated with differentiation into pavement cells. In spacing divisions, the BASL peripheral crescent relocates to maintain its position distal to the newly formed meristemoid.

4.2. Phosphorylation links the MAPK module to the transcriptional network

While most of the protein domains of SPCH, MUTE, and FAMA are highly conserved, SPCH contains a unique region that neither MUTE, nor FAMA, nor any other bHLH possesses. This domain, named the MAPK target domain (MPKTD), is enriched with predicted MAPK phosphorylation target sequences. An *in vitro* kinase assay confirmed that the SPCH MPKTD was phosphorylated by MPK3 and MPK6. Deletion of this domain removed all detectable phosphorylation *in vitro* and *in vivo* and resulted in the accumulation of SPCH protein, excess stomatal lineage divisions, and overproduction of stomatal lineage cells (Lampard *et al.*, 2008). It was hypothesized that in the plant epidermis, phosphorylation of SPCH by MPK3/6 would reduce the level of active SPCH (Fig. 9.3). To test this, SPCH variants bearing mutations in some of the predicted phosphorylation sites ("weakly overactive") were monitored in the background of different MAPK and receptor mutations; in these backgrounds, SPCH overactivity was enhanced, indicating that the previously identified stomatal signaling components could work upstream of SPCH and modulate its activity. Disruption of one specific phosphorylation site resulted in a form of SPCH that was incapable of producing stomata, while not affecting the protein's ability to promote asymmetric division. It appears, therefore, that MAPK-mediated phosphorylation not only generally downregulates SPCH, but it may also be required to activate SPCH's cell fate promoting activity. How this complex regulation is achieved is not yet known (Lampard *et al.*, 2008), but it is interesting to consider in light of the opposite effects MAPK signaling can have on the different stomatal transitions described in the previous section.

Could the YDA MAPK module regulate stomatal transcription factors other than SPCH? Based on *in vitro* phosphorylation data, the conclusion is probably yes (Fig. 9.3). ICE1/SCRM, FLP, and MYB88 contain consensus MAPK phosphorylation sites (Blom *et al.*, 1999). On protein array-based kinase assays, MUTE and MYB88 are *in vitro* targets of MAPKs, though interestingly, MUTE is a target of MAPKs other than MPK3/6 (Feilner *et al.*, 2005; Popescu *et al.*, 2009). It should be noted that these protein arrays do not include all the transcription factors known to be involved in stomatal development (for example, SPCH is not present). Although promising, thus far, the biological relevance of phosphorylation has not been demonstrated yet. Confirmation of phosphorylation sites by mass spectrometry and *in vivo* assays on variants of SCRM/SCRM2, MUTE, and MYB88 containing point mutations that disrupt the putative phosphorylation sites will be necessary to determine whether phosphorylation affects their function and whether the effects seen by altering MAPK signaling in different stomatal cell types can be explained by phosphoregulation of these known transcription factors.

4.3. How does phosphorylation affect transcription factor function?

In general, phosphorylation has the potential to change many properties of proteins including their stability, subcellular localization, activity, and protein–protein or protein–DNA interactions. One interesting model to consider for stomatal development is how phosphorylation could modulate the relationship between SPCH and SCRM/SCRM2. The phenotype of *scrm-D* homozygous mutant plants—overproduction of stomata and stomatal lineage cells—mimics that generated by expressing MPKTD-deleted SPCH in plants. Consistent with this, the double mutant *scrm; scrm2* exhibits no stomatal production in the epidermis, resembling the *spch* mutant (Kanaoka *et al.*, 2008). SCRM and SCRM2 appear to weakly bind to SPCH, but the SCRM-D mutant protein shows a strikingly increased affinity for SPCH (Kanaoka *et al.*, 2008). If a partnership between SCRM/SCRM2 and SPCH promotes initiation of stomatal development and SPCH is normally downregulated by MAPK phosphorylation, then the stimulatory effect of SCRM/SCRM2 on SPCH could be achieved by these proteins also serving as MAPK substrates. SCRM/SCRM2 could (1) serve as competitive substrates (shielding SPCH from phosphorylation) and (2) change their activity upon phosphorylation such that they bind SPCH and stabilize it, or interfere with the MPK3/6-directed SPCH phosphorylation and degradation. Of course, these models require more rigorous tests of whether SCRM/SCRM2 phosphorylation takes place *in vivo*, and better assays to determine the effects of phosphorylation on stomatal lineage transcription factors.

Although many transcription factors have now been identified, the targets of these transcription factors are not yet known. Based on their ability to "reprogram" cell fate and functional conservation between *Arabidopsis* and rice proteins (Ohashi-Ito *et al.*, 2006; Pillitteri and Torii, 2007; Kanaoka *et al.*, 2008; Liu *et al.*, 2009), it is likely that the bHLHs will be high-level controllers, regulating many other transcriptional regulators as well as cell cycle and differentiation genes. Several outstanding questions are as follows:

- What are the targets of each transcription factor at each stomatal transition phase? Do putative partners (e.g., SPCH and SCRM/SCRM2) have overlapping regulatory profiles? Information from large-scale transcriptomics, proteomics, and phosphoproteomics may be especially useful, in combination with forward genetics and cellular or biochemical analysis, to establish biologically relevant targets.
- What other transcription factors classes participate in stomatal cell fate, and how are they regulated? For example, the MADS-box protein AGL16 appears to affect stomatal development and it is targeted by a microRNA (Kutter *et al.*, 2007). Conventional forward genetics will still

be valuable for isolating new mutants, and screening for enhancers or suppressors of the existing stomatal mutants may identify regulators with more subtle effects.

5. Regulation of Stomatal Asymmetric Cell Division

We have described the individual stages in stomatal development and some of the signals and intrinsic factors that regulate them. We now turn to the mechanism of ACD that establishes different identities and division behaviors in these different cell types. ACD refers to divisions that generate two daughter cells with distinctive cell fates; in the case of meristematic or stem cells, one of the daughter cells adopts the fate of its mother, while the other takes on a divergent fate. ACD is a primary mechanism that leads to cell-type diversity and is fundamental for growth and development of multicellular organisms including animals, fungi, and plants (Horvitz and Herskowitz, 1992; Caussinus and Hirth, 2007; Gonczy, 2008).

5.1. Models of asymmetric cell division

In general, ACD requires the generation of cellular asymmetry in (or polarization of) the progenitor cell; this includes physical changes in cell shape and asymmetric distribution of cellular components such as organelles, proteins, and RNAs. Following cellular polarization, a precisely oriented division plane will be established and daughter cell fates specified. In animals, two basic mechanisms are thought to confer distinguishable daughter cell fates. Differential distribution of protein complexes, most notably the conserved PAR–aPKC membrane kinase complex, can unequally segregate internal fate determinants (Goldstein and Macara, 2007). Alternatively, extrinsic mechanisms may be employed in which initially equivalent daughters obtain unequal cell fates depending on their physical location post division. In many cases, a combination of both intrinsic and extrinsic mechanisms may be used (Munro, 2006; Knoblich, 2008).

Extensive effort has been applied to study the molecular mechanisms underlying ACD in *Arabidopsis*, with emphasis on early embryogenesis, stomatal production, root morphogenesis, and male gametophyte generation (reviewed in Heidstra, 2007). Both internal factors and external cues, especially for stomatal development, are thought to play crucial roles. For example, in early leaf development, MMCs divide asymmetrically and generate two physically and functionally distinct cells, but their division plane is not obviously oriented by any external information (Serna *et al.*, 2002). However, later, when the secondary meristemoids are generated

(Fig. 9.1), external signals from GCs, GMCs, or meristemoids orient the direction of the new ACDs, so that the newly formed meristemoid is placed distal from them (Nadeau and Sack, 2002). In this review, we will point out where signals and transcription factors regulate stomatal lineage ACDs, but will consider in more depth two newly identified regulators that appear to be critical internal polarity factors.

5.2. Stomatal asymmetric cell division

Many genes affect the number and position of stomatal lineage ACDs, such as the ligands (EPF1, EPF2, CHAL, STOMAGEN), membrane receptors (TMM and ERf), MAPK module, and bHLH transcription factors (SPCH, MUTE, and SCRM1/2) previously described. This reflects an integration of external signaling and internal control in stomatal development (Petricka *et al.*, 2009). While it is indeed true that all these genes regulate stomatal ACDs, most do so by specifying cell fates in which asymmetric divisions are part of their identities. None of these genes, from a cellular perspective, is likely to be part of the machinery that creates a physically and functionally asymmetric division. Based on studies in plant roots and embryos, it appeared that plants relied primarily on extrinsic information to generate ACDs (Song *et al.*, 2006; Cui *et al.*, 2007; Willemsen *et al.*, 2008; Bayer *et al.*, 2009). Recently, however, identification of two polarly localized proteins, PANGLOSS1 (PAN1) and BREAKING OF ASYMMETRY IN THE STOMATAL LINEAGE (BASL), suggests that plants might also utilize polarized protein localization as a mechanism to generate ACDs (Cartwright *et al.*, 2009; Dong *et al.*, 2009).

5.3. Polarized receptor protein: PAN1 from maize

In grasses, both GC and SC formation require asymmetric divisions (Fig. 9.4; Gallagher and Smith, 2000). The maize genes controlling the first asymmetric division that creates GMCs are still unknown, but several mutations that affect SMC divisions have been identified (Gallagher and Smith, 2000; Cartwright *et al.*, 2009; Wright *et al.*, 2009). Two of these genes, *PAN1* and *PAN2*, are involved in SMC ACD and SC fate specification (Gallagher and Smith, 2000; Cartwright *et al.*, 2009). *PAN1* encodes a LRR-RLK and is expressed in SMCs. Immunolocalization of PAN1 in maize leaves indicated that this protein forms strikingly polarized "patches" at sites where the SMCs contact GMCs (Fig. 9.4). In a correctly patterned ACD, actin patches form and SMC nuclei migrate to these same contact sites. Absence of PAN1 patches in the *pan1* mutant leads to defects in both actin and nuclear recruitment to contact site, eventually leading to incorrectly placed division planes and incorrectly specified SCs (Cartwright *et al.*, 2009). PAN1, as a receptor or a co-receptor in the SMCs, may respond to the GMC-derived signals that promote SMC polarization in preparation for asymmetric divisions. The molecular identity or nature of the GMC-derived cue that positions PAN1 protein in SMCs is still unknown, but the ability to detect PAN1 will provide an excellent tool to identify genes or processes required for its polarization in the future. Once PAN1 is polarized, how it transmits this information to downstream processes like actin patch formation also needs further investigation. Although annotated as a receptor-like kinase, the PAN1 sequence does not contain the key residue for kinase activity and PAN1 failed to phosphorylate a generic substrate in *in vitro* kinase assays (Cartwright *et al.*, 2009).

Although they all encode LRR-containing transmembrane proteins, the asymmetric localization of PAN1 distinguishes this protein from *Arabidopsis* TMM and the ERf LRR-RLKs. A TMM–GFP fusion expressed under the TMM promoter was generally plasma membrane localized in stomatal lineage cells (Nadeau and Sack, 2002), and the subcellular localizations of the Erf members have not yet been described. Considering the large collection of LRR-RLKs (~220) in the *Arabidopsis* genome (Torii, 2004), it is plausible that other LRR-RLKs function like PAN1 by accumulating asymmetrically in response to external cues and by polarizing cells in advance of asymmetric division. It is also possible that since the specific developmental event, the recruitment of neighboring cells to form SCs of the stomatal complexes, is not found in *Arabidopsis*, there will not be LRR-RLKs with comparable polarized expression or functions.

5.4. Polarized and segregated novel protein: BASL in *Arabidopsis*

PAN1 is clearly a polarized protein in plant cells; however, there is no evidence that it is segregated during an ACD to yield daughter cells of different fates. A novel protein that does appear to be segregated is *Arabidopsis* BASL. Like PAN1, BASL regulates asymmetric divisions in stomatagenesis and exhibits a strikingly polarized subcellular distribution (Dong *et al.*, 2009). The *basl* mutant loses asymmetries associated with stomatal divisions; marker expression is inappropriately segregated, and the daughter cell sizes and identities are similar to each other (Dong *et al.*, 2009). BASL protein has no predicted functional domains and is encoded only in the genomes of dicot plants. GFP–BASL exhibits a dynamic subcellular localization pattern and is found in a crescent at the cell periphery, a localization reminiscent of the PAR/aPKC complex in animals (Fig. 9.4). In asymmetrically dividing stomatal lineage cells, BASL first appears in the interphase nucleus and begins to accumulate in a crescent at the cell periphery prior to the cell exhibiting any morphological asymmetry. After a typical asymmetric division, the BASL peripheral crescent is always inherited by only the larger daughter cell (the SLGC, Fig. 9.4). Immediately after division, both daughter cells have BASL–GFP in their nuclei, but nuclear BASL does not always persist. Disappearance of BASL from the nucleus correlates with cell differentiation; if the cell did not inherit peripheral BASL, it will become a GMC and eventually GCs, but if the cell does possess BASL at the cell periphery, loss of nuclear BASL correlates with adoption of pavement cell fate (Fig. 9.4). Interestingly, peripheral BASL is more than a convenient asymmetrically segregated marker. This region is the site of BASL function as demonstrated by the expression of an N-terminal-deleted variant of BASL (BASL-IC) that is localized to a crescent in the cell periphery (but not the nucleus), yet it fully rescues the *basl* mutant phenotypes (Dong *et al.*, 2009).

In contrast to PAN1, BASL seems to be an internal regulator of stomatal ACDs. The orientation of the BASL crescent does not appear to respond to external cues such as EPF1 coming from GCs, nor does BASL pattern change in a *tmm* mutant. It was hypothesized that BASL may trigger local cell expansion to induce local physical asymmetries. Indeed, when ectopically expressed in nonstomatal lineage cells of the hypocotyl, the BASL peripheral crescent overlays areas of abnormal cell outgrowth (Dong *et al.*, 2009). The BASL crescent, to some extent, also coordinates cell fate determination; it is invariably in the larger daughter cell resulting from an asymmetric division and the division and fate of each sister can be predicted by monitoring BASL in the nucleus and at the periphery (Dong *et al.*, 2009). Together with evidence that BASL is polarized before any outward physical asymmetries can be detected in stomatal lineage cells, it appears that BASL plays a role very similar to that of the PAR–aPKC complexes in animal cells.

Discovery of PAN1 and BASL, polarized proteins that regulate asymmetric division, suggested that the establishment of cell polarity by accumulation of proteins in distinct subcellular regions is a mechanism used not only by animals, but by plants, for asymmetric division control. Still, there are differences between plants and animals; the molecular identities of BASL and PAN1 as well as the lack of genes encoding homologs of animal polarity proteins indicate that each group recruited a different set of proteins.

It seems likely that the way that BASL (and possibly PAN1) promotes asymmetric division must also be different from what is known from animal systems. Classical morphology and cell biology already showed that there are different mechanisms for division plane specification in animals and plants. In animal cells, the position of the centrosomes and the mitotic spindle determine the orientation of the division plane (Cowan and Hyman, 2004). In an animal asymmetric division, polarity proteins (such as the PAR/aPKC complex), coordinating with microtubules, adjust the spindle position and the division plane is set when the membrane pinches down from the periphery to the midzone of the spindle in a "purse-string closure" cytokinesis. By contrast, the division plane of a plant cell is predicted by a specialized ring of cortical cytoskeleton elements (microtubules and F-actin) named the preprophase band (PPB). The PPB appears during interphase and is transient, being largely disassembled before mitosis; however, a reminder of its position is maintained and the phragmoplast and new cell wall dividing the daughter cells will form at this site. What is the connection between PAN1 and BASL and PPB orientation? Does either protein orient PPB placement directly, and if so, how? PAN1 appears to establish cellular asymmetry by recruiting actin, so it is possible that it creates a platform for orientation of the PPB through this polymer. However, the placement of the actin patch in SMCs is not where the PPB would assemble, so it is unlikely that PAN1 is a direct scaffold. Interphase nuclear position also influences PPB placement (Murata and Wada, 1991), though the precise mechanism is still unknown. Since BASL is expressed in the nucleus before PPB formation, might nuclear BASL direct the placement of the PPB? Alternatively, nuclear BASL might act in coordination with polarized peripheral BASL to facilitate PPB formation at a specific site.

Polarity generation, though known from decades of morphological studies to be a critical part of stomatal development and pattern, is just now beginning to be studied at a molecular level. We anticipate many new discoveries in the coming decade and that many of these will translate to polarity issues outside of the stomatal lineage. Some of the outstanding questions are as follows:

- How are the PAN1 and BASL proteins brought to and maintained at their polarized peripheral locations? Much work on plant cell polarity has followed the localization of the transmembrane auxin transporters of the PIN family. Will the trafficking elements required for PIN localization also be used for PAN1 and BASL, or are the mechanisms different? Experiments comparing PIN and BASL localization in roots suggest that they are different (Dong *et al.*, 2009), leaving open the question of what is required for the dynamic localization of BASL.
- What is the link between the presence of polarized proteins and the generation of cellular asymmetry? Do PAN1 and BASL serve as scaffolds for other proteins with enzymatic functions to become localized to one region? Or do these proteins have (as yet unknown) abilities to directly organize differential growth and division plane orientation?
- What is the functional or regulatory connection between the cell fate regulators, including the ligands, receptors, MAPK signaling components, and transcription factors described in previous sections and the (presumably downstream) polarity proteins PAN1 and BASL?

6. Concluding Remarks

In the last decade, stomatal development and pattern has emerged from a phase where most of our information was purely descriptive to become a maturing field in which we know the molecular identities of key players and many of the regulatory principles. We have sought in this review to provide a comprehensive picture of signals, receptors, transcription factors, and polarity proteins currently known and how they connect with each other. We summarize these key points below. In describing these recent advances, we have also tried to highlight areas in which we think there will be new discoveries in the near future and areas where technical advances are needed to drive the field forward.

7. Key Conclusions

1. Stomatal development has emerged as an ideal system for study of intercellular and intracellular signaling and the underlying molecular mechanisms for cell fate determination and patterning during plant tissue formation and morphogenesis.
2. The intercellular communication in the stomatal lineage appears to be mediated by a set of LRR receptor and kinases that must interpret multiple extracelluar signals. Peptide ligands secreted from neighboring cells (not necessarily limited to the stomatal lineage) deliver messages to membrane receptors that can result in either negative or positive regulation of stomatal fates.
3. Coordinated activity of transcription factors is required to regulate three sequential cell fate transitions during stomatal development. Promoters of some of these transcription factors, due to their cell-type specificity, are proving useful for manipulating gene activity in restricted stomatal lineage cell types.
4. The YDA MAPK cascade appears to link signaling initiated by interaction of ligands with the LRR-RLKs at the plasma membrane to responses in the nucleus by phosphorylation of transcription factors.
5. Polarly localized and segregated proteins regulate ACD in the stomatal lineage, hinting at conserved mechanisms among plants and animals.

References

Abrash, E. B., and Bergmann, D. C. (2010). Regional specification of stomatal production by the putative ligand CHALLAH. *Development* **137**, 447–455.

Baena-Gonzalez, E., Rolland, F., Thevelein, J. M., and Sheen, J. (2007). A central integrator of transcription networks in plant stress and energy signalling. *Nature* **7156**, 938–942.

Baena-Gonzalez, E., and Sheen, J. (2008). Convergent energy and stress signaling. *Trends Plant Sci.* **9**, 474–482.

Barton, M. K. (2007). Making holes in leaves: promoting cell state transitions in stomatal development. *Plant Cell* **4**, 1140–1143.

Bayer, M., Nawy, T., Giglione, C., Galli, M., Meinnel, T., and Lukowitz, W. (2009). Paternal control of embryonic patterning in *Arabidopsis thaliana*. *Science* **5920**, 1485–1488.

Berger, D., and Altmann, T. (2000). A subtilisin-like serine protease involved in the regulation of stomatal density and distribution in *Arabidopsis thaliana*. *Genes Dev.* **9**, 1119–1131.

Bergmann, D. C. (2004). Integrating signals in stomatal development. *Curr. Opin. Plant Biol.* **1**, 26–32.

Bergmann, D. C., Lukowitz, W., and Somerville, C. R. (2004). Stomatal development and pattern controlled by a MAPKK kinase. *Science* **5676**, 1494–1497.

Bhave, N. S., Veley, K. M., Nadeau, J. A., Lucas, J. R., Bhave, S. L., and Sack, F. D. (2009). TOO MANY MOUTHS promotes cell fate progression in stomatal development of Arabidopsis stems. *Planta* **2**, 357–367.

Blom, N., Gammeltoft, S., and Brunak, S. (1999). Sequence and structure-based prediction of eukaryotic protein phosphorylation sites. *J. Mol. Biol.* **5**, 1351–1362.

Byrne, M. E., Kidner, C. A., and Martienssen, R. A. (2003). Plant stem cells: divergent pathways and common themes in shoots and roots. *Curr. Opin. Genet. Dev.* **5**, 551–557.

Cartwright, H. N., Humphries, J. A., and Smith, L. G. (2009). PAN1: a receptor-like protein that promotes polarization of an asymmetric cell division in maize. *Science* **5914**, 649–651.

Caussinus, E., and Hirth, F. (2007). Asymmetric stem cell division in development and cancer. *Prog. Mol. Subcell. Biol.* **45**, 205–225.

Chinnusamy, V., Ohta, M., Kanrar, S., Lee, B. H., Hong, X., Agarwal, M., and Zhu, J. K. (2003). ICE1: a regulator of cold-induced transcriptome and freezing tolerance in Arabidopsis. *Genes Dev.* **8**, 1043–1054.

Colcombet, J., and Hirt, H. (2008). Arabidopsis MAPKs: a complex signalling network involved in multiple biological processes. *Biochem. J.* **2**, 217–226.

Cowan, C. R., and Hyman, A. A. (2004). Asymmetric cell division in *C. elegans*: cortical polarity and spindle positioning. *Annu. Rev. Cell Dev. Biol.* **20**, 427–453.

Cui, H., Levesque, M. P., Vernoux, T., Jung, J. W., Paquette, A. J., Gallagher, K. L., Wang, J. Y., Blilou, I., Scheres, B., and Benfey, P. N. (2007). An evolutionarily conserved mechanism delimiting SHR movement defines a single layer of endodermis in plants. *Science* **5823**, 421–425.

Dong, J., MacAlister, C. A., and Bergmann, D. C. (2009). BASL controls asymmetric cell division in Arabidopsis. *Cell* **7**, 1320–1330.

Feilner, T., Hultschig, C., Lee, J., Meyer, S., Immink, R. G., Koenig, A., Possling, A., Seitz, H., Beveridge, A., Scheel, D., Cahill, D. J., Lehrach, H., *et al.* (2005). High throughput identification of potential Arabidopsis mitogen-activated protein kinases substrates. *Mol. Cell. Proteomics* **10**, 1558–1568.

Fiers, M., Ku, K. L., and Liu, C. M. (2007). CLE peptide ligands and their roles in establishing meristems. *Curr. Opin. Plant Biol.* **1**, 39–43.

Fletcher, J. C., Brand, U., Running, M. P., Simon, R., and Meyerowitz, E. M. (1999). Signaling of cell fate decisions by CLAVATA3 in Arabidopsis shoot meristems. *Science* **5409**, 1911–1914.

Gallagher, K., and Smith, L. G. (2000). Roles for polarity and nuclear determinants in specifying daughter cell fates after an asymmetric cell division in the maize leaf. *Curr. Biol.* **19**, 1229–1232.

Geisler, M. J., Deppong, D. O., Nadeau, J. A., and Sack, F. D. (2003). Stomatal neighbor cell polarity and division in Arabidopsis. *Planta* **4**, 571–579.

Goldstein, B., and Macara, I. G. (2007). The PAR proteins: fundamental players in animal cell polarization. *Dev. Cell.* **5**, 609–622.

Gomez-Gomez, L., and Boller, T. (2002). Flagellin perception: a paradigm for innate immunity. *Trends Plant Sci.* **6**, 251–256.

Gonczy, P. (2008). Mechanisms of asymmetric cell division: flies and worms pave the way. *Nat. Rev. Mol. Cell Biol.* **5**, 355–366.

Gray, J. E., Casson, S., and Hunt, L. (2008). Intercellular peptide signals regulate plant meristematic cell fate decisions. *Sci. Signal.* **49**, pe53.

Hara, K., Kajita, R., Torii, K. U., Bergmann, D. C., and Kakimoto, T. (2007). The secretory peptide gene EPF1 enforces the stomatal one-cell-spacing rule. *Genes Dev.* **14**, 1720–1725.

Hara, K., Yokoo, T., Kajita, R., Onishi, T., Yahata, S., Peterson, K. M., Torii, K. U., and Kakimoto, T. (2009). Epidermal cell density is autoregulated via a secretory peptide, EPIDERMAL PATTERNING FACTOR 2 in Arabidopsis leaves. *Plant Cell Physiol.* **6**, 1019–1031.

Heidstra, R. (2007). Asymmetric cell division in plant development. *Prog. Mol. Subcell. Biol.* 1–37.

Hernandez, M. L., Passas, H. J., Smith, L. G. (1999). Clonal analysis of epidermal patterning during maize leaf development. *Dev. Biol.* **216**, 646–658.

Horvitz, H. R., and Herskowitz, I. (1992). Mechanisms of asymmetric cell division: two Bs or not two Bs, that is the question. *Cell* **2**, 237–255.

Hunt, L., and Gray, J. E. (2009). The signaling peptide EPF2 controls asymmetric cell divisions during stomatal development. *Curr. Biol.* **10**, 864–869.

Ito, Y., Nakanomyo, I., Motose, H., Iwamoto, K., Sawa, S., Dohmae, N., and Fukuda, H. (2006). Dodeca-CLE peptides as suppressors of plant stem cell differentiation. *Science* **5788**, 842–845.

Kanaoka, M. M., Pillitteri, L. J., Fujii, H., Yoshida, Y., Bogenschutz, N. L., Takabayashi, J., Zhu, J. K., and Torii, K. U. (2008). SCREAM/ICE1 and SCREAM2 specify three cell-state transitional steps leading to Arabidopsis stomatal differentiation. *Plant Cell* **7**, 1775–1785.

Knoblich, J. A. (2008). Mechanisms of asymmetric stem cell division. *Cell* **4**, 583–597.

Kondo, T., Kajita, R., Miyazaki, A., Hokoyama, M., Nakamura-Miura, T., Mizuno, S., Masuda, Y., Irie, K., Tanaka, Y., Takada, S., Kakimoto, T., and Sakagami, Y. (2009). Stomatal density is controlled by a mesophyll-derived signaling molecule. *Plant Cell Physiol.* **51**, 1–8.

Kondo, T., Sawa, S., Kinoshita, A., Mizuno, S., Kakimoto, T., Fukuda, H., and Sakagami, Y. (2006). A plant peptide encoded by CLV3 identified by in situ MALDI-TOF MS analysis. *Science* **5788**, 845–848.

Kutter, C., Schob, H., Stadler, M., Meins, F., Jr., and Si-Ammour, A. (2007). MicroRNA-mediated regulation of stomatal development in Arabidopsis. *Plant Cell* **8**, 2417–2429.

Lai, L. B., Nadeau, J. A., Lucas, J., Lee, E. K., Nakagawa, T., Zhao, L., Geisler, M., and Sack, F. D. (2005). The Arabidopsis R2R3 MYB proteins FOUR LIPS and MYB88 restrict divisions late in the stomatal cell lineage. *Plant Cell* **10**, 2754–2767.

Lampard, G. R., Lukowitz, W., Ellis, B. E., and Bergmann, D. C. (2009). Novel and expanded roles for MAPK signaling in Arabidopsis stomatal cell fate revealed by cell type-specific manipulations. *Plant Cell*, **21**, 3506–3517.

Lampard, G. R., Macalister, C. A., and Bergmann, D. C. (2008). Arabidopsis stomatal initiation is controlled by MAPK-mediated regulation of the bHLH SPEECHLESS. *Science* **5904**, 1113–1116.

Liu, T., Ohashi-Ito, K., and Bergmann, D. C. (2009). Orthologs of *Arabidopsis thaliana* stomatal bHLH genes and regulation of stomatal development in grasses. *Development* **13**, 2265–2276.

Lukowitz, W., Roeder, A., Parmenter, D., and Somerville, C. (2004). A MAPKK kinase gene regulates extra-embryonic cell fate in Arabidopsis. *Cell* **1**, 109–119.

MacAlister, C. A., Ohashi-Ito, K., and Bergmann, D. C. (2007). Transcription factor control of asymmetric cell divisions that establish the stomatal lineage. *Nature* **445**, 537–540.

Müller, R., Bleckmann, A., Simon R. (2008). The receptor kinase CORYNE of Arabidopsis transmits the stem cell–limiting signal CLAVATA3 independently of CLAVATA1. *Plant Cell* **20**, 934–946.

Munro, E. M. (2006). PAR proteins and the cytoskeleton: a marriage of equals. *Curr. Opin. Cell Biol.* **1**, 86–94.

Murata, T., and Wada, M. (1991). Effects of centrifugation on preprophase-band formation in *Adiantum protonemata. Planta* **183**, 391–398.

Nadeau, J. A., and Sack, F. D. (2002). Control of stomatal distribution on the Arabidopsis leaf surface. *Science* **5573**, 1697–1700.

Nadeau, J. A., and Sack, F. D. (2003). Stomatal development: cross talk puts mouths in place. *Trends Plant Sci.* **6**, 294–299.

Ogawa, M., Shinohara, H., Sakagami, Y., and Matsubayashi, Y. (2008). Arabidopsis CLV3 peptide directly binds CLV1 ectodomain. *Science* **5861**, 294.

Ohashi-Ito, K., and Bergmann, D. C. (2006). Arabidopsis FAMA controls the final proliferation/differentiation switch during stomatal development. *Plant Cell* **10**, 2493–2505.

Payne, C. T., Zhang, F., and Lloyd, A. M. (2000). *GL3* Encodes a bHLH protein that regulates trichome development in Arabidopsis through interaction with GL1 and TTG1. *Genetics* **156**, 1349–1362.

Petricka, J. J., Van Norman, J. M., and Benfey, P. N. (2009). Symmetry breaking in plants: molecular mechanisms regulating asymmetric cell division in Arabidopsis. *Cold Spring Harb. Perspect. Biol.* **1**(5), a000497. Cold Spring Harbor Laboratory Press.

Pillitteri, L. J., and Torii, K. U. (2007). Breaking the silence: three bHLH proteins direct cell-fate decisions during stomatal development. *Bioessays* **9**, 861–870.

Popescu, S. C., Popescu, G. V., Bachan, S., Zhang, Z., Gerstein, M., Snyder, M., and Dinesh-Kumar, S. P. (2009). MAPK target networks in *Arabidopsis thaliana* revealed using functional protein microarrays. *Genes Dev.* **1**, 80–92.

Serna, L., Torres-Contreras, J., and Fenoll, C. (2002). Clonal analysis of stomatal development and patterning in Arabidopsis leaves. *Dev. Biol.* **1**, 24–33.

Shpak, E. D., McAbee, J. M., Pillitteri, L. J., and Torii, K. U. (2005). Stomatal patterning and differentiation by synergistic interactions of receptor kinases. *Science* **5732**, 290–293.

Song, S. K., Lee, M. M., and Clark, S. E. (2006). POL and PLL1 phosphatases are CLAVATA1 signaling intermediates required for Arabidopsis shoot and floral stem cells. *Development* **23**, 4691–4698.

Stahl, Y., and Simon, R. (2009). Plant primary meristems: shared functions and regulatory mechanisms. *Curr. Opin. Plant Biol.* **13**, 53–58.

Sugano, S. S., Shimada, T., Imai, Y., Okawa, K., Tamai, A., Mori, M., and Hara-Nishimura, I. (2009). Stomagen positively regulates stomatal density in Arabidopsis. *Nature* **463**, 241–244.

Torii, K. U. (2004). Leucine-rich repeat receptor kinases in plants: structure, function, and signal transduction pathways. *Int. Rev. Cytol.* **234**, 1–46.

van Zanten, M., Snoek, L. B., Proveniers, M. C., and Peeters, A. J. (2009). The many functions of ERECTA. *Trends Plant Sci.* **4**, 214–218.

Von Groll, U., Berger, D., and Altmann, T. (2002). The subtilisin-like serine protease SDD1 mediates cell-to-cell signaling during Arabidopsis stomatal development. *Plant Cell* **7**, 1527–1539.

Wang, H., Ngwenyama, N., Liu, Y., Walker, J. C., and Zhang, S. (2007). Stomatal development and patterning are regulated by environmentally responsive mitogen-activated protein kinases in Arabidopsis. *Plant Cell* **1**, 63–73.

Widmann, C., Gibson, S., Jarpe, M. B., and Johnson, G. L. (1999). Mitogen-activated protein kinase: conservation of a three-kinase module from yeast to human. *Physiol. Rev.* **1**, 143–180.

Willemsen, V., Bauch, M., Bennett, T., Campilho, A., Wolkenfelt, H., Xu, J., Haseloff, J., and Scheres, B. (2008). The NAC domain transcription factors FEZ and SOMBRERO control the orientation of cell division plane in Arabidopsis root stem cells. *Dev. Cell.* **6**, 913–922.

Woodward, F. (1987). Stomatal numbers are sensitive to increases in CO_2 from pre-industrial levels. *Nature* **327**, 617–618.

Wright, A. J., Gallagher, K., and Smith, L. G. (2009). Discordia1 and alternative Discordia1 function redundantly at the cortical division site to promote preprophase band formation and orient division planes in maize. *Plant Cell* **1**, 234–247.

Yang, M., and Sack, F. D. (1995). The too many mouths and four lips mutations affect stomatal production in Arabidopsis. *Plant Cell* **12**, 2227–2239.

CHAPTER TEN

Trichome Patterning in *Arabidopsis thaliana*: From Genetic to Molecular Models

Rachappa Balkunde, Martina Pesch, *and* Martin Hülskamp

Contents

Abstract

The aerial organs of plants typically produce trichomes that may adopt various functions, including light, wind, frost, and herbivore protection. Trichomes are of epidermal origin regularly distributed on the surface. The mechanism by which trichome differentiation is triggered in individual cells in a field of protodermal cells is best studied in *Arabidopsis thaliana*. The genetic analysis has revealed a number of key genes controlling this patterning process, and

Botanical Institute III, University of Cologne, Köln, Germany

Current Topics in Developmental Biology, Volume 91
ISSN 0070-2153, DOI 10.1016/S0070-2153(10)91010-7

further molecular analysis has enabled the in-depth cell-biological and biochemical analysis. The established models explain trichome patterning by the mutual interaction between positive and negative factors. Three activators, a bHLH (helix–loop–helix), a R2R3 MYB-related transcription factor, and a WD40 domain protein, form an active complex. The activity of this complex is counteracted by R3 MYB factors that compete with the R2R3 MYB for binding to the bHLH factor. The R3 MYBs can move between cells and thereby mediate cellular interactions. This general model cannot explain all genetic observations and recent data suggest the existence of several parallel patterning mechanisms. In this chapter we aim to summarize the current data and sketch possible alternative, not mutually exclusive theoretical models.

1. Introduction

During plant development, cells successively adopt a distinct cell fate and undergo the respective differentiation. The temporal and spatial coordination of this process is called pattern formation. In plants pattern formation is mainly based on cellular interactions, which may occur between immediate neighbor cells or over long distances by plant hormones.

Most of our current knowledge on pattern formation is derived from the genetic analysis in the model plant *Arabidopsis thaliana*. In particular, patterning processes in the epidermis are well studied, mainly because the epidermis is readily accessible for visual inspection and for experimental manipulations. In addition, the aerial epidermis consists of only a few cell types, the pavement cells, the stomata, and the trichomes. Trichomes and stomata are both regularly distributed on the leaf surface between the pavement cells; however, the mechanism underlying stomata patterning is fundamentally different (chapter 9; Bergmann and Sack, 2007).

In this chapter we focus on the trichome patterning system. The underlying patterning mechanism is shared to a large extent with the root hair pattering system in *Arabidopsis* (Ishida *et al.*, 2008; Schellmann *et al.*, 2007); however, this system will not be considered in this chapter. We will provide an overview of the genetic and molecular data and discuss how theoretical modeling helps to interpret the molecular and cell biological data.

2. Trichome Form and Function in Plants

The function of trichomes is mostly inferred from their form and physiology. They are considered to be relevant for the protection of the plant against insects and UV radiation, the control of leaf surface temperature, and to prevent dehydration by increasing reflection and by reducing air exchange near the surface (Johnson, 1975; Wagner *et al.*, 2004). Trichome form, size, and density vary considerably between plants and they may be unicellular or multicellular, secretory glandular or nonglandular (Esau, 1977; Theobald *et al.*, 1979; Uphof, 1962). Although some scattered molecular data are available for various plants, a mechanistic understanding of trichome patterning and development is only available for *Arabidopsis thaliana*.

3. Trichome Development in *Arabidopsis thaliana*

In *Arabidopsis* trichomes are found on most aerial organs, including the rosette leaves, the stem, cauline leaves, and sepals. They are not found on the hypocotyl and the cotyledons. Trichome patterning genes function downstream of the organ and tissue layer-specific signals. This was concluded from experiments in which overexpression of the trichome-promoting gene *GLABRA1* (*GL1*) in the absence of the trichome-suppressing gene *TRIPTYCHON* (*TRY*) was shown to be sufficient to cause ectopic trichome formation on all organs and even in the subepidermis (Schnittger *et al.*, 1998; Szymanski and Marks, 1998).

Most studies on the origin of the trichome pattern have focused on rosette leaves. Trichomes are unicellular and branched cells covering the whole leaf surface. *Arabidopsis* trichomes develop from single protodermal cells at the leaf base where they are typically separated by three to four cells (Hulskamp *et al.*, 1994). Concomitant with further leaf growth new trichomes are initiated at the leaf base and the already existing trichomes are separated due to cell divisions of the intervening epidermal cells (Hulskamp *et al.*, 1994; Larkin *et al.*, 1996). Trichome development is characterized by a stereotyped series of steps (Hulskamp, 2004). The first recognizable change from a protodermal cell to an incipient trichome cell is an increase of the nuclear size due to the initiation of endoreduplication cycles, which are cell cycles lacking cell and nuclear divisions (Hulskamp *et al.*, 1994). On average, trichomes proceed through four endoreduplication cycles resulting in a mature trichome with a DNA content of 32C (Hulskamp *et al.*, 1994; Melaragno *et al.*, 1993). Trichome cell morphogenesis is characterized by an initial outgrowth, two successive branching events with a stereotype arrangement of the branches (Folkers *et al.*, 1997; Hulskamp *et al.*, 1994). At maturity the trichome is about 0.5 mm high and they are typically separated from the neighboring trichome cells by about 0.5–1 mm.

4. Random Versus Regulated Trichome Pattern

The distribution of trichomes could in principle be due to a random decision of individual cells to develop into trichomes or an underlying system controlling their spatial arrangement. Two sets of experiments suggest a controlled mechanism. The first experimental set up was based on the consideration that a random choice of protodermal cells to develop into trichomes should lead to the formation of two trichomes immediately next to each other with a certain probability that should depend on their density. The statistical analysis of trichome distribution revealed that the number of clustered trichomes is much less frequently found than expected for a random distribution, indicating the existence of a patterning system (Larkin *et al.*, 1996). These data were confirmed using more sophisticated methods for the mapping of trichomes on leaves. The three-dimensional reconstruction of whole leaves by optical projection tomography showed a regular distribution with distances ranging from 54 to 299 μm and a mean distance of 164 μm (Lee *et al.*, 2006). More detailed statistical evaluation of micro X-ray computed tomography reconstructions of whole leaves also confirmed a nonrandom distribution of trichomes on mature leaves (Kaminuma *et al.*, 2008). Kaminuma and co-workers also confirmed the previous observation that trichome number is higher at the leaf border as compared to the rest of the leaf blade (Hauser *et al.*, 2001; Marks, 1997).

The question as to which type of patterning mechanism is operating was addressed by genetic mosaic experiments. In principle, the trichome distribution pattern could be explained by two alternative mechanisms. First, stereotyped asymmetric cell divisions could segregate trichomes and pavement cells in a manner preventing their formation next to each other (see chapter 9). Second, the interaction between initially equivalent cells could prevent the formation of cells in the near neighborhood of already existing trichomes. The *triptychon* (*try*) mutant, which displays trichome clusters, was used to distinguish between the two possibilities (Schnittger *et al.*, 1999). In the first scenario, all trichomes within a cluster should be daughter cells from the same mother cell. In the second scenario, trichomes in a cluster may have different ancestor cells. A clonal analysis of *try* mutants revealed that trichome clusters can have different mother cells, which excludes a cell lineage scenario.

5. Genetic Analysis of the Core Patterning Machinery

Various screens for mutants affecting trichome development have revealed candidate genes for the regulation of trichome patterning. They generally fall into two classes (Fig. 10.1). One class of mutants shows fewer

Figure 10.1 Phenotypes of trichome patterning mutants.(A) Wild-type plant with regularly distributed trichomes on the leaf surfaces. (B) *gl1* mutant plant lacking trichomes. (C) *try* mutant plant exhibiting trichome clusters (arrow). (D) *try cpc* mutant plant with large trichome clusters containing up to 40 trichomes.

or no trichomes and they are therefore formally positive regulators of trichome formation. The second class of mutants has either more trichomes or trichome clusters and the corresponding genes are therefore considered to act as negative regulators of trichome development.

5.1. Positive Regulators of Trichome Initiation

Two mutants were identified that are characterized by the general absence of trichomes (Fig. 10.1B). Mutations in the *TRANSPARENT TESTA GLABRA1* (*TTG1*) gene result in the complete loss of trichomes on all organs. Strikingly, the weak *ttg1* mutant phenotype creates a paradox genetic situation. They develop trichomes that are arranged in clusters suggesting that *TTG1* is not only an activator, but functions also as a negative regulator of trichome initiation. In addition to the trichome phenotype, several other phenotypes are expressed in *ttg1* mutants, including the formation of ectopic root hairs, the lack of seed coat mucilage, and reduced anthocyanin production (Galway *et al.*, 1994; Koornneef, 1981). *TTG1* encodes a protein with WD40 domains that generally function as protein–protein interaction domains (Walker *et al.*, 1999).

In *glabra1* (*gl1*) mutants, trichomes on the leaf surface are absent and only a few trichomes are found at the leaf margin. The corresponding *GL1* gene encodes an R2R3 MYB-related transcription factor (Oppenheimer *et al.*, 1991). In support of its function as a transcription factor, GL1 protein localizes to the nucleus (Szymanski *et al.*, 1998). A partially redundant function to *GL1* was reported for the R2R3 MYB encoding gene *MYB23*. *MYB23* overexpression supports a role in trichome patterning (Kirik *et al.*, 2001), and a comparison of the single *gl1* and *myb23* mutant phenotypes with that of the double mutant revealed that *MYB23* is required for the production of the leaf margin trichomes in the *gl1* mutant background (Kirik *et al.*, 2005).

A third activator was identified as a mutant with a subtle reduction in trichome number and cell size, the *glabra3* (*gl3*) mutant (Hulskamp *et al.*, 1994; Koornneef *et al.*, 1982). The *GL3* gene encodes a bHLH-like transcription factor (Payne *et al.*, 2000) and was shown to be a nuclear protein (Esch *et al.*, 2003). The function of the *GL3* gene is redundantly encoded by the closely related bHLH gene *ENHANCER OF GLABRA3* (*EGL3*). This is evident from the *egl3 gl3* double mutant, which is completely glabrous, while trichome patterning in the *egl3* single mutant is indistinguishable from wild type (Zhang *et al.*, 2003).

5.2. Negative Regulators of Trichome Initiation

The *TRY* gene was the first identified negative regulator of trichome initiation (Hulskamp *et al.*, 1994). In *try* mutants trichomes are occasionally arranged in clusters of two or three trichomes suggesting that *TRY* represses trichome initiation immediately next to another trichome (Fig. 10.1C). *TRY* encodes an R3 single-repeat MYB protein lacking a transcriptional activation domain. This suggested that TRY could inhibit the R2R3 MYB factor GL1 by competition for DNA binding (Schellmann *et al.*, 2002). Its general role as a negative regulator of trichome initiation was demonstrated by showing that its overexpression results in a glabrous phenotype (Schellmann *et al.*, 2002). *TRY* functions redundantly with other R3 MYB genes. First, a redundancy with *CAPRICE* (*CPC*) was documented. *CPC* was initially identified as an important regulator of root hair development (Wada *et al.*, 1997). Later a higher trichome density was reported, indicating an additional role in trichome patterning (Schellmann *et al.*, 2002).

To date, four additional R3 MYB homologs of *TRY* and *CPC* have been found. They are *TRICHOMELESS1* (*TCL1*), *ENHANCER OF TRIPTYCHON AND CAPRICE 1, 2*, and *3* (*ETC1, 2 and 3*) (Wang *et al.*, 2007; Kirik *et al.*, 2004a, b; Wester *et al.*, 2009). A phylogenetic study places *TRY* and *ETC2* in one group and *CPC, TCL1, ETC1*, and *ETC3* in a separate group (Wang *et al.*, 2007). The six genes share partially redundant functions in root hair and trichome formation. Five of the six R3 MYB genes are important during root hair formation. *CPC* has the most prominent role, exhibiting a reduced root hair number in single mutants (Wada *et al.*, 1997). In contrast to the root hair system, all six R3 MYB genes are important for trichome initiation and act in a partially redundant manner. While *try* is the only mutant displaying a cluster phenotype (Hulskamp *et al.*, 1994) a higher trichome density is also found in *etc2* and *etc3* mutants (Kirik *et al.*, 2004b; Schellmann *et al.*, 2002; Tominaga *et al.*, 2008; Wester *et al.*, 2009). Combinations of multiple mutants revealed further redundancies. The *try cpc* double mutant exhibits large clusters containing up to 40 trichomes (Fig. 10.1D), indicating an additional function of *CPC* in the local patterning events (Schellmann *et al.*, 2002). Though less pronounced, a similar contribution was found for *ETC3* and *ETC2* in the corresponding *etc3 try* and *etc2 try* double mutants (Kirik *et al.*, 2004b; Wester *et al.*, 2009).

Apart from redundancies, also a functional diversification was noted for several R3 MYB genes. This is most clearly seen for *TCL1*. The *tcl1* single mutant develops ectopic trichomes on the inflorescence revealing its relevance for organ-specific trichome regulation (Wang *et al.*, 2007), whereas a redundant role in trichome initiation was only detected in the *tcl1 cpc etc1 etc3* mutant (Wang *et al.*, 2010, 2008). A region-specific role for *ETC1* and *ETC2* in the control of trichome formation on petioles was concluded from *etc1 try cpc* and *etc2 try cpc* triple mutants (Kirik *et al.*, 2004a, b). A redundancy between *TCL1* and *CPC* was demonstrated in the *tcl1 cpc* double mutant that shows enhanced trichome formation on pedicels and stems (Wang *et al.*, 2007).

6. Genetic Interactions Between the Trichome Initiation Genes

The general logic underlying the interaction network of positive and negative regulators of trichome initiation was elegantly addressed in various genetic studies in which mutants and transgenic lines modulating the activity of patterning genes were combined. Attempts to rescue the *gl1* mutant by overexpression of *GL3* and *gl3* mutants by overexpression of *GL1* failed, indicating that these proteins cannot substitute for each other (Larkin *et al.*, 1994; Lloyd *et al.*, 1994). Overexpression of *GL3* and *GL1* caused a phenotype much stronger than expected from the simple addition of the two single overexpression phenotypes (Payne *et al.*, 2000). Together these data indicate that the two genes act at the same hierarchical level in a joined manner. TTG1 seems to act upstream of GL3 and GL1. This is suggested by the finding that overexpression of *GL3* and *EGL3* or *GL3* and

GL1 can, to some extend, rescue the *ttg1* mutant trichome phenotype (Payne *et al.*, 2000; Zhang *et al.*, 2003). These genetic data are equally consistent with TTG1 directly activating GL3/GL1 targets or alternatively with TTG1 modulating/enhancing the activity of GL3/GL1 (Zhao *et al.*, 2008).

The relationships between positive and negative regulators was also addressed genetically. In one set of experiments, the functional gene dosage of a positive regulator and of *TRY* were both reduced to 50% in transheterozygous plants. These plants exhibit trichome clusters, although a wild-type copy of both genes is present. The finding that the combinatorial reduction of their gene expression levels leads to patterning defects indicates that the activators and inhibitors act together in the same process. Similar conclusions were derived from the finding that plants overexpressing *GL1* or the maize R gene (a functional *GL3* homolog from maize) show a much stronger phenotype in the absence of *TRY*, indicating that *TRY*, represses more or less directly (Schnittger *et al.*, 1998; Szymanski *et al.*, 1998).

7. Expression Patterns of Patterning Genes: An Apparent Paradox

Analysis of the temporal and spatial activities of genes during patterning processes is essential for understanding the underlying mechanisms. For trichome patterning, one could expect that the activator genes and inhibitor genes show complementary patterns of expression, with the activators being expressed in trichomes and the inhibitors in nontrichome cells. Expression of *GL1* changes substantially during leaf development. *GL1* is expressed in all cells of young leaf primordia. Slightly later, trichome initials show elevated expression levels, and in later staged leaves, trichomes continue to express *GL1* while epidermal cells lose the expression completely (Larkin *et al.*, 1993). Its close homolog, *MYB23*, lacks the initial ubiquitous expression and is expressed exclusively in developing trichomes (Kirik *et al.*, 2005). The expression patterns of *GL3* and *EGL3* are similar to that of *GL1* (Zhang *et al.*, 2003). Using a *pGL3:GUS* reporter construct, *GL3* was shown to be ubiquitously expressed in the lateral regions of the basal part of young leaves (Zhao *et al.*, 2008). Expression is elevated in trichomes, and in mature leaves expression is found exclusively in trichomes. *EGL3* is expressed ubiquitously in the whole basal region of young leaves with elevated levels in trichomes. In contrast to *GL3*, expression of *EGL3* persists in pavement cells as well as in trichome, but at low levels (Zhao *et al.*, 2008). *TTG1* expression differs from *GL1, GL3*, and *EGL3* in

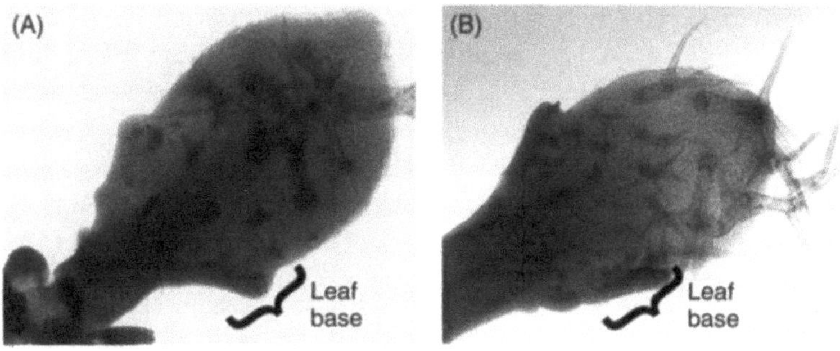

Figure 10.2 Expression pattern of GL1 and TRY. (A) GL1:GUS leaf showing the wild-type expression pattern of GL1. (B) TRY:GUS leaf showing the wild-type expression pattern of TRY. Trichomes are initiated at the leaf base. Note that both genes exhibit ubiquitous expression in the basal region of the leaf and elevated levels in trichomes. (See Color Insert.)

that it is ubiquitously expressed in all epidermal cells and no obvious elevated expression in trichomes is found (Bouyer *et al.*, 2008; Zhao *et al.*, 2008).

The expression patterns of the inhibitors were unexpected. Rather than showing a complementary expression pattern to that of the activators, *TRY* and the homologs studied so far are expressed initially in all cells of the developing leaf and expression becomes exclusively restricted to trichomes in mature leaves (Kirik *et al.*, 2004a, b; Koshino-Kimura *et al.*, 2005; Schellmann *et al.*, 2002; Tominaga *et al.*, 2008; Wester *et al.*, 2009). Thus trichome activators and inhibitors are coexpressed (Fig. 10.2). This raises the question, by which mechanism this apparent paradox can be explained.

8. Protein–Protein Interactions Between Patterning Components

As the genetic experiments suggest an intimate interaction among all patterning genes, it is likely that the corresponding proteins interact biochemically. This was tested thoroughly using the yeast two-hybrid assay. Interactions were found between GL1 and GL3/EGL3 (Payne *et al.*, 2000; Zhang *et al.*, 2003). The interaction sites on both proteins were mapped using smaller fragments. These experiments revealed that the N-terminus including the MYB repeats of GL1 and the first 96 amino acids of GL3 are relevant for the interaction. GL3 can also interact with itself, suggesting the formation of homodimers. Here the C-terminal region containing the bHLH domains is relevant. In addition, the GL3 protein can bind to TTG1. As TTG1 does not interact with GL1 and as the protein domains of GL3 responsible for GL1 and TTG binding are different, the formation of a GL1/GL3/TTG1 trimer was suggested (Payne *et al.*, 2000). In support of this, immunoprecipitation of GL1 from plant extracts co-precipitates TTG1 (Zhao *et al.*, 2008). As GL1 does not bind to TTG1 in the yeast two-hybrid system, these data suggest that GL3 mediates the precipitation of TTG1 due to the formation of a linear trimeric complex. Taking into account that *ttg1* mutants can be rescued by *GL1* and *GL3* overexpression it is likely that TTG1 has a more accessory function, such as the modulation of activity or the stabilization of the complex (Payne *et al.*, 2000).

Also the negative interactors bind to GL3 in yeast two-hybrid assays (Esch *et al.*, 2003). Strikingly, TRY binds to the same domain of GL3 as GL1 and competition studies in the yeast three-hybrid system demonstrated that TRY can compete with GL1 for the binding to GL3. This finding leads to the model that TRY inhibits the activators by creating an inactive TRY/TTG/GL3 complex. Later pull-down experiments revealed a direct binding of GL1 and TRY (Digiuni *et al.*, 2008). Thus TRY can, in principle, inhibit the activator complex due to binding to either GL1 or GL3.

Binding to GL3 and the competition with GL1 for binding to GL3 has also been shown for most other inhibitors (Tominaga *et al.*, 2008; Wester *et al.*, 2009). Strikingly, the binding and competition behavior differs tremendously between the five inhibitors analyzed. In binding assays, TRY shows the strongest binding affinity, followed by CPC, ETC2, ETC1, and ETC3 (Tominaga *et al.*, 2008). In the yeast three-hybrid competition assay, CPC is the most potent competitor for binding of GL1 to GL3, followed by ETC1, TRY, ETC3, and ETC2 (Wester *et al.*, 2009). While there is only a weak correlation between the results from the two experimental setups, differences between the five inhibitors with respect to GL3 binding and GL1 competition is clearly evident.

9. Transcriptional Cross-Regulation Between Patterning Genes

To analyze of the transcriptional regulation of a given gene by a second gene, its expression pattern is typically studied in the relevant mutant background. An intrinsic problem of this type of analysis for dissecting trichome patterning is that the genes involved act in various feedback loops. As a result, it is not possible to judge how direct or indirect the expression changes are or whether compensation mechanisms step into place. Most studies have centered on the questions: How the activators regulate each other's expression and how the activators and inhibitors affect each other? The regulation of the inhibitors by the activators is best studied for *TRY*. Here it has been shown that the TRY:GUS reporter is not expressed in *gl1* and *gl3* mutants, indicating that *TRY* is activated by these activators (Digiuni *et al.*, 2008). Conversely, it was shown that the expression of *GL1*, but not *GL3* and *TTG1*, is strongly reduced in plants overexpressing *TCL1* (Wang *et al.*, 2007) indicating a selective repression of activators by inhibitors. *GL1* expression was also found to be unchanged in *ttg1* and *gl1* mutants (Kirik *et al.*, 2005).

One possibility of unraveling the complex regulatory interactions between genes involved in trichome patterning is to determine which proteins act at which promoters to regulate activity. One big step toward this goal comes from ChiP (chromatin-immuno-precipitation)-based studies that assess the association of proteins with defined promoter regions. Three elegant studies assessed the binding behavior of GL1, GL3, EGL3, and TTG1 to promoters using the following general strategy (Morohashi and Grotewold, 2009; Morohashi *et al.*, 2007; Zhao *et al.*, 2008): Proteins of interest were expressed as glucocorticoid receptor fusions in the respective mutant backgrounds. Upon the application of a steroid hormone, the protein fusions are transported into the nucleus and become active. This strategy enables the temporal analysis of the activation of downstream genes. Chromatin was immunoprecipitated using an antibody against the tagged proteins (either GR or YFP fusions) and the associated DNA analyzed by PCR techniques or micro-array analysis (called ChiP chip). The expression of candidate genes was studied to confirm that the binding proteins trigger expression. The use of the protein synthesis inhibitor cycloheximide enables to distinguish between indirect and direct responses. With this strategy it was shown that GL3 binds and activates *CPC, ETC1, CPL3*, and *TRY* (Morohashi and Grotewold, 2009; Morohashi *et al.*, 2007). Binding was also found to its own promoter. Here, however, an auto-inhibition was reported (Morohashi *et al.*, 2007). The activation of the *CPC* and *ETC3* promoters seemed to require GL3 and GL1, as GL1 also binds to the *CPC* and *ETC3* promoters and its presence is required for GL3 binding. By contrast, the auto-inhibition of *GL3* seems to be independent of GL1 (Morohashi and Grotewold, 2009; Morohashi *et al.*, 2007; Zhao *et al.*, 2008). TTG1 binding and activation was found for the *CPC* and *ETC1* promoters (Zhao *et al.*, 2008). Together these data clearly show that the three activators bind to a common set of promoters.

It is difficult to extract a stringent logic from these data sets on the interaction scheme of the patterning genes. One problem is that the binding behavior of a protein varies depending on the exact methodology used. For example, binding of GL3 to the *TRY* promoter was not found in the *p35S:GL3:GR* and *pGL3:GL3:GR* plants but was successful in *pGL3:GL3:YFP* lines (Morohashi *et al.*, 2007). A second problem is that the use of whole plants combines trichome cells and nontrichome cells, which are expected to contain contrasting expression and protein-association patterns. As a result, it is not possible to decide whether a certain *in vivo* binding pattern corresponds to the minority of trichome cells or the more abundant epidermal pavement cells.

10. Cellular Interactions are Mediated by Intercellular Protein Movement

An essential requirement for a patterning system is that the cells communicate in order to make cell fate decisions in a temporally and spatially coordinated manner. During trichome patterning, this is governed by the movement of proteins through plasmodesmata. Plasmodesmata are membrane-coated channels, which traverse the cell walls between cells. The size exclusion limits vary tremendously depending on the cell type and developmental situation. In addition, proteins and RNAs can actively open the plasmodesmata far beyond the actual size exclusion limit, thereby enabling the transfer of large proteins or RNAs from one cell to another in a controlled manner (Lucas and Lee, 2004).

Initially, it was shown that CPC can move between cells in the root epidermis. Here *CPC* is expressed in nonroot hair cells and it was shown that a CPC:GFP fusion protein expressed under its own promoter can move into the root hair cells (Wada *et al.*, 2002). A protein domain was mapped that is required for CPC movement, indicating that the movement occurs via a regulated process rather than passive diffusion (Kurata *et al.*, 2005). The ability of CPC to move was confirmed for the leaf epidermis by particle bombardment (Fig. 10.3). Moreover, TRY and ETC3 were shown to move between cells, suggesting that intercellular mobility is a general feature of all six inhibitors (Digiuni *et al.*, 2008; Wester *et al.*, 2009; Zhao *et al.*, 2008).

Strikingly, mobility properties seem to differ between the inhibitors. Theoretical considerations predict that inhibitors with a high affinity for GL3 show a reduced mobility rate whereas a relatively low binding affinity results in an increased mobility rate. Quantitative measurements of mobility rates showed that coexpression of GL3 affects the movement rate of inhibitors as predicted (Wester *et al.*, 2009). Thus the movement rates of the inhibitors are inversely correlated with their inhibitory function.

The three activators, GL1, GL3, and TTG1, were shown to be nonmobile in particle bombardment experiments on leaves (Fig. 10.3; Digiuni *et al.*, 2008; Zhao *et al.*, 2008). A more detailed analysis revealed, however, that TTG1 can move between cells (Bouyer *et al.*, 2008). This was demonstrated in several

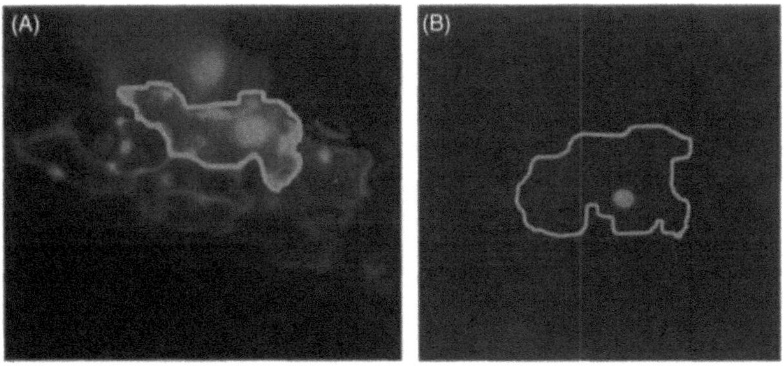

Figure 10.3 Mobility of trichome patterning proteins. Single cells in Arabidopsis leaves were transformed by particle bombardment. The cell border of the transformed cell is marked by an orange line. (A) TRY:YFP is found after transformation in the transformed cell and in surrounding cells indicating its movement between cells. (B) GL3:YFP does not move into adjacent cells and is only found in the transformed cell. This demonstrates that GL3:YFP cannot move between cells.

different ways. The Cre-Lox system was used to create genetic mosaics in which mutant *ttg1* sectors are formed immediately next to wild-type tissues. In such leaves, the phenotypes of *ttg1* mutant regions are fully rescued even at large distances from the wild-type sectors, indicating that TTG1 functions non-cell autonomously. A non-cell-autonomous function was also demonstrated by creating *ttg1* mutant plants expressing *TTG1* in the subepidermis. These plants are fully rescued indicating a movement of TTG1 activity between cell layers. Mobility of TTG1 protein or RNA was demonstrated by expressing *TTG1:YFP* in the epidermal pavement cells. Fluorescence was observed not only in the pavement cells but also in the trichomes, proving a transfer between pavement cells and trichomes. Finally, TTG1 protein was injected into tobacco mesophyll cells along with markers that normally do not move between cells. In these experiments, TTG1 and the marker were found to move from the injected cell into the neighboring cells. The observation that not only TTG1 but also the marker moves in these experiments is taken as evidence that movement occurs through plasmodesmata.

11. Immediate Downstream Genes—or Additional Patterning Components?

Downstream genes have been identified with different methods. Genetic analysis has identified approximately 70 genes that are involved in the development of trichomes by regulating various processes, including the regulation of endoreduplication, growth, and branching (Schellmann *et al.*, 2007). ChiP chip experiments with the core patterning proteins identified 513 genes as a minimum number of genes involved in trichome development (Morohashi and Grotewold, 2009). Many of the genetically identified genes were also recovered in these ChiP chip experiments. Several studies used micro-arrays to identify trichome-specific genes. The transcriptome comparison of mature trichomes and trichomeless shoot tissues revealed 516 trichome genes that are expressed specifically in mature wild-type trichomes (Marks *et al.*, 2009). In this study, wild-type trichomes were compared with those of a *gl3-sst siamese* double mutant. This combination of the *gl3-sst* allele and the cell cycle mutant *siamese* exhibits severe trichome morphogenesis defects and therefore was considered to be ideal for the identification of cell morphogenesis genes. This approach revealed several new genes involved in normal trichome development. Another study identified 3231 trichome-specific genes when comparing trichomes with leaf tissues devoid of trichomes (Jakoby *et al.*, 2008). Additional genes were identified that are specifically up- or downregulated in *gl3* or *try* mutants (Jakoby *et al.*, 2008). Similarly, the analysis of single cell transcription profiles of epidermal, basal, and trichome cells that were collected by glass capillaries revealed about 1000 genes expressed specifically in one of these cell types (Lieckfeldt *et al.*, 2008). A further refinement of this type of single cell analyses enabled a comparison of trichome initials and mature trichomes (Kryvych *et al.*, 2008).

Out of the potential downstream genes, two genes in particular that encodes for, the homeo domain transcription factor GLABRA2 (GL2) and the WRKY transcription factor TRANSPARENT TESTA GLABRA2 (TTG2), are considered candidates to function as immediate downstream genes of the core patterning machinery. The *gl2* and *ttg2* phenotypes are pleiotropic. Mutations in *GL2* cause a reduction of seed coat mucilage, the overproduction of root hairs, and a reduction of trichome growth (DiCristina *et al.*, 1996; Johnson *et al.*, 2002; Rerie *et al.*, 1994). The *ttg2* mutants have defects in trichome development, proanthocyanin and mucilage accumulation in seed coats, and integument cell elongation (Garcia *et al.*, 2005; Johnson *et al.*, 2002). The trichome phenotype of both mutants is characterized by a wide range of morphogenesis phenotypes suggesting that cell differentiation is effected in many ways. This has been taken as a genetic argument to consider both genes as integrators of the positional information provided by the patterning machinery to trigger the expression of downstream trichome differentiation genes. In support of this, *TTG2* and *GL2* show a very similar expression profile and become activated most strongly 24–48 h after trichome initiation (Morohashi and Grotewold, 2009).

Several lines of evidence support the idea that *TTG2* is a direct target of the patterning genes. *TTG2* is expressed in young leaves ubiquitously and with elevated levels in developing trichomes (Johnson *et al.*, 2002). No expression is found in *gl3 egl3* double mutants and in *p35S:CPC* plants, indicating that the core genes transcriptionally regulate *TTG2* (Ishida *et al.*, 2007). In addition, it was shown that TTG1, GL1, and GL3 bind to the *TTG2* promoter in ChiP experiments (Morohashi and Grotewold, 2009; Zhao *et al.*, 2008).

Similar experiments indicate a direct activation of *GL2* by the core patterning genes. The analysis of *pGL2:GUS* reporter lines in *gl1, ttg1* and *gl3 egl3* revealed a strong reduction or absence of expression (Szymanski *et al.*, 1998; Zhang *et al.*, 2003). In *Arabidopsis* protoplast transient expression assays, the co-transfection of *GL1* and *GL3* was sufficient to activate the *GL2* promoter (Wang and Chen, 2008). Activation of the *GL2* promoter was also reported for stable transgenic lines expressing a GL1–GL3 fusion protein (Wang and Chen, 2008). Binding of all three activators, GL3/EGL3, TTG1, and GL1 in ChiP experiments was reported and the induction of expression after trichome initiation was shown even in the presence of a protein synthesis inhibitor (Morohashi and Grotewold, 2009; Morohashi *et al.*, 2007; Zhao *et al.*, 2008). Together these data indicate that *GL2* is a direct target gene of the trimeric complex.

One experiment provides evidence for an additional involvement of *GL2* in trichome patterning. The introduction of additional copies of *GL2* under its own promoter resulted in an increased number of trichomes and the formation of clusters. These findings suggest that *GL2* can promote trichome formation and hence can be considered a positive regulator of trichome initiation (Ohashi *et al.*, 2002).

12. The Role of Chromatin Structure in Patterning

Although most of the current results on the role of chromatin structure in patterning were obtained in the root hair system, we will briefly sketch the data and implications. *In situ* hybridization studies with a *GL2* bacterial artificial chromosome had shown that root epidermal cells expressing *GL2* have an open chromatin state around the *GL2* locus whereas the nonexpressing cells have a closed state (Costa and Shaw, 2006). Strikingly, positional information regulates the reorganization of the chromatin state at the *GL2* locus in the G1 phase of the cell cycle. A gene identified as a *GL2* expression modifier, the *GEM-1* gene, seems to regulate the chromatin state through methylation. Trimethylated K9 of histone H3 (H3K9me3) marks an open chromatin state whereas mono- and dimethylated H3K9 (H3K9me2) is associated with a closed state (Caro *et al.*, 2007). In the *gem-1* mutant, H3K9me3 increases and H3K9me2 decreases at the *GL2* promoter, indicating that *GEM-1* gene inhibits the formation of the open chromatin state. As *gem-1* mutants show an increased trichome density, *GEM-1* seems to act as a negative regulator of trichome initiation.

Interestingly, *GEM-1* connects patterning with cell cycle control through the DNA replication component CDT1. It was shown that TTG1 can disrupt the interaction of GEM-1 with CDT1 in the yeast three-hybrid system. While the relevance of these findings is still unclear, it is tempting to speculate that this machinery might be involved in the switch from mitotic cell divisions to endoreduplication cycles after trichome cell fate determination.

13. Modeling the Patterning System—Starting With Two Components

The initial step toward a model explaining how cells are singled out to become trichomes in a field of initially equivalent cells was guided by theoretical principles formulated by Meinhardt and co-workers (Koch and Meinhardt, 1994; Meinhardt and Gierer, 1974). In an attempt to identify general principles underlying biological patterning processes, two models have been developed.

The first principle is called the activator–inhibitor model. This model explains patterning by two counteracting activities, an activator and an inhibitor. The activator activates its own inhibitor. Because the inhibitor moves faster than the activator, the inhibitor concentration is always lower than the activator concentration at places where the activator is produced. In addition, the activator upregulates its own expression (positive feedback loop). This is important to enable the system to amplify random fluctuations and thereby to break the initial balance between cells (Meinhardt and Gierer, 1974; Turing, 1952).

The second principle is called the activator–substrate model (Koch and Meinhardt, 1994). In this model a substrate required for the autocatalysis is depleted. As a result, regions around a region with high activator concentrations have successively less substrate and therefore produce less activator.

The current data are consistent with the idea that both systems are operating during trichome formation. Most predictions made by the activator–inhibitor model were experimentally demonstrated. For the activator–inhibitor model, *GL1, GL3,* and *TTG1* are collectively considered to represent the activator and the six R3 single repeat MYBs *TRY, CPC, ETC1, ETC2, ETC3,* and *TCL1* the inhibitor (Fig. 10.4). The activation of some inhibitors by the activators has been shown in transgenic lines expressing a *pTRY:GUS* reporter construct in wild type and mutants (Digiuni *et al.*, 2008). Besides, the mobility of the inhibitors and the nonmobility of GL1 and GL3 were demonstrated as described above.

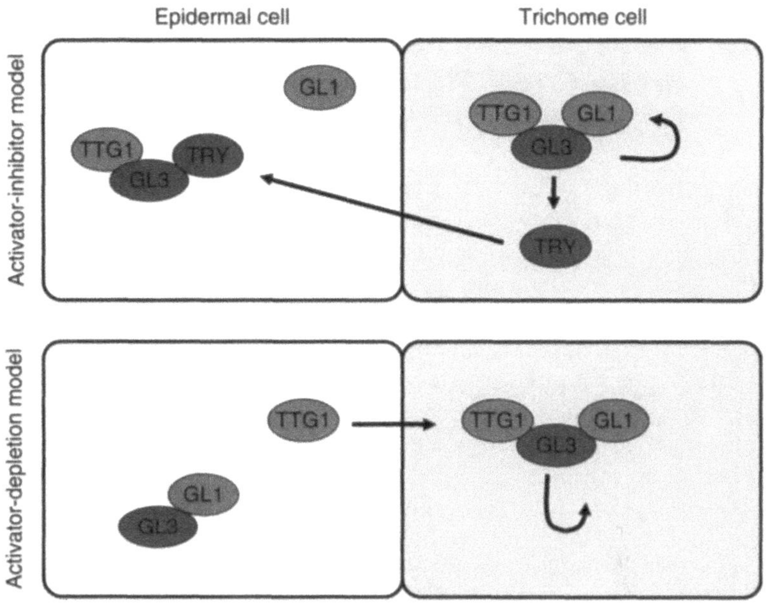

Figure 10.4 Trichome patterning models. The top diagram shows the activator–inhibitor model. TTG1/GL1/GL3 forms an active complex that triggers trichome development. This trimeric complex activates its own inhibitor TRY that moves into the neighboring cell where it replaces GL1, rendering the complex inactive. In addition, the activator complex is autoactivating. The bottom diagram shows the activator-depletion Model. TTG1 can freely move between cells. Its binding to GL3, which is strongly expressed in trichomes, results in a trapping of TTG1 in trichome cells and a depletion around trichome cells.

The only prediction that has not been experimentally confirmed is the self-enhancement of the activators.

In parallel, a second mechanism seems to operate during trichome patterning, the activator–substrate mechanism—though in a slightly different form (Fig. 10.4). The comparison of the expression pattern of *TTG1* and the localization of the TTG1:YFP fusion protein revealed that TTG1 proteins is depleted around incipient trichome cells (Bouyer *et al.*, 2008). This depletion was not found in *gl3* mutants. These observations led to a model in which TTG1 can move in young tissues between cells and accumulate in cells containing elevated levels of GL3 due to its binding to GL3. As a consequence, incipient trichomes accumulate TTG1 whereas the neighboring cells have reduced TTG1 levels. These data provide an explanation for the paradox that *TTG1* functions as a negative (cluster formation in weak alleles) as well as a positive regulator (no trichomes in strong alleles). According to the activator-depletion model, *TTG1* is an essential component for the activation of trichome fate and an inhibitor, because its depletion in neighboring cells represents as equivalent to an inhibitory function.

14. Modeling the Patterning System: Global Interaction Models versus Solution of Discrete Problems

Apart from the intuitive modeling approaches, two strategies toward a comprehensive patterning model are used. One strategy is to incorporate all genes and their regulatory interactions into one model. The second strategy is to theoretically evaluate discrete problems in sufficient detail to understand it with all parameters—ideally in a quantitative manner.

Several approaches have been published that integrate all patterning genes into a gene regulatory network. Initially reduced networks were analyzed that contain the three core activators and one or two inhibitors representing their combined action with the underlying logic formulated by the activator–inhibitor model (Benitez *et al.*, 2007; Dupuy *et al.*, 2008). Simulations of these models revealed patterns reminiscent to the trichome pattern on leaves. A more sophisticated model, including all observed interactions, the mobility, and the regulatory interactions, was published recently (Benitez *et al.*, 2008). In this study, the model was validated by showing that the simulation of some selected mutants revealed the expected results.

There are also approaches aiming to understand small regulatory circuits in more detail and to test experimentally the theoretically predicted properties or even their relevance. The latter was successfully studied for the question, whether and which interactions of TRY to components of the trimeric activator complex are relevant. This question came up when a new interaction between TRY and GL1 was detected in pull-down experiments (Digiuni *et al.*, 2008). This new finding led to three possible inhibition scenarios: (i) TRY binds to GL3 and replaces GL1; (ii) TRY binds to GL3 or GL1 and prevents the formation of an activator complex; and (iii) TRY inactivates the activator complex by binding to the complex. Although the simulation of all three scenarios successfully revealed the correct pattern, differences were predicted for the phenotype of plants overexpressing *GL3* ubiquitously under the 35S promoter or the *GL2* promoter. The phenotype found for these two transgenic lines was only consistent for the first scenario. This suggests that not all experimentally found protein–protein interactions are biologically relevant.

A second example is the simulation of the TTG1/GL3 activator-depletion model (Bouyer *et al.*, 2008). Modeling the behavior of the two proteins enabled the authors to understand the properties of the regulatory circuit and revealed that it is sufficient to generate a *de novo* pattern. It is important to note that in this case any experimental functional approach is virtually impossible because the same components are also relevant in the activator–inhibitor patterning circuit. Thus the theoretical in-depth analysis of the properties of each regulatory circuit might be the only possibility to understand the patterning machinery as a whole.

15. Perspective

The genetic, molecular, and cellular analysis has revealed a detailed view about the function and interaction of the involved genes, yet the mechanistic understanding of the patterning process is very speculative and general. Quantitative biology combined with theoretical approaches will be necessary to unravel the molecular principles underlying trichome patterning.

REFERENCES

Benitez, M., Espinosa-Soto, C., Padilla-Longoria, P., and Alvarez-Buylla, E. R. (2008). Interlinked nonlinear subnetworks underlie the formation of robust cellular patterns in Arabidopsis epidermis: a dynamic spatial model. *BMC Syst. Biol.* **2**, 98.

Benitez, M., Espinosa-Soto, C., Padilla-Longoria, P., Diaz, J., and Alvarez-Buylla, E. R. (2007). Equivalent genetic regulatory networks in different contexts recover contrasting spatial cell patterns that resemble those in Arabidopsis root and leaf epidermis: a dynamic model. *Int. J. Dev. Biol.* **51**, 139–155.

Bergmann, D. C., and Sack, F. D. (2007). Stomatal development. *Annu. Rev. Plant Biol.* **58**, 163–181.

Bouyer, D., Geier, F., Kragler, F., Schnittger, A., Pesch, M., Wester, K., Balkunde, R., Timmer, J., Fleck, C., and Hulskamp, M. (2008). Two-dimensional patterning by a trapping/depletion mechanism: the role of TTG1 and GL3 in Arabidopsis trichome formation. *PLoS Biol.* **6**, e141.

Caro, E., Castellano, M. M., and Gutierrez, C. (2007). A chromatin link that couples cell division to root epidermis patterning in Arabidopsis. *Nature* **447**, 213–217.

Costa, S., and Shaw, P. (2006). Chromatin organization and cell fate switch respond to positional information in Arabidopsis. *Nature* **439**, 493–496.

DiCristina, M., Sessa, G., Dolan, L., Linstead, P., Ruberti, S., and Morelli, G. (1996). The Arabidopsis Athb-10 (GLABRA2) is an HD-Zip protein required for regulation of root hair development. *Plant J.* **10**, 393–402.

Digiuni, S., Schellmann, S., Geier, F., Greese, B., Pesch, M., Wester, K., Dartan, B., Mach, V., Srinivas, B. P., Timmer, J., Fleck, C., and Hulskamp, M. (2008). A competitive complex formation mechanism underlies trichome patterning on Arabidopsis leaves. *Mol. Syst. Biol.* **4**, 217.

Dupuy, L., Mackenzie, J., Rudge, T., and Haseloff, J. (2008). A system for modelling cell–cell interactions during plant morphogenesis. *Ann. Bot.* **101**, 1255–1265.

Esau, K. (1977). *Anatomy of Seed Plants.* John Wiley and Sons, New York.

Esch, J. J., Chen, M., Sanders, M., Hillestad, M., Ndkium, S., Idelkope, B., Neizer, J., and Marks, M. D. (2003). A contradictory GLABRA3 allele helps define gene interactions controlling trichome development in Arabidopsis. *Development* **130**, 5885–5894.

Folkers, U., Berger, J., and Hulskamp, M. (1997). Cell morphogenesis of trichomes in Arabidopsis: differential control of primary and secondary branching by branch initiation regulators and cell growth. *Development* **124**, 3779–3786.

Galway, M. E., Masucci, J. D., Lloyd, A. M., Walbot, V., Davis, R. W., and Schiefelbein, J. W. (1994). The *TTG* gene is required to specify epidermal cell fate and cell patterning in the *Arabidopsis* root. *Dev. Biol.* **166**, 740–754.
Garcia, D., Fitz Gerald, J. N., and Berger, F. (2005). Maternal control of integument cell elongation and zygotic control of endosperm growth are coordinated to determine seed size in Arabidopsis. *Plant Cell* **17**, 52–60.
Hauser, M. T., Harr, B., and Schlotterer, C. (2001). Trichome distribution in *Arabidopsis thaliana* and its close relative *Arabidopsis lyrata*: molecular analysis of the candidate gene GLABROUS1. *Mol. Biol. Evol.* **18**, 1754–1763.
Hulskamp, M. (2004). Plant trichomes: a model for cell differentiation. *Nat. Rev. Mol. Cell Biol.* **5**, 471–480.
Hulskamp, M., Misera, S., and Jürgens, G. (1994). Genetic dissection of trichome cell development in *Arabidopsis*. *Cell* **76**, 555–566.
Ishida, T., Hattori, S., Sano, R., Inoue, K., Shirano, Y., Hayashi, H., Shibata, D., Sato, S., Kato, T., Tabata, S., Okada, K., and Wada, T. (2007). Arabidopsis TRANSPARENT TESTA GLABRA2 is directly regulated by R2R3 MYB transcription factors and is involved in regulation of GLABRA2 transcription in epidermal differentiation. *Plant Cell* **19**, 2531–2543.
Ishida, T., Kurata, T., Okada, K., and Wada, T. (2008). A genetic regulatory network in the development of trichomes and root hairs. *Annu. Rev. Plant Biol.* **59**, 364–386.
Jakoby, M. J., Falkenhan, D., Mader, M. T., Brininstool, G., Wischnitzki, E., Platz, N., Hudson, A., Hulskamp, M., Larkin, J., and Schnittger, A. (2008). Transcriptional profiling of mature Arabidopsis trichomes reveals that NOECK encodes the MIXTA-like transcriptional regulator MYB106. *Plant Physiol.* **148**, 1583–1602.
Johnson, H. B. (1975). Plant pubescence: an ecological perspective. *Bot. Rev.* **41**, 233–258.
Johnson, C. S., Kolevski, B., and Smyth, D. R. (2002). TRANSPARENT TESTA GLABRA2, a trichome and seed coat development gene of Arabidopsis, encodes a WRKY transcription factor. *Plant Cell* **14**, 1359–1375.
Kaminuma, E., Yoshizumi, T., Wada, T., Matsui, M., and Toyoda, T. (2008). Quantitative analysis of heterogeneous spatial distribution of Arabidopsis leaf trichomes using micro X-ray computed tomography. *Plant J.* **56**, 470–482.
Kirik, V., Lee, M. M., Wester, K., Herrmann, U., Zheng, Z., Oppenheimer, D., Schiefelbein, J., and Hulskamp, M. (2005). Functional diversification of MYB23 and GL1 genes in trichome morphogenesis and initiation. *Development* **132**, 1477–1485.
Kirik, V., Schnittger, A., Radchuk, V., Adler, K., Hulskamp, M., and Baumlein, H. (2001). Ectopic expression of the Arabidopsis AtMYB23 gene induces differentiation of trichome cells. *Dev. Biol.* **235**, 366–377.
Kirik, V., Simon, M., Hulskamp, M., and Schiefelbein, J. (2004a). The ENHANCER OF TRY AND CPC1 (ETC1) gene acts redundantly with TRIPTYCHON and CAPRICE in trichome and root hair cell patterning in Arabidopsis. *Dev. Biol.* **268**, 506–513.
Kirik, V., Simon, M., Wester, K., Schiefelbein, J., and Hulskamp, M. (2004b). ENHANCER of TRY and CPC 2 (ETC2) reveals redundancy in the region-specific control of trichome development of Arabidopsis. *Plant Mol. Biol.* **55**, 389–398.
Koch, A. J., and Meinhardt, H. (1994). Biological pattern formation: from basic mechanisms to complex structures. *Rev. Mod. Phys.* **66**, 1481–1507.
Koornneef, M. (1981). The complex syndrome of *ttg* mutants. *Arab. Info. Ser.* **18**, 45–51.
Koornneef, M., Dellaert, L. W. M., and VanderVeen, J. H. (1982). EMS- and radiation-induced mutation frequencies at individual loci in *Arabidopsis thaliana*. *Mutat. Res.* **93**, 109–123.
Koshino-Kimura, Y., Wada, T., Tachibana, T., Tsugeki, R., Ishiguro, S., and Okada, K. (2005). Regulation of CAPRICE transcription by MYB proteins for root epidermis differentiation in Arabidopsis. *Plant Cell Physiol.* **46**, 817–826.

Kryvych, S., Nikiforova, V., Herzog, M., Perazza, D., and Fisahn, J. (2008). Gene expression profiling of the different stages of *Arabidopsis thaliana* trichome development on the single cell level. *Plant Physiol. Biochem.* **46**, 160–173.

Kurata, T., Ishida, T., Kawabata-Awai, C., Noguchi, M., Hattori, S., Sano, R., Nagasaka, R., Tominaga, R., Koshino-Kimura, Y., Kato, T., Sato, S., Tabata, S., *et al.* (2005). Cell-to-cell movement of the CAPRICE protein in Arabidopsis root epidermal cell differentiation. *Development* **132**, 5387–5398.

Larkin, L. C., Oppenheimer, D. G., Lloyd, A. M., Paparozzi, E. T., and Marks, M. D. (1994). The roles of the *GLABROUS1* and *TRANPARENT TESTA GLABRA* genes in *Arabidopsis* trichome development. *Plant Cell* **6**, 1065–1076.

Larkin, J. C., Oppenheimer, D. G., Pollock, S., and Marks, M. D. (1993). *Arabidopsis* GLABROUS1 gene requires downstream sequences for function. *Plant Cell* **5**, 1739–1748.

Larkin, J. C., Young, N., Prigge, M., and Marks, M. D. (1996). The control of trichome spacing and number in *Arabidopsis*. *Development* **122**, 997–1005.

Lee, K., Avondo, J., Morrison, H., Blot, L., Stark, M., Sharpe, J., Bangham, A., and Coen, E. (2006). Visualizing plant development and gene expression in three dimensions using optical projection tomography. *Plant Cell* **18**, 2145–2156.

Lieckfeldt, E., Simon-Rosin, U., Kose, F., Zoeller, D., Schliep, M., and Fisahn, J. (2008). Gene expression profiling of single epidermal, basal and trichome cells of *Arabidopsis thaliana*. *J. Plant Physiol.* **165**, 1530–1544.

Lloyd, A. M., Schena, M., Walbot, V., and Davis, R. W. (1994). Epidermal cell fate determination in *Arabidopsis*: patterns defined by steroid-inducible regulator. *Science* **266**, 436–439.

Lucas, W. J., and Lee, J. Y. (2004). Plasmodesmata as a supracellular control network in plants. *Nat. Rev. Mol. Cell Biol.* **5**, 712–726.

Marks, M. D. (1997). Molecular genetic analysis of trichome development in *Arabidopsis*. *Annu. Rev. Plant Physiol. Plant Mol. Biol.* **48**, 137–163.

Marks, M. D., Wenger, J. P., Gilding, E., Jilk, R., and Dixon, R. A. (2009). Transcriptome analysis of Arabidopsis wild-type and gl3-sst sim trichomes identifies four additional genes required for trichome development. *Mol. Plant* **2**, 803–822.

Meinhardt, H., and Gierer, A. (1974). Applications of a theory of biological pattern formation based on lateral inhibition. *J. Cell Sci.* **15**, 321–346.

Melaragno, J. E., Mehrotra, B., and Coleman, A. W. (1993). Relationship between endopolyploidy and cell size in epidermal tissue of *Arabidopsis*. *Plant Cell* **5**, 1661–1668.

Morohashi, K., and Grotewold, E. (2009). A systems approach reveals regulatory circuitry for Arabidopsis trichome initiation by the GL3 and GL1 selectors. *PLoS Genet.* **5**, e1000396.

Morohashi, K., Zhao, M., Yang, M., Read, B., Lloyd, A., Lamb, R., and Grotewold, E. (2007). Participation of the Arabidopsis bHLH factor GL3 in trichome initiation regulatory events. *Plant Physiol.* **145**, 736–746.

Ohashi, Y., Ruberti, I., Morelli, G., and Aoyama, T. (2002). Entopically additive expression of GLABRA2 alters the frequency and spacing of trichome initiation. *Plant J.* **21**, 5036–5046.

Oppenheimer, D. G., Herman, P. L., Sivakumaran, S., Esch, J., and Marks, M. D. (1991). A *myb* gene required for leaf trichome differentiation in *Arabidopsis* is expressed in stipules. *Cell* **67**, 483–493.

Payne, C. T., Zhang, F., and Lloyd, A. M. (2000). GL3 encodes a bHLH protein that regulates trichome development in Arabidopsis through interaction with GL1 and TTG1. *Genetics* **156**, 1349–1362.

Rerie, W. G., Feldmann, K. A., and Marks, M. D. (1994). The glabra 2 gene encodes a homeo domain protein required for normal trichome development in *Arabidopsis*. *Genes Dev.* **8**, 1388–1399.

Schellmann, S., Hulskamp, M., and Uhrig, J. (2007). Epidermal pattern formation in the root and shoot of Arabidopsis. *Biochem. Soc. Trans.* **35**, 146–148.

Schellmann, S., Schnittger, A., Kirik, V., Wada, T., Okada, K., Beermann, A., Thumfahrt, J., Jurgens, G., and Hulskamp, M. (2002). TRIPTYCHON and CAPRICE mediate lateral inhibition during trichome and root hair patterning in Arabidopsis. *EMBO J.* **21**, 5036–5046.

Schnittger, A., Folkers, U., Schwab, B., Jürgens, G., and Hulskamp, M. (1999). Generation of a spacing pattern: the role of *TRIPTYCHON* in trichome patterning in Arabidopsis. *Plant Cell* **11**, 1105–1116.

Schnittger, A., Jurgens, G., and Hulskamp, M. (1998). Tissue layer and organ specificity of trichome formation are regulated by GLABRA1 and TRIPTYCHON in Arabidopsis. *Development* **125**, 2283–2289.

Szymanski, D. B., Jilk, R. A., Pollock, S. M., and Marks, M. D. (1998). Control of GL2 expression in *Arabidopsis* leaves and trichomes. *Development* **125**, 1161–1171.

Szymanski, D. B., and Marks, M. D. (1998). *GLABROUS1* overexpression and *TRIPTYCHON* alter the cell cycle and trichome cell fate in Arabidopsis. *Plant Cell* **10**, 2047–2062.

Theobald, W. L., Krahulik, J. L., and Rollins, R. C. (1979). Trichome description and classification. In: Metcalfe, C. R., and Chalk, L. (Eds.), *Anatomy of the Dicotyledons*, Vol. 1, 2nd ed. Clarendon press, Oxford, pp. 40–53.

Tominaga, R., Iwata, M., Sano, R., Inoue, K., Okada, K., and Wada, T. (2008). Arabidopsis CAPRICE-LIKE MYB 3 (CPL3) controls endoreduplication and flowering development in addition to trichome and root hair formation. *Development* **135**, 1335–1345.

Turing, A. (1952). The chemical basis of morphogenesis. *Philos. Trans. R. Soc. Lond., B* **237**, 37–72.

Uphof, J. C. T. (1962). *Plant Hairs.* Gebr. Bornträger, Berlin.

Wada, T., Kurata, T., Tominaga, R., Koshino-Kimura, Y., Tachibana, T., Goto, K., Marks, M. D., Shimura, Y., and Okada, K. (2002). Role of a positive regulator of root hair development, CAPRICE, in Arabidopsis root epidermal cell differentiation. *Development* **129**, 5409–5419.

Wada, T., Tachibana, T., Shimura, Y., and Okada, K. (1997). Epidermal cell differentiation in Arabidopsis determined by a Myb homolog, CPC. *Science* **277**, 1113–1116.

Wagner, G. J., Wang, E., and Shepherd, R. W. (2004). New approaches for studying and exploiting an old protuberance, the plant trichome. *Ann. Bot. (Lond.)* **93**, 3–11.

Walker, A. R., Davison, P. A., Bolognesi-Winfield, A. C., James, C. M., Srinivasan, N., Blundell, T. L., Esch, J. J., Marks, M. D., and Gray, J. C. (1999). The TTG1 (transparent testa, glabra1) locus which regulates trichome differentiation and anthocyanin biosynthesis in Arabidopsis encodes a WD40-repeat protein. *Plant Cell* **11**, 1337–1350.

Wang, S., Barron, C., Schiefelbein, J., and Chen, J.-G. (2010). Distinct relationships between GLABRA2 and single-repeat R3 MYB transcription factors in the regulation of trichome and root hair patterning in Arabidopsis. *New Phytol.* **185**, 387–400.

Wang, S., and Chen, J. G. (2008). Arabidopsis transient expression analysis reveals that activation of GLABRA2 may require concurrent binding of GLABRA1 and GLABRA3 to the promoter of GLABRA2. *Plant Cell Physiol.* **49**, 1792–1804.

Wang, S., Hubbard, L., Chang, Y., Guo, J., Schiefelbein, J., and Chen, J. G. (2008). Comprehensive analysis of single-repeat R3 MYB proteins in epidermal cell patterning and their transcriptional regulation in Arabidopsis. *BMC Plant Biol.* **8**, 81.

Wang, S., Kwak, S. H., Zeng, Q., Ellis, B. E., Chen, X. Y., Schiefelbein, J., and Chen, J. G. (2007). TRICHOMELESS1 regulates trichome patterning by suppressing GLABRA1 in Arabidopsis. *Development* **134**, 3873–3882.

Wester, K., Digiuni, S., Geier, F., Timmer, J., Fleck, C., and Hulskamp, M. (2009). Functional diversity of R3 single-repeat genes in trichome development. *Development* **136**, 1487–1496.

Zhang, F., Gonzalez, A., Zhao, M., Payne, C. T., and Lloyd, A. (2003). A network of redundant bHLH proteins functions in all TTG1-dependent pathways of Arabidopsis. *Development* **130**, 4859–4869.

Zhao, M., Morohashi, K., Hatlestad, G., Grotewold, E., and Lloyd, A. (2008). The TTG1-bHLH-MYB complex controls trichome cell fate and patterning through direct targeting of regulatory loci. *Development* **135**, 1991–1999.

CHAPTER ELEVEN

Comparative Analysis of Flowering in Annual and Perennial Plants

Maria C. Albani *and* George Coupland

Contents

Abstract

In plants the switch from vegetative growth to flowering involves a major transition in the development of the shoot apex. This transition can occur once, in annual species, or repeatedly, in perennial plants. In annuals, flowering is associated with senescence and death of the whole plant, whereas perennials flower in consecutive years and maintain vegetative development after flowering. The perennial life strategy depends on differential behavior of meristems on a single plant so that some remain in the vegetative state while others undergo the floral transition. *A. thaliana* provides a powerful model system for

Department of Plant Developmental Biology, Max Planck Institute for Plant Breeding Research, Cologne, Germany

Current Topics in Developmental Biology, Volume 91
ISSN 0070-2153, DOI 10.1016/S0070-2153(10)91011-9

understanding the mechanisms of flowering in annuals. Here we review the events that occur in the meristem of *A. thaliana* during the floral transition and compare these with our understanding of flowering in perennial systems.

1. Introduction

After embryogenesis, plants pass through successive phases of vegetative and reproductive development. During these phases, organs are formed on the shoot. These organs derive from the shoot apical meristem (SAM), a group of undifferentiated cells present at the apex of the plant. During vegetative development the SAM gives rise to vegetative organs such as leaves or vegetative shoots. At a time determined by the environmental conditions and the genotype of the individual plant it undergoes the transition to reproductive development. The SAM changes shape, becoming an inflorescence meristem, and gives rise to flowers or flowering shoots. In annual plants this transition is irreversible, whereas the majority of perennial plants utilize different developmental strategies to repeatedly pass through phases of vegetative and reproductive development.

The floral transition is also an important life history trait that adapts plants to particular environments by ensuring that flowering occurs in favorable conditions. The number of flowering events plants undergo during their lifetime depends on the life strategy they follow. All annuals and a few perennial species such as bamboo and agave follow the monocarpic life strategy (Battey and Tooke, 2002). This life strategy is also known as semelparous and is characterized by a massive flowering episode which is followed by senescence and death of the whole plant. However, the majority of perennials follow a polycarpic or iteroparous life strategy in which they flower many times during their lifetime. Perennials have the ability to resume vegetative growth every year after flowering whereas annuals die after a single flowering event. Differences in flowering behavior between annuals and perennials contribute to differences in their life strategy (Grillo *et al.*, 2009; Thomas *et al.*, 2000; Wang *et al.*, 2009b). For example in annuals, all meristems behave the same way and upon flower induction become reproductive. Perennials on the other hand do not commit all meristems to flower but retain some vegetative meristems which will support vegetative growth the following year (Foster *et al.*, 2003; Wang *et al.*, 2009b). Moreover, some perennials can cycle back and forth between reproductive and vegetative development because reproductive meristems revert to the vegetative state after flowering (Diomaiuto, 1988; Tooke *et al.*, 2005).

Here we will concentrate on the molecular mechanisms known in *A. thaliana* to regulate flowering in the meristem and then broaden the discussion to the control of flowering in polycarpic perennials.

2. Control of Flowering in the Annual Model *Arabidopsis thaliana*

Flowering of Arabidopsis is promoted by a range of environmental cues. Some of these vary in a predictable way with the changing seasons such as day length and sustained cold winter temperatures, while others reflect the local environment including the light quality associated with shading by other plants or fluctuations in ambient temperature. In addition, the age of the plant and other endogenous factors influence flowering time. These distinct effects on flowering time are conferred by different genetic pathways that ultimately converge to regulate the expression of a set of floral integrator genes (Baurle and Dean, 2006; Mouradov *et al.*, 2002; Simpson *et al.*, 1999). Several reviews describe the activity of individual pathways such as the photoperiod pathway (Kobayashi and Weigel, 2007; Turck *et al.*, 2008), vernalization pathway (Kim *et al.*, 2009), autonomous pathway (Farrona *et al.*, 2008), age-related pathway (Fornara and Coupland, 2009), and the GA pathway (Mutasa-Gottgens and Hedden, 2009).

Floral development occurs at the SAM, but some of the environmental response pathways that regulate flowering act in the leaves. For example, photoperiod perception occurs in the leaves and in response to promotive day lengths a signal is transmitted to the meristem that induces changes in gene expression and ultimately flower development at the shoot meristem (Kobayashi and Weigel, 2007; Turck *et al.*, 2008). Here we mainly summarize the events that occur in the meristem (Fig. 11.1), as these provide the background required for comparison with perennial systems.

Exposure to environmental cues can be used to provide a temporal framework for events in the meristem. For example, Arabidopsis plants grown under short days (8- or 10-h long) remain vegetative, but after shifting to long days (16-h long) genes involved in flowering are expressed at the meristem within 24 h and floral primordia are detectable within 3 or 5 days depending on the accession and precise conditions. We have organized the following sections according to this temporal framework starting with the earliest events that are detected in the meristem and associated with flowering and progressing to the development of floral primordia. Expression of the *APETALA1* (*AP1*) gene, which encodes a MADS box transcription factor, in developing flower primordia is usually assumed to be the end of the floral induction process and the first stage of floral development (Hempel *et al.*, 1997). We will therefore summarize the events leading to *AP1* expression but will not proceed further into floral development.

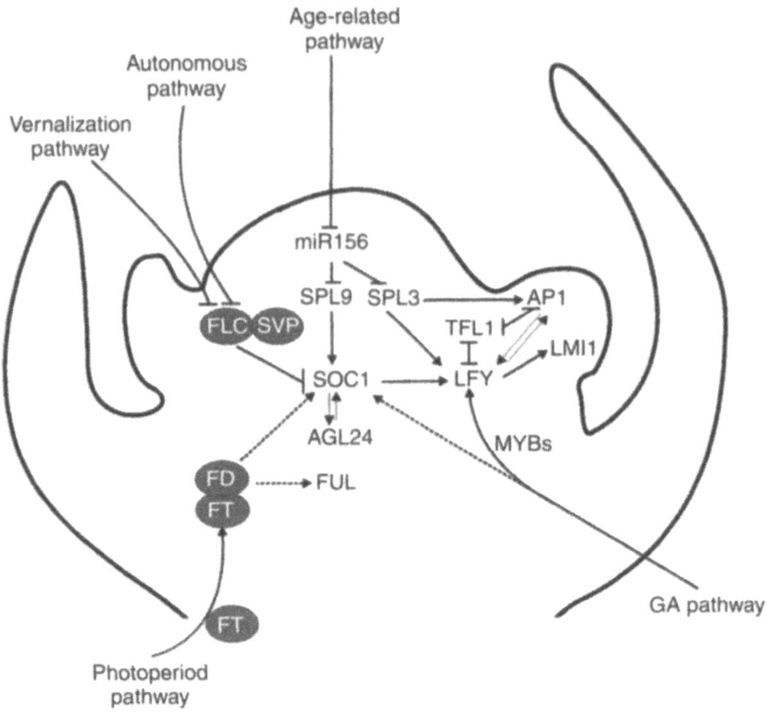

Figure 11.1 Regulation of flowering in the *A. thaliana* meristem. Environmental and endogenous signals are integrated by *SOC1* which is one of the first genes activated in the meristem during floral induction. SOC1 increases the expression of *AGL24* by directly binding to *AGL24* promoter, but also SOC1 and AGL24 proteins interact in a positive feedback loop increasing each others expression. In response to long photoperiods *FT* mRNA level is increased in the leaf. The FT protein is transmitted to the meristem. FT interacts with FD and both are required for transcriptional activation of *SOC1* in the meristem. The vernalization and autonomous pathways regulate the expression of *FLC* which encodes a floral repressor. FLC interacts with SVP and the FLC–SVP complex represses *SOC1* transcription. miR156 represses the expression of the floral promoters *SPL3* and *SPL9* that increase with age. SPL9 binds directly to the *SOC1* promoter whereas SPL3 binds to the promoters of the meristem identity genes *LFY* and *AP1*. Gibberellin promotes flowering either by activating the expression of *SOC1* or by regulating *LFY* transcription through MYB transcription factors. *FUL* acts redundantly with *SOC1* to promote flowering and is also upregulated by FT and FD. *TFL1* represses *LFY* and *AP1* expression in the shoot apical meristem while *LFY* and *AP1* enhance each others expression within floral primordia. Additionally, LFY activates expression of other genes conferring floral identity such as LMI1. PENNYWISE and PENNY-FOOLISH are not included in the diagram as their precise interactions with other genes are not known. Arrows indicate flower promoting events whereas T symbols represent repression events. Solid lines indicate direct interactions and dotted lines indicate interactions that might be indirect. Circles demonstrate protein interactions. This schematic diagram is not intended to indicate the spatial distribution of events within the meristem. *AGL24, AGAMOUS-LIKE 24; AP1, APETALA 1; LMI1, LATE MERISTEM IDENTITY 1; LFY, LEAFY;* miR156, microRNA 156; *FLC, FLOWERING LOCUS C; FT, FLOWERING LOCUS T; FUL, FRUITFUL; SOC1, SUPPRESSOR OF OVEREXPRESSION OF CONSTANS; SPL3, SUAMOSA-PROMOTER BINDING PROTEIN LIKE 3; SPL9, SQUAMOSA-PROMOTER BINDING PROTEIN LIKE 9; SVP, SHORT VEGETATIVE PHASE; TFL1, TERMINAL FLOWER 1*

2.1. Promoters of flowering acting at the shoot apical meristem early during the floral transition

The transition to flowering in Arabidopsis is most readily synchronized by shifting plants from short to long days (Hempel *et al.*, 1997; Schmid *et al.*, 2003; Searle *et al.*, 2006). Two-week-old plants grown in SD are not induced to flower, but on transfer to LDs, *FLOWERING LOCUS T* (*FT*) mRNA is expressed in the leaf during the first day (Corbesier *et al.*, 2007). The FT protein is a member of the CETS (CEN1; TFL1; FT) family (Pnueli *et al.*, 2001) and is related to phosphatidylethanolamine binding proteins (Kardailsky *et al.*, 1999; Kobayashi *et al.*, 1999). This small protein is expressed in the phloem companion cells and then transmitted through the phloem sieve elements to the meristem where it is involved in transcriptional reprogramming (Corbesier *et al.*, 2007; Jaeger and Wigge, 2007; Lin *et al.*, 2007; Mathieu *et al.*, 2007; Notaguchi *et al.*, 2008; Tamaki *et al.*, 2007). The mechanism by which FT alters gene expression is through interaction with the FLOWERING LOCUS D (FD) bZIP transcription factor (Abe *et al.*, 2005; Wigge *et al.*, 2005). FD is already expressed in the meristem of 2-week-old SD-grown plants, but its level of expression rises in the meristem after the shift to LDs. Mutations in *FT* or *FD* delay flowering, and after transfer of plants to LDs these mutations also delay the expression of *SUPPRESSOR OF OVEREXPRESSION OF CONSTANS 1* (*SOC1*), which encodes a MADS box transcription factor. In wild-type plants *SOC1* mRNA increases in the meristem within 1 day of activation of *FT* expression in the leaf, but is greatly delayed in *ft* or *fd* mutants (Borner *et al.*, 2000; Lee *et al.*, 2000; Samach *et al.*, 2000; Searle *et al.*, 2006). *SOC1* is the earliest-acting gene known to respond to FT/FD in the SAM. Although *SOC1* is expressed both in the leaves and the shoot meristem, it promotes flowering much more strongly when ectopically expressed in the meristem compared to the leaf (Searle *et al.*, 2006). These observations suggest that FT and FD initiate flowering and reprogramme the behavior of the meristem at least in part by activating *SOC1* transcription, but it is still not known whether this is a direct or indirect effect.

SOC1 is an important integrator of floral signals at the shoot meristem. The photoperiod pathway promotes flowering in response to LDs, and within this pathway the expression of *FT* and the related gene *TWIN SISTER OF FT* (*TSF*) is activated in the leaf specifically in LDs by the zinc finger protein CONSTANS (Jang *et al.*, 2009; Michaels *et al.*, 2005; Samach *et al.*, 2000; Yamaguchi *et al.*, 2005). Activation of *SOC1* at the meristem occurs in response to transport of FT, and probably TSF, protein to the meristem. Similarly, *SOC1* expression is regulated by other promotive pathways. Gibberellins (GAs) promote flowering of Arabidopsis, particularly under SDs. GA accumulates at the meristem prior to flowering under SDs and mutations with a severe effect on GA biosynthesis prevent flowering under SD (Eriksson *et al.*, 2006; Reeves and Coupland., 2001; Wilson *et al.*, 1992). Furthermore, GA acts redundantly with the photoperiod pathway under LDs, so that the late-flowering phenotype of mutants impaired in the photoperiod pathway is strongly enhanced by GA mutants under LDs. The GA pathway also acts through *SOC1*, so that treatment with GA leads to an increase in *SOC1* mRNA levels, although the mechanism by which this occurs is still not clear (Moon *et al.*, 2003). Finally *SOC1* transcription is repressed by FLOWERING LOCUS C (FLC), which binds to the promoter and intron regions of *SOC1* (Hepworth *et al.*, 2002; Searle *et al.*, 2006; Sheldon *et al.*, 2006). FLC plays important roles in the vernalization pathway, which controls the response to low temperatures, and the autonomous pathway, which regulates flowering independently of environmental cues. Overexpression of *SOC1* from a heterologous promoter can overcome the repressive effect of FLC on flowering (Lee *et al.*, 2000). The function of FLC is discussed in more detail in the next section on repressors of flowering. In summary, *SOC1* is the earliest promoter of flowering whose expression rises in the meristem during the transition to flowering, and it acts as an integrator of the photoperiod, GA, autonomous, and vernalization pathways.

FRUITFULL (*FUL*) is expressed in the meristem only slightly later during floral induction than *SOC1* and encodes a closely related MADS box transcription factor. Mutations in *FUL* also delay flowering (Gu *et al.*, 1998), and *soc1 ful* double mutants flower significantly later than either single mutant, suggesting that these two related proteins are partially redundant (Melzer *et al.*, 2008). Furthermore, the floral transition in the double mutant is unstable so that it reverts to vegetative growth, indicating that SOC1 and FUL have roles in maintenance of inflorescence development. The extreme early flowering caused by *FT* overexpression is largely suppressed in the *soc1 ful* double mutant indicating that the effect of FT on flowering time is mainly through these two MADS box proteins (Melzer *et al.*, 2008). Consistent with this conclusion, expression of *SOC1* and *FUL* in the meristem is strongly delayed in the *ft tsf* or *fd* mutants (Jang *et al.*, 2009; Searle *et al.*, 2006; Wang *et al.*, 2009a). In addition, the inflorescence apical meristem in the *soc1 ful* double mutant has a longer lifespan than the wild type. This observation together with the reversion to vegetative growth after flowering led to the suggestion that *soc1 ful* double mutants showed characteristics of perennial plants. Taken together these analyses demonstrate important and partially redundant functions for *SOC1* and *FUL* in flowering time as well as inflorescence development.

A third MADS box transcription factor expressed in the meristem early during the floral transition is AGAMOUS-LIKE 24 (AGL24). The *agl24* mutants are late flowering and show reduced *SOC1* expression in the meristem at an early stage in the floral transition (Liu *et al.*, 2008; Yu *et al.*, 2002). Furthermore, overexpression of *AGL24* causes early flowering and it binds directly to the *SOC1* promoter. *AGL24* is increased in expression by the vernalization and photoperiod pathways, suggesting that it might be a major integrator of environmental information involved in the early activation of *SOC1* expression. The SOC1 protein also binds directly to the *AGL24* promoter. Furthermore, SOC1 and AGL24 proteins interact and are co-localized to the nucleus (Lee *et al.*, 2008). Therefore, these two MADS box proteins appear to be involved in a positive-feedback loop that increases the expression of each gene early during the floral transition.

In summary, these observations indicate that three MADS box transcription factors, SOC1, AGL24, and FUL, play important roles in the meristem during the earliest stages of the floral transition in Arabidopsis.

2.2. Repressors of flowering acting at the shoot apical meristem during the floral transition

Many mutations have been described that cause early flowering and define genes that play a role in repressing flowering. Here we focus on those genes that act as repressors of flowering within the SAM. Two transcriptional repressors with central roles in this process are the MADS box transcription factors FLC and SHORT VEGETATIVE PHASE (SVP). The *FLC* and *SVP* mRNAs are both expressed in the SAM and in the leaf. In the SAM they act as direct repressors of *SOC1* transcription, while in the leaf they repress *FT* and *TSF* transcription (Jang *et al.*, 2009; Lee *et al.*, 2000; Liu *et al.*, 2009). FLC and SVP interact to form a protein complex; however, they can probably also act independently because the *svp flc* double mutant is much earlier flowering than either single mutant (Fujiwara *et al.*, 2008; Li *et al.*, 2008). The mechanism by which they repress transcription likely involves histone modifications as they physically interact in protein complexes with two proteins TERMINAL FLOWER 2 (TFL2)/LIKE HETEROCHROMATIN PROTEIN 1 (LHP1) and SIN3-ASSOCIATED POLYPEPTIDE OF 18 kDa (SAP18) which are implicated in influencing transcription through different histone modifications (Liu *et al.*, 2009). TFL2 binds trimethylated lysine 27 on histone 3, while SAP18 is a member of a histone deacetylase complex.

FLC plays central roles in the vernalization and autonomous pathways. In early-flowering summer annual accessions of Arabidopsis such as Columbia and Landsberg *erecta* which are widely used in the laboratory, *FLC* mRNA is expressed at a low level or the gene carries mutations that prevent expression of an active protein (Michaels and Amasino, 1999; Sheldon *et al.*, 1999). However, in almost all late-flowering accessions *FLC* is active and the mRNA is expressed at high levels. Flowering of most of these accessions can be accelerated by vernalization, extended exposure to low temperatures that mimic winter temperatures. Typically plants are exposed to 4°C for 4–8 weeks. Under these conditions *FLC* mRNA is progressively reduced, so that when the plant is returned to normal growth temperatures it will rapidly undergo the transition to flowering (Bastow *et al.*, 2004; Finnegan and Dennis, 2007; Michaels and Amasino, 1999; Sheldon *et al.*, 1999; Sung and Amasino, 2004; Sung *et al.*, 2006). The reduction of *FLC* expression in low temperatures involves expression of an antisense RNA (Swiezewski *et al.*, 2009) and machinery that modifies the tails of histone 3 at the *FLC* gene, particularly trimethylation of lysine 27 (Finnegan and Dennis, 2007). These processes result in reduction of *FLC* mRNA expression in the cold and stable repression of *FLC* when plants are returned to normal growth temperatures. Thus *FLC* plays a central role in vernalization response, repressing flowering prior to vernalization and being repressed in expression during vernalization.

SVP is expressed in early-flowering summer annual accessions as well as late-flowering accessions. Mutations in *SVP* accelerate flowering under both LDs and SDs, but have their most severe effect under SD (Hartmann *et al.*, 2000). In Columbia, *SVP* is expressed in the SAM of 10-day-old LD-grown plants, so prior to the floral transition (Jang *et al.*, 2009). However, *SVP* expression is reduced during the floral transition as *SOC1* mRNA starts to rise. So that soon after *SOC1* is strongly expressed in the meristem, *SVP* mRNA is not detectable. The mechanism by which *SVP* is repressed in the meristem is unclear.

Another important repressor of flowering time that acts in the meristem is the CETs family protein TERMINAL FLOWER 1 (TFL1). The *TFL1* mRNA is expressed weakly in the center of the meristem of vegetative plants and more strongly in the inflorescence meristem (Bradley *et al.*, 1997; Conti and Bradley, 2007; Ratcliffe *et al.*, 1998). TFL1 is related to FT, and can move from the domain of *TFL1* mRNA expression throughout the meristem (Conti and Bradley, 2007). Mutations in *tfl1* cause early flowering under long and short days while overexpressor plants are late flowering, indicating that in contrast to FT the TFL1 protein is a repressor of flowering. One possible mechanism by which TFL1 represses flowering is through competition with FT for binding to FD (Ahn *et al.*, 2006). *In situ* hybridizations demonstrated that the floral meristem identity genes *AP1* and *LFY* are ectopically expressed throughout the meristem in *tfl1* mutant plants rather than predominately in floral primordia (Bradley *et al.*, 1997; Gustafson-Brown *et al.*, 1994; Liljegren *et al.*, 1999). This ectopic expression is associated with termination of inflorescence development by conversion of the SAM into a flower (Alvarez *et al.*, 1992; Shannon and Meeks-Wagner, 1991). Similarly, axillary meristems of *tfl1* mutants are converted into flowers. *AP1* is also a target gene for the FT/FD complex, and its increased expression in *tfl1* mutants would be consistent with enhanced FT activity.

Clearly repressors of flowering play important roles in the meristem by modulating the floral transition and ensuring the proper temporal and spatial patterns of expression of floral promoting genes.

2.3. Promoters of flowering that act at the shoot apical meristem later during the floral transition

Several regulatory proteins that promote flowering are expressed in the SAM subsequent to the activation of *SOC1*. Indeed this stage is likely to include a dramatic reprogramming of transcription in the shoot meristem (Schmid *et al.*, 2003). Important roles for specific transcription factors have been described at this stage.

Members of the SQUAMOSA-PROMOTER BINDING PROTEIN LIKE (SPL) family of transcription factors activate expression of a number of key genes in the meristem at different stages of the floral transition. Loss-of-function mutations in *SPL* genes that have been described do not show an obvious flowering-time phenotype, probably because of functional redundancy between members of the family. Of the 16 SPL transcription factors encoded in the Arabidopsis genome, 10 are regulated by the microRNA miR156. Overexpression of this microRNA causes a delay in flowering, consistent with the promotive effect of the SPLs on flowering (chapter 12; Schwab *et al.*, 2005). Also overexpression of single *SPL* genes, notably *SPL3*, causes early flowering. Furthermore, SPL3 binds directly to the promoters of *FUL* and the meristem identity genes *AP1* and *LEAFY* (Wang *et al.*, 2009a; Yamaguchi *et al.*, 2009), while SPL9 binds directly to *SOC1*. Binding of the SPLs is probably associated with activation of transcription, because an inducible version of SPL9 rapidly activated *FUL* transcription (Wang *et al.*, 2009a). The *SPL9* and *SPL3* mRNAs slowly increase with the age of the plant and are expressed at the shoot meristem, correlating with the gradual reduction in miR156 levels. Thus SPLs are believed to function in the meristem within an age-related flowering pathway that would make the plant more likely to flower as it becomes older.

Two homeodomain transcription factors, PENNYWISE (PNY) and PENNY-FOOLISH (PNF), are also essential for the floral transition (Kanrar *et al.*, 2008; Smith *et al.*, 2004). Double mutants in which both of these genes are inactivated never flowered and continued to produce leaves for at least 200 days under LD conditions. The meristem of the double mutant increased in size as if it entered the floral transition but did not expand to the same extent as wild-type plants. *SOC1* was also expressed but floral primordia were not produced. The *PNY* and *PNF* mRNAs are expressed at the SAM during flowering; however, because they are also expressed in the vegetative meristem they are assumed not to be sufficient to induce flowering. Nevertheless, PNY and PNF are essential for expression of key floral meristem identity genes as in *pny pnf* double mutants *LFY* and *AP1* expression is strongly reduced or abolished, while the response to FT is also attenuated (Kanrar *et al.*, 2008; Smith *et al.*, 2004). Thus these homeodomain transcription factors seem to play an essential role in the meristem during the floral transition in facilitating changes in the morphology of the meristem and the induction of expression of genes that act downstream of SOC1.

As mentioned earlier the growth regulator GA promotes flowering of Arabidopsis by activating expression of *SOC1*. However, GA has also been assigned roles later in the floral transition. In particular *LFY* is proposed to be activated by MYB transcription factors related to GAMYB, which acts in the barley aleurone to activate transcription of target genes in response to GA (Woodger *et al.*, 2003b). In Arabidopsis, three MYB transcription factors, AtMYB33, AtMYB101, and AtMYB65, are related in sequence to GAMYB (Gocal *et al.*, 2001), and AtMYB33 as well as AtMYB65 are expressed in the early inflorescence meristem. One of these, AtMYB33, binds *in vitro* to elements present in the *LFY* promoter. All three of the Arabidopsis MYBs are regulated by a microRNA, miR159. Overexpression of miR159 delays flowering and reduces LFY expression, consistent with a role for the AtGAMYBs in the promotion of flowering (chapter 12; Achard *et al.*, 2004). Therefore, although lack of mutant alleles prevents analysis of the contribution of individual AtGAMYBs, the late-flowering phenotype caused by downregulation of the entire family using miR159 supports the idea that these transcription factors have a significant role in the floral transition.

In this section we have summarized the function in the meristem of three families of transcription factors that broadly appear to act in the floral transition after *SOC1* expression but before induction of the floral meristem identity genes. In temporal terms this would roughly correspond to the period between 2 and 4 days subsequent to shifting plants from short to long days. The complex circuitry linking these different groups of proteins and their spatial expression patterns in the meristem during the floral transition have only just begun to be described. Furthermore, there are likely many more transcription factors and other regulatory proteins with important roles during this period still to be discovered.

2.4. The floral meristem identity genes *LFY* and *AP1*

Flowers develop from primordia produced on the flanks of the inflorescence meristem. A small set of genes confer floral identity on these primordia. Among these floral identity genes *LEAFY* (*LFY*) and *APETALA1* (*AP1*) have the most dramatic effects. *LFY* encodes a plant-specific transcription factor (Hames *et al.*, 2008; Weigel *et al.*, 1992) and *AP1* encodes a MADS box transcription factor. Both genes are expressed strongly in young floral primordia. Mutations in *LFY* cause many of these primordia to develop as shoots rather than flowers, and overexpression of *LFY* is sufficient to confer floral identity on the SAM (Weigel and Nilsson, 1995). Similarly, the flowers of *ap1* mutants have an indeterminate character related to inflorescences, and again overexpression of *AP1* is sufficient to confer floral identity on the SAM (Mandel and Yanofsky, 1995). *AP1* becomes expressed around 5 days after shifting 2-week-old short-day-grown plants to long days, and returning plants to short days does not alter their flowering behavior. Therefore, when *AP1* is expressed plants are committed to flower. Thus this can define the end of the floral induction process and the beginning of floral development.

Strong *LFY* expression occurs rather late in the meristem during floral induction and is activated by several classes of transcription factors. Both SOC1 and SPL3 have been shown to bind directly to *LFY in vivo* using chromatin immunoprecipitation (Lee *et al.*, 2008; Yamaguchi *et al.*, 2009). Also, as described above, at least *in vitro*, the AtGAMYB AtMYB33 binds directly to LFY. Activation of *LFY* by these transcription factors can explain the observation that LFY acts downstream of several flowering pathways (Blazquez and Weigel, 2000). LFY directly activates *AP1* expression (Wagner *et al.*, 1999), so that both major floral meristem identity genes are expressed. In addition LFY activates other genes with roles in conferring floral meristem identity such as *LATE MERISTEM IDENTITY 1* (LMI1) (Saddic *et al.*, 2006).

AP1 can also feed back to activate *LFY*, probably redundantly with the related proteins CAULIFLOWER (CAL) and FUL (Ferrandiz *et al.*, 2000). In this way LFY and AP1 enhance each others expression reinforcing floral identity (Bowman *et al.*, 1993) and initiating floral organ specification and development. At this stage the flower development programme in the developing primordia is established and the plant is committed to flower, ending the phase of floral transition.

3. Flowering in Perennials

Flowering control in perennials determines the perennial life strategy and is integrated with other perennial traits. The molecular mechanisms controlling flowering in perennials have not been studied as extensively as in annual plants. A major factor hampering detailed studies is that often only a part of the plant responds to flower inductive signals, because only a subset of meristems become reproductive. In a perennial plant, therefore, reproductive competence varies between meristems so that when they are exposed to favorable environmental conditions only competent meristems perceive flower inductive signals and differentiate into inflorescences and flowers (Battey and Tooke, 2002; Martin-Trillo and Martinez-Zapater, 2002; Walton *et al.*, 1997). Flower initiation in many temperate species is interrupted by winter, which seems to be involved in retaining some meristems in the vegetative state (Battey, 2000). In these species, flowering spans 2 successive years. Flower induction and often also flower initiation is followed by a period of winter dormancy, before flowers fully develop the following year (Abbott, 1970; Battey *et al.*, 1998; Mullins *et al.*, 1992; Walton *et al.*, 1997). Vegetative growth resumes in the spring together with the emerging flowers from the meristems that did not become reproductive the previous year. The molecular mechanisms involved in these processes are poorly understood.

Studies in kiwifruit and grapevine, which follow this pattern of flowering over 2 years, associated flowering with expression of homologs of characterized *A. thaliana* genes, which act as early markers of floral development (Carmona *et al.*, 2002; Walton *et al.*, 2001). The grapevine *LFY* homolog (*VFL*), shows a bimodal expression pattern which coincides with the time of flower induction and inflorescence initiation during the first year and the formation and development of flowers during the second year (Carmona *et al.*, 2002). The *AP1* homolog (*VAP1*) in grapevine increases in expression during the second year as floral meristems develop, although it is also involved in the formation of other structures such as tendrils (Calonje *et al.*, 2004).

Floral reversion has been suggested as a mechanism that maintains perennial behavior (Tooke *et al.*, 2005). Seasonal environmental changes during the year can cause inflorescence meristems to revert from producing axillary flowers to forming axillary vegetative shoots, thereby switching back from reproductive to vegetative development. The molecular mechanisms of floral reversion in the context of perenniality have not been studied in detail. However, the flowering patterns of some trees suggest that inflorescence reversion might take place every year regulating the cyclical changes between flowering and vegetative development (Diomaiuto, 1988; Tooke *et al.*, 2005).

Perennial flowering is complex and cannot be described as a single trait. Studies in perennials so far have targeted several distantly related species and described the involvement of homologs of *A. thaliana* genes in the regulation of perennial-specific traits that are related to flowering. In the following sections we will summarize our current knowledge on the physiology and molecular mechanisms involved in the regulation of some perennial-specific traits and the events that occur in the meristem.

3.1. Juvenility

Juvenility is the stage early in a plant's life cycle during which it is unable to respond to flower inductive signals (Baurle and Dean, 2006; Poethig, 2003). The duration of the juvenile period varies among species and is most pronounced in perennial plants where it can last from a few weeks to several years.

Overexpression of genes that regulate flowering in *A. thaliana* can overcome the juvenile phase in distantly related perennial species. Constitutive expression of the Arabidopsis *LFY* and *AP1* meristem identity genes in citrus reduce the juvenile phase and transgenic plants flower within 12–20 months when grown in greenhouse conditions (Pena *et al.*, 2001). Similarly, overexpression in apple of the silver birch gene *BpMADS4*, which is the *A. thaliana FUL* homolog, causes premature flowering (Flachowsky *et al.*, 2007). These studies based on gene overexpression do not provide knowledge of the mechanisms regulating the juvenile-to-adult phase transition in wild-type plants. Nevertheless, they demonstrate that the flowering time pathways described in *A. thaliana* are likely to be at least partially conserved in distantly related species.

Overexpression of homologs from genes that promote flowering in *A. thaliana* in trees can also cause premature flowering. For example, constitutive overexpression of the citrus *FT* (*CiFT*) homolog caused transgenic plants to flower within 3–22 months (Endo *et al.*, 2005). Similarly, poplar plants overexpressing the poplar orthologs of *FT* (*PtFT1* or *FT2*) show precocious flowering (Bohlenius *et al.*, 2006; Hsu *et al.*, 2006). Expression of *PtFT1* and *FT2* increase as plants get older suggesting that temporal changes in *FT* expression during development might enable the juvenile-to-adult transition to flowering (Bohlenius *et al.*, 2006; Hsu *et al.*, 2006).

The involvement of floral repressors in the regulation of juvenility has been studied in many perennial species. The citrus *TFL1* homolog (*CsTFL1*) is expressed in juvenile stems but not in stems of adult plants (Pillitteri *et al.*, 2004). In a similar way, the apple *TFL1* (*MdTFL1*) homolog is expressed during vegetative development and *MdTFL1* mRNA levels are downregulated during floral bud differentiation at about the same time that the apple *LFY*-like gene *AFL1* is upregulated in the bud (Kotoda and Wada, 2005). *MdTFL1* is a floral repressor as overexpression of *MdTFL1* in *A. thaliana* causes late flowering (Kotoda and Wada, 2005). Moreover, apple transgenic plants expressing antisense RNA of *MdTFL1* showed precocious flowering suggesting that *MdTFL1* represses the transition to flowering and maintains the juvenile phase in apple (Kotoda *et al.*, 2006, 2003).

In summary, overexpression of genes that act in *A. thaliana* as floral integrators or meristem identity genes and downregulation of floral repressors can overcome the repression of flowering during the juvenile phase in several perennial species.

3.2. "Seasonal flowering" in perennials, limiting the duration of flowering

Flowering during favorable environmental conditions ensures successful reproduction. Annuals senesce and die after flowering, thereby completing their life cycle after the production of seeds. Perennials do not die after flowering and they can either flower continuously throughout the year or show seasonal flowering, in which flowering occurs for a relatively short period between phases of vegetative growth. Seasonal flowering has an adaptive advantage and ensures that seeds and fruits will successfully develop before the winter. Some perennials flower as long as the environmental conditions are favorable and show longer duration of the flowering season, which is called perpetual flowering. The duration of flowering (seasonal vs. perpetual) directly determines the amount of fruit/seed production, which is of commercial interest.

Most perennials flower seasonally; however, within the Rosaceae family several strawberry and rose species follow the perpetual flowering habit (Hanccock, 1999; Semeniuk, 1971). In strawberry, the duration of flowering is correlated with differences in requirement for flower induction. Seasonal-flowering strawberry plants (also known as June-bearers) flower during 2 successive years, in which flowers are initiated in the autumn and emerge the following spring (Battey *et al.*, 1998). Perpetual flowering plants (everbearers) are day neutral and do not only initiate flower buds in the autumn, but also in the spring. A study of day neutrality (and therefore duration of flowering season) in cultivated strawberries demonstrated that the inheritance of this trait is under polygenic control (Weebadde *et al.*, 2008). In *Fragaria vesca*, which is a wild relative of the cultivated strawberry, *SEASONAL FLOWERING LOCUS* (*SFL*) segregates as a single Mendelian locus in a cross between accessions that follow the seasonal and perpetual flowering habit (Albani *et al.*, 2004; Brown and Wareing, 1965). Similarly, in rose *RECURRENT BLOOMING* (*RB*) regulates the recurrent blooming trait (Crespel *et al.*, 2002; Semeniuk, 1971). Neither *SFL* nor *RB* has been isolated and therefore the mechanisms regulating seasonal flowering in these species are not known.

Recently, the rose homolog of the *A. thaliana* flowering time gene *SPINDLY* (*RoSPY*) was mapped close to the *RB* locus (Remay *et al.*, 2009). *SPY* in *A. thaliana* is a negative regulator of GA signalling, indicating that GAs might regulate seasonal flowering. In rose, but also in many woody perennials, GAs play an opposite role than in *A. thaliana* and inhibit flower initiation by promoting vegetative development (Lenahan *et al.*, 2006; Lord and Eckard, 1987; Salazar-Garcia and Lovatt, 1998). Comparisons of endogenous GA levels between rose plants exhibiting seasonal and perpetual flowering habit indicate that GAs might play a role in the duration of flowering season by regulating the return to vegetative development. In seasonal flowering plants, GA1 and GA3 levels increase in the shoot apices just after flower initiation, whereas in perpetual flowering plants GA levels are low throughout the year (Roberts *et al.*, 1999).

Recently, studies in *Arabis alpina,* a member of the Brassicaceae related to *Arabidopsis thaliana,* demonstrated the involvement of a floral repressor in the regulation of the seasonal flowering (Wang *et al.*, 2009b). *PERPETUAL FLOWERING 1* (*PEP1*) is the *A. alpina* ortholog of the *A. thaliana FLC* gene which encodes a MADS box transcription factor that represses flowering before vernalization (Kim *et al.*, 2009). *A. alpina* wild-type plants require exposure to a prolonged period of cold in order to flower whereas *pep1* mutant plants (similar to the *A. thaliana flc* mutant) have lost the requirement for vernalization. Moreover, *pep1* has other flowering phenotypes indicating that the *PEP1* gene has additional roles in the perennial life cycle of *A. alpina.* Mutant *pep1* plants flower perpetually suggesting that *PEP1* restricts the duration of flowering season. This additional role of *PEP1* in the *A. alpina* life cycle is due to the unstable repression of *PEP1* by vernalization. *PEP1* mRNA levels are downregulated during vernalization but when plants are returned to warm temperatures, *PEP1* expression levels rise again and repress flowering in the shoots that have not yet become reproductive. Therefore, in wild-type *A. alpina*, flowers are not initiated after the winter and, therefore, the plants will flower only for a restricted period.

Poplar trees also flower seasonally (Yuceer *et al.*, 2003). The SAM of each branch never terminates and remains vegetative (Yuceer, personal communication). However, within 1 year branches develop forming vegetative axillary buds (vegetative zone I), floral axillary buds (floral zone), and then revert to vegetative development producing new vegetative axillary buds (vegetative zone II) (Yuceer *et al.*, 2003). Expression of *FT2* in the leaves is increased when branches produce floral axillary buds (floral zone). Moreover, branches in transgenic plants overexpressing *FT2* transcript terminate with an inflorescence indicating that seasonal fluctuation of *FT2* mRNA levels might regulate the reversion to vegetative development and therefore seasonal flowering (Hsu *et al.*, 2006).

3.3. Polycarpy

A plant is considered to be polycarpic if it is able to reproduce more than once before dying. Perennial plants follow such a polycarpic growth habit by retaining at least one meristem in the vegetative state after flowering (Thomas *et al.*, 2000). In most species, vegetative development is maintained by axillary meristems and their developmental fate can influence the plant's life strategy (Foster *et al.*, 2003; Onishi *et al.*, 2003; Wang *et al.*, 2009b; Yuceer *et al.*, 2003). Comparisons between annual and perennial *Lolium* species, for example, demonstrated that axillary meristems of annual *Lolium* species die after flowering whereas the perennial species maintain their potential to actively grow the following year (Onishi *et al.*, 2003).

Studies in perennial species investigated the involvement of floral repressors and the reproductive competence of different branches (Jensen *et al.*, 2001; Mimida *et al.*, 2009). As mentioned before *TFL1* homologs in apple act as floral repressors. Transgenic plants that have reduced *TFL1* expression exhibit reduced juvenile period, and also show disturbed polycarpic behavior in which more branches become reproductive than in wild-type plants. *MdTFL1* and *MdTFL1a* are expressed during vegetative development in branches that are destined to stay vegetative as well as in branches that are committed to flowering. Therefore, these *TFL1* homologs in apple cannot explain differences in reproductive competence between shoots (Mimida *et al.*, 2009). In *A. alpina* temporal changes in expression of *PEP1* enhance polycarpic growth habit (Wang *et al.*, 2009b). Vegetative development after flowering is maintained from axillary branches on the axis of the leaves very close to the SAM. In *pep1* mutant plants, more branches become reproductive compared to the wild-type plants. However, as *pep1* plants do not die after flowering, additional factors must be involved in maintaining polycarpic behavior.

The developmental status of a meristem depends both on endogenous factors and environmental signals. A meristem can only commit to reproductive development under favorable environmental conditions when it has reached a certain developmental state. In citrus, overexpression of the *A. thaliana* meristem identity genes *LFY* and *AP1* that act downstream of the flowering pathway causes premature flowering, but transgenic plants still maintain polycarpic development and show seasonal flowering (Pena *et al.*, 2001). This suggests that meristems in a perennial plant might have different levels of reproductive competence (Martin-Trillo and Martinez-Zapater , 2002).

The position of an axillary meristem on the plant, its age/size, and time of outgrowth might also determine its developmental fate. Vegetative development in apple and *A. alpina* is maintained by axillary meristems close to the SAM (Fig. 11.2; Foster *et al.*, 2003; Wang *et al.*, 2009b). Additionally, in kiwifruit the size of the meristem prior to bud break in the spring can determine the fate of the second-order axillary meristems (Walton *et al.*, 1997).

As mentioned before, GAs in many perennials repress flowering and promote vegetative development suggesting that they might be involved in the regulation of polycarpic growth habit. Indeed, grapevine plants that carry mutations in the homolog of the Arabidopsis gene *GA INSENSITIVE* (*GAI*) show reduced polycarpic behavior with transformation of tendrils into inflorescences (Boss and Thomas, 2002). However, *vvgai1* mutant plants are still polycarpic suggesting that the regulation of flowering through *VvGAI1* acts downstream of meristem competence or additional factors might be involved in the regulation of polycarpic growth habit.

3.4. Biennial bearing

Biennial bearing is a trait in which trees alternate in producing many flowering branches 1 year and fewer or no flowering branches the following year (Jonkers, 1979; Lavee *et al.*, 1983; Scholefield *et al.*, 1985). In cases

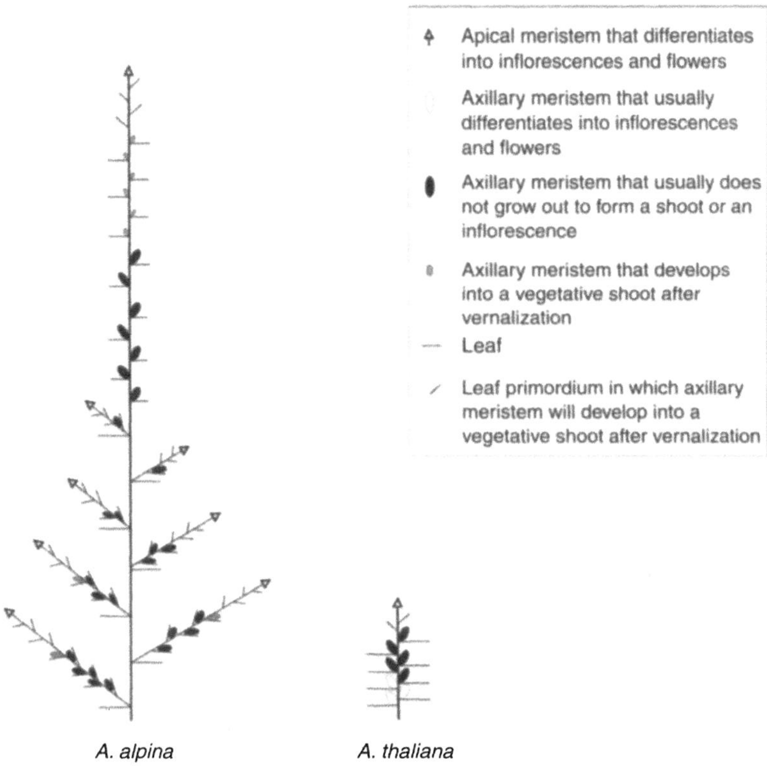

Figure 11.2 Schematic representation of the predicted behavior of meristems in *A. alpina* compared to *A. thaliana*. In the herbaceous perennial *A. alpina* not all meristems respond to flower inductive signals. An *A. alpina* plant grown for 8 weeks in long days comprises of the main shoot as well as axillary branches in the axils of the lower leaves and meristems in the axils of upper (younger) leaves. When such a plant is vernalized only the shoot apical meristem of the main axis and of the developed axillary branches will flower. Meristems in the axils of the leaves do not respond to flower inductive signals but behave in a predictable way after vernalization depending on their position on the plant. New vegetative axillary branches will grow in the axils of leaves located close to the shoot apical meristem. Meristems located in the axils of lower (older) leaves abort. In *A. thaliana* the shoot apical meristem and all meristems that form axillary shoots flower.

where flowering spans 2 years, the time of flower bud formation usually overlaps with the time of embryo development in the fruits produced from the buds initiated the previous year. Thus the presence of developing fruits on a tree appears to inhibit the potential of a bud to differentiate into a flower bud. Although the molecular mechanisms regulating this trait are not understood, GAs produced in the developing seeds of apples have been proposed to inhibit formation of flower buds in neighboring meristems (Marino and Greene, 1981). However, measurements of endogenous GAs between biennial- and non-biennial-bearing cultivars did not detect differences in GA levels between these two types (Tu, 2000).

4. The Evolution of Plant Life Strategies

The annual and perennial life strategies were adopted by many groups of plants. Phylogenetic studies support the idea that annuals derived from perennial ancestors and that the shift between these two strategies is unidirectional (Andreasen and Baldwin, 2001; Church, 2003; Datson *et al.*, 2008). However, in some genera it is believed that a switch back to perennial behavior might have also occurred (Bena *et al.*, 1998; Tank and Olmstead, 2008). Annual species are usually randomly distributed within the phylogeny, indicating independent shifts in life strategy during evolution (Andreasen and Baldwin, 2001; Church, 2003; Datson *et al.*, 2008). Annuals make up a large fraction of desert floras and in general predominate in areas where annual seed banks can survive more easily than adult perennial plants. Therefore, shifts to annual life strategy might have occurred as an adaptive response to environmental conditions such as heat and drought (Barbier *et al.*, 1991; Ciereszko *et al.*, 2001; Datson *et al.*, 2008; Evans *et al.*, 2005).

Closely related species within a genus can be annuals or perennials, but variation in life strategies can also occur between different populations within a species. Therefore, crosses between annual and perennial relatives can be a useful tool in understanding the genetic basis of annual or perennial life strategies. In rice and maize, crosses between annual and perennial species associate perennial growth habit with the formation of rhizomes and the presence of more axillary tillers (Hu *et al.*, 2003; Lammer *et al.*, 2004; Westerbergh and Doebley, 2004). Additionally, crosses between *Oryza nivara*, which is an annual, with *Oryza rufipogon*, which is a perennial, revealed that the genetic basis of the life strategy is complex (Grillo *et al.*, 2009).

The return to vegetative development after flowering is a major determinant for the number of reproductive events that a plant will undergo during its lifetime and for the life strategy that it will follow. Recently, comparative studies between *A. alpina* and *A. thaliana* demonstrated that differences in the regulation of a flowering time gene, *FLC/PEP1*, are associated with their life strategies (Finnegan and Dennis, 2007; Wang *et al.*, 2009b). In this case, differences in epigenetic regulation of these orthologs contribute to phenotypic variation observed between these species. At present, it is still unknown whether the epigenetic repression of *FLC* has been gained in *A. thaliana* or lost in *A. alpina*. However, the comparative analysis between *A. alpina* and *A. thaliana* demonstrated that perennial relatives of *A. thaliana* can be a useful tool to increase our understanding of annual and perennial flowering.

5. Concluding Remarks

The complexity and diversity of the mechanisms underlying perennial flowering as well as the long-generation times of perennial plants have hampered progress in understanding this trait. However, recent developments promise much more rapid progress. Focus on a small number of model species that are genetically tractable and susceptible to transformation will allow identification of crucial genes and the testing of hypotheses (Bohlenius *et al.*, 2006; Grillo *et al.*, 2009; Hsu *et al.*, 2006; Wang *et al.*, 2009b). Furthermore, the availability of the genome sequences of perennial species will facilitate gene isolation and analysis of regulatory sequences (Tuskan *et al.*, 2006). In addition, the increasing depth of our understanding of the mechanisms of flowering in *A. thaliana* will provide hypotheses for testing in perennial species (Fig. 11.1; Melzer *et al.*, 2008). This combination of approaches seems well poised to provide a molecular understanding of perennial flowering in the coming years.

REFERENCES

Abbott, D. L. (1970). The role of budscales in the morphogenesis and dormancy of the apple fruit bud. In: Luckwill, L. C., and Cuttings, C. V. (Eds.), *Physiology of Tree Crops*. Academic Press, New York, pp. 64–82.

Abe, M., Kobayashi, Y., Yamamoto, S., Daimon, Y., Yamaguchi, A., Ikeda, Y., Ichinoki, H., Notaguchi, M., Goto, K., and Araki, T. (2005). FD, a bZIP protein mediating signals from the floral pathway integrator FT at the shoot apex. *Science* **309**, 1052–1056.

Achard, P., Herr, A., Baulcombe, D. C., and Harberd, N. P. (2004). Modulation of floral development by a gibberellin-regulated microRNA. *Development* **131**, 3357–3365.

Ahn, J. H., Miller, D., Winter, V. J., Banfield, M. J., Lee, J. H., Yoo, S. Y., Henz, S. R., Brady, R. L., and Weigel, D. (2006). A divergent external loop confers antagonistic activity on floral regulators FT and TFL1. *EMBO J.* **25**, 605–614.

Albani, M. C., Battey, N. H., and Wilkinson, M. J. (2004). The development of ISSR-derived SCAR markers around the SEASONAL FLOWERING LOCUS (SFL) in *Fragaria vesca*. *Theor. Appl. Genet.* **109**, 571–579.

Alvarez, J., Guli, C. L., Yu, X. H., and Smyth, D. R. (1992). Terminal-flower—a gene affecting inflorescence development in *Arabidopsis-thaliana*. *Plant J.* **2**, 103–116.

Andreasen, K., and Baldwin, B. G. (2001). Unequal evolutionary rates between annual and perennial lineages of checker mallows (Sidalcea, Malvaceae): Evidence from 18S–26S rDNA internal and external transcribed spacers. *Mol. Biol. Evol.* **18**, 936–944.

Barbier, P., Morishima, H., and Ishihama, A. (1991). Phylogenetic-relationships of annual and perennial wild-rice—probing by direct DNA sequencing. *Theor. Appl. Genet.* **81**, 693–702.

Bastow, R., Mylne, J. S., Lister, C., Lippman, Z., Martienssen, R. A., and Dean, C. (2004). Vernalization requires epigenetic silencing of FLC by histone methylation. *Nature* **427**, 164–167.

Battey, N. H. (2000). Aspects of seasonality. *J. Exp. Bot.* **51**, 1769–1780.

Battey, N. H., Le Miere, P., Tehranifar, A., Cekic, C., Taylor, S., Shrives, K. J., Hadley, P., Greenland, A. J., Darby, J., and Wilkinson, M. J. (1998). *Genetic and Environmental Control of Flowering in Strawberry*. CAB International, Wallingford, UK.

Battey, N. H., and Tooke, F. (2002). Molecular control and variation in the floral transition. *Curr. Opin. Plant Biol.* **5**, 62–68.

Baurle, I., and Dean, C. (2006). The timing of developmental transitions in plants. *Cell* **125**, 655–664.

Bena, G., Lejeune, B., Prosperi, J. M., and Olivieri, I. (1998). Molecular phylogenetic approach for studying life-history evolution: the ambiguous example of the genus Medicago L. *Proc. R. Soc. Lond., B, Biol. Sci.* **265**, 1141–1151.

Blazquez, M. A., and Weigel, D. (2000). Integration of floral inductive signals in Arabidopsis. *Nature* **404**, 889–892.

Bohlenius, H., Huang, T., Charbonnel-Campaa, L., Brunner, A. M., Jansson, S., Strauss, S. H., and Nilsson, O. (2006). CO/FT regulatory module controls timing of flowering and seasonal growth cessation in trees. *Science* **312**, 1040–1043.

Borner, R., Kampmann, G., Chandler, J., Gleissner, R., Wisman, E., Apel, K., and Melzer, S. (2000). A MADS domain gene involved in the transition to flowering in Arabidopsis. *Plant J.* **24**, 591–599.

Boss, P. K., and Thomas, M. R. (2002). Association of dwarfism and floral induction with a grape "green revolution" mutation. *Nature* **416**, 847–850.

Bowman, J. L., Alvarez, J., Weigel, D., Meyerowitz, E. M., and Smyth, D. R. (1993). Control of flower development in *Arabidopsis-thaliana* by APETALA1 and interacting genes. *Development* **119**, 721–743.

Bradley, D., Ratcliffe, O., Vincent, C., Carpenter, R., and Coen, E. (1997). Inflorescence commitment and architecture in Arabidopsis. *Science* **275**, 80–83.

Brown, T., and Wareing, P. F. (1965). Genetical control of everbearing habit and 3 other characters in varieties of Fragaria vesca. *Euphytica* **14**, 97–112.

Calonje, M., Cubas, P., Martinez-Zapater, J. M., and Carmona, M. J. (2004). Floral meristem identity genes are expressed during tendril development in grapevine. *Plant Physiol.* **135**, 1491–1501.

Carmona, M. J., Cubas, P., and Martinez-Zapater, J. M. (2002). VFL, the grapevine FLORICAULA/LEAFY ortholog, is expressed in meristematic regions independently of their fate. *Plant Physiol.* **130**, 68–77.

Church, S. A. (2003). Molecular phylogenetics of Houstonia (Rubiaceae): descending aneuploidy and breeding system evolution in the radiation of the lineage across North America. *Mol. Phylogenet. Evol.* **27**, 223–238.

Ciereszko, I., Johansson, H., Hurry, V., and Kleczkowski, L. A. (2001). Phosphate status affects the gene expression, protein content and enzymatic activity of UDP-glucose pyrophosphorylase in wild-type and pho mutants of Arabidopsis. *Planta (Berlin)* **212**, 598–605.

Conti, L., and Bradley, D. (2007). TERMINAL FLOWER1 is a mobile signal controlling Arabidopsis architecture. *Plant Cell* **19**, 767–778.

Corbesier, L., Vincent, C., Jang, S. H., Fornara, F., Fan, Q. Z., Searle, I., Giakountis, A., Farrona, S., Gissot, L., Turnbull, C., and Coupland, G. (2007). FT protein movement contributes to long-distance signaling in floral induction of Arabidopsis. *Science* **316**, 1030–1033.

Crespel, L., Chirollet, M., Durel, C. E., Zhang, D., Meynet, J., and Gudin, S. (2002). Mapping of qualitative and quantitative phenotypic traits in Rosa using AFLP markers. *Theor. Appl. Genet.* **105**, 1207–1214.

Datson, P. M., Murray, B. G., and Steiner, K. E. (2008). Climate and the evolution of annual/perennial life-histories in Nemesia (Scrophulariaceae). *Plant Syst. Evol.* **270**, 39–57.

Diomaiuto, J. (1988). Periodic flowering or continual flowering as a function of temperature in a perennial species: the Ravenelle wallflower (Cheiranthus Cheiri L.). *Phytomorphology* **38**, 163–171.

Endo, T., Shimada, T., Fujii, H., Kobayashi, Y., Araki, T., and Omura, M. (2005). Ectopic expression of an FT homolog from Citrus confers an early flowering phenotype on trifoliate orange (*Poncirus trifoliata* L. Raf.). *Transgenic Res.* **14**, 703–712.

Eriksson, S., Bohlenius, H., Moritz, T., and Nilsson, O. (2006). GA(4) is the active gibberellin in the regulation of LEAFY transcription and Arabidopsis floral initiation. *Plant Cell* **18**, 2172–2181.

Evans, M. E. K., Hearn, D. J., Hahn, W. J., Spangle, J. M., and Venable, D. L. (2005). Climate and life-history evolution in evening primroses (Oenothera, Onagraceae): A phylogenetic comparative analysis. *Evolution* **59**, 1914–1927.

Farrona, S., Coupland, G., and Turck, F. (2008). The impact of chromatin regulation on the floral transition. *Semin. Cell Dev. Biol.* **19**, 560–573.

Ferrandiz, C., Gu, Q., Martienssen, R., and Yanofsky, M. F. (2000). Redundant regulation of meristem identity and plant architecture by FRUITFULL, APETALA1 and CAULIFLOWER. *Development* **127**, 725–734.

Finnegan, E. J., and Dennis, E. S. (2007). Vernalization-induced trimethylation of histone H3 lysine 27 at FLC is not maintained in mitotically quiescent cells. *Curr. Biol.* **17**, 1978–1983.

Flachowsky, H., Peil, A., Sopanen, T., Elo, A., and Hanke, V. (2007). Overexpression of BpMADS4 from silver birch (Betula pendula Roth.) induces early-flowering in apple (Malus × domestica Borkh). *Plant Breed.* **126**, 137–145.

Fornara, F., and Coupland, G. (2009). Plant phase transitions make a SPLash. *Cell* **138**, 625–627.

Foster, T., Johnston, R., and Seleznyova, A. (2003). A morphological and quantitative characterization of early floral development in apple (Malus × domestica Borkh. *Ann. Bot.* **92**, 199–206.

Fujiwara, S., Oda, A., Yoshida, R., Niinuma, K., Miyata, K., Tomozoe, Y., Tajima, T., Nakagawa, M., Hayashi, K., Coupland, G., and Mizoguchi, T. (2008). Circadian clock proteins LHY and CCA1 regulate SVP protein accumulation to control flowering in Arabidopsis. *Plant Cell* **20**, 2960–2971.

Gocal, G.F.W., Sheldon, C. C., Gubler, F., Moritz, T., Bagnall, D. J., MacMillan, C. P., Li, S. F., Parish, R. W., Dennis, E. S., Weigel, D., and King, R. W. (2001). GAMYB-like genes, flowering, and gibberellin signaling in Arabidopsis. *Plant Physiol.* **127**, 1682–1693.

Grillo, M. A., Li, C. B., Fowlkes, A. M., Briggeman, T. M., Zhou, A. L., Schemske, D. W., and Sang, T. (2009). Genetic architecture for the adaptive origin of annual wild rice, *Oryza nivara*. *Evolution* **63**, 870–883.

Gu, Q., Ferrandiz, C., Yanofsky, M. F., and Martienssen, R. (1998). The FRUITFULL MADS-box gene mediates cell differentiation during Arabidopsis fruit development. *Development* **125**, 1509–1517.

Gustafson-Brown, C., Savidge, B., and Yanofsky, M. F. (1994). Regulation of the Arabidopsis floral homeotic gene APETALA1. *Cell* **76**, 131–143.

Hames, C., Ptchelkine, D., Grimm, C., Thevenon, E., Moyroud, E., Gerard, F., Martiel, J. L., Benlloch, R., Parcy, F., and Muller, C. W. (2008). Structural basis for LEAFY floral switch function and similarity with helix-turn-helix proteins. *EMBO J.* **27**, 2628–2637.

Hanccock, J. F. (1999). *Strawberries*. CABI Publishing, New York.

Hartmann, U., Hoehmann, S., Nettesheim, K., Wisman, E., Saedler, H., and Huijser, P. (2000). Molecular cloning of SVP: a negative regulator of the floral transition in Arabidopsis. *Plant J.* **21**, 351–360.

Hempel, F. D., Weigel, D., Mandel, M. A., Ditta, G., Zambryski, P. C., Feldman, L. J., and Yanofsky, M. F. (1997). Floral determination and expression of floral regulatory genes in Arabidopsis. *Development* **124**, 3845–3853.

Hepworth, S. R., Valverde, F., Ravenscroft, D., Mouradov, A., and Coupland, G. (2002). Antagonistic regulation of flowering-time gene SOC1 by CONSTANS and FLC via separate promoter motifs. *EMBO J.* **21**, 4327–4337.

Hsu, C. Y., Liu, Y. X., Luthe, D. S., and Yuceer, C. (2006). Poplar FT2 shortens the juvenile phase and promotes seasonal flowering. *Plant Cell* **18**, 1846–1861.

Hu, F. Y., Tao, D. Y., Sacks, E., Fu, B. Y., Xu, P., Li, J., Yang, Y., McNally, K., Khush, G. S., Paterson, A. H., and Li, Z. K. (2003). Convergent evolution of perenniality in rice and sorghum. *Proc. Natl. Acad. Sci. U.S.A.* **100**, 4050–4054.

Jaeger, K. E., and Wigge, P. A. (2007). FT protein acts as a long-range signal in Arabidopsis. *Curr. Biol.* **17**, 1050–1054.

Jang, S., Torti, S., and Coupland, G. (2009). Genetic and spatial interactions between FT, TSF and SVP during the early stages of floral induction in Arabidopsis. *Plant J.* **60**, 614–625.

Jensen, C. S., Salchert, K., and Nielsen, K. K. (2001). A TERMINAL FLOMER1-Like gene from perennial ryegrass involved in floral transition and axillary meristem identity. *Plant Physiol.* **125**, 1517–1528.

Jonkers, H. (1979). Biennial bearing in apple and pear—literature survey. *Sci. Hortic.* **11**, 303–317.

Kanrar, S., Bhattacharya, M., Arthur, B., Courtier, J., and Smith, H.M.S. (2008). Regulatory networks that function to specify flower meristems require the function of homeobox genes PENNYWISE and POUND-FOOLISH in Arabidopsis. *Plant J.* **54**, 924–937.

Kardailsky, I., Shukla, V. K., Ahn, J. H., Dagenais, N., Christensen, S. K., Nguyen, J. T., Chory, J., Harrison, M. J., and Weigel, D. (1999). Activation tagging of the floral inducer FT. *Science* **286**, 1962–1965.

Kim, D. H., Doyle, M. R., Sung, S., and Amasino, R. M. (2009). Vernalization: winter and the timing of flowering in plants. *Annu. Rev. Cell Dev. Biol.* **25**, 277–299.

Kobayashi, Y., Kaya, H., Goto, K., Iwabuchi, M., and Araki, T. (1999). A pair of related genes with antagonistic roles in mediating flowering signals. *Science* **286**, 1960–1962.

Kobayashi, Y., and Weigel, D. (2007). Move on up, it's time for change—mobile signals controlling photoperiod-dependent flowering. *Genes Dev.* **21**, 2371–2384.

Kotoda, N., Iwanami, H., Takahashi, S., and Abe, K. (2006). Antisense expression of MdTFL1, a TFL1-like gene, reduces the juvenile phase in apple. *J. Am. Soc. Hortic. Sci.* **131**, 74–81.

Kotoda, N., and Wada, M. (2005). MdTFL1, a TFL1-like gene of apple, retards the transition from the vegetative to reproductive phase in transgenic Arabidopsis. *Plant Sci.* **168**, 95–104.

Kotoda, N., Wada, M., Masuda, T., and Soejima, J. (2003). The break-through in the reduction of juvenile phase in apple using transgenic approaches. *Acta Horticultural* **625**, 337–343.

Lammer, D., Cai, X. W., Arterburn, M., Chatelain, J., Murray, T., and Jones, S. (2004). A single chromosome addition from *Thinopyrum elongatum* confers a polycarpic, perennial habit to annual wheat. *J. Exp. Bot.* **55**, 1715–1720.

Lavee, S., Haskal, A., and Bental, Y. (1983). Girdling olive trees, a partial solution to biennial bearing .1. methods, timing and direct tree response. *J. Hortic. Sci.* **58**, 209–218.

Lee, J., Oh, M., Park, H., and Lee, I. (2008). SOC1 translocated to the nucleus by interaction with AGL24 directly regulates LEAFY. *Plant J.* **55**, 832–843.

Lee, H., Suh, S. S., Park, E., Cho, E., Ahn, J. H., Kim, S. G., Lee, J. S., Kwon, Y. M., and Lee, I. (2000). The AGAMOUS-LIKE 20 MADS domain protein integrates floral inductive pathways in Arabidopsis. *Genes Dev.* **14**, 2366–2376.

Lenahan, O. M., Whiting, M. D., and Elfving, D. C. (2006). Gibberellic acid inhibits floral bud induction and improves "Bing" sweet cherry fruit quality. *Hortscience* **41**, 654–659.

Li, D., Liu, C., Shen, L., Wu, Y., Chen, H., Robertson, M., Helliwell, C. A., Ito, T., Meyerowitz, E., and Yu, H. (2008). A repressor complex governs the integration of flowering signals in Arabidopsis. *Dev. Cell* **15**, 110–120.

Liljegren, S. J., Gustafson-Brown, C., Pinyopich, A., Ditta, G. S., and Yanofsky, M. F. (1999). Interactions among APETALA1, LEAFY, and TERMINAL FLOWER1 specify meristem fate. *Plant Cell* **11**, 1007–1018.

Lin, M. K., Belanger, H., Lee, Y. J., Varkonyi-Gasic, E., Taoka, K. I., Miura, E., Xoconostle-Cazares, B., Gendler, K., Jorgensene, R. A., Phinney, B., Lough, T. J., and Lucas, W. J. (2007). FLOWERING LOCUS T protein may act as the long-distance florigenic signal in the cucurbits. *Plant Cell* **19**, 1488–1506.

Liu, C., Chen, H., Er, H. L., Soo, H. M., Kumar, P. P., Han, J. H., Liou, Y. C., and Yu, H. (2008). Direct interaction of AGL24 and SOC1 integrates flowering signals in Arabidopsis. *Development* **135**, 1481–1491.

Liu, C., Xi, W. Y., Shen, L. S., Tan, C. P., and Yu, H. (2009). Regulation of floral patterning by flowering time genes. *Dev. Cell* **16**, 711–722.

Lord, E. M., and Eckard, K. J. (1987). Shoot development in *Citrus-sinensis* L. (Washington Navel Orange) .2. Alteration of developmental fate of flowering shoots after GA3 treatment. *Bot. Gaz.* **148**, 17–22.

Mandel, M. A., and Yanofsky, M. F. (1995). A gene triggering flower formation in Arabidopsis. *Nature* **377**, 522–524.

Marino, F., and Greene, D. W. (1981). Involvement of gibberellins in the biennial bearing of "Early Mcintosh" apples. *J. Am. Soc. Hortic. Sci.* **106**, 593–596.

Martin-Trillo, M., and Martinez-Zapater, J. M. (2002). Growing up fast: manipulating the generation time of trees. *Curr. Opin. Biotechnol.* **13**, 151–155.

Mathieu, J., Warthmann, N., Kuttner, F., and Schmid, M. (2007). Export of FT protein from phloem companion cells is sufficient for floral induction in Arabidopsis. *Curr. Biol.* **17**, 1055–1060.

Melzer, S., Lens, F., Gennen, J., Vanneste, S., Rohde, A., and Beeckman, T. (2008). Flowering-time genes modulate meristem determinacy and growth form in *Arabidopsis thaliana*. *Nat Genet.* **40**, 1489–1492.

Michaels, S. D., and Amasino, R. M. (1999). FLOWERING LOCUS C encodes a novel MADS domain protein that acts as a repressor of flowering. *Plant Cell* **11**, 949–956.

Michaels, S. D., Himelblau, E., Kim, S. Y., Schomburg, F. M., and Amasino, R. M. (2005). Integration of flowering signals in winter-annual Arabidopsis. *Plant Physiol.* **137**, 149–156.

Mimida, N., Kotoda, N., Ueda, T., Igarashi, M., Hatsuyama, Y., Iwanami, H., Moriya, S., and Abe, K. (2009). Four TFL1CEN-like genes on distinct linkage groups show different expression patterns to regulate vegetative and reproductive development in apple (Malusdomestica Borkh). *Plant Cell Physiol.* **50**, 394–412.

Moon, J., Suh, S. S., Lee, H., Choi, K. R., Hong, C. B., Paek, N. C., Kim, S. G., and Lee, I. (2003). The SOC1 MADS-box gene integrates vernalization and gibberellin signals for flowering in Arabidopsis. *Plant J.* **35**, 613–623.

Mouradov, A., Cremer, F., and Coupland, G. (2002). Control of flowering time: interacting pathways as a basis for diversity. *Plant Cell* **14**, S111–S130.

Mullins, M. G., Bouquet, A., and Williams, L. E. (1992). *Biology of the Grapevine*. Cambridge University Press, Cambridge, UK.

Mutasa-Gottgens, E., and Hedden, P. (2009). Gibberellin as a factor in floral regulatory networks. *J. Exp. Bot.* **60**, 1979–1989.

Notaguchi, M., Abe, M., Kimura, T., Daimon, Y., Kobayashi, T., Yamaguchi, A., Tomita, Y., Dohi, K., Mori, M., and Araki, T. (2008). Long-distance, graft-transmissible action of Arabidopsis FLOWERING LOCUS T protein to promote flowering. *Plant Cell Physiol.* **49**, 1922–1922.

Onishi, K., Sano, Y., and Nakashima, H. (2003). Developmental fates of axillary for the pattern of life buds as a major determinant history in Lolium. *Plant Prod. Sci.* **6**, 179–184.

Pena, L., Martin-Trillo, M., Juarez, J., Pina, J. A., Navarro, L., and Martinez-Zapater, J. M. (2001). Constitutive expression of Arabidopsis LEAFY or APETALA1 genes in citrus reduces their generation time. *Nat. Biotechnol.* **19**, 263–267.

Pillitteri, L. J., Lovatt, C. J., and Walling, L. L. (2004). Isolation and characterization of a TERMINAL FLOWER homolog and its correlation with juvenility in citrus. *Plant Physiol.* **135**, 1540–1551.

Pnueli, L., Gutfinger, T., Hareven, D., Ben-Naim, O., Ron, N., Adir, N., and Lifschitz, E. (2001). Tomato SP-interacting proteins define a conserved signaling system that regulates shoot architecture and flowering. *Plant Cell* **13**, 2687–2702.

Poethig, R. S. (2003). Phase change and the regulation of developmental timing in plants. *Science* **301**, 334–336.

Ratcliffe, O. J., Amaya, I., Vincent, C. A., Rothstein, S., Carpenter, R., Coen, E. S., and Bradley, D. J. (1998). A common mechanism controls the life cycle and architecture of plants. *Development* **125**, 1609–1615.

Reeves, P. H., and Coupland, G. (2001). Analysis of flowering time control in Arabidopsis by comparison of double and triple mutants. *Plant Physiol.* **126**, 1085–1091.

Remay, A., Lalanne, D., Thouroude, T., Le Couviour, F., Hibrand-Saint Oyant, L., and Foucher, F. (2009). A survey of flowering genes reveals the role of gibberellins in floral control in rose. *Theor. Appl. Genet.* **119**, 767–781.

Roberts, A. V., Blake, P. S., Lewis, R., Taylor, J. M., and Dunstan, D. I. (1999). The effect of gibberellins on flowering in roses. *J. Plant Growth Regul.* **18**, 113–119.

Saddic, L. A., Huvermann, B. R., Bezhani, S., Su, Y. H., Winter, C. M., Kwon, C. S., Collum, R. P., and Wagner, D. (2006). The LEAFY target LMI1 is a meristem identity regulator and acts together with LEAFY to regulate expression of CAULIFLOWER. *Development* **133**, 1673–1682.

Salazar-Garcia, S., and Lovatt, C. J. (1998). GA(3) application alters flowering phenology of "Hass" avocado. *J. Am. Soc. Hortic. Sci.* **123**, 791–797.

Samach, A., Onouchi, H., Gold, S. E., Ditta, G. S., Schwarz-Sommer, Z., Yanofsky, M. F., and Coupland, G. (2000). Distinct roles of CONSTANS target genes in reproductive development of Arabidopsis. *Science* **288**, 1613–1616.

Schmid, M., Uhlenhaut, N. H., Godard, F., Demar, M., Bressan, R., Weigel, D., and Lohmann, J. U. (2003). Dissection of floral induction pathways using global expression analysis. *Development* **130**, 6001–6012.

Scholefield, P. B., Sedgley, M., and Alexander, D. M. (1985). Carbohydrate cycling in relation to shoot growth, floral initiation and development and yield in the Avocado. *Sci. Hortic.* **25**, 99–110.

Schwab, R., Palatnik, J. F., Riester, M., Schommer, C., Schmid, M., and Weigel, D. (2005). Specific effects of MicroRNAs on the plant transcriptome. *Dev. Cell* **8**, 517–527.

Searle, I., He, Y. H., Turck, F., Vincent, C., Fornara, F., Krober, S., Amasino, R. A., and Coupland, G. (2006). The transcription factor FLC confers a flowering response to vernalization by repressing meristem competence and systemic signaling in Arabidopsis. *Genes Dev.* **20**, 898–912.

Semeniuk, P. (1971). Inheritance of recurrent and nonrecurrent blooming in Goldilocks by Rosa-Wichuraiana Progeny. *J. Hered.* **62**, 319.

Shannon, S., and Meeks-Wagner, D. R. (1991). A mutation in the Arabidopsis Tfl1 gene affects inflorescence meristem development. *Plant Cell* **3**, 877–892.

Sheldon, C. C., Burn, J. E., Perez, P. P., Metzger, J., Edwards, J. A., Peacock, W. J., and Dennis, E. S. (1999). The FLF MADS box gene: A repressor of flowering in Arabidopsis regulated by vernalization and methylation. *Plant Cell* **11**, 445–458.

Sheldon, C. C., Finnegan, E. J., Dennis, E. S., and Peacock, W. J. (2006). Quantitative effects of vernalization on FLC and SOC1 expression. *Plant J.* **45**, 871–883.

Simpson, G. G., Gendall, A. R., and Dean, C. (1999). When to switch to flowering. *Annu. Rev. Cell Dev. Biol.* **15**, 519.

Smith, H. M. S., Campbell, B. C., and Hake, S. (2004). Competence to respond to floral inductive signals requires the homeobox genes PENNYWISE and POUND-FOOLISH. *Curr. Biol.* **14**, 812–817.

Sung, S. B., and Amasino, R. M. (2004). Vernalization in *Arabidopsis thaliana* is mediated by the PHD finger protein VIN3. *Nature* **427**, 159–164.

Sung, S. B., He, Y. H., Eshoo, T. W., Tamada, Y., Johnson, L., Nakahigashi, K., Goto, K., Jacobsen, S. E., and Amasino, R. M. (2006). Epigenetic maintenance of the vernalized state in *Arabidopsis thaliana* requires LIKE HETEROCHROMATIN PROTEIN 1. *Nat. Genet.* **38**, 706–710.

Swiezewski, S., Liu, F. Q., Magusin, A., and Dean, C. (2009). Cold-induced silencing by long antisense transcripts of an Arabidopsis Polycomb target. *Nature* **462**, 799–U122.

Tamaki, S., Matsuo, S., Wong, H. L., Yokoi, S., and Shimamoto, K. (2007). Hd3a protein is a mobile flowering signal in rice. *Science* **316**, 1033–1036.

Tank, D. C., and Olmstead, R. G. (2008). From annuals to perennials: phylogeny of subtribe Castillernae (Orobanchaceae). *Am. J. Bot.* **95**, 608–625.

Thomas, H., Thomas, H. M., and Ougham, H. (2000). Annuality, perenniality and cell death. *J. Exp. Bot.* **51**, 1781–1788.

Tooke, F., Ordidge, M., Chiurugwi, T., and Battey, N. (2005). Mechanisms and function of flower and inflorescence reversion. *J. Exp. Bot.* **56**, 2587–2599.

Tu, Y. (2000). Endogenous gibberellins in developing apple seeds in relation to alternate bearing, Vol. MSc. Purdue University.

Turck, F., Fornara, F., and Coupland, G. (2008). Regulation and identity of florigen: FLOWERING LOCUS T moves center stage. *Annu. Rev. Plant Biol.* **59**, 573–594.

Tuskan, G. A., DiFazio, S., Jansson, S., Bohlmann, J., Grigoriev, I., Hellsten, U., Putnam, N., Ralph, S., Rombauts, S., Salamov, A., Schein, J., Sterck, L., *et al.* (2006). The genome of black cottonwood, *Populus trichocarpa* (Torr. & Gray). *Science* **313**, 1596–1604.

Wagner, D., Sablowski, R. W. M., and Meyerowitz, E. M. (1999). Transcriptional activation of APETALA1 by LEAFY. *Science* **285**, 582–584.

Walton, E. F., Fowke, P. J., Weis, K., and McLeay, P. L. (1997). Shoot axillary bud morphogenesis in kiwifruit (*Actinidia deliciosa*). *Ann. Bot.* **80**, 13–21.

Walton, E. F., Podivinsky, E., and Wu, R. M. (2001). Bimodal patterns of floral gene expression over the two seasons that kiwifruit flowers develop. *Physiol. Plant.* **111**, 396–404.

Wang, J. W., Czech, B., and Weigel, D. (2009a). miR156-Regulated SPL transcription factors define an endogenous flowering pathway in *Arabidopsis thaliana*. *Cell* **138**, 738–749.

Wang, R. H., Farrona, S., Vincent, C., Joecker, A., Schoof, H., Turck, F., Alonso-Blanco, C., Coupland, G., and Albani, M. C. (2009b). PEP1 regulates perennial flowering in *Arabis alpina*. *Nature* **459**, 423–U138.

Weebadde, C. K., Wang, D., Finn, C. E., Lewers, K. S., Luby, J. J., Bushakra, J., Sjulin, T. M., and Hancock, J. F. (2008). Using a linkage mapping approach to identify QTL for day-neutrality in the octoploid strawberry. *Plant Breed.* **127**, 94–101.

Weigel, D., Alvarez, J., Smyth, D. R., Yanofsky, M. F., and Meyerowitz, E. M. (1992). Leafy controls floral Meristem identity in Arabidopsis. *Cell* **69**, 843–859.

Weigel, D., and Nilsson, O. (1995). A developmental switch sufficient for flower initiation in diverse plants. *Nature* **377**, 495–500.

Westerbergh, A., and Doebley, J. (2004). Quantitative trait loci controlling phenotypes related to the perennial versus annual habit in wild relatives of maize. *Theor. Appl. Genet.* **109**, 1544–1553.

Wigge, P. A., Kim, M. C., Jaeger, K. E., Busch, W., Schmid, M., Lohmann, J. U., and Weigel, D. (2005). Integration of spatial and temporal information during floral induction in Arabidopsis. *Science* **309**, 1056–1059.

Wilson, R. N., Heckman, J. W., and Somerville, C. R. (1992). Gibberellin is required for flowering in *Arabidopsis-thaliana* under short days. *Plant Physiol.* **100**, 403–408.

Woodger, F. J., Gubler, F., Pogson, B. J., and Jacobsen, J. V. (2003a). A Mak-like kinase is a repressor of GAMYB in barley aleurone. *Plant J.* **33**, 707–717.

Woodger, F. J., Millar, A., Murray, F., Jacobsen, J. V., and Gubler, F. (2003b). The role of GAMYB transcription factors in GA-regulated gene expression. *J. Plant Growth Regul.* **22**, 176–184.

Yamaguchi, A., Kobayashi, Y., Goto, K., Abe, M., and Araki, T. (2005). TWIN SISTER OF FT (TSF) acts as a floral pathway integrator redundantly with FT. *Plant Cell Physiol.* **46**, 1175–1189.

Yamaguchi, A., Wu, M. F., Yang, L., Wu, G., Poethig, R. S., and Wagner, D. (2009). The MicroRNA-regulated SBP-box transcription factor SPL3 is a direct upstream activator of LEAFY, FRUITFULL, and APETALA1. *Dev. Cell* **17**, 268–278.

Yu, H., Xu, Y. F., Tan, E. L., and Kumar, P. P. (2002). AGAMOUS-LIKE 24, a dosage-dependent mediator of the flowering signals. *Proc. Natl. Acad. Sci. U.S.A.* **99**, 16336–16341.

Yuceer, C., Land, S. B., Kubiske, M. E., and Harkess, R. L. (2003). Shoot morphogenesis associated with flowering in *Populus deltoides* (Salicaceae). *Am. J. Bot.* **90**, 196–206.

CHAPTER TWELVE

Sculpting the Flower; the Role of microRNAs in Flower Development

Anwesha Nag *and* Thomas Jack

Contents

Abstract

microRNAs (miRNAs) are small approximately 21-nucleotide RNAs that function posttranscriptionally to regulate gene activity. miRNAs function by binding to complementary sites in target genes causing mRNA degradation and/or translational repression of the target. Since the discovery of miRNAs in plants in 2002 much has been learned about the function of these small regulatory RNAs. miRNAs function broadly to control many aspects of plant biology and plant development. This review focuses on the role of miRNAs in flower development. miRNAs function throughout flower development, from the earliest stages (floral induction) to very late stages (floral organ cell type specification). miRNAs such as miR156 and miR172 play a key role in vegetative phase change and in the vegetative to reproductive transition in both Arabidopsis and maize. miR172 in Arabidopsis and maize and miR169 in Petunia and

Department of Biological Sciences, Dartmouth College, Hanover, New Hampshire, USA

Current Topics in Developmental Biology, Volume 91
ISSN 0070-2153, DOI 10.1016/S0070-2153(10)91012-0

Antirrhinum function to control floral organ identity fate during the early stages of flower development by regulating the spatial boundaries of expression of target genes. miR164, miR319, miR159, and miR167 function to specify particular cell types during later stages of flower development. Although much has been learned about the role of miRNAs in flower development in the last 8 years, many challenges remain to fully elucidate the function of these important regulatory molecules.

1. miRNAs in Plant Development

miRNAs are 21–22-nucleotide noncoding RNAs that regulate eukaryotic gene activity posttranscriptionally. miRNAs were discovered in the early 1990s as the product of a gene that controls developmental timing in *Caenorhabditis elegans* (Lee *et al.*, 1993). However, it was not until 2001 that miRNAs were shown to be broadly distributed in the animal kingdom (Lagos-Quintana *et al.*, 2001; Lau *et al.*, 2001; Lee and Ambros, 2001). Shortly thereafter, numerous miRNA families were discovered in the plant kingdom (Llave *et al.*, 2002; Park *et al.*, 2002; Reinhart *et al.*, 2002). Over the past 8 years plant miRNAs have been shown to be involved in regulating many aspects of plant biology, including metabolism, hormone response (Liu and Chen, 2009), and biotic (Voinnet, 2008) and abiotic stress (Sunkar *et al.*, 2007). Despite the wide range of processes affected by plant miRNAs, the majority of plant miRNAs target genes that function in cell differentiation and development, including genes that function in flower development. This review will focus on what has been learned to date about the role of miRNAs in flower development.

The broad role of plant miRNAs in development is evident from the phenotype of mutants defective in miRNA biogenesis components (Fernandez *et al.*, 1999; Golden *et al.*, 2002; Jacobsen *et al.*, 1999). Before describing the function of various miRNAs involved in flower development, we will first briefly review how plant miRNAs are produced and the molecular details of how they function. (Fig. 12.1) (Chen, 2008). miRNAs are most frequently transcribed by RNA polymerase II and are synthesized as long primary transcripts called pri-miRNAs. Pri-miRNAs are stabilized by the nuclear RNA binding protein DAWDLE (DDL) (Yu *et al.*, 2008) before being processed by the RNase III-like enzyme DICER-LIKE1 (DCL1) into a hairpin structure that contains the 21-nucleotide miRNA (Fig. 12.1) (Kurihara and Watanabe, 2004; Qi *et al.*, 2005). There are two enzymes that assist DCL1 in this process: the double-stranded RNA binding protein HYPONASTIC LEAVES1 (HYL1) (Han *et al.*, 2004) and the C2H2 zinc finger protein SERRATE (SE) (Grigg *et al.*, 2005; Kurihara *et al.*, 2006; Yang *et al.*, 2006a; Laubinger *et al.*, 2008). The pre-miRNA is

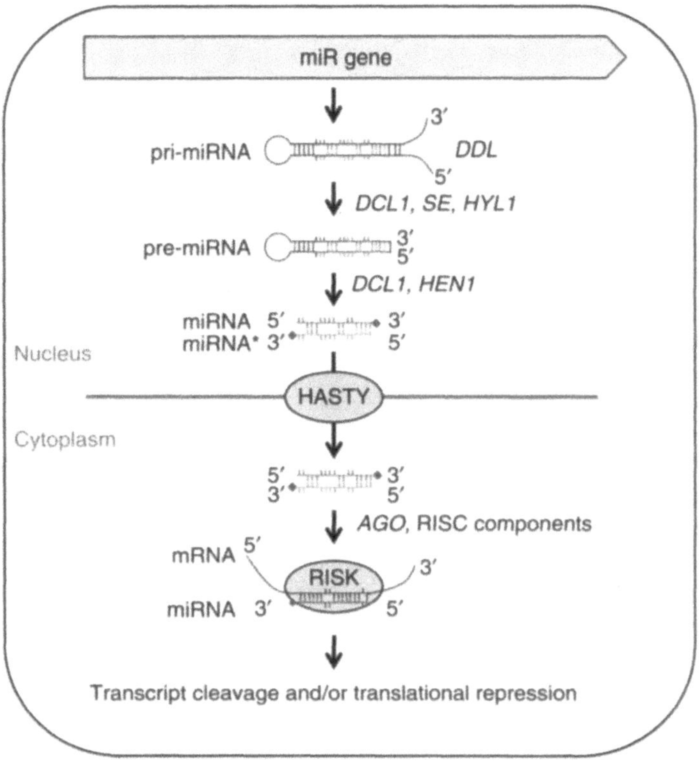

Figure 12.1 miRNA biogenesis pathway. A primary transcript (pri-miRNA) is produced from a miRNA gene by RNA polymerase II. The pri-mRNA adopts a stem-loop structure that is stabilized by DWADLE (DDL). The primary transcript gets processed by the RNAse III-like enzyme DICER-LIKE1 (DCL1) to produce the hairpin-like structure called the pre-mRNA. This processing is aided by the C2H2 zinc finger protein SERRATE (SE) and double-stranded RNA binding protein HYPONASTIC LEAVES 1 (HYL1). The pre-miRNA gets processed by DCL1 a second time to lead to the miRNA/miRNA* complex. miRNA/miRNA* gets methylated by HUA ENHANCER 1 (HEN1) at the 3′ ends. This complex then gets transported from the nucleus to the cytoplasm with the help of the exportin-5 homolog HASTY. In the cytoplasm, miRNA* gets degraded and the miRNA gets incorporated in an RNA induced silencing complex (RISC). The RISC complex then interacts with target mRNA leading to either transcript cleavage and/or translational repression.

processed a second time by DCL1 to give rise to a structure that is composed of the mature miRNA and the antisense strand called miRNA*(Dong *et al.*, 2008). The miRNA/miRNA* duplex gets methylated at the 2′OH of the 3′ terminal nucleotide by the S-adenosyl methionine-dependent HUA ENHANCER 1 (HEN1) protein (Fig. 12.1) (Park *et al.*, 2005; Yang *et al.*, 2006b; Yu *et al.*, 2005). HEN1 functions to stabilize the miRNA; if the miRNA is not methylated by HEN1, the miRNA is uridylated and subsequently degraded (Li *et al.*, 2005; Ramachandran and Chen, 2008). Following methylation, the miRNA/miRNA* duplex is assembled into an RNA induced silencing complex (RISC). One of the key protein components of RISC is an ARGONAUTE (AGO) protein that contains both an RNA binding PAZ domain and a PIWI domain similar to the endonuclease domain of RNase H. In Arabidopsis, the key AGO protein for miRNA function is AGO1, which possesses the ability to cleave target mRNAs by using catalytic residues in the PIWI domain (Vaucheret, 2008). miRNA transcription and processing takes place in the nucleus, but miRNAs ultimately function in the cytoplasm. The HASTY protein, the Arabidopsis homolog of exportin 5, plays a role in miRNA nuclear export of at least some miRNAs (Fig. 12.1) (Park *et al.*, 2005). Once incorporated into RISC, the miRNA binds to partially or completely complementary sequences referred to as miRNA binding sites that are present in target mRNAs (Fig. 12.1). In general, a single miRNA recognizes multiple targets, ranging from several to several dozen (Jones-Rhoades and Bartel, 2004; Jones-Rhoades *et al.*, 2006). Binding of the miRNA to the miRNA binding site leads most often either to target mRNA degradation or to translational repression of the target. The details of the mechanism of mRNA degradation are well understood; miRNAs direct AGO1 to catalyze a single endonucleolytic cleavage of the mRNA (i.e. "slicer" activity) approximately in the middle of the duplex miRNA–mRNA region. By contrast, in the cases where the miRNA leads to translational repression, the mechanistic details remain very unclear.

The crucial role that miRNAs play in plant development is evident from the embryonic- and seedling-lethal phenotypes of null alleles of *dcl1* and *ago1*, respectively (Bohmert *et al.*, 1998; Fernandez *et al.*, 1999; Golden *et al.*, 2002; Jacobsen *et al.*, 1999; Kidner and Martienssen, 2005; Schauer *et al.*, 2002; Vaucheret *et al.*, 2004). A partial loss-of-function allele of *dcl1* is more informative of organism-wide miRNA function as these *dcl1* mutants exhibit phenotypic defects in leaf morphology, axillary meristem maintenance, flowering time, inflorescence determinacy, floral organ patterning, and ovule development (Park *et al.*, 2002). Similarly, partial loss-of-function alleles of *ago1* are defective in lateral organ polarity, and morphology of leaves and flowers (Kidner and Martienssen, 2004; Morel *et al.*, 2002; Vaucheret *et al.*, 2004). *ago1* and *dcl1* mutants exhibit decreased levels of global miRNA accumulation with a resulting increase in the level of miRNA target expression. Loss-of-function alleles of other components of the miRNA machinery such as *hen1* (Chen *et al.*, 2002), *hyl1* (Lu and Fedoroff, 2000), and *hasty* (Telfer and Poethig, 1998; Park *et al.*, 2005) also exhibit pleiotropic mutant phenotypes that are similar to the phenotypes observed in *ago1* and *dcl1* mutants. In summary, the pleiotropic developmental phenotypes observed in mutants in the miRNA biogenesis machinery provide strong evidence that miRNAs play a critical role in multiple aspects of plant development.

As mentioned above, miRNAs most often act via two molecular mechanisms: mRNA degradation and translational repression. To date, only one plant miRNA has been demonstrated to act via methylation-induced gene silencing (Bao *et al.*, 2004) though at present it is unclear whether this is a general mechanism of miRNA function. All mechanisms involve, as a first step, the miRNA binding to the miRNA binding site of the target. Until recently, the dogma in the field was that the most widespread and important mechanism of function of plant miRNAs was via mRNA degradation and that translational repression was infrequent. However, the idea that miRNA-mediated translational repression is rare in plants has been challenged (Brodersen *et al.*, 2008; Dugas and Bartel, 2008; Gandikota *et al.*, 2007). It now seems likely that most plant miRNAs function both in mRNA degradation and translational repression. When considered in a developmental context these two modes of regulation offer different potential benefits to the organism. For example, miRNA-directed mRNA cleavage provides an irreversible way to remove accumulated transcripts. Such a mechanism can be useful, for example, for specifying how one cell type becomes stably distinguished from another. The second miRNA mechanism, translational repression, is reversible, and likely serves to fine-tune levels of gene activity. This type of regulation could be important in cases where the level of gene activity needs to be carefully controlled, i.e. where too much or too little gene activity results in negative consequences.

miRNAs have been demonstrated to function by regulating both the spatial pattern and the level of expression of target mRNAs. In some cases, the miRNA and target are expressed in mutually exclusive non-overlapping expression patterns. In this scenario, the miRNA functions to define the spatial boundary of expression of the target gene. miR172 and its cognate target *APETALA2* (*AP2*) are postulated to function in this way (Chen, 2004). In other cases, the miRNA and target exhibit overlapping expression patterns. In this model, the function of the miRNA is to regulate the level of transcript and/or protein accumulation rather than to reduce transcript/protein levels to low or undetectable levels. miRNAs function in this case to fine-tune levels of expression. miR164 regulation of *CUP SHAPED COTYLEDON1* (*CUC1*) and *CUC2* in the flower is postulated to function in this way (Sieber *et al.*, 2007). Overlapping expression of the miRNA and the target mRNA may suggest that the primary mode of regulation is via translational repression rather than mRNA degradation. Unfortunately, for most miRNAs, translational repression has been neither ruled out nor confirmed because protein levels have not been carefully analyzed for most miRNA targets (Chen, 2004; Chuck *et al.*, 2007b; Gandikota *et al.*, 2007).

Most plant miRNAs have been studied using a standard two-pronged approach. The first approach involves construction of transgenic lines that constitutively express the miRNA, resulting in downregulation of all target genes that contain a miRNA binding site. Most often, miRNA overexpression lines resemble loss-of-function alleles in the miRNA target genes. The second approach involves studying miRNA target genes that contain mutations in the miRNA binding site that dramatically reduce or eliminate miRNA binding without affecting the protein coding sequence; these miRNA-resistant mutants result in the target being insensitive to regulation by all members of a particular miRNA family. miRNA-resistant mutants are dominant to wild type, and the resulting phenotypes yield information about the developmental processes that are controlled by the miRNA. In an ideal case, miRNA-resistant mutants resemble loss-of-function alleles of the miRNA genes themselves. Since most plant miRNAs are redundantly encoded in plant genomes, single loss-of-function miRNA alleles are often aphenotypic and thus are not informative. In only a very small number of cases to date have loss-of-function alleles for all members of a miRNA family been rigorously studied (e.g., miRNA164 (Sieber *et al.*, 2007)). It is time consuming to construct the multiple mutants required for these analyses, so for now, miRNA-resistant mutants are used as a proxy to infer miRNA function.

2. miR172 Functions to Control Genes that Function in both Flowering Time Control and Floral Organ Identity in Arabidopsis

The five genes that comprise the miR172 family in Arabidopsis affect multiple aspects of flower development including flowering time control (Aukerman and Sakai, 2003), floral organ identity (Chen, 2004), and floral determinacy (Fig. 12.2, Table 12.1) (Zhao *et al.*, 2007). Unlike most plant miRNAs, miR172 was initially postulated to exert much of its function via translational repression rather than by RNA degradation (Aukerman and Sakai, 2003; Chen, 2004). However, recent evidence suggests that miR172 also functions to cause degradation of target mRNAs (Jung *et al.*, 2007; Schwab *et al.*, 2005).

During the vegetative to reproductive transition, miR172 regulates a set of four target AP2 transcription factor genes that contain a miR172 binding site: *TARGET OF EAT1* (*TOE1*), *TOE2, SCHNARCHZAPFEN*, and *SCHLAFMÜTZE* (Aukerman and Sakai, 2003; Schmid *et al.*, 2003). Constitutive expression of any of these four AP2-family genes results in a late-flowering phenotype while either overexpression of miR172 or *toe1* loss-of-function mutants exhibit an early-flowering phenotype (Aukerman and Sakai, 2003; Jung *et al.*, 2007). As plants proceed through vegetative development, the level of miR172 increases, resulting in a concomitant decrease in *TOE* gene activity. Downregulation of *TOE1* and *TOE2* is postulated to be necessary for the vegetative to reproductive transition to occur. The low levels of miR172 during early stages of vegetative development is indirectly regulated, at least in part, by miR156 (Wu *et al.*, 2009) (see the section on miR156 below for more details).

Many years of effort by a large number of labs have clearly demonstrated that control of flowering time by photoperiod in Arabidopsis functions via the genes *GIGANTEA* (*GI*) and *CONSTANS* (*CO*), the latter of which integrates day length and circadian clock signals to ensure that flowering is promoted under long-day growth conditions (Kobayashi and Weigel, 2007). Once activated, *CO* functions to activate the key floral inducer *FLOWERING LOCUS T* (*FT*) (chapter 11). miR172 is postulated to function on the photoperiod pathway based on the fact that levels of miR172 in Arabidopsis are responsive to day length; a higher level of miR172 accumulates in long days compared to short days (Schmid *et al.*, 2003). Although miR172 functions to regulate the levels of *FT* RNA, surprisingly miR172 appears to function via a *GI*-dependent but *CO*-independent pathway. The fact that the level of miR172 is regulated by *GI* suggests that *GI* has separable functions in the photoperiod pathway, one of which leads to activation of miR172, the other of which leads to activation of *CO* (Jung *et al.*, 2007).

miR172 also plays a critical role in floral organ identity (Fig. 12.2). The ABC model has been the dominant paradigm for studies of floral organ identity since it was postulated almost 20 years ago (Bowman, 1991; Coen and Meyerowitz, 1991). The ABC model was initially postulated based on genetic experiments; specifically on phenotypic analysis of single, double, and triple mutant combinations among a group of floral organ identity mutants. The ABC model has several major tenets. First, it postulates three activities, A, B, and C, function in adjacent whorls of the flower: A class activity in whorls 1 and 2, B class activity in whorls 2 and 3, and C class activity in whorls 3 and 4. The model is combinatorial in that different combinations of activities specify organ identity; A alone specifies whorl 1 sepals, A plus B specifies whorl 2 petals, B plus C specifies whorl 3 stamens, and C alone specifies whorl 4 carpels. Another major aspect of the model is the mutual repression of A and C classes; in the absence of A class activity, C activity is present in all four whorls, and in the absence of C class activity, A activity is present throughout the flower. In Arabidopsis, the key A class genes are *APETALA1* (*AP1*) and *AP2*, the B class genes are *APETALA3* (*AP3*) and *PISTILLATA* (*PI*), and the C class gene is *AGAMOUS* (*AG*). All of these genes, except AP2, encode transcription factors in the MADS family (Jack, 2001). AP2 encodes a transcription factor in a different family, the AP2 transcription factor family (Kim *et al.*, 2006).

The first hint that miRNAs might function in floral organ identity came from a genetic screen designed to isolate enhancers of the weak *ag* allele *ag-4* (Chen and Meyerowitz, 1999). Strong loss-of-function *ag* mutants like *ag-3* are indeterminate and exhibit homeotic transformations of both stamens to petals and carpels to sepals. The weak loss-of-function *ag-4* mutant is

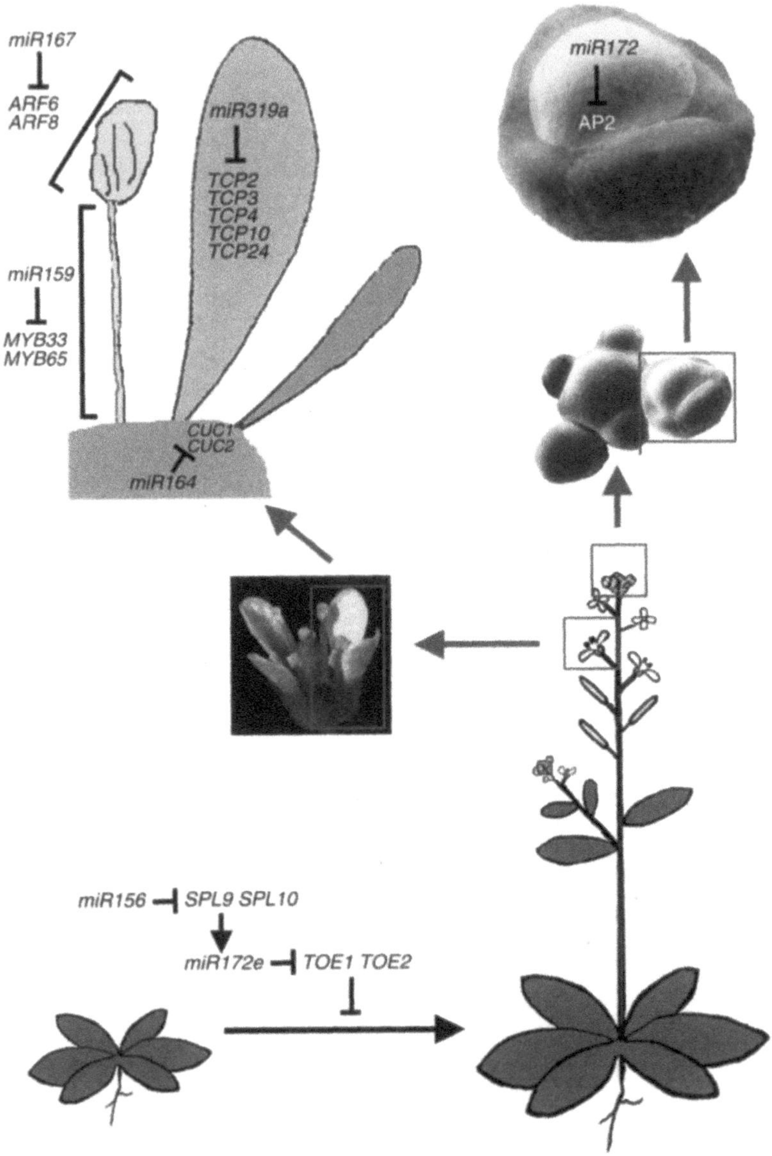

Figure 12.2 *(Continued)*

indeterminate, but the organ identity transformations are incomplete in that *ag-4* mutants produce stamens (Sieburth *et al.*, 1995). A gene named *HEN1* was identified as a result of a series of genetic enhancer/suppressor screens that began with *ag-4*. *HEN1* encodes a methyltransferase involved in miRNA processing and maturation (Fig. 12.1) (Boutet *et al.*, 2003; Park *et al.*, 2002; Yu *et al.*, 2005). This discovery suggested that a miRNA might be involved in regulating floral organ identity. Simultaneously, large-scale miRNA sequencing efforts were progressing in Arabidopsis (Reinhart *et al.*, 2002; Rhoades *et al.*, 2002), and one miRNA, miR172, was shown to have a binding site in the floral organ identity gene *AP2*. A role for miR172 in regulating *AP2* and floral organ identity came from the demonstration that transgenic lines that constitutively express miR172 develop flowers that resemble *ap2* mutant flowers suggesting that miR172 functions to downregulate *AP2* activity in the flower (Chen, 2004). Further support for the importance of miR172 regulation of *AP2* came from the study of miRNA-resistant versions of *AP2* (e.g. *mAP2*); for example, transgenic Arabidopsis that constitutively express *mAP2* develops flowers that resemble *ag* mutants in that stamens and carpels develop as petals and sepals, respectively, and the flowers are indeterminate (Chen, 2004; Zhao *et al.*, 2007). miR172 appears to regulate *AP2* primarily at the level of translational repression based on the fact that *AP2* RNA accumulates to approximately equal levels throughout the flower while AP2 protein accumulates to high levels in sepals and petals, but not stamens and carpels (Chen, 2004). Early papers describing this work suggested that miR172 functioned solely via translational repression (Aukerman and Sakai, 2003; Chen, 2004), but subsequent studies have shown that miR172 also can lead to degradation of *AP2* mRNA (Jung *et al.*, 2007; Schwab *et al.*, 2005).

The finding that miR172 can regulate *AP2* helped to explain a long-standing puzzle in Arabidopsis flower development. The well-characterized

Figure 12.2 miRNAs in flower development in Arabidopsis. This schematic shows some activities of miRNAs in Arabidopsis flower development. Note that many miRNAs function in multiple cell types and organs, and this figure does not accurately reflect all of the activities of all miRNAs that function in flower development; instead, this figure is designed to provide a sampling. As Arabidopsis plants mature, they proceed from a vegetative rosette (bottom left) to a reproductive form that possesses a flowering stalk or inflorescence (bottom right). miR156 and miR172 function as negative and positive regulators, respectively, of the vegetative to floral transition. During early floral stages (top right), miR172 functions to restrict AP2 activity in the center of the flower, in whorls 3 and 4, which will develop into the reproductive organs, the stamens and carpels. In later-staged flowers (upper left), miR164 functions to negatively regulate the *CUC* genes in intrawhorl and interwhorl boundaries, miR159 functions to negatively regulate *MYB33* and *MYB59* in stamens, miR167 targets *ARF6* and *ARF8* in stamens, and miR319 targets *TCP* RNAs in the petal.

Table 12.1 miRNAs in flower development

miRNA	Species	Number of miRNA genes[a]	Loss-of-function alleles	Target genes	Function in flowers
miR156	Arabidopsis	8	–	10 members of the *SPL* family (*SPL2, 3, 4, 5, 6, 9, 10, 11, 13, 15*)	Control of vegetative to reproductive transition, adult epidermal cell patterning
	Maize	11	–	*SPL* genes	Control of juvenile to adult phase transition, adult epidermal cell patterning
miR159	Arabidopsis	3	miR159a, miR159b	*MYB33, MYB65, MYB101*	Stamen filament length, male fertility
miR164	Arabidopsis	3	*miR164c/early extra petals 1, miR164a-1, miR164b-1*	*CUC1, CUC2*	Intrawhorl boundary specification
miR167	Arabidopsis	4	–	*ARF6, ARF8*	Stamen filament length, male and female fertility
miR169	Antirrhinum	–	*miR169/fistulata*	*NF-YA*	Repression of C activity in whorls 1 and 2
	Petunia	–	*miR169/blind*	*NF-YA*	

					Repression of C activity in whorls 1 and 2
miR172	Arabidopsis	5	–	*AP2*	Repression of AP2 in whorls 3 and 4
				TOE1, TOE2, SMZ, SNZ	Control of vegetative to reproductive transition
	Maize	6[b]	*miR172e/tasselseed 4*	*IDS1/TASSELSEED6, SID1*	Carpel abortion in tassel, spikelet meristem determinacy
			–	*GLOSSY15*	Control of vegetative to reproductive transition, control of juvenile to adult phase change
miR319	Arabidopsis	3	*miR319a*[129]	*TCP2, TCP3, TCP4, TCP10, TCP24*	Petal and stamen morphogenesis

[a] Numbers based on miRBase release 14 (September 2009) (http://www.mirbase.org/), unless indicated otherwise.
[b] http://bioinformatics.cau.edu.cn/cgi-bin/PMRD/further/detail.cgi?identifier=zma-MIR172f.

floral organ identity genes *AP3, PI*, and *AG* all exhibit persistent RNA expression patterns in the region of the flower that exhibits defects in mutants. For example, *AG* RNA accumulates in third-whorl stamens and fourth-whorl carpels; in *ag* mutants there are homeotic transformations of third-whorl stamens to petals and fourth-whorl carpels to sepals. Similarly, *AP3* and *PI* RNA are persistently expressed in petals and stamens; *ap3* and *pi* mutants exhibit homeotic transformations of petals to sepals and stamens to carpels. *ap2* mutants exhibit organ identity defects in first-whorl sepal and second-whorl petal development. The prediction had been that *AP2* RNA, like *PI, AP3*, and *AG*, would be confined to the region of the flower that exhibited defects in mutants, i.e. in whorls 1 and 2. Thus, it was surprising in 1994 when *AP2* RNA was shown to accumulate in all four whorls of the flower (Jofuku *et al.*, 1994). The discovery that miR172 regulates AP2 provides a potential explanation for this long-standing dilemma.

Although miR172 explains some aspects of this dilemma, it leaves other aspects unanswered and raises additional questions. If miR172 is the key to confining *AP2* activity to whorls 1 and 2, then the prediction is that miR172 expression should be confined to whorls 3 and 4. Although miR172 accumulates at highest levels in whorls 3 and 4 during mid-to-late stages of flower development, during early stages, when floral organ identity is being specified, miR172 is expressed throughout the flower primordium. At this point, it seems that miR172 alone does not entirely explain why AP2 activity is confined to whorls 1 and 2; other factors must also be involved. In addition, nothing is known about the factors that function upstream of miR172 to establish expression throughout the flower during early floral stages, and later to confine expression to whorls 3 and 4.

The simple model that comes out of the ABC model is that C class activity antagonizes A class activity in whorls 3 and 4. *AG* is known to directly repress expression of the A class gene *AP1* in whorls 3 and 4. Although *AG* does not define the spatial limits of the RNA expression domain of *AP2, AG*, in formal terms, it is still a negative regulator of *AP2* because in *ag* mutants reproductive organs develop as perianth organs. This phenotypic effect is not caused by *AP1* misexpression because *ag ap1* double mutants also develop whorl 3 and whorl 4 organs that are largely sepals and petals. Thus, *AG* functions to negatively regulate *AP2* activity in whorls 3 and 4, but the molecular nature of this negative regulation is unclear. One question is whether miR172 functions independently of *AG*; that is, does *miR172* play any role in whorls 3 and 4 when *AG* is active? The answer is yes, as revealed by careful experiments performed by Zhao *et al.* (Zhao *et al.*, 2007) that demonstrate that when miRNA-resistant versions of *AP2* (e.g. 35S:AP2-m3) are constitutively expressed in whorls 3 and 4, even in the presence of functional *AG*, dramatic defects in flower development result. The take-home message from these studies it that normal flower development requires regulation of A class activity by both *AG* and miR172. The fact that 35S:AP2-m3 *ag* doubles have an enhanced phenotype provides additional evidence that *AG* and miR172 act independently on *AP2,* and that both are required for proper floral patterning (Zhao *et al.*, 2007).

3. miR172 Functions to Regulate Inflorescence Development in Maize

miR172 has also been shown to be a key regulator of inflorescence development in maize. Upon transition to flowering a maize plant produces two types of inflorescences, a male inflorescence called the tassel and a female inflorescence called the ear. The tassel is produced from the shoot apical meristem and the ears are produced from the axillary meristems (Fig. 12.3A). The maize tassel is comprised of a main spike and several lateral branches both of which bear smaller branches comprised of a pair of spikelets. Each spikelet contains two florets enclosed by two leaf-like organs called glumes. The maize tassel produces several branch meristems each of which produces several spikelet pair meristems. In turn, each spikelet pair meristem produces spikelet meristems. Each spikelet meristem gives rise to two floral meristems. The developmental pattern of the ear is very similar to that of the tassel except that instead of giving rise to branch meristems the axillary meristems directly produce spikelet pair meristems; that is, there are no branch meristems in the ear. Surrounding the reproductive structures are lodicules (the functional equivalent of petals in dicots (Whipple *et al.*, 2004)) as well as the palea and lemma, which are sepal/bract-like structures. During early stages of development, each floret is bisexual and is comprised of both male and female reproductive organs. As flower development proceeds, stamens arrest in the female flowers and the carpels abort in male flowers (Laudencia-Chingcuanco and Hake, 2002; McSteen *et al.*, 2000; Thompson and Hake, 2009).

The two *tasselseed* (*ts*) mutants, the recessive *ts4* and the dominant *Ts6*, fail to abort the carpel from the male inflorescence. *ts4* mutants also fail to develop stamens. In addition, *ts4* and *Ts6* mutants exhibit increased indeterminate meristem branching resulting in supernumerary spikelet meristems and florets (Chuck *et al.*, 2007b; Irish, 1997). *ts4* encodes *miR172e* (Chuck *et al.*, 2007b). As in Arabidopsis, the targets of miR172 in maize include members of the AP2 transcription factor family such as *INDETERMINATE SPIKELET 1* (*IDS1*). In *ids1* mutants, the spikelet meristem is indeterminate in that it is not completely consumed when it gives rise to two floral meristems; instead, the spikelet meristem persists between the florets and produces two additional florets. This indeterminacy can occur several times giving rise to many florets which have only

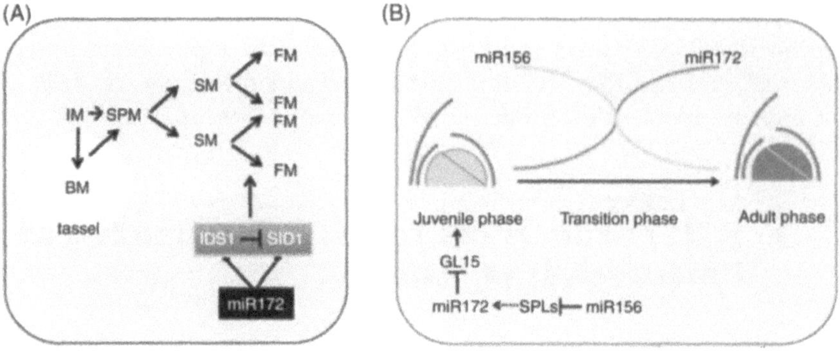

Figure 12.3 miR156 and miR172 in maize flower development. (A) The development of the maize tassel (male inflorescence) and ear (female inflorescence) follow similar patterns. In the tassel the inflorescence meristem (IM) gives rise to branch meristems (BM) that subsequently gives rise to spikelet pair meristems (SPM). In the ear, the IM directly gives rise to SPMs. SPMs give rise to two spikelet meristems (SM), which subsequently form two floral meristems (FM). miR172 and its targets *IDS1* and *SID1* function to influence the SM to FM conversion. *IDS1* is also a negative regulator of its homolog *SID1*. Both *IDS1* and *SID1* are positive regulators of the SM to FM transition. Thus, in double *ids1 sid1* mutants floral meristems do not form. By contrast, mutants such as *ts4* that ectopically express *IDS1* exhibit supernumerary floral meristem initiation. miR172 also influences floral organ identity as evidenced by failure of carpel abortion from the male inflorescence. (B) The juvenile to adult phase transition in maize is controlled by miR156 and miR172. Several *SPL* genes are targeted by miR156 in maize. In Arabidopsis it has been clearly demonstrated that *SPL9* and *SPL10* positively regulate miR172e transcription. In maize, miR156 is postulated to indirectly activate miR172 via the *SPL* genes. miR172 has been shown to negatively regulate *GL15*, which functions to promote the maintenance of the juvenile state. Early in development, the level of miR156 is high which results in low levels of the SPL genes and miR172, and high levels of *GL15*. As development proceeds, the level of miR156 decreases, ultimately leading to downregulation of *GL15* and the development of adult leaf characteristics.

minor defects in reproductive organ differentiation (Chuck *et al.*, 1998). The branching and sex determination phenotypes of *ts4* are almost completely suppressed by loss-of-function mutations in *IDS1* (Chuck *et al.*, 2008) suggesting that *IDS1* is a key target of miR172e. Additional evidence that *IDS1* is a key target comes from molecular analysis of *Ts6*. *Ts6* is a dominant allele of *IDS1* that contains a mutation in miR172 binding site that renders it insensitive to regulation by miR172 (Chuck *et al.*, 2007b). Thus, in both *ts4* and *Ts6*, *IDS1* fails to be properly regulated by *miR172* resulting in defects in sex determination and branching. Examination of the RNA and protein expression patterns of *IDS1* in *ts4* compared to wild type indicates that *miR172e* likely controls *IDS1* activity primarily by translational repression (Chuck *et al.*, 2007b), similar to how *miR172* is postulated to function in Arabidopsis to regulate *AP2*.

A second AP2 gene, *SISTER OF IDS1* (*SID1*) (Chuck *et al.*, 2008), which also contains a miR172 binding site, is a secondary target of miR172 in controlling reproductive meristem identity in maize. Although *sid1* mutants alone do not suppress the *ts4* phenotype, the *sid1 ids1 ts4* triple mutant exhibits enhanced suppression of the *ts4* phenotype compared to the *ids1 ts4* double mutant. Although single mutants of *sid1* are largely aphenotypic, *sid1* enhances the *ids1* phenotype; *ids1 sid1* double mutants fail to initiate floral meristems and instead produce multiple bract-like organs in their place (Chuck *et al.*, 1998, 2008). The phenotype of *ids1 sid1* double mutants suggests that both *ids1* and *sid1* are necessary for the spikelet meristem to floral meristem transition. Initially it was puzzling as to why both loss-of-function and gain-of-function mutations of *ids1* resulted in the same phenotype, namely the development of extra florets. However, the demonstration that the *ids1 sid1* double mutant has a phenotype (no floral meristems) opposite that of *Ts6* (supernumerary floral meristems) presents a more consistent picture of how *ids1* and *sid1* function. The indeterminate meristem phenotype of *ids1* single loss-of-function mutants can be explained by postulating that *ids1* functions as a negative regulator of *sid1* so that in *ids1* mutants, *sid1* levels increase, similar to what is observed in *Ts6* mutants (Chuck *et al.*, 1998; Thompson and Hake, 2009).

4. miR156 Regulates Phase Change and Floral Induction

In maize and Arabidopsis, both miR172 and miR156 are critical for two interrelated processes: the juvenile to adult phase change, and the vegetative to reproductive transition. As plants proceed through vegetative development, they gradually undergo a transition from producing leaves with juvenile characteristics to producing leaves with adult characteristics. During normal development, floral induction depends on plants having proceeded to the adult phase of vegetative development. Although phase change and floral induction are related processes, these two processes are clearly separable in at least some aspects as it is possible to isolate mutants that affect one of these processes but not the other, i.e. mutants that affect only phase change but not flowering time and vice versa. However, many mutants affect both phase change and floral induction. Evidence from maize and Arabidopsis suggests that miR156 and miR172 and their target genes are members of the group that affect both phase change and floral induction.

The juvenile to adult transition affects a number of traits including leaf epidermal cell characteristics such as the formation of trichomes on the abaxial surface of leaves (Fornara and Coupland, 2009; Moose and Sisco,

1994; Telfer *et al.*, 1997). In maize, a loss-of-function mutation in the AP2 family gene *glossy15 (gl15)* leads to replacement of juvenile leaf epidermal characteristics with adult leaf epidermal characteristics (Moose and Sisco, 1994) demonstrating that *gl15* is required for maintenance of juvenile leaf epidermal traits (Moose and Sisco, 1996) (Fig. 12.3B). By contrast, increasing *GL15* activity leads to an increase in the number of leaves with juvenile characteristics and delays flowering (Lauter *et al.*, 2005). Thus, *GL15* is a key determinant in the specification of juvenile epidermal features. *GL15* contains a miR172 binding site. In maize, miR172 is expressed at very low levels during early stages of vegetative development but is detected at increasing levels as plants proceed through the vegetative development (Lauter *et al.*, 2005). miR172 is critical for degradation of *GL15* mRNA, and downregulation of *GL15* is required for the maize plant to proceed from the juvenile to adult phase, i.e. to express adult leaf characteristics (Lauter *et al.*, 2005) (Fig. 12.3). While *gl15* mutants develop leaves that precociously exhibit adult epidermal cell fate, the dominant *Corngrass1 (Cg1)* mutant has the opposite phenotype, namely the retention of juvenile characteristics in adult plants (Chuck *et al.*, 2007a). This phenotype is reminiscent of plants that ectopically express *GL15* or exhibit reduced expression of miR172. The molecular lesion in *Cg1* results in overexpression in the shoot meristem and lateral organs of two tandemly arranged miR156 genes. In *Cg1* mutants, the level of miR172 is reduced (Chuck *et al.*, 2007a) suggesting that one or more of the miR172 genes might be a target of miR156. However, the miR172 genes do not appear to be direct targets; instead, the key targets of miR156 are seven members of the *SQUAMOSA PROMOTER BINDING PROTEIN LIKE* (*SPL*) family that contain a miR156 binding site. In maize, several *SPL* genes are downregulated in *Cg1*, but the precise details of the indirect regulation of miR172 by miR156 have not been elucidated. However, in Arabidopsis, the regulation of miR172 by miR156 has been conclusively shown to be indirect and is mediated via the *SPL* genes (Gandikota *et al.*, 2007; Wang *et al.*, 2009; Wu and Poethig, 2009, 2006; Yamaguchi *et al.*, 2009). In both maize and Arabidopsis, miR156 is expressed at high levels at early stages of vegetative development and levels decrease as vegetative development proceeds. In Arabidopsis, ten SPL genes contain a miR156 binding site. A decrease in miR156 expression correlates with an increase in expression of the SPL target genes, and a concomitant transition to the adult phase and an early-flowering phenotype. By contrast, overexpression of miR156 leads to a prolonged juvenile phase and a late-flowering phenotype. Similarly, constitutive expression of miR156-resistant *SPL* targets leads to a premature transition to the adult phase and an early-flowering phenotype (Wu *et al.*, 2009, Wu and Poethig, 2006).

Thus, miR156 and the SPL targets exhibit a complementary temporal expression pattern during vegetative development. Similarly, miR156 and miR172 exhibit opposite expression profiles; miR156 is expressed at high levels during juvenile stages and levels decrease as plants age while miR172 levels increase throughout vegetative development. In Arabidopsis the regulatory connection between these two miRNAs is mediated by the miR156 targets *SPL9* and *SPL10*, both of which are capable of binding to sequences in the regulatory region of *miR172e* (Fig. 12.2) (Wu *et al.*, 2009). Thus, as vegetative development proceeds, the levels of miR156 decrease, leading to an increase in *SPL9* and *SPL10* which in turn results in an increase in the level of transcription of *miR172e*. miR172 then targets genes such as *TOE1* and *TOE2*, which need to be downregulated for the vegetative to reproductive transition to occur. *TOE1* and *TOE2* also function to promote juvenile epidermal identity during vegetative development (Wu *et al.*, 2009).

In summary, results from Arabidopsis suggest that two miRNAs, miR156 and miR172, function sequentially to coordinate both the juvenile to adult phase transition and the vegetative to reproductive transition. The data from maize, although incomplete at this point, suggests a similar mechanism of phase change and flowering time control likely exists in maize as well.

5. miR164 Regulates Genes that Function in Organ Boundary Formation

The three members of the miR164 family play an important role in lateral organ separation, organ boundary formation, and lateral organ proliferation (Fig. 12.2, Table 12.1). The key targets of miR164 are *CUC1* and *CUC2*, which function in both meristem maintenance and in lateral (including floral) organ separation (Aida *et al.*, 1997). Single *cuc1* mutants have only very subtle sepal fusion phenotypes, but *cuc1 cuc2* double mutants are seedling lethal; the two cotyledons fuse to form a single cup-shaped structure and the shoot apical meristem arrests. Based on the loss-of-function phenotype, the *CUC* genes are postulated to be important for formation of lateral organ boundaries by suppressing cell growth between lateral organs.

CUC1 and *CUC2* are members of the NAC transcription factor family. Early in flower development, *CUC2* is expressed at the boundary between the shoot meristem and the developing flower primordia (Takada *et al.*, 2001). At later stages of flower development, *CUC2* is expressed at the boundaries between floral organs within a whorl (i.e., the intrawhorl boundary). Based on this, *CUC2* was postulated to function in the suppression of cell proliferation, thus allowing separate floral organs to develop.

miR164 functions by targeting for degradation the *CUC1* and *CUC2* mRNAs (Laufs *et al.*, 2004; Mallory *et al.*, 2004). That the three miR164 genes are not completely functionally redundant comes from the finding that a loss-of-function mutation in *miR164c* called *early extra petals 1* (*eep1*) results in plants that have supernumerary petals in the first flowers that form on the inflorescence (Baker *et al.*, 2005). This flower phenotype is enhanced when *miR164c-eep1* is combined with loss-of-function mutants in *miR164a* and *miR164b*; in the *miR164a miR164b miR164c* triple mutant, carpels are unfused, petal and sepal number is increased, and stamen number is decreased (Sieber *et al.*, 2007). There are also changes in floral organ size and phyllotaxy in these triple mutants (Peaucelle *et al.*, 2007). A similar extra floral organ phenotype is observed when miRNA-resistant versions of *CUC1* and *CUC2* are expressed in the flower (Baker *et al.*, 2005; Mallory *et al.*, 2004).

In some ways it was surprising to find that a mutation in one member of a redundant miRNA family results in a loss-of-function phenotype. The three miR164 genes encode very similar miRNA products and are postulated to regulate the same set of target genes. One possibility is that *eep1* exhibits a phenotype because the three miR164 genes exhibit unique spatial and temporal expression patterns. In an elegant analysis, Sieber *et al.* (Sieber *et al.*, 2007) demonstrated that the expression patterns of the three miR164 genes are not completely overlapping. In this study, the expression patterns were determined via two different methods. First, miR164 promoter-GFP fusion transgenic lines were constructed and expression patterns were analyzed. Second, in situ hybridization was performed. In general, in situ hybridization is challenging for miRNAs that are members of a redundant gene family because it is very difficult to design probes that are specific for the closely related family members. Sieber *et al.* got around this problem by analyzing expression patterns in double miRNA mutants, for example, by examining *miR164a* expression in a *miR164b miR164c* double mutant etc. These experiments definitively showed that the three miR164 genes have unique expression patterns (Sieber *et al.*, 2007), and that these differences explain the functional diversification of the three miR164 genes. The experiments done on miR164 serve as a model for how expression analyses of redundant miRNA families should be carried out. To date, no other miRNA family has been as carefully analyzed mostly due to the lack of availability of loss-of-function miRNA mutants.

Although loss-of-function of *miR164c* (i.e., *eep1*) results in a phenotype indicating some unique functions for miR164c, the fact that the miR164 triple mutant has a dramatically enhanced phenotype demonstrates that there is functional redundancy among the miR164 genes. This is best exemplified by the fact that all three miR164 single mutants have a normal phyllotaxis, but the miR164 triple mutant exhibits severe defects in phyllotaxis (Peaucelle *et al.*, 2007).

miR164 also serves as the best-studied example of how miRNAs fine-tune spatial and temporal expression patterns. miR164, *CUC1*, and

CUC2 RNAs are expressed in overlapping domains in interwhorl and intrawhorl boundary regions of flowers. In the miR164 triple mutant, the expression domain of *CUC1* and *CUC2* is, at most, only subtly spatially expanded, but the level of expression of the CUC targets is dramatically increased (Sieber *et al.*, 2007). The most likely model is that miR164 functions primarily to modulate the level of the *CUC1* and *CUC2* RNAs, rather than by defining the limits of the spatial expression domain of these RNAs. Another possibility that has not been carefully analyzed is that miR164 functions to translationally repress *CUC1* and *CUC2.*

Recently, the long-held hypothesis that *CUC1* and *CUC2* function primarily to restrict cell growth and proliferation has been called into question (Nikovics *et al.*, 2006; Peaucelle *et al.*, 2007; Sieber *et al.*, 2007). Larue *et al.* (Larue *et al.*, 2009) isolated a *cuc2* mutant with a mutation in the miR164 binding site (*cuc2-1D*) that exhibits larger leaf and petal area and larger pedicels suggesting that *CUC2* functions to promote, rather than to restrict, cell proliferation in some cells (Larue *et al.*, 2009).

6. miR169 Represses C Class Activity in the Perianth in Antirrhinum and Petunia

In Arabidopsis, repression of C class activity in whorls 1 and 2 is carried out by *AP2* as well as by a number of other repressors including LEUNIG, SEUSS, and CURLY LEAF (Goodrich *et al.*, 1997; Liu and Meyerowitz, 1995; Franks *et al.*, 2002). This latter group of repressors is important for *AG* repression, but plays no role in specifying organ identity. In Antirrhinum, the two *AP2* homologs, *LIPLESS1* (*LIP1*) and *LIP2,* function redundantly to specify perianth organ identity, but they fail to repress C class activity in the perianth based on the fact that in double loss-of-function *lip1 lip2* mutants conversions of perianth organs to reproductive organs, which would be indicative of C class gene misexpression, are not observed (Keck *et al.*, 2003). Instead, in Antirrhinum and Petunia, miR169 functions to repress C class activity in the perianth. Both the Antirrhinum *blind* (*bl*) mutant and the Petunia *fistulata* (*fis*) mutant (Motte *et al.*, 1998) exhibit staminoid petals in whorl 2 suggesting that these mutants have defects in repressing C class activity in whorl 2. Both *bl* and *fis* exhibit elevated levels of transcripts of *AG* orthologs, specifically *PLENA* (*PLE*) in Antirrhinum and *pMADS3* in Petunia. Both *BL* and *FIS* encode *miR169* (Cartolano *et al.*, 2007), which targets members of the NF-YA (also called *HAP2* or *CBF-B*) family. NF-YA transcription factors bind to CCAAT box sequences found in the second intron of many *AG* orthologs; NF-YA binding has been demonstrated to be necessary for enhancement and maintenance of *AG* transcription (Hong *et al.*, 2003). Thus, the obvious prediction would be that in *bl* and *fis* mutants expansion of C class activity is caused by the failure of miR169 to repress NF-YA leading to ectopic expression of *NF-YA* and subsequently *PLE/pMADS3* in the perianth. But contrary to this hypothesis *NF-YA* (and *miRBL, miRFIS*) transcripts are detected throughout all four floral whorls of wild-type flowers making it less likely that these two miRNAs function to completely eliminate *NF-YA* transcripts from the outer whorls of the flower. Instead, the demonstration that the steady-state level of *PhYA1* expression in a *bl* mutant was elevated makes it more likely that miR169 functions by fine-tuning the levels of NF-YA RNA. It is also possible that miR169 functions via a translational repression mechanism, but this has not yet been rigorously tested (Cartolano *et al.*, 2007). In Arabidopsis miR169 has been demonstrated to regulate the *NFYA5* gene, which is involved in drought resistance (Li *et al.*, 2008). To date, there is no evidence that miR169 regulation of NF-YA genes plays a role in C class repression in Arabidopsis.

7. *miR319A* Controls Floral Organ Size and Shape in Arabidopsis

The miR319 family consists of three members: *miR319a, miR319b,* and *miR319c*. *miR319a* was initially suggested to play a role in leaf development (Palatnik *et al.*, 2003), but recent work suggests that *miR319a* is critical for controlling the size and shape of the floral organs (Nag *et al.*, 2009). *miR319a* was discovered as the gene that is ectopically expressed in an activation-tagged line (called *jaw-D*) that resulted in plants with large leaves that exhibit extremely curled and buckled leaf margins (Palatnik *et al.*, 2003). The key targets of miR319 are five members of the TCP transcription factor family that contain a miR319 binding site. In *jaw-D* the five TCP genes *TCP2, TCP3, TCP4, TCP10,* and *TCP24* are down-regulated in seedlings (Palatnik *et al.*, 2003). These five TCP genes are postulated to be negative regulators of cell growth and/or promoters of cell differentiation (chapter 7; Crawford *et al.*, 2004; Efroni *et al.*, 2008; Luo *et al.*, 1999, 1996; Nath *et al.*, 2003).

A role for *miR319a* in flower development comes from the analysis of a loss-of-function *miR319a* mutant that exhibits a reduction of petal width and decreased petal and stamen length (Fig. 12.2) (Nag *et al.*, 2009). This *miR319a* allele contains a mutation in the middle of the mature miRNA. Careful analysis of the spatial and temporal expression patterns of the three miR319 family members in Arabidopsis demonstrates that these genes have a largely non-overlapping expression pattern in seedlings and inflorescences (Nag *et al.*, 2009). For example, *miR319a* is not expressed in leaves, but *miR319c* is expressed in the proximal region of young leaves. Based on this, it appears likely that the phenotypes of miR319 overexpression alleles, like *jaw-D*, is due to targeting of genes that are normally targets of *miR319c* in leaves. Although all three miR319 genes are expressed in the inflorescence, *miR319a* is expressed most strongly throughout developing petals and stamens during early and mid stages of flower development; this expression pattern is consistent with the loss-of-function *miR319a* phenotype which is strongest in petals and stamens. *TCP4* was shown to be a key target of *miR319a* by utilizing a *tcp4* allele that contains a mutation in the *miR319a* binding site that is complementary to the change that occurs in the *miR319a* loss-of-function mutation (Palatnik *et al.*, 2007). By itself, *tcp4* mutants exhibit normal flowers, but in *tcp4 miR319a* double mutants, the floral defects of the *miR319* mutant are largely suppressed (Nag *et al.*, 2009) strongly suggesting that *TCP4* is a key target of *miR319a* in flowers.

8. miR159 Regulates Genes with Broad Roles in Flower Development

The miR159 family members, *miR159a, miR159b*, and *miR159c* target seven MYB genes that contain close matches to the miR159 binding site. To date, only three targets of miR159 have been confirmed: *MYB33, MYB65*, and *MYB101/DUO POLLEN1* (*DUO1*) (Millar and Gubler, 2005). Transgenic lines of Arabidopsis lines that overexpress *miR159a* develop flowers with stunted stamens and reduced fertility (Fig. 12.2) (Achard *et al.*, 2004; Allen *et al.*, 2007). Single loss-of-function *miR159a* and *miR159b* mutants fail to exhibit a dramatic flower phenotype, but double *miR159a miR159b* mutants have pleiotropic morphological defects that include dwarfing, reduced apical dominance, hyponastic leaves, reduced fertility, and seeds with irregular shape. The fact that a miRNA-resistant version of *MYB33*, when expressed under its endogenous promoter, has a phenotype that strongly resembles *miR159a miR159b* double mutants suggests that *MYB33* is a key target of *miR159a* and *miR159b*. Further support comes from analysis of the *miR159a miR159b myb33* triple mutant, which largely, but not completely, suppresses the developmental defects observed in the *miR159a miR159b* double mutant. *MYB65*, a close homolog of *MYB33*, also contains a miR159 binding site. Not surprisingly, *MYB33* and *MYB65* function redundantly in anther development. The best evidence for this comes from analysis of the *miR159a miR159b myb33 myb65* quadruple mutant, which fully rescues the developmental defects of the *miR159a miR159b* double mutant. Together, these data indicate that miR159 exerts its role on flower development by regulating both *MYB33* and *MYB65* (Allen *et al.*, 2007).

RNA for both *MYB33* and *MYB65* is detected in roots, leaves, flowers, and siliques. In flowers *MYB33* and *MYB65* are expressed in a nearly identical pattern in developing anthers, pollen grains, and ovules (Gocal *et al.*, 2001). Analysis of promoter:GUS fusions indicates that *miR159a* and *miR159b* have nearly identical expression pattern; both miRNA genes are expressed in root tips, lateral roots, the shoot apex, and inflorescences. In flowers, *miR159a* and *miR159b* are expressed in the receptacles, carpels, sepals, and stamen filaments but are not detected in anthers. Thus, the absence of *miR159a* and *miR159b* in the anther allows for *MYB33* and *MYB65* expression in the anther, which is necessary for proper anther development (Allen *et al.*, 2007). A third target of *miR159* is *MYB101*. *MYB101* RNA can be cleaved by *miR159* (Palatnik *et al.*, 2007; Reyes and Chua, 2007), and this targeting likely has biological relevance based on the fact that overexpression of a miRNA-resistant version of *MYB101* leads to numerous developmental defects including severe seedling patterning defects, twisted and curly leaves, flower defects such as short petals and stamens and partial sterility (Palatnik *et al.*, 2007).

Interestingly, the miR319 family is very closely related in sequence to the miR159 family; the two families possess sequence identity in 17 of the 21 nucleotides of the mature miRNA (Palatnik *et al.*, 2007). The interrelatedness and function of these two miRNA families has been studied in detail (Palatnik *et al.*, 2007). While miR319 has the ability to regulate the miR159 targets *MYB33, MYB65*, and *MYB101*, the restricted expression domain and low level of miR319 accumulation prevent miR319 from regulating the MYB targets in a biologically significant manner. By contrast, *miR159a* and *miR159b* do not efficiently target *TCP2, TCP3, TCP4, TCP10*, and *TCP24* due to sequence differences between the miR159 and the miRNA binding sites in the TCP targets. In summary, the miR319 family has the ability to target both TCP and MYB targets, but *in vivo* only the TCP genes are targeted. By contrast, the miR159 family is able to efficiently target only the MYB genes but not TCP genes (Palatnik *et al.*, 2007).

9. miR167 is Critical for Male and Female Fertility

miR167 targets two members of the auxin response factor family (ARF), *ARF6* and *ARF8*. The ARFs are sequence-specific DNA binding proteins that transcriptionally activate or repress genes that mediate the auxin response. *ARF6* and *ARF8* function largely redundantly in flower maturation as single mutants have only subtle floral defects while double mutants have stronger phenotypes. In *arf6 arf8* double mutants, stamen filaments are short and anthers do not properly release pollen (Nagpal *et al.*, 2005). *arf6 arf8* double mutants also exhibit carpel defects that result in female sterility. Constitutive expression of miR167 throughout the plant phenocopies the *arf6 arf8* double mutant phenotype (Wu *et al.*, 2006), which is consistent with the hypothesis that *ARF6* and *ARF8* are the key targets of miR167. Expression of miRNA-resistant versions of *ARF6* and *ARF8* under the control of endogenous *ARF6* and *ARF8* control sequences results in plants that develop sterile flowers and small leaves. The flowers of these plants are both male and female sterile. The male sterility results from a failure of anther dehiscence, and the female sterility results from premature arrest of ovule development. Careful analyses of the expression patterns of both miR167 and the *ARF6/ARF8* mRNAs demonstrate that in both the ovule and anther, the expression patterns of miR167 and *ARF6/ARF8* are mutually exclusive (Wu *et al.*, 2006). Thus, miR167 is critical for defining the correct spatial expression domains of *ARF6* and *ARF8* in stamens and ovules to ensure proper ovule and anther development and male and female fertility.

miRNAs in addition to miR167 also affect the auxin response and thus likely have wide-ranging effects on plant development including flower development. For example, miR393 targets the auxin receptor family members including *TRANSPORT INHIBITOR RESPONSE 1* (*TIR1*) and the *AUXIN SIGNALING F-BOX* (*AFB*) genes (Parry *et al.*, 2009). The precise roles of the miR393 family in flower development have yet to be elucidated.

10. Regulation of Floral Organ Polarity by miR166

Adaxial–abaxial organ polarity, including floral organ polarity, is regulated by the miR166 family (Husbands *et al.*, 2009). The targets of miR166 include the HD-ZIPIII transcription factor genes *PHABULOSA*, *PHAVOLUTA*, *REVOLUTA*, *CORONA*, and *ATHB-8*. A second type of small RNA, called tasiRNAs, also influences organ polarity by regulating *ARF3* and *ARF4* (Chitwood *et al.*, 2009). Please refer to the chapter in this volume by Kidner and Timmermans (Chapter 5) for a detailed description of the mechanisms of action of the miRNAs and tasiRNAs that mediate organ polarity in plants.

11. Perspective and Future Challenges

Much has been learned about the function of plant miRNAs since they were first discovered in plants in 2002, but significant gaps in our knowledge and understanding remain. Although many miRNAs have been characterized genetically or biochemically, there are no examples of plant miRNA families that are completely characterized both genetically and biochemically; that is to say there are no miRNAs for which (1) loss-of-function mutants in all members of a miRNA gene family have been carefully analyzed and (2) protein expression patterns of the target genes have been analyzed to ascertain the degree of translational repression of targets. Even in Arabidopsis, where miRNAs have been studied in the most detail, there are a number of miRNA families that have been studied only very superficially or not at all (Rajagopalan *et al.*, 2006). There is much still left to accomplish; plant miRNAs will remain the focus of intensive research for some time to come.

An obvious gap in knowledge is a lack of detailed understanding of the factors that control expression of miRNA genes themselves. For most plant miRNAs, little is known about the details of the spatial and temporal expression patterns of the miRNA genes, and the endogenous, biotic, and abiotic inputs that affect miRNA gene expression. The lessons of both miR164 (Sieber *et al.*, 2007) and miR319 (Nag *et al.*, 2009) demonstrate that understanding the details of the expression patterns of the miRNA are critical to unraveling their function. Because the spatial and temporal expression patterns of the redundant miRNA family members can be different, researchers need to exercise caution in interpreting experiments analyzing transgenic lines that constitutively overexpress miRNAs as the results can often be deceiving. More emphasis needs to be placed on analysis of loss-of-function alleles of miRNA genes. But because single loss-of-function mutants are most often aphenotypic and because miRNAs are small targets for mutagenesis, loss-of-function alleles have been slow to be isolated and characterized. Also, since single miRNA mutants fail to exhibit a phenotype, multiple mutants need to be constructed and analyzed, which is a time-consuming process. The discovery of miRNA target mimics may provide a way to systematically engineer plant lines that resemble loss-of-function miRNA mutants (Franco-Zorrilla *et al.*, 2007). A miRNA target mimic is a miRNA that interacts with the miRNA binding site of the target in a nonproductive way such that miRNA:target RNA duplex cannot be cleaved by the slicer activity of AGO. In this way, the miRNA target mimic resembles a loss-of-function mutant in the miRNA.

The plant miRNA field also needs to pay more attention to translational repression as a mechanism of miRNA function. In only a small number of cases (Chen, 2004; Chuck *et al.*, 2007b; Gandikota *et al.*, 2007) have protein levels of miRNA targets been analyzed in any detail. Since the miRNA genes are expressed in discrete and dynamic spatial and temporal patterns, both the RNA and protein expression patterns of targets need to analyzed in detail to get a complete picture of the degree to which miRNAs are directing mRNA degradation vs. mRNA translational repression.

ACKNOWLEDGMENTS

We thank Aubrey Frank for comments on the manuscript. Work in the Jack lab on miRNAs and flower development is supported by the United States National Science Foundation Grant IOS-0926347.

REFERENCES

Achard, P., Herr, A., Baulcombe, D. C., and Harberd, N. P. (2004). Modulation of floral development by a gibberellin-regulated microRNA. *Development* **131**, 3357–3365.

Aida, M., Ishida, T., Fukaki, H., Fujusawa, H., and Tasaka, M. (1997). Genes involved in organ separation in Arabidopsis: an analysis of the *cup-shaped cotyledon* mutant. *Plant Cell* **9**, 841–857.

Allen, R. S., Li, J., Stahle, M. I., Dubroue, A., Gubler, F., and Millar, A. A. (2007). Genetic analysis reveals functional redundancy and the major target genes of Arabidopsis miR159 family. *Proc. Natl. Acad. Sci. USA* **104**, 16371–16376.

Aukerman, M. J., and Sakai, H. (2003). Regulation of flowering time and floral organ identity by a microRNA and its *APETALA2*-like target genes. *Plant Cell* **15**, 2730–2741.

Baker, C. C., Sieber, P., Wellmer, F., and Meyerowitz, E. M. (2005). The *early extra petals1* mutant uncovers a role for microRNA *miR164c* in regulating petal number in Arabidopsis. *Curr. Biol.* **15**, 303–315.

Bao, N., Lye, K. W., and Barton, M. K. (2004). MicroRNA binding sites in Arabidopsis class III HD-ZIP mRNAs are required for methylation of the template chromosome. *Dev. Cell* **7**, 653–662.

Bohmert, K., Camus, I., Bellini, C., Bouchez, D., Caboche, M., and Benning, C. (1998). *AGO1* defines a novel locus of Arabidopsis controlling leaf development. *EMBO J.* **17**, 170–180.

Boutet, S., Vazquez, F., Liu, J., Beclin, C., Fagard, M., Gratias, A., Morel, J. B., Crete, P., Chen, X., and Vaucheret, H. (2003). Arabidopsis *HEN1*: A genetic link between endogenous miRNA controlling development and siRNA controlling transgene silencing and virus resistance. *Curr. Biol.* **13**, 843–848.

Bowman, J. L. (1991). Genetic interactions among floral homeotic genes in Arabidopsis. *Development* **112**, 1–20.

Brodersen, P., Sakvarelidze-Achard, L., Bruun-Rasmussen, M., Dunoyer, P., Yamamoto, Y. Y., Sieburth, L., and Voinnet, O. (2008). Widespread translational inhibition by plant miRNAs and siRNAs. *Science* **320**, 1185–1190.

Cartolano, M., Castillo, R., Efremova, N., Kuckenberg, M., Zethof, J., Gerats, T., and Schwartz-Sommer, Z. (2007). A conserved microRNA module exerts homeotic control over *Petunia hybrida* and *Antirrhinum majus* floral organ identity. *Nat. Genet.* **39**, 901–905.

Chen, X. (2004). A microRNA as a translational repressor of *APETALA2* in Arabidopsis flower development. *Science* **303**, 2022–2025.

Chen, X. (2008). MicroRNA metabolism in plants. *Current Topics Micro. Immuno.* **320**, 117–136.

Chen, X., Liu, J., and Jia, D. (2002). *HEN1* functions pleiotropically in Arabidopsis development and acts in C function in the flower. *Development* **129**, 1085–1094.

Chen, X., and Meyerowitz, E. M. (1999). *HUA1* and *HUA2* are two members of the floral homeotic *AGAMOUS* pathway. *Mol. Cell* **3**, 349–360.

Chitwood, D. H., Nogueira, F.T.S., Howell, M. D., Montgomery, T. A., Carrington, J. C., and Timmermans, M.C.P. (2009). Pattern formation via small RNA mobility. *Genes Dev.* **23**, 549–554.

Chuck, G., Cigan, A. M., Saeteurn, K., and Hake, S. (2007a). The heterochronic maize mutant *Corngrass1* results from overexpression of a tandem microRNA. *Nat. Genet.* **39**, 544–549.
Chuck, G., Meeley, R. B., and Hake, S. (1998). The control of maize spikelet meristem fate by the *APETALA2*-like gene *indeterminate spikelet1*. *Genes Dev.* **12**, 1145–1154.
Chuck, G., Meeley, R., and Hake, S. (2008). Floral meristem initiation and meristem fate are regulated by the maize AP2 genes *ids1* and *sid1*. *Development* **135**, 3013–3019.
Chuck, G., Meeley, R., Irish, E., Sakai, H., and Hake, S. (2007b). The maize *tasselseed4* microRNA controls sex determination and meristem cell fate by targeting *Tasselseed6/ indeterminate spikelet1*. *Nat. Genet.* **39**, 1517–1521.
Coen, E. S., and Meyerowitz, E. M. (1991). The war of the whorls: genetic interactions controlling flower development. *Nature* **353**, 31–37.
Crawford, B. C., Nath, U., Carpenter, R., and Coen, E. (2004). *CINNCINNATA* controls both cell differentiation and growth in petal lobes and leaves of Antirrhinum. *Plant Physiol.* **135**, 244–253.
Dong, Z., Han, M. H., and Fedoroff, N. V. (2008). The RNA-binding proteins HYL1 and SE promote accurate in vitro processing of pri-mRNA by DCL1. *Proc. Natl. Acad. Sci. USA* **105**, 9970–9975.
Dugas, D. V., and Bartel, B. (2008). Sucrose induction of Arabidopsis miR398 represses two Cu/Zn superoxide dismutases. *Plant Mol. Biol.* **67**,
Efroni, I., Blum, E., Goldshmidt, A., and Eshed, Y. (2008). A protracted and dynamic maturation schedule underlies Arabidopsis leaf development. *Plant Cell* **20**, 2293–2306.
Fernandez, L. K., Aida, M., Sedbrook, J., Tasaka, M., Masson, P., and Barton, M. K. (1999). The *PINHEAD/ZWILLE* gene acts pleiotropically in Arabidopsis development and has overlapping functions with the *ARGONAUTE1* gene. *Development* **126**, 469–481.
Fornara, F., and Coupland, G. (2009). Plant phase transitions make a SPLash. *Cell* **138**, 625–627.
Franco-Zorrilla, J. M., Valli, A., Todesco, M., Mateos, I., Puga, M. I., Rubio-Somoza, I., Leyva, A., Weigel, D., Garcia, J. A., and Paz-Ares, J. (2007). Target mimicry provides a new mechanism for regulation of microRNA activity. *Nat. Genet.* **39**, 1033–1037.
Franks, R. G., Wang, C., Levin, J. Z., and Liu, Z. (2002). SEUSS, a member of a novel family of plant regulatory proteins, represses floral homeotic gene expression with LEUNIG. *Development* **129**, 253–263.
Gandikota, M., Birkenbihl, R. P., Hohmann, S., Cardon, G. H., Saedler, H., and Huijser, P. (2007). The miR156/157 recognition element in the 3' UTR of the Arabidopsis SBP box gene *SPL3* prevents early flowering by translational inhibition. *Plant J.* **49**, 683–693.
Gocal, G. F., Sheldon, C. C., Gubler, F., Moritz, T., Bagnall, D. J., MacMillan, C. P., Li, S. F., Parish, R. W., Dennis, E. S., Weigel, D., and King, R. W. (2001). GAMYB-like genes, flowering, and gibberellin signaling in Arabidopsis. *Plant Physiol.* **127**, 1682–1693.
Golden, T. A., Schauer, S. E., Lang, J. D., Pien, S., Mushegian, A. R., Grossniklaus, U., Meinke, D. W., and Ray, A. (2002). *SHORT INTEGUMENTS1/SUSPENSOR1/ CARPEL FACTORY*, a Dicer homolog, is a maternal effect gene required for embryo development in Arabidopsis. *Plant Physiol.* **130**, 808–822.
Goodrich, J., Puangsomiee, P., Martin, M., Long, D., Meyerowitz, E. M., and Coupland, G. (1997). A polycomb-group gene regulates homeotic gene expression in Arabidopsis. *Nature* **386**, 44–51.
Grigg, S. P., Canales, C., Hay, A., and Tsiantis, M. (2005). *SERRATE* coordinates shoot meristem function and axial leaf patterning in Arabidopsis. *Nature* **437**, 1022–1026.
Han, M. H., Goud, S., Song, L., and Fedoroff, N. (2004). The Arabidopsis double-stranded RNA-binding protein HYL1 plays a role in microRNA-mediated gene regulation. *Proc. Natl. Acad. Sci. USA* **101**, 1093–1098.

Hong, R. L., Hamaguchi, L., Busch, M. A., and Weigel, D. (2003). Regulatory elements of the floral homeotic gene *AGAMOUS* identified by phylogenetic footprinting and shadowing. *Plant Cell* **15**, 1296–1309.

Husbands, A., Chitwood, D., Plavskin, Y., and Timmermans, M. (2009). Signals and prepatterns: new insights into organ polarity in plants. *Genes Dev.* **23**, 1986–1997.

Irish, E. E. (1997). Experimental analysis of tassel development in the maize mutant *Tassel Seed 6*. *Plant Physiol.* **114**, 817–825.

Jack, T. (2001). Plant development going MADS. *Plant Mol. Biol.* **46**, 515–520.

Jacobsen, S. E., Running, M. P., and Meyerowitz, E. M. (1999). Disruption of an RNA helicase/RNAse III gene in Arabidopsis causes unregulated cell division in floral meristems. *Development* **126**, 5231–5243.

Jofuku, K. D., den Boer, B.G.W., Van Montagu, M., and Okamuro, J. K. (1994). Control of Arabidopsis flower and seed development by the homeotic gene *APETALA2*. *Plant Cell* **6**, 1211–1225.

Jones-Rhoades, M. W., and Bartel, D. (2004). Computational identification of plant microRNAs and their targets, including a stress-induced miRNA. *Mol. Cell* **14**, 787–799.

Jones-Rhoades, M. W., Bartel, D. P., and Bartel, B. (2006). MicroRNAs and their regulatory roles in plants. *Ann. Rev. Plant Biol.* **57**, 19–53.

Jung, J.-H., Seo, J.-H., Reyes, J. L., Yun, J., Chua, N. H., and Park, C.-M. (2007). The *GIGANTEA*-regulated microRNA172 mediates photoperiodic flowering independent of *CONSTANS* in Arabidopsis. *Plant Cell* **19**, 2736–2748.

Keck, E., McSteen, P., Carpenter, R., and Coen, E. (2003). Separation of genetic functions controlling organ identity in flowers. *EMBO J.* **22**, 1058–1066.

Kidner, C. A., and Martienssen, R. A. (2004). Spatially restricted microRNA directs leaf polarity through *ARGONAUTE1*. *Nature* **428**, 81–84.

Kidner, C. A., and Martienssen, R. (2005). The role of *ARGONAUTE1* (*AGO1*) in meristem formation and identity. *Dev. Biol.* **280**, 504–517.

Kim, S., Soltis, P. S., Wall, K., and Soltis, D. E. (2006). Phylogeny and domain evolution in the APETALA2-like genes family. *Mol. Biol. Evol.* **23**, 107–120.

Kobayashi, Y., and Weigel, D. (2007). Move on up, it's time for change-mobile signals controlling photoperiod-dependent flowering. *Genes Dev.* **21**, 2371–2384.

Kurihara, Y., Takashi, Y., and Watanabe, Y. (2006). The interaction between DCL1 and HYL1 is important for efficient and precise processing of pri-mRNA in plant microRNA biogenesis. *RNA* **12**, 206–212.

Kurihara, Y., and Watanabe, K. (2004). Arabidopsis micro-RNA biogenesis through Dicer-like 1 protein functions. *Proc. Natl. Acad. Sci. USA* **101**, 12753–12758.

Lagos-Quintana, M., Rauhut, R., Lendeckel, W., and Tuschl, T. (2001). Identification of novel genes coding for small expressed RNAs. *Science* **294**, 853–858.

Larue, C. T., Wen, J., and Walker, J. C. (2009). A microRNA-transcription factor module regulates lateral organ size and patterning in Arabidopsis. *Plant J.* **58**, 450–463.

Lau, N. C., Lim, L. P., Weinstein, E. G., and Bartel, D. P. (2001). An abundant class of tiny RNAs with probable regulatory roles in *Caenorhabditis elegans*. *Science* **294**, 858–862.

Laubinger, S., Sachsenberg, T., Zeller, G., Busch, W., Lohmann, J. U., Ratsch, G., and Weigel, D. (2008). Dual roles of the nuclear cap-binding complex and SERRATE in pre-mRNA splicing and miRNA processing in *Arabidopsis thaliana*. *Proc. Natl. Acad. Sci. USA* **105**, 8795–8800.

Laudencia-Chingcuanco, D., and Hake, S. (2002). The *indeterminate floral apex1* gene regulates meristem determinacy and identity in the maize inflorescence. *Development* **129**, 2629–2638.

Laufs, P., Peaucelle, A., Morin, H., and Traas, J. (2004). MicroRNA regulation of the *CUC* genes is required for boundary size control in Arabidopsis meristems. *Development* **131**, 4311–4322.

Lauter, N., Kampani, A., Carlson, S., Goebel, M., and Moose, S. P. (2005). microRNA172 down-regulates *glossy15* to promote vegetative phase change in maize. *Proc. Natl. Acad. Sci. USA* **102**, 9412–9417.

Lee, R. C., and Ambros, V. (2001). An extensive class of small RNAs in *Caenorhabditis elegans*. *Science* **294**, 862–864.

Lee, R. C., Feinbaum, R. L., and Ambros, V. (1993). The *C. elegans* heterochronic gene *lin-4* encodes small RNAs with antisense complimentarity to *lin-14*. *Cell* **75**, 843–854.

Li, W.-X., Oono, Y., Zhu, J., He, X.-J., Wu, J.-M., Iida, K., Lu, X.-Y., Cui, X., Jin, H., and Zhu, J.-K. (2008). The Arabidopsis NFYA5 transcription factor is regulated transcriptionally and posttranscriptionally to promote drought resistance. *Plant Cell* **20**, 2238–2251.

Li, J., Yang, Z., Yu, B., Liu, J., and Chen, X. (2005). Methylation protects miRNAs and siRNAs from 3'-end uridylation activity in Arabidopsis. *Curr. Biol.* **15**, 1501–1507.

Liu, Q., and Chen, Y. Q. (2009). Insights into the mechanism of plant development: Interactions of miRNA pathways with phytohormone response. *Biochem. Biophys. Res. Commun.* **384**, 1–5.

Liu, Z., and Meyerowitz, E. M. (1995). *LEUNIG* regulates *AGAMOUS* expression in Arabidopsis flowers. *Development* **121**, 975–991.

Llave, C., Kasschau, K. D., Rector, M. A., and Carrington, J. C. (2002). Endogenous and silencing-associated small RNAs in plants. *Plant Cell* **14**, 1605–1619.

Lu, C., and Fedoroff, N. (2000). A mutation in the Arabidopsis *HYL1* gene encoding a dsRNA binding protein affects responses to abscisic acid, auxin, and cytokinin. *Plant Cell* **12**, 2351–2366.

Luo, D., Carpenter, R., Copsey, L., Vincent, C., and Coen, E. (1999). Control of organ asymmetry in flowers in Antirrhinum. *Cell* **99**, 367–376.

Luo, D., Carpenter, R., Vincent, C., Copsey, L., and Coen, E. (1996). Origin of floral asymmetry in Antirrhinum. *Nature* **383**, 794–799.

Mallory, A. C., Dugas, D. V., Bartel, D. P., and Bartel, B. (2004). Micro-RNA regulation of NAC-domain targets is required for proper formation and separation of adjacent embryonic, vegetative, and floral organs. *Curr. Biol.* **14**, 1035–1046.

McSteen, P., Laudencia-Chingcuanco, D., and Colasanti, J. (2000). A floret by any other name: control of meristem identity in maize. *Trends Plant Sci.* **5**, 61–66.

Millar, A. A., and Gubler, F. (2005). The Arabidopsis *GAMYB*-like genes, *MYB33* and *MYB65*, are microRNA-regulated genes that redundantly facilitate anther development. *Plant Cell* **17**, 705–721.

Moose, S. P., and Sisco, P. H. (1994). *Glossy15* controls the epidermal juvenile-to-adult transition in maize. *Plant Cell* **6**, 1343–1355.

Moose, S. P., and Sisco, P. H. (1996). *Glossy15*, an *APETALA2*-like gene from maize that regulates leaf epidermal cell identity. *Genes Dev.* **10**, 3018–3027.

Morel, J.-B., Godon, C., Mourain, P., Beclin, C., Boutet, S., Feuerbach, F., Proux, F., and Vaucheret, H. (2002). Fertile hypomorphic *ARGONAUTE* (*ago1*) mutants impaired in post-transcriptional gene silencing and virus resistance. *Plant Cell* **14**, 629–639.

Motte, P., Saedler, H., and Schwartz-Sommer, Z. (1998). *STYLOSA* and *FISTULATA*: regulatory components of the homeotic control of Antirrhinum floral organogenesis. *Development* **125**, 71–84.

Nag, A., King, S., and Jack, T. (2009). *miR319a* targeting of *TCP4* is critical for petal growth and development in Arabidopsis. *Proc. Natl. Acad. Sci. USA* **106**, 22534–22539.

Nagpal, P., Ellis, C. M., Weber, H., Ploense, S. E., Barkawi, L. S., Guilfoyle, T. J., Hagen, G., Alonso, J. M., Cohen, J. D., Farmer, E. E., Ecker, J. R., and Reed, J. W. (2005). Auxin response factors ARF6 and ARF8 promote jasmonic acid production and flower maturation. *Development* **132**, 4107–4118.

Nath, U., Crawford, B.C.W., Carpenter, R., and Coen, E. (2003). Genetic control of surface curvature. *Science* **299**, 1404–1407.

Nikovics, K., Blein, T., Peaucelle, A., Ishida, T., Morin, H., Aida, M., and Laufs, P. (2006). The balance between the *MIR164A* and *CUC2* genes controls leaf margin serration in Arabidopsis. *Plant Cell* **18**, 2929–2945.

Palatnik, J. F., Allen, E., Wu, X., Schommer, C., Schwab, R., Carrington, J. C., and Weigel, D. (2003). Control of leaf morphogenesis by microRNAs. *Nature* **425**, 257–263.

Palatnik, J. F., Wollmann, H., Schommer, C., Schwab, R., Boisbouvier, J., Rodriguez, R., Warthmann, N., Allen, E., Dezulian, T., Huson, D., Carrington, J. C., and Weigel, D. (2007). Sequence and expression differences underlie functional specialization of Arabidopsis microRNAs miR159 and miR319. *Dev. Cell* **13**, 115–125.

Park, W., Li, J., Song, R., Messing, J., and Chen, X. (2002). CARPEL FACTORY, a Dicer homolog, and HEN1, a novel protein, act in microRNA metabolism in *Arabidopsis thaliana. Curr. Biol.* **12**, 1484–1495.

Park, M. Y., Wu, G., Gonzalez-Suiser, A., Vaucheret, H., and Poethig, R. S. (2005). Nuclear processing and export of microRNAs in Arabidopsis. *Proc. Natl. Acad. Sci. USA* **102**, 3691–3696.

Parry, G., Calderon-Villalobos, L. I., Prigge, M., Peret, B., Dharmasiri, S., Itoh, H., Lechner, E., Gray, W. M., Bennett, M., and Estelle, M. (2009). Complex regulation of the TIR1/AFB family of auxin receptors. *Proc. Natl. Acad. Sci. USA* **106**, 22540–22545.

Peaucelle, A., Morin, H., Traas, J., and Laufs, P. (2007). Plants expressing a miR164-resistant *CUC2* gene reveal the importance of post-meristematic maintenance of phyllotaxy in Arabidopsis. *Development* **134**, 1045–1050.

Qi, Y., Denli, A. M., and Hannon, G. J. (2005). Biochemical specialization within Arabidopsis RNA silencing pathways. *Mol. Cell* **19**, 421–428.

Rajagopalan, R., Vaucheret, H., Trejo, J., and Bartel, D. P. (2006). A diverse and evolutionarily fluid set of microRNAs in *Arabidopsis thaliana. Genes Dev.* **20**, 3407–3425.

Ramachandran, V., and Chen, X. (2008). Degradation of microRNAs by a family of exoribonucleases in Arabidopsis. *Science* **321**, 1490–1492.

Reinhart, B. J., Weinstein, E. G., Rhoades, M. W., Bartel, B., and Bartel, D. P. (2002). MicroRNAs in plants. *Genes Dev.* **16**, 1616–1626.

Reyes, J. L., and Chua, N. H. (2007). ABA induction of miR159 controls transcript levels of two MYB factors during Arabidopsis seed germination. *Plant J.* **49**, 592–606.

Rhoades, M. W., Reinhart, B. J., Lim, L. P., Burge, C. B., Bartel, B., and Bartel, D. P. (2002). Prediction of plant microRNA targets. *Cell* **110**, 513–520.

Schauer, S. E., Jacobsen, S. E., Meinke, D. W., and Ray, A. (2002). DICER-LIKE1: blind men and elephants in Arabidopsis development. *Trends Plant Sci.* **7**, 487–491.

Schmid, M., Uhlenhaut, N. H., Godard, F., Demar, M., Bressan, R., Weigel, D., and Lohmann, J. U. (2003). Dissection of floral induction pathways using global expression analysis. *Development* **130**, 6001–6012.

Schwab, R., Palatnik, J. F., Riester, M., Schommer, C., Schmid, M., and Weigel, D. (2005). Specific effects of microRNAs on the plant transcriptome. *Dev. Cell* **8**, 517–527.

Sieber, P., Wellmer, F., Gheyselinck, J., Riechmann, J. L., and Meyerowitz, E. M. (2007). Redundancy and specialization among plant microRNAs: role of the MIR164 family in developmental robustness. *Development* **134**, 1051–1060.

Sieburth, L. E., Running, M. P., and Meyerowitz, E. M. (1995). Genetic separation of third and fourth whorl functions of *AGAMOUS. Plant Cell* **7**, 1249–1258.

Sunkar, R., Chinnusamy, V., Zhu, J., and Zhu, J. K. (2007). Small RNAs as big players in plant abiotic stress responses and nutrient deprivation. *Trends Plant Sci.* **12**, 301–309.

Takada, S., Hibara, K. I., Ishida, T., and Tasaka, M. (2001). The *CUP-SHAPED COTYLEDON1* gene of Arabidopsis regulates shoot apical meristem formation. *Development* **128**, 1127–1135.

Telfer, A., Bollman, K. M., and Poethig, R. S. (1997). Phase change and the regulation of trichome distribution in *Arabidopsis thaliana*. *Development* **124**, 645–654.

Telfer, A., and Poethig, R. S. (1998). *HASTY*: a gene that regulates the timing of shoot maturation in *Arabidopsis thaliana*. *Development* **125**, 1889–1898.

Thompson, B. E., and Hake, S. (2009). Translational biology: from Arabidopsis flowers to grass inflorescence architecture. *Plant Physiol.* **149**, 38–45.

Vaucheret, H. (2008). Plant ARGONAUTES. *Trends Plant Sci.* **13**, 350–358.

Vaucheret, H., Vazquez, F., Crete, P., and Bartel, D. P. (2004). The action of *ARGONAUTE1* in the miRNA pathway and its regulation by the miRNA pathway is crucial for plant development. *Genes Dev.* **18**, 1187–1197.

Voinnet, O. (2008). Post-transcriptional RNA silencing in plant–microbe interactions: a touch of robustness and versatility. *Curr. Opin. Plant Biol.* **11**, 464–470.

Wang, J. W., Czech, B., and Weigel, D. (2009). miR156-regulated SPL transcription factors define an endogenous flowering pathway in Arabidopsis. *Cell* **138**, 738–749.

Whipple, C. J., Ciceri, P., Padilla, C. M., Ambrose, B. A., Bandong, S. L., and Schmidt, R. J. (2004). Conservation of B-class floral homeotic gene function between maize and Arabidopsis. *Development* **131**, 6083–6091.

Wu, G., Park, M. Y., Conway, S. R., Wang, J. W., Weigel, D., and Poethig, R. S. (2009). The sequential action of miR156 and miR172 regulates developmental timing in Arabidopsis. *Cell* **138**, 750–759.

Wu, G., and Poethig, R. S. (2006). Temporal regulation of shoot development in *Arabidopsis thaliana* by miR156 and its target *SPL3*. *Development* **133**, 3539–3547.

Wu, M. F., Tian, Q., and Reed, J. W. (2006). Arabidopsis microRNA167 controls patterns of *ARF6* and *ARF8* expression, and regulates both female and male reproduction. *Development* **133**, 4211–4218.

Yamaguchi, A., Wu, M. F., Yang, L., Wu, G., Poethig, R. S., and Wagner, D. (2009). The microRNA-regulated SBP-box transcription factor SPL3 is a direct upstream activator of *LEAFY, FRUITFULL*, and *APETALA1*. *Dev. Cell* **17**, 268–278.

Yang, Z., Ebright, Y. W., Yu, B., and Chen, X. (2006b). HEN1 recognizes 21–24 nt small RNA duplexes and deposits a methyl group onto the 2' OH of the 3' terminal nucleotide. *Nuc. Acids Res.* **34**, 667–675.

Yang, L., Liu, Z., Lu, F., Dong, A., and Huang, H. (2006a). *SERRATE* is a novel nuclear regulator in primary microRNA processing in Arabidopsis. *Plant J.* **47**, 841–850.

Yu, B., Bi, L., Zheng, B., Ji, L., Chevalier, D., Agarwal, M., Ramachandran, V., Li, W., Lagrange, T., Walker, J. C., and Chen, X. (2008). The FHA domain proteins DAWDLE in Arabidopsis and SNIP1 in humans act in small RNA biogenesis. *Proc. Natl. Acad. Sci. USA* **105**, 10073–10078.

Yu, B., Yang, Z., Li, J., Minakhina, S., Yang, M., Padgett, R. W., Steward, R., and Chen, X. (2005). Methylation as a crucial step in plant microRNA biogenesis. *Science* **307**, 932–935.

Zhao, L., Kim, Y.-J., Dinh, T. T., and Chen, X. (2007). miR172 regulates stem cell fate and defines the inner boundary of *APETALA3* and *PISTILLATA* expression domain in Arabidopsis floral meristems. *Plant J.* **51**, 840–849.

CHAPTER THIRTEEN

Development of Flowering Plant Gametophytes

Hong Ma[*] *and* Venkatesan Sundaresan[†]

Contents

Abstract

Plant reproduction occurs through the production of gametes by a haploid generation, the gametophyte. Flowering plants have highly reduced male and female gametophytes, called pollen grains and embryo sacs, respectively, consisting of only a few cells. Gametophytes are critical for sexual reproduction, but detailed understanding of their development remains poor as compared to

[*] State Key Laboratory of Genetic Engineering and School of Life Sciences, Institute of Plant Biology, Fudan University, Shanghai, China, and Department of Biology, Pennsylvania State University, University Park, Pennsylvania, USA
[†] Departments of Plant Biology and Plant Sciences, University of California, Davis, California, USA

Current Topics in Developmental Biology, Volume 91
ISSN 0070-2153, DOI 10.1016/S0070-2153(10)91013-2

the diploid sporophyte. This article reviews recent progress in understanding the mechanisms underlying gametophytic development and function in flowering plants. The focus is on genes and molecules involved in the processes of initiation, growth, cell specification, and fertilization of the male and female gametophytes derived primarily from studies in model systems.

1. Introduction

The life cycle of land plants involves an alternation of generations between a haploid gametophyte and a diploid sporophyte. Whereas animal gametes are formed directly after meiosis, plant gametes are produced only after growth of the multicellular haploid gametophyte. The morphological complexity of the haploid generation ranges from the lower plant gametophytes which form the main plant body and support the much smaller sporophytes, to the highly reduced three-celled male gametophyte (pollen) and seven-celled female gametophyte (embryo sac) that are characteristic of most angiosperms. Despite the crucial importance of the gametophytic generation to the reproduction of flowering plants, their small size, short life-span, and inaccessibility during development has resulted in their receiving comparatively little attention, leading to the description "the forgotten generation" (Brukhin *et al.*, 2005). Nevertheless, the past few years have seen some exciting progress in the study of flowering plant gametophytes, which forms the basis for this review. Here, we discuss advances in understanding the development of the male and the female gametophytes, and their interactions during the process of double fertilization, with an emphasis on genes from *Arabidopsis thaliana.*

2. Development of the Male Gametophyte or Pollen Grain

2.1. Signaling pathways and transcriptional control of microsporogenesis

During flower development, cells in the anther primordium divide and differentiate to form several cell types (Fig. 13.1; Ma, 2005; Wilson and Zhang, 2009). At stage 1, an anther primodium is formed from the floral meristem. Cell division leads to more cells in each of the three cell layers and some cells in L2 become archesporial cells (ACs) at stage 2 (Fig. 13.1A). The ACs then divide periclinally at stage 3 to form inner primary sporogenous cells and outer primary parietal cells, which further divide at stage 4 to form secondary parietal cells. At stage 5, there are four

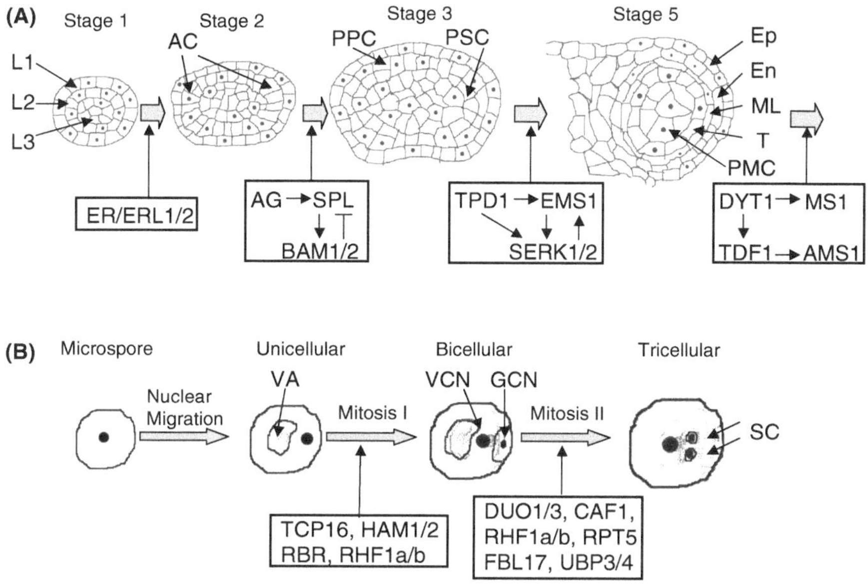

Figure 13.1 Morphological features of microsporogenesis and microgametogenesis. (A) Early stages of anther development (from left): Stage 1: the anther primordium has three cell layers derived from the floral meristem: L1, L2, and L3; Stage 2, Archesporial cells (AC) elongate in preparation for periclinal division; Stage 3, Periclinal division of archesporial cells give rise to primary parietal cells (PPC, outer) and primary sporogenous cells (PSC, inner); Stage 4 (not shown), further division of parietal cells; Stage 5, one of four anther lobes is shown with epidermis (Ep) derived from the L1, and four other cell types derived from the L2: endothecium (En), middle layer (ML), tapetum (T), and pollen mother cells (PMC). Genes acting at different stages are shown in the boxes, with arrows between genes indicating proposed positive interactions, and the "T" shape representing repression of *SPL* by *BAM1/2* (see text). (B) Pollen development (from left): microspore, with a nucleus; the nucleus migrates to one side, and a large vacuole forms (VA); pollen mitosis I, yielding the vegetative cell nucleus (VCN) and generative cell nucleus (GCN) in the binucleate pollen; pollen mitosis II (of the generative cell) produces sperm cells, forming the mature trinucleate pollen. Genes required for normal mitoses I and II are indicated in two boxes, respectively.

anther lobes (one is shown in Fig. 13.1A) in a typical anther; each anther lobe contains four somatic cell types, the epidermis, endothecium, middle layer, and the tapetum, which encircle a group of pollen mother cells (Fig. 13.1A). Therefore, the epidermis results from cell division in the L1 layer, whereas the other cells types are largely progeny of the L2 layer. Molecular genetic studies in recent years have uncovered evidence for cell–cell communication that contributes to cell fate determination during the cell division and differentiation process that results in the formation of the five cell types in the anther lobe. One of the earliest events in anther development is the formation in the L2 layer of the ACs, which give rise to the precursors of pollen mother cells and also the inner three somatic cell layers of the anther lobe.

Recent genetic studies in Arabidopsis revealed that the ER/ERL1/ERL2 (ERECTA/ERECTA-LIKE1/2) trio of receptor-linked protein kinases and MPK3/6 (MAP KINASE3/6) are both important for the formation of anther lobes. Mutants defective in these genes often form anthers that lack one or more anther lobes, sometimes with only a filament (Hord *et al.*, 2008). The observations that the entire anther lobe can fail to form strongly suggest that these genes are crucial for the specification of the ACs, or the normal function of these cells. Further studies using AC-specific markers are needed to test these ideas. The fact that ER/ERL1/2 proteins are members of the predicted leucine-rich repeat receptor-linked protein kinases (LRR-RLKs) suggests that cell–cell communication is important for formation of anther lobes. Other studies have found that both the ER/ERL1/ERL2 and MPK3/6 (MAP KINASE3/6) and other components of the MAP kinase cascade are important for other developmental processes, such as guard cell development (chapter 9; Lampard *et al.*, 2009; Shpak *et al.*, 2005; Wang *et al.*, 2007). Therefore, their functions in anther development might also be connected somehow. It is not unreasonable to hypothesize that these two signaling modules are involved in the same pathway controlling AC development and/or function.

As the progeny of the AC differentiates into the primary parietal cells and primary sporogenous cells, the precursors of somatic cells and pollen mother cells, respectively, it is important for these two kinds of descendants to develop in a coordinated way. One of the earliest genes discovered to be essential for anther development is the *SPL/NZZ* (*SPOROCYTELESS/NOZZLE*) gene, which encodes a putative transcription factor (Schiefthaler *et al.*, 1999; Yang *et al.*, 1999c). In the *spl* mutant, anther lobes are formed, but they lack the normal cells interior of the epidermis. The *SPL/NZZ* gene is also needed for megasporogenesis (Schiefthaler *et al.*, 1999; Yang *et al.*, 1999a; see below for additional discussion), suggesting that its primary role might be to promote the formation of sporogenous cells, and the failure to form L2-derived somatic cells in the *spl* anther might be a secondary effect. *SPL* was shown to be a direct target of the floral homeotic protein *AG* (*AGAMOUS*) and *AG*-independent expression of *SPL* results in microspore production in the *ag* mutant (Ito *et al.*, 2004). Recently, a microarray analysis using the *spl* mutant anther showed that over 1900 genes were expressed at levels at least twofold different from those in the wild-type anther, suggesting that at least some of these genes might be directly regulated by *SPL/NZZ* (Wijeratne *et al.*, 2007).

Another recent study has shed some light on the regulation of cell differentiation during the early stages of another development, particularly the role of cell–cell communication in this regulation. *BAM1* (*BARELY ANY MERISTEM1*) and *BAM2* are two close paralogs of *CLV1*

(*CLAVATA1*), which encodes an LRR-RLK and is key regulator of the stem cell pool in the shoot apical meristem (Clark *et al.*, 1997). Single mutations in either *BAM1* or *BAM2* cause no obvious defects, but *bam1 bam2* double mutants have meristem defects and are male and female sterile (DeYoung *et al.*, 2006; Hord *et al.*, 2006). Detailed phenotypic analysis revealed that the *bam1 bam2* anther fails to form the three L2-derived somatic cell layers: the endothecium, middle layer, and tapetum (Hord *et al.*, 2006). Further cytological and molecular tests indicated that the cells occupying the position of these somatic cells have properties of meiotic cells. Therefore, the *BAM1/2* genes are needed for the normal differentiation of these somatic cells and, in *bam1/2*, extra meiotic cells are formed instead, suggesting that cell–cell signaling is needed for proper somatic cell differentiation. Furthermore, expression analysis of *BAM1/2* and *SPL/NZZ* in *spl* and *bam1/2* mutant anthers, respectively, provides support for a possible positive–negative feedback loop between *SPL/NZZ* and *BAM1/2* (Hord *et al.*, 2006), similar to that between *CLV1/3* and *WUS* (*WUSCHEL*) (Carles and Fletcher, 2003).

As the primary parietal cells divide to form secondary parietal cells, and then eventually generating three L2-derived somatic cell layers (Fig. 13.1A), contributions of several components of a signaling pathway, including *EMS1/EXS* (*EXCESS MALE SPOROCYTE1/EXCESS SPOROGENOUS CELLS*), *TPD1* (*TAPETUM DETERMINANT1*), and *SERK1/SERK2* (*SOMATIC EMBRYO RECEPTOR KINASE1/2*) are required. (Albrecht *et al.*, 2005; Canales *et al.*, 2002; Yang *et al.*, 2003a, 2005; Zhao *et al.*, 2002). *EMS1/EXS* was discovered in separate mutant screens and found to be required for the formation of the tapetum (Canales *et al.*, 2002; Zhao *et al.*, 2002). In *ems1/exs* mutant anthers, the tapetum is absent and extra pollen mother cells are formed. Two alternative models have been proposed to explain this phenotype: either cells that normally form the tapetum develop into pollen mother cells, or cell division in the tapetum precursors is reduced whereas cell division for pollen mother cells is elevated. Regardless of these options for cell division/differentiation, the fact that *EMS1/EXS* encodes a member of the LRR-RLK family and that the protein is localized to the cell surface in an onion epidermal cell assay strongly suggest that cell–cell signaling is important for the development of the tapetum.

The idea that cell–cell signaling is important is further supported by the finding that *TPD1* is required for the same process. The *tpd1* mutant shows a phenotype identical to that of the *ems1/exs* mutant, and the TPD1 protein is predicted to be a secreted ligand (Yang *et al.*, 2003a). If extra pollen mother cells are formed when the tapetum is absent, it seems to make sense that the signal should be produced in the pollen mother cells, and that the receptor should be found in the cells that are forming the tapetum. Indeed, *EMS1* expression was detected in the precursor of both the tapetum and pollen mother cells, at the highest levels in cells of the nascent tapetum, whereas the highest expression of *TPD1* was found in the pollen mother cells (Yang *et al.*, 2003a; Zhao *et al.*, 2002). Further studies have shown that *EMS1/EXS* and *TPD1* act in the same genetic pathway and that the proteins also interact physically, providing additional evidence for them being the receptor and ligand, respectively, for signaling between the pollen mother cell and the early tapetum cells (Yang *et al.*, 2005; Jia *et al.*, 2008).

Two additional studies have added another component to this signaling module, *SERK1/2*, which are functionally similar and can be lost singly without causing dramatic phenotypes in anther development (Albrecht *et al.*, 2005; Colcombet *et al.*, 2005). The *serk1/2* double mutant, however, is male sterile and forms anthers that are similar to those of *ems1/exs* and *tpd1* mutants. *SERK1* and *SERK2* also encode members of the LRR-RLK family, suggesting that they also are involved in cell–cell signaling. Although there is no direct evidence, there is some hint that SERK1/2 might physically interact with EMS1/EXS. It was shown that a close paralog of SERK1 and SERK2, BAK1 (BRI1 ASSOCIATED KINASE)/SERK3, can interact with the receptor for the hormone brassinosteroid, BRI1, (BR-INSENSITIVE1) suggesting that SERK3/BAK1 and BRI1 form heteromeric receptors (Li *et al.*, 2002; Nam and Li, 2002). Although EMS1/EXS lacks the island domain required for binding of brassinolides, EMS1/EXS is in the same clade/subfamily as BRI1. Therefore, it is possible that SERK1/2 and EMS1/EXS form heteromeric receptors for the specification of tapetum fate. Although pollen mother cells are formed in the *ems1* mutant and can undergo meiotic nucleic divisions, meiotic cytokinesis fails to occur in these mutants, suggesting that the tapetum is required for normal meiotic cytokinesis (Zhao *et al.*, 2002).

In addition to the *SPL/NZZ* gene mentioned earlier, several other genes encoding transcription factors have been found to be crucial for anther development. One of the earliest male-specific regulators is the *DYT1* (*DYSFUNCTIONAL TAPETUM1*) gene, which encodes a member of the bHLH family (Zhang *et al.*, 2006); *dyt1* mutant anthers contain a slightly deformed tapetum with precociously large vacuoles and show abnormally low expression of several genes important for post-meiotic anther development (Zhang *et al.*, 2006), including *AMS* (*ABORTED MICROSPORES*) and *MS1* (*MALE STERILITY1*), both encoding putative transcription factors (Ito *et al.*, 2007; Sorensen *et al.*, 2003; Xu *et al.*, 2010; Yang *et al.*, 2007). AMS, also a member of the bHLH family, is required for post-meiotic microspore development (Sorensen *et al.*, 2003) and interacts with the SET domain protein ASHR3, which is a putative histone methylase and is involved in anther development (Thorstensen *et al.*, 2008). A recent study showed that AMS regulates the expression of many genes in the flower and that it can bind to the promoters of many genes *in vivo*, suggesting that AMS is a direct regulator of transcription of these genes (Xu *et al.*, 2010). In addition, it was found through genetic and expression studies that *TDF1* (*TAPETUM DEVELOPMENT AND FUNCTION1*)/*MYB35* is activated by *DYT1* (*DYSFUNCTIONAL TAPETUM1*) directly or indirectly and that it activates *AMS* expression (Wijeratne *et al.*, 2007; Zhu *et al.*, 2008). *MS1* encodes a putative transcriptional regulator with a PHD domain, and is required for normal pollen wall formation (Ito *et al.*, 2007; Yang *et al.*, 2007). Analysis using a fusion of MS1 with a repressor motif indicates that MS1 normally activates transcription (Ito *et al.*, 2007). Furthermore, gene expression studies using an inducible MS1 protein and microarray analysis revealed that MS1 is likely a direct activator of the *MYB99* gene (Ito *et al.*, 2007; Yang *et al.*, 2007), which is expressed in the tapetum and might have redundant functions with other MYB genes in the regulation of phenylpropanoid metabolism during pollen development (Alves-Ferreira *et al.*, 2007). From these studies, a transcriptional cascade or network is emerging that is essential for anther development (Fig. 13.1A; Wijeratne *et al.*, 2007; Wilson and Zhang, 2009; Zhu *et al.*, 2008). Furthermore, studies in rice showed that several genes are conserved and play similar roles, including the homologs of *EMS1/EXS, DYT1,* and AMS (Wilson and Zhang, 2009).

Plant hormones are known to be important for male fertility, including jasmonate, gibberellic acid, and brassinosteroids (e.g., Cheng *et al.*, 2004; Suzuki *et al.*, 2004; von Malek *et al.*, 2002). A recent study showed that BR promotes anther and pollen development in several ways, including cell division and expansion. Moreover, mutants defective in brassinosteroid synthesis and response show reduced pollen numbers and viability (Ye *et al.*, 2010). It was shown that these mutants have reduced expression of several key anther development genes, including *SPL, TDF1, AMS,* and *MS1,* and *in vivo* binding experiments revealed that BR-dependent transcription factor BES1 (BRI1 EMS SUPPRESSOR1) is enriched at the promoter regions of these genes, suggesting that BES1 is a direct transcriptional regulator of these genes (Ye *et al.*, 2010).

2.2. Regulation of male meiotic processes by multiple mechanisms

Male meiosis is an essential step for pollen production, and has been studied extensively to understand the key meiotic processes. To achieve the purpose of dividing the genome into two haploid chromosome sets homologous chromosomes (homologs) must recognize and pair with each other, synapse together, and undergo recombination, to maintain stable association until they are ready to separate at the transition between metaphase I and anaphase I. Because of space limitations, this review will focus on several regulatory mechanisms; for meiotic processes of homolog pairing, synapsis, recombination, and chromosome segregation, please be referred to recent reviews (e.g., Hamant *et al.*, 2006; Ma, 2006; Muyt *et al.*, 2009).

Although transcriptional regulation within the meiotic cell is likely important, there is relatively little information on genes important for the regulation of meiotic transcription. Nevertheless, one such potential regulator, called *MMD1/DUET*, was found by genetic studies (Reddy *et al.*, 2003; Yang *et al.*, 2003b). The Arabidopsis *mmd1* (*male meiocyte death1*) mutant was identified based on its male sterility phenotype; further cell biological analysis showed that meiosis can initiate, but becomes abnormal soon afterwards, during prophase I. Meiotic cells then undergo programmed cell death, showing chromosome fragmentation. This is a very unusual phenotype because most plant meiotic mutants showing abnormal chromosomal morphology or behavior can complete meiosis, and even initiate microspore development. *MMD1/DUET* encodes a putative transcriptional regulator with a PHD domain, which is associated with chromatin remodeling. Therefore, it is possible that MMD1/DUET is involved in the reprogramming of transcriptional processes for normal meiosis, and a failure in such a regulatory mechanism causes defects beyond those directly associated with chromosomal interactions or segregation, which are processes affected in most plant meiotic mutants.

For successful meiosis, homolog pairing, synapsis, and recombination must be closely coordinated spatially and temporally; this coordination requires very tight regulation. Although there is just limited knowledge, genetic studies have revealed several key regulators of these meiotic processes. One of these regulators is called *SDS* (*SOLO DANCERS*), which was isolated as a nearly sterile Arabidopsis mutant and is required for normal synapsis and recombination, probably pairing as well (Azumi *et al.*, 2002). The *SDS* gene encodes a putative cyclin that is somewhat similar to the A and B type cyclins, but is not a true member of the A and B types according to phylogenetic analysis (Wang *et al.*, 2004), which also identified a putative rice ortholog, suggesting that this gene is conserved in plants. Indeed, analysis using transgenic rice plants indicates that the rice *SDS* homolog has similar functions (Chang *et al.*, 2009). The SDS protein can interact with known cyclin-dependent protein kinases (CDKs) in a yeast two-hybrid assay, supporting the idea that it is an authentic cyclin (Azumi *et al.*, 2002). In situ RNA hybridization experiments indicate that *SDS* is specifically expressed in the male and female meiocytes; this, together with the only detected mutant phenotype being in meiosis, strongly suggests that SDS is a meiosis-specific cyclin. It is possible that SDS and its cognate CDK(s) together phosphorylate key proteins, thereby regulating the progression of meiotic processes, analogous to the functions of cyclins/CDKs during the mitotic cell cycle.

During meiosis, proteins that have served their functions would need to be removed to prevent abnormally prolonged activity, similar to those for mitosis and many other processes. In Arabidopsis, the *ASK1* (*Arabidopsis SKP1-LIKE1*) gene, which encodes a subunit of the SCF (SKP1-Cullin-F-box)-type ubiquitin ligases, was found to be essential for male meiosis, revealing the importance of regulated protein degradation by the ubiquitin-mediated proteasome-dependent pathway (Yang *et al.*, 1999a, b; Zhao *et al.*, 2003, 2006). In the *ask1* mutant, a number of chromosomal processes are defective, including condensation and reorganization, and release from the nuclear envelope. In addition, homolog pairing and separation are also abnormal (Yang *et al.*, 1999a, 2006b). Because the removal of cohesin is required for homolog separation, it is possible that ASK1-containing SCF(s) is(are) needed for cohesin removal, as supported by persistent localization of cohesin in the *ask1* mutant meiotic chromosomes (Zhao *et al.*, 2006). This is a novel role for a putative SCF in regulating meiosis, and preliminary analysis suggests that it is conserved in other plants. Further investigation is needed to uncover the F-box subunit of the SCF and the substrate(s) of ubiquitination. These studies suggest that plant meiosis is regulated by a diverse array of mechanisms and we are only just beginning to get a glimpse of such regulation.

2.3. Regulation of the asymmetric first pollen mitosis

Following meiosis during early pollen development, the microspore nucleus first migrates to one side of the cell, with the formation of a large vacuole; then an asymmetric mitotic cell division produces a large vegetative cell and a small generative cell, which is the male germline (Fig. 13.1B). Subsequently, the vegetative cell completely encloses the generative cell, which then divide to form two sperm cells (Fig. 13.1B). Following pollination, the vegetative cell produces a pollen tube, which grows through female tissues and delivers the sperm cells to the female gametophyte. Large-scale mutant screens and transcriptomics have identified mutations and genes involved in pollen development (Honys and Twell, 2003; Lalanne *et al.*, 2004). Expression studies showed that the *TCP16* (*TB1-CYC-PCF16*) gene, encoding a putative transcription factors, is expressed when microspores are released from the tetrad and before the first mitosis (Takeda *et al.*, 2006). Using a GUS reporter gene, *TCP16* expression was detected at low levels at the tetrad stage and found to be much stronger afterwards in microspores with a single nucleus. *TCP16* expression subsequently decreases gradually, becoming undetectable in the mature pollen. Its requirement for normal early microspore development is supported by RNAi analysis that showed defects at the unicellular stage (Takeda *et al.*, 2006), including anucleated microspores soon after release from the tetrad, with an increased frequency at the uninucleate stage.

An important mechanism for regulating gene expression is the control of chromatin structure via histone modification. Recently, it was found that *HAM1* (*HISTONE ACETYLTRANSFERASE OF THE MYST FAMILY1*) and *HAM2*, two members of the MYST family of histone acetylases, are essential for normal male gametogenesis (Latrasse *et al.*,

2008). *HAM1* and *HAM2* belong to a plant-specific clade of MYST genes, suggesting that they might have evolved plant-specific functions, such as those for multicellular gametophyte development. It was found that *ham1* or *ham2* single mutants develop without obvious morphological defects. However, *ham1 ham2* double mutant plants could not be obtained, indicating that these two genes have similar functions and are redundant under laboratory conditions. Furthermore, in *ham1/ham1 HAM2/ham2* and *HAM1/ham1 ham2/ham2* plants, pollen development is abnormal in a fraction of the pollen, consistent with the *ham1 ham2* double mutant pollen being defective. The defective pollen grains contain a single large nucleus, indicating that there is a failure prior to the first pollen mitosis and that the HAM1/2-mediated histone modification is needed for normal gene expression during early pollen development.

Analysis of the Arabidopsis homolog (*RBR, RETINOBLASTOMA RELATED*) of the Retinoblastoma protein, a negative regulator of cell proliferation (Chen *et al.*, 2009; Johnston *et al.*, 2008), revealed that *RBR* is required for normal differentiation of the vegetative cell and the male germline. In normal pollen grains, the vegetative cell is a terminally differentiated cell, without further cell proliferation; however, in many *rbr* mutant pollen grains there is a further mitotic division of the vegetative cell into two cells, each containing a nucleus resembling that of the normal vegetative cell (Chen *et al.*, 2009; Johnston *et al.*, 2008). A more complex phenotype was observed in a small fraction of the *rbr* mutant pollen grains; in addition to two vegetative-like cells, there were two small generative-like cells, at the stage corresponding to the normal two-cell stage, suggesting that the mutant generative cell sometimes also divided once more than normal (Chen *et al.*, 2009). Nevertheless, cell-type-specific markers seem to be expressed normally in the *rbr* mutant pollen, suggesting that *RBR* is not needed for cell fate determination (Chen *et al.*, 2009). Therefore, normal *RBR* function is needed to specifically prevent excess cell proliferation, thereby allowing the formation of single vegetative and generative cells at the two-cell stage.

Not surprisingly, the first pollen mitosis also requires proper cell cycle control, as supported by molecular genetic studies (Liu *et al.*, 2008). The mitotic cell cycle depends on the activity of cyclins and cyclin-dependent protein kinases (CDKs), which are inhibited by CDK inhibitors to prevent overly active cell cycles. To allow the cell cycle to progress, CDK inhibitors must be removed at specific points in the cell cycle. Molecular studies of two highly similar RING-finger E3 ubiquitin ligases, RHF1a (RING H2 group F1a) and RHF1b, indicate that they play redundant roles in the degradation of the CDK inhibitor KRP6 (Kip RELATED PROTEIN6) and promote the progression of the first pollen mitosis (Liu *et al.*, 2008). In 30–40% of the *rhf1a rhf1b* double mutant pollen grains, pollen mitosis I fails to occur, leading to the formation of uninucleate pollen.

2.4. Control of the second pollen mitosis and sperm cell differentiation

The second pollen mitosis, that of the generative cell to produce two sperm cells, is the only male-germline-specific mitosis in flowering plants; therefore, it offers an unique opportunity to study the regulation of germline development in plants. Molecular genetic studies found that the *Arabidopsis* *DUO1* (*DUO POLLEN1*) gene is expressed in the generative cell and subsequently also in the sperm cells (Rotman *et al.*, 2005). In addition, *duo1* mutant pollen fails to undergo the second pollen mitosis, resulting in the formation of a single diploid sperm-like cell. *DUO1* encodes a MYB protein, suggesting that it is important for gene expression needed for the second pollen mitosis. At least part of this requirement is fulfilled by the activation by DUO1 of expression of the cell cycle regulator cyclin B1:1 in the generative cell (Brownfield *et al.*, 2009a). This is consistent with the finding that the cyclin-dependent kinase CDKA is required for pollen mitosis II (Harashima *et al.*, 2007; Iwakawa *et al.*, 2006). In addition, the *DUO3* gene, which encodes a conserved plant protein that is related to a regulator of *C. elegans* gonadogenesis, also has important roles in the regulation of second pollen mitosis and sperm cell differentiation (Brownfield *et al.*, 2009a, b). The second pollen mitosis in the *duo3* mutant either is delayed or fails to occur, suggesting that DUO3 acts to promote the division of the generative cell to form the two sperm cells. It was observed that cyclin B1:1 is expressed in the *duo3* mutant pollen indicating that the *duo3* defect is not due to a reduction in the level of cyclin B1:1, suggesting that DUO3 affects a distinct aspect of cell cycle control from that of DUO1.

Transcriptomic analysis indicates that the sperm cell expresses a large number of genes (Borges *et al.*, 2008). A recent study further showed that DUO1 is also needed for sperm-specific gene expression and normal sperm cell differentiation (Brownfield *et al.*, 2009a), consistent with the observation that the diploid sperm-like cell of the *duo1* mutant pollen fails undergo fertilization (Rotman *et al.*, 2005). Like DUO1, DUO3 is also required for normal expression of sperm cell genes and for sperm differentiation; specifically, DUO3 promotes the expression of a subset of genes that depend on DUO1 activity (Brownfield *et al.*, 2009b). The analyses of *DUO1* and *DUO3* suggest the second pollen mitosis is coupled with sperm cell differentiation and that the DUO1 and DUO3 proteins play related yet distinct roles in these crucial regulatory processes.

However, cell cycle regulation can also be decoupled from sperm differentiation. The chromatin remodeling factor CAF1 (CHROMATIN ASSEMBLY FACTOR1) was found to be important for the second pollen mitosis (Chen *et al.*, 2008). Pollen grains that are deficient in CAF1 often fail to undergo the second mitosis and contain a vegetative cell and a single sperm-like cell. Unlike the diploid sperm cells in the *duo1* and *duo3* mutants, the single *caf1* mutant sperm cell can express cell-specific markers and is able to fertilize either the egg or central cells, suggesting that CAF1 is important for the second pollen mitosis, but not for sperm cell fate determination (Chen *et al.*, 2008). Therefore, the specification of sperm cell differentiation is not dependent on normal cell cycle regulation. The differentiation of the single generative cell into a functional diploid sperm was also observed when the translational inhibitor diphtheria toxin A was expressed from the Arabidopsis *HAP2/GCS1* (*HAPLESS2/GENERATIVE CELL SPECIFIC1*) promoter, indicating that protein synthesis is needed for the second pollen mitosis (Frank and Johnson, 2009).

Another piece of evidence for the importance of cell cycle regulation came from the study of RHF1a and RHF1b (Liu *et al.*, 2008), which were mentioned earlier for their role in the first pollen mitosis. Among the *rhf1a rhf1b* double mutant pollen grains that are able to carry out the first pollen mitosis, up to 30% fail the second pollen mitosis. The importance of removal of KRP6 was further demonstrated by the analysis of a mutant in the gene encoding an F-box protein, FBL17 (F-BOX-LIKE17; Kim *et al.*, 2008). In the *fbl17* mutant, a fusion of KRP6 with GFP persists in the generative nucleus and the *fbl17* mutant generative nucleus fails to divide. Furthermore, normal regulation of protein degradation requires the cleavage of ubiquitin moiety from the ubiquitin-substrate conjugates. The ubiquitin proteases UBP3/UBP4 (UBIQUITIN SPECIFIC PEPTIDASE3/4) were found to be important for pollen mitosis II; *ubp3 ubp4* double mutant pollen grains have a vegetative cell and a single generative-like cell, suggesting that the second pollen mitosis is defective (Doelling *et al.*, 2007). Finally, double mutant in the two paralogs encoding the RPN5/RPT5 (REGULATORY PARTICLE5) subunit of the 19S regulatory particle (RP) of the proteasome are defective in male gametogenesis, consistent with an essential role for the proteasome in pollen mitosis I and/or II (Book *et al.*, 2009; Gallois *et al.*, 2009).

2.5. Function of small RNAs in regulating sperm maturation and protecting the germline genome

Recent studies have shown that the pollen express diverse small RNAs, including microRNAs, tasiRNAs, and siRNAs (Chambers and Shuai, 2009; Grant-Downton *et al.*, 2009a; Slotkin *et al.*, 2009). In addition, genes involved in small RNA biogenesis are expressed during pollen development (Grant-Downton *et al.*, 2009b). Multiple microRNAs were detected in the pollen, perhaps to render their target genes inactive in the mature pollen (Chambers and Shuai, 2009; Grant-Downton *et al.*, 2009a, b). To protect the genome of the male germline, it seems that it is particularly important to repress the activity of transposons; this idea is supported by the finding that pollen express a large number of siRNAs that match known transposons (Slotkin *et al.*, 2009). However, transposon-encoded genes are expressed and can transpose in the vegetative cell, whose genome is not transmitted to the next generation. It was proposed that siRNAs generated in the vegetative cell can be transported into the sperm, where they silence the transposons to prevent the generation of mutations (Slotkin *et al.*, 2009).

3. Development of the Female Gametophyte or Embryo SAC

3.1. Initiation of female gametophyte development

In angiosperms, the female gametophytes, known as megagametophytes or embryo sacs, develop within the ovule (Grossniklaus and Schneitz, 1998). The process begins with the ovule primordium, where the central region called the nucellus will give rise to a single pre-meiotic cell known as the Megasporocyte or the Megaspore Mother Cell (MMC; Fig. 13.2A). The MMC is specified uniquely in most flowering plants from a single hypodermal cell at the distal end of the nucellus, called the AC, which enlarges and becomes conspicuously polarized with some callose deposition. Meiotic division of the MMC results in four haploid spores called megaspores. In most flowering plants, only one megaspore survives per ovule, typically this is the most proximal megaspore, while the remaining three undergo programmed cell death (Fig. 13.2A). The surviving megaspore is called the Functional Megaspore (FM) and it gives rise to the embryo sac after mitotic divisions (see below).

Sporophytic genes are known to affect specification of the AC. In maize, *mac1* (*multiple ACs*) mutants result in ovules that contain several ACs that can develop into embryo sacs, although partially sterile (Sheridan *et al.*, 1996); this mutant also has excessive ACs in the anther (Sheridan *et al.*, 1999). A very similar phenotype was observed in the *msp1* (*multiple sporocyte1*) mutant of rice (Nonomura *et al.*, 2003). The *MSP1* gene has been shown to encode an LRR receptor kinase (Nonomura *et al.*, 2003), related to the *EMS1/EXS* receptor kinase that controls microspore and tapetum specification in Arabidopsis together with its presumptive ligand, *TPD1*, as discussed earlier in this review. Interestingly, knock-down of the rice homolog of *TPD1,* called *OsTPD1a,* also results in a phenotype resembling the *msp1* mutant (Zhao *et al.*, 2008). Thus specification of ACs in both the anther and the ovule might involve cell–cell signaling by receptor kinase–ligand interactions that are common to all angiosperms. Subsequent to specification of the AC, the development of the AC into the pre-meiotic MMC requires the activity of the *SPL/NZZ* gene (Schiefthaler *et al.*, 1999; Yang *et al.*, 1999b), which is a target of the floral organ identity gene *AG* (Ito *et al.*, 2004), as discussed in the section on anther development. In

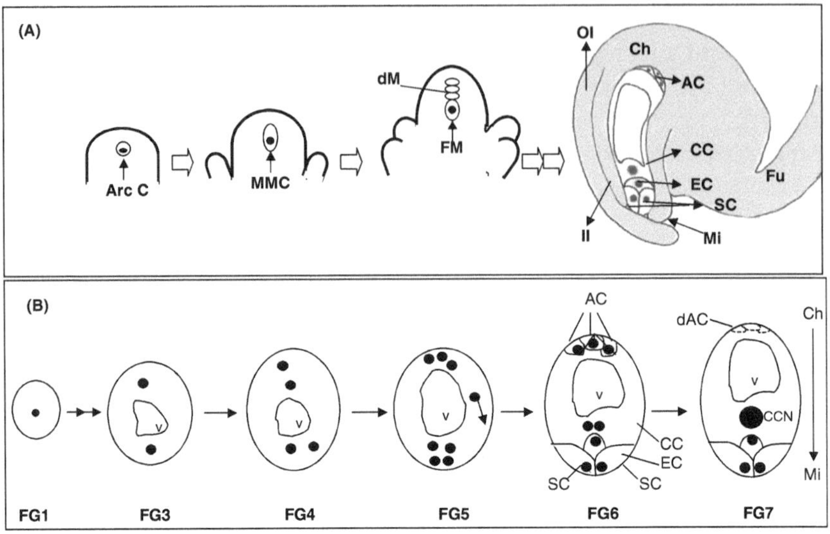

Figure 13.2 Schematic of female gametophyte initiation and development within the Arabidopsis ovule. (A) A hypodermal cell in the ovule primordium is specified as an Archesporial cell, which differentiates into a Megaspore Mother Cell and undergoes meiosis generating four spores. A single spore at the proximal (chalazal) position survives and becomes the Functional megaspore, which gives rise eventually to the mature female gametophyte within the fully developed ovule on the right. (B). Female gametophyte stages FG1–FG7 showing the sequence of syncitial mitotic divisions that lead to cellularization and formation of the seven-cell eightnucleate embryo sac at FG6. The two polar nuclei of the Central cell fuse and the antipodals degenerate by FG7. AC, Antipodal cell; Arc C, Archesporial cell; CC, Central cell; CCN, Central cell nucleus; Ch, Chalaza; dAC, Degenerating antipodal cells; dM, Degenerating megaspores; EC, Egg cell; FM, Functional megaspore; Fu, Funiculus; II, Inner integument; Mi, Micropyle; MMC, Megaspore mother cell; OI, Outer integument; SC, Synergid cell; V, Vacuole. See text for genes known to act at different developmental stages.

addition to these interactions, recent findings have uncovered a surprising role for the small RNA pathway in specification of germline cells.

Mutations in the *ARGONAUTE9* (*AGO9*) gene result in the formation of ectopic MMCs in the ovule; only the normally positioned MMC will form an embryo sac, but an ectopic MMC can differentiate without undergoing meiosis into a diploid megaspore that arrests at the one-nuclear stage (Olmedo-Monfil *et al.*, 2010). The *AGO9* gene is not expressed in the AC or MMC but in the surrounding cells of the ovule, so it must act non-cell-autonomously to limit MMC formation to a single cell. *AGO9* seems to be primarily functioning in the siRNA-mediated transposon silencing pathway, and gametes in *ago9* mutants exhibit active transcription of normally silent transposons, pointing to an interesting parallel with transposon silencing by small RNAs in the germline of animals (Klattenhoff and Theurkauf, 2008). The mechanism by which *AGO9* acts to restrict MMC fate in the ovule is completely unknown. However, mutants in other genes in the 24nt siRNA transposon silencing pathway, including RDR2 (RNA-dependent RNA polymerase 2), DCL3 (Dicer-Like 3), and DNA-dependent RNA Polymerases IV and V, have also been shown to have the same effect, suggesting that the mechanism is directly related to the function of *AGO9* in RNA-directed silencing and is not a separate or distinct function (Olmedo-Monfil *et al.*, 2010). In rice, a different *ARGONAUTE* gene, called *MEIOSIS ARRESTED AT LEPTOTENE1* (*MEL1*), has been found to be required for progression of the MMC through meiosis, although the precise mechanism is unclear (Nonomura *et al.*, 2007). *MEL1* is expressed in the MMC, and appears not to be orthologous to the Arabidopsis *AGO* genes; it is most closely related to Arabidopsis *AGO1* which is primarily involved in miRNA regulation. Again, these findings point to the importance of RNA silencing for normal initiation of gametophytic development, and the mechanisms by which these processes are connected will be of great interest for future studies.

3.2. Development of the female gametophyte

There has been over a century of studies on female gametophyte development in a large number of diverse plant species, and many variations in cell numbers and organization of the embryo sac occur in nature (Friedman *et al.*, 2008). For this review we will consider only the Polygonum-type (Maheshwari, 1950), containing seven cells of four distinct cell types, which is found in the large majority of flowering plants, including *Arabidopsis*, maize, and rice. A schematic outline of the development of a seven-celled embryo sac based on *Arabidopsis* is depicted in Fig. 13.2A. After the MMC is formed in the ovule, it undergoes meiosis to generate four spores, three of which undergo programmed cell death, leaving only the proximal (chalazal) megaspore as the functional megaspore (FM stage or female gametophyte stage 1 (FG1)) in each ovule. The functional megaspore then undergoes three sequential mitotic nuclear divisions, designated as stages FG2–FG5 (Christensen *et al.*, 1997), to generate the eight nuclei that will contribute to the mature embryo sac, four at each pole. Subsequent cellularization results in the formation of seven cells, owing to the nuclear migration of two of these nuclei called the polar nuclei (Fig. 13.2B), which will constitute the central cell nucleus. These seven cells are the egg cell, which gives rise to the embryo, two accessory cells called synergid cells that are required for pollen tube attraction and entry, the central cell which gives rise to the endosperm, and three cells of uncertain function called antipodal cells. The entire embryo sac is enclosed within diploid sporophytic tissues called integuments, which will constitute the seed coat in the mature seed.

An opening in the integuments at the distal end of the ovule, called the micropyle, permits entry of the pollen tube into the embryo sac during fertilization. The micropylar (distal) end of the ovule is adjacent to the egg cell and the two synergid cells, whereas the chalazal (proximal) end of the ovule is adjacent to the three antipodal cells. All of the cells within the embryo sac possess distinctive morphologies (Huang and Russell, 1992). The homo-diploid (2N) central cell is the largest cell of the embryo sac and contains a large central vacuole.

The mitotic divisions leading to the formation of the eight-nucleate embryo sac before cellularization are marked by their fixed nuclear positioning and migration. After the first mitotic division, the two daughter nuclei migrate to opposite poles of the embryo sac at stage FG3. The next mitotic division gives rise to a pair of nuclei at each pole at FG4, and two pairs at each pole at the beginning of stage FG5. The nuclear positioning at the eight-nucleate FG5 stage is correlated with cell fate specification (Huang and Sheridan, 1994; Webb and Gunning, 1994). At the micropylar pole, the two most distal nuclei will form the future synergids. Of the remaining two nuclei, the more distal nucleus of the pair will form the future egg cell, whereas the more central nucleus will become one of the two polar nuclei of the central cell. The most centrally located nucleus of the four nuclei from the chalazal pole will constitute the other polar nucleus of the central cell. This nucleus migrates towards the micropylar pole and becomes positioned close to the first polar nucleus as the embryo sac undergoes cellularization to form the seven-celled embryo sac (Webb and Gunning, 1994). In *Arabidopsis*, the polar nuclei fuse prior to fertilization to form the homodiploid central cell nucleus, and the antipodal cells will have disappeared at the time of fertilization (Fig. 13.2B). The end of mitotic division at the eight-nucleate stage requires the *RBR* (*RETINOBLASTOMA RELATED*) gene, as *rbr* mutants generate additional nuclei that eventually cellularize (Ebel *et al.*, 2004). Although the mechanism of cellularization is not well understood, it has been shown recently that at least in maize, the migration and arrangement of the nuclei during embryo sac development depends upon the activity of the *ZmDSUL* (Zea mays di-Small ubiquitin-related modifier) gene (Srilunchang *et al.*, 2010). Down-regulation of *ZmDSUL* results in mis-positioning of all eight nuclei to the center of the embryo sac, and mutant gametophytes arrest prior to maturation and degenerate. *ZmDSUL* may be involved in the regulation of spindle elongation and asymmetry during female gametophyte development (Srilunchang *et al.*, 2010).

In cereals, the overall development of the embryo sac is very similar to that in *Arabidopsis* and most other eudicots, but with some notable differences at the late stages. The two polar nuclei are generally unfused until fertilization, when they will fuse with a sperm nucleus to form the triploid endosperm (Mol *et al.*, 1994). Another difference is that the three antipodal cells do not degenerate, but instead begin to proliferate by nuclear division and cytokinesis. In maize, at the time of fertilization, proliferation results in up to 40 antipodal cells (Huang and Sheridan, 1994). The antipodal cells in cereals have cell walls resembling those of plant cells involved in nutrient transfer (known as transfer cells); it has been hypothesized that the antipodal cells might perform an essential function in the transfer of nutrients from maternal tissues to the cereal endosperm (Diboll and Larson, 1996; Maeda and Miyake, 1997). However, in the majority of flowering plants, including *Arabidopsis*, the antipodals appear to degenerate before fertilization, and their function if any remains uncertain.

3.3. Genes and mutants affecting female gametophyte development

Genetic studies of embryo sac development have been limited by the small size of the female gametophyte and its inaccessibility, being embedded within the sporophytic tissues of the ovule. Genetic screens using insertional mutagens with selectable markers, such as T-DNA or transposons, have generally utilized the principle that a female gametophyte mutation will result in reduced transmission through the female gametes of a marker linked to that mutation (Brink, 1925). Such screens have identified hundreds of mutants affecting female gametophyte development (Bonhomme *et al.*, 1998; Christensen *et al.*, 1998; Feldmann *et al.*, 1997; Howden *et al.*, 1998; Pagnussat *et al.*, 2005). A comprehensive review of mutants affecting early embryo sac development has been published recently (Yang *et al.*, 2010). In the majority of cases, the mutations that affect early embryo sac development and growth appear to be in genes that control essential functions, such as cellular metabolism, ribosomal RNA processing, or the cell cycle. It is worth noting that none of the characterized mutants affecting early embryo sac development appear to be specific for gametophytic growth, based on expression pattern as well as predicted cellular function. It is likely that these genes are required for sporophytic growth as well, but their sporophytic functions cannot be easily assessed, as homozygous mutant individuals are not found due to the lethality of the mutation on the female gametophyte. In this respect, screens for gametophytic mutants have yielded valuable mutants that would have been missed in standard mutant screens.

When multiple redundant genes are required for a developmental process, mutants might not be found in forward genetic screens. Expression profiling studies can overcome this limitation, but it is problematic to isolate whole embryo sacs without contamination with sporophytic tissues. It has been possible to isolate embryo sacs and egg cells from maize and wheat to construct cDNA libraries, leading to the identification of specifically expressed genes in those plants (Le *et al.*, 2005; Sprunck *et al.*, 2005; Yang *et al.*, 2006a). An alternative approach used with the much smaller embryo sac of *Arabidopsis* is comparative expression profiling, in which embryo sac-specific genes are identified by comparing the expression profiles of whole wild-type ovules with those of mutants that do not form embryo sacs. This approach has been employed by several labs and has led to the identification of hundreds of genes that are specifically expressed or enriched in the embryo sac, many of which display cell-specific patterns of expression (Johnston *et al.*, 2007; Jones-Rhoades *et al.*, 2007; Steffen *et al.*, 2007; Yu *et al.*, 2005). Recently laser-assisted micro-dissection has been successfully adapted to the Arabidopsis embryo sac to generate expression profiles of individual cell-types, specifically the egg cell, central cell, and synergids (Wuest *et al.*, 2010). This approach identified ~8850 Arabidopsis genes expressed in the embryo sac, of which at least 431 were specific to one of the three cell types examined. Interesting parallels with animal germline transcriptomes included the enrichment of genes for dsRNA binding proteins in the egg cell, possibly involved in transposon silencing, as well as RNAs for basic cellular processes in the egg cell that could be utilized during embryo development. The latter finding might explain why nearly half the female gametophytic mutants appear to be defective in early embryogenesis (Pagnussat *et al.*, 2005). This comprehensive dataset of single-cell expression profiles is certain to be valuable for future characterization of female gametophytic genes and functions.

For most of the genes identified in these studies, elucidating functions in embryo sac development will require systematic reverse genetic studies using loss-of-function mutants. Thus the AtMYB98 transcription factor identified by its synergid-specific expression has been shown to be required for synergid cell differentiation, specifically the formation of a structure called the filiform apparatus required for pollen tube entry (Punwani *et al.*, 2007, 2008). *AGL80* and *AGL61*, two *Arabidopsis* genes of the type I MADS-box family of transcription factors expressed specifically in the central cell after cellularization, appear to be important for the central cell differentiation and function, specifically fertilization of the central cell and initiation of endosperm development (Bemer *et al.*, 2008; Portereiko *et al.*, 2006). Because these genes are expressed relatively late after cellularization, it is likely that they establish central cell identity and function after cell specification.

3.4. Patterning, cell fate specification, and maintenance of cell identities in the embryo sac

There are only a few mutants described that alter cell fate specification in the embryo sac. The *eostre* mutation results in an extra functional egg cell in place of a synergid cell (Pagnussat *et al.*, 2007). This is a gain-of-function mutant in which the KNOX-TALE homeodomain gene *BELL-LIKE*

HOMEODOMAIN1 (*BLH1*) is ectopically expressed in the embryo sac. The mutation is suppressed by loss-of-function mutation in *KNAT3*, encoding a putative KNOX heterodimerization partner of BLH1, and is phenocopied by loss-of-function of *OVATE5*, a presumptive negative regulator of *BLH1* (Pagnussat *et al.*, 2007). In the *eostre* mutant, alteration of cell fates is correlated with mis-positioning of nuclei prior to cellularization, suggesting that the cell fate specification is dependent on a location-specific mechanism within the syncytial female gametophyte. In maize, mutation of *Indeterminate gametophyte1*, which is closely related to the Arabidopsis *ASYMMETRIC LEAVES2* gene coding for a transcription factor (Evans, 2007), results in the formation of supernumerary nuclei in the embryo sac that, upon cellularization, appear to assume specific cell fates correlated with their position (Guo *et al.*, 2004). These observations are consistent with a model in which positional information within the syncytium might rely on the asymmetric distribution of a morphogenetic determinant, similar to the positional information provided during the development of the *Drosophila* embryo by gradients of transcription factors (Lewis, 2008).

Recent work provides an overall mechanism for the patterning of the female gametophyte with auxin as a morphogenetic determinant (Pagnussat *et al.*, 2009). In this study, the distribution of auxin was followed by the synthetic auxin-responsive reporter *DR5::GFP* (Ulmasov *et al.*, 1997), and a highly asymmetric distribution of auxin in the developing embryo sac was observed (Fig. 13.3B). At the FG1 stage of female gametophyte development, the auxin signal is strong in the nucellus, outside the gametophyte and distal to it (Fig. 13.3A). As the female gametophyte develops, a strong signal was localized within the embryo sac at the micropylar end up to the FG5 stage, when the third mitotic division has been completed and eight nuclei are positioned prior to cellularization (Fig. 13.3C). The asymmetric distribution of auxin in the syncitial embryo sac indicated that specification of synergids occurs in nuclei exposed to the highest levels of auxin, while specification of the antipodal cells occurs in nuclei at the lowest auxin, and intermediate higher and lower concentrations for the egg cell and central cell, respectively. Increasing the levels of auxin by expression of the auxin biosynthetic gene *YUCCA1* (Zhao *et al.*, 2001) in the embryo sac resulted in the ectopic acquisition of synergid cell fates, and in some cases, egg cell fates. Therefore, high auxin levels throughout the embryo sac alter cell identities, resulting in the conversion of chalazal (proximal) cell identities to micropylar (distal) cell identities. Importantly, auxin over-production did not result in any abnormalities in nuclear positioning during syncytial embryo sac development, indicating that the nuclear positioning pathway is auxin-independent.

Conversely, attenuation of auxin signaling by either down-regulation of the *AUXIN RESPONSE FACTOR* genes or *ARFs* (Guilfoyle and Hagen, 2007), or mutation of the auxin receptor genes *TIR1* and *AFB1-3* (Dharmasiri *et al.*,

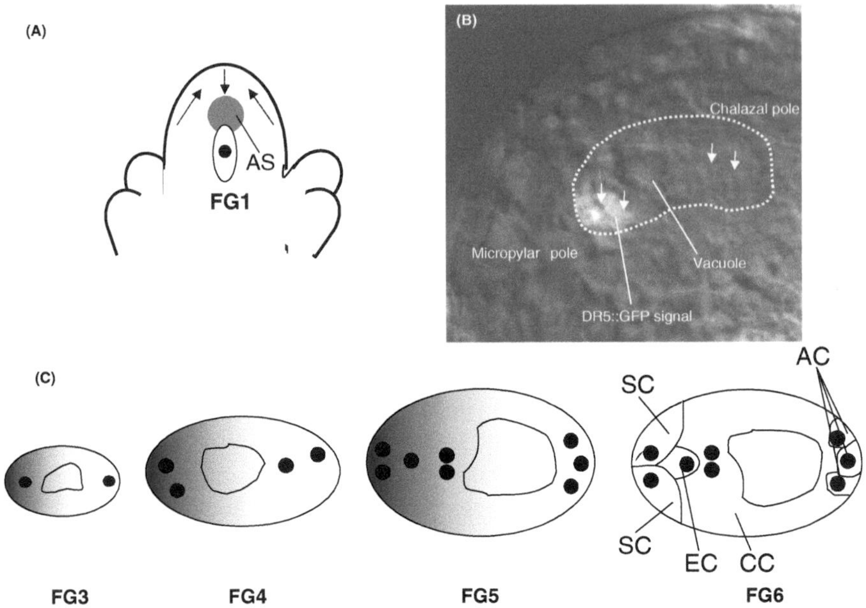

Figure 13.3 Auxin in the developing female gametophyte. (A) Auxin from polar auxin transport in the epidermal layer of the primordium accumulates in the sporophytic ovule cells distal to the one nuclear female gametophyte FG1, and a sporophytic auxin source AS is specified. (B) Four-nucleate embryo sac at stage FG4, displaying internal asymmetric auxin distribution. The *DR5::GFP* fluorescent reporter shows that the highest auxin is at the micropylar pole, and lowest at the chalazal pole. Auxin biosynthesis by the embryo sac, monitored by *YUCCA* gene expression, is also similarly asymmetric. (C) Schematic drawing of auxin gradients in the developing female gametophyte prior to cellularization indicated by shading. Darkest shading corresponds to highest auxin, which is at the micropylar pole corresponding to the nuclei of the future synergids. AC, Antipodal cell; AS, Auxin source; CC, Central cell; EC, Egg cell; FG1, Female gametophyte stage 1; SC, Synergid cell.

2005), results in shift of cell fate specification away from synergids towards more chalazal identities. For example, expression in the embryo sac of an artificial microRNA (*amiR-ARFa*) that targets a subset of the *ARF*s, specifically *ARF1-8* and *ARF19*, results in defective embryo sacs in which all three micropylar cells acquired egg cell attributes. These embryo sacs do not attract any pollen tubes, consistent with the loss of synergid identities (Pagnussat *et al.*, 2009).

These observations indicate that differences in auxin concentration specify cell fates. An auxin gradient is formed in the syncytial embryo sac, with a maximum at the micropylar pole and a minimum at the chalazal pole. Each nucleus would then select a different cell fate as a read-out of the auxin signal (Fig. 13.3C). Thus, synergid fate would correspond to the highest auxin concentration and antipodal fate to the lowest, with egg cell and central cell fates representing different intermediate auxin levels.

In sporophytic development, auxin maxima arise from the polar transport of auxin, which occurs through the action of auxin efflux carriers encoded by the *PIN-FORMED* (*PIN*) gene family (Galweiler *et al.*, 1998). However, in the female gametophyte auxin appears to arise from localized synthesis by the auxin-biosynthetic products of the *YUCCA* (*YUC*) genes (Cheng *et al.*, 2006), because the expression of *YUC* genes overlaps with the auxin signal in the ovules. *YUC* expression appears first at the micropylar region outside of the embryo sac at FG1 (Fig. 13.3A), and then localizes to the micropylar pole of the gametophyte from FG3 through to the later developmental stages, while being undetectable in the sporophyte. Interestingly, the *PIN1* gene is expressed in the diploid tissues of the ovule until the FG1 stage, but disappears soon thereafter (Pagnussat *et al.*, 2009). These observations suggest a sequential source mechanism for the formation of the gametophytic auxin gradient. The early auxin flux in the nucellus, as indicated by PIN1 localization, might be involved in establishing an initial auxin maximum, which could then provide the trigger for events that establish gametophyte patterning by specifying a sporophytic auxin source. Auxin from the sporophytic source leads to the specification of secondary auxin sources within the gametophyte at the micropylar pole, which, in turn, result in a gradient of auxin along the micropylar–chalazal axis. The outcome is the graded specification of different cell types according to nuclear position (Fig 13.3C).

The maintenance of an auxin gradient within the developing embryo sac is somewhat unexpected. The syncytial embryo sac is partitioned into cytoplasmic domains (Brown and Lemmon, 1991, 1992; Webb and Gunning, 1994), so that the highly asymmetric expression of GFP or GUS reporters can be accounted for by restriction of the translation of mRNAs to the domain that contains the nucleus and limiting subsequent diffusion of the proteins. However, it seems doubtful that auxin, a small molecule, could be restricted from diffusion even with cytoplasmic partitioning, which suggests that other mechanisms might be functioning to maintain the gradient. Possibilities include the export of auxin from the embryo sac at the chalazal pole by unidentified non-PIN auxin carriers, or alternatively the inactivation of auxin by conjugation or degradation at the chalazal pole.

While auxin appears to be a primary determinant for cell specification, it has been found that maintenance of cell fates in the embryo sac requires interactions between the gametic cells, *i.e.* the egg cell and central cell, and the non-gametic accessory cells, the synergids and antipodals, respectively. Three genes, *LACHESIS* (*LIS*), *GAMETOPHYTIC FACTOR1/ CLOTHO* (*GFA1/CLO*), and *ATROPUS* (*ATO*) have been found to be required to restrict gametic cell fate after cellularization (Coury *et al.*, 2007; Gross-Hardt *et al.*, 2007; Moll *et al.*, 2008). Mutants in these genes show changes in cell identities inside the embryo sac after initial normal development and cell-specification (Gross-Hardt *et al.*, 2007; Moll *et al.*, 2008). In the embryo sacs of *lis, gfa1/clo*, and *ato* mutants, cells that are normally specified as synergid cells can express egg cell-specific markers, and cells that are normally specified as antipodal cells can express central cell-specific markers. It has been proposed that a mechanism of lateral inhibition by the gametic cells, i.e. the egg cell and central cell, operates to prevent the formation of excess gametic cells inside the female and to maintain the cell fate of the accessory synergids and antipodals (Gross-Hardt *et al.*, 2007). All three genes, *LIS, GFA1/CLO*, and *ATO*, encode pre-mRNA splicing factors that are conserved among all eukaryotes (Bartels *et al.*, 2003; Horowitz *et al.*, 1997) and are expressed throughout the *Arabidopsis* plant. It is not clear why these specific cell fate phenotypes arise from mutations in genes that encode components of the core splicing machinery. It is possible that the splicing defects might disrupt signaling mechanisms required for the maintenance of cell fates specified at cellularization. Alternatively, the splicing of auxin response factors involved in cell-fate specification might be affected in these mutants, leading to abnormal auxin responses and to the mis-specification of cell fates. Further studies should reveal the mechanistic basis for these intriguing mutant phenotypes.

4. Double Fertilization: Signaling and Reception between Male and Female

In flowering plants, both sperm cells in the pollen are used to fertilize cells of the female gametophyte, in a process unique to angiosperms called double fertilization. The entire embryo sac is enclosed within diploid sporophytic tissues called integuments, which will constitute the seed coat in the mature seed. When pollen lands on the stigma it forms a structure called the pollen tube that penetrates the embryo sac through an opening in the integuments called the micropyle, and delivers two sperm cells by entering a synergid cell, which subsequently degenerates (Fig. 13.4A, B). Eventually, the other synergid cell will also degenerate. The two gametic cells of the embryo sac, the egg cell and the central cell, will undergo double fertilization by the two sperm cells of the pollen (Fig. 13.4B). The fertilization of the egg cell within the embryo sac by one of the sperm cells delivered by the pollen tube results in the diploid zygote, which will initiate the next sporophyte generation. The fertilization of the central cell by the second sperm cell results in the triploid endosperm, which functions as a nutritive source for the embryo or the germinating seedling.

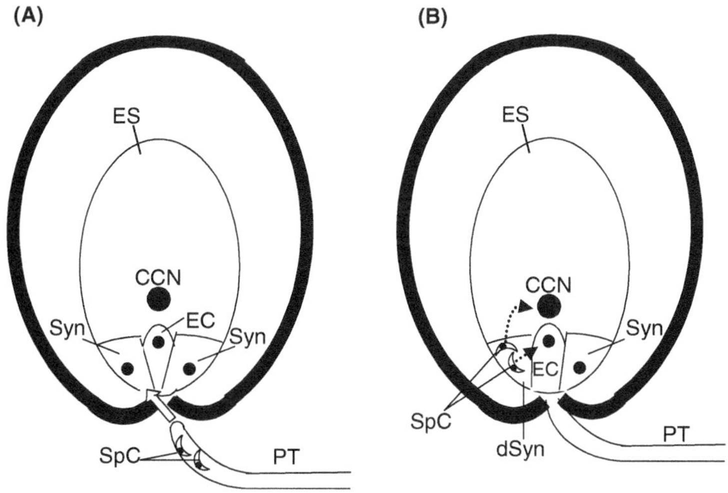

Figure 13.4 Pollen tube guidance, entry and double fertilization. (A) Embryo sac within an ovule attracting a Pollen tube carrying two Sperm cells. The attracting signal is secreted by the Synergid cells. (B) Pollen tube enters through the micropyle into one of the Synergid cells, ruptures and discharges the two Sperm cells. Dotted arrows show that one Sperm cell will fertilize the Egg cell, and the other will fertilize the Central cell. CCN, Central cell nucleus; dSyn, Degenerating synergid; EC, Egg cell; ES, Embryo sac; PT, Pollen tube; SpC, Sperm cells; Syn, Synergid cell.

During the growth through the female tissues of the pistil, the pollen tube must navigate through a complex passageway, requiring cues and guidance from the female. The isolation and analysis of the *pop2* mutant showed that the chemical gamma-amino butyric acid (GABA) might play an important role in pollen tube guidance (Palanivelu *et al.*, 2003). It was observed that GABA forms a gradient that increases along the pathway of the pollen tube from the stigma to the style and then septum, and finally to the integuments of the ovule. However, in the *pop2* mutant, pollen tubes fail to grow in the proper direction and often miss the destination of pollen tube growth, the micropyle of the ovule. It turns out that *POP2* encodes an enzyme that degrades GABA, but in the *pop2* mutant, the failure to degrade GABA results in abnormal GABA distribution, and mis-directed growth of pollen tubes.

The mature female gametophyte provides essential functions required for fertilization, including pollen tube attraction and the release of the male gametes (sperm cells) from the pollen tube. As the pollen tube grows closer to the female gametophyte, synergid cells play an essential role in pollen tube guidance and pollen tube reception (Higashiyama *et al.*, 2001; Rotman *et al.*, 2003; Sandaklie-Nikolova *et al.*, 2007). The synergids have characteristic cell wall ingrowths at the micropylar end termed the filiform apparatus, which are located at the site of pollen tube entry (Huang and Russell, 1992). In maize, a small protein, EA1, which is synthesized in the egg apparatus (i.e. the synergids and the egg cell), is required for micropylar pollen tube guidance (Marton *et al.*, 2005). In *Arabidopsis*, the synergid-specific transcription factor *MYB98* seems to regulate a battery of genes that encode small peptides, although these have not yet been demonstrated to be pollen tube attractants (Kasahara *et al.*, 2005; Punwani *et al.*, 2007, 2008). Recently, a significant advance was reported in *Torenia fournieri*, showing that small cysteine-rich polypeptides called LUREs, which are synthesized in the synergids, function as pollen tube attractants in that plant species (Okuda *et al.*, 2009).

After entry into the synergid cell, the process of pollen tube reception and male gamete release requires the plasma membrane receptor kinase *FER* (*FERONIA*), which might play a role in species recognition (Escobar-Restrepo *et al.*, 2007), as well as *LORELEI*, a membrane GPI (glucosylphophatidylinositol)-anchored protein (Capron *et al.*, 2008). Also required for successful pollen tube reception is the expression of the peroxisomal protein encoded by ABSTINENCE BY MUTUAL CONSENT in both the male and the female gametophytes, which serves to illustrate the active participation of the male gametophyte in gamete release (Boisson-Dernier *et al.*, 2008). The Arabidopsis mutants *fer* and *lorelei* are female semi-sterile, and the pollen tube fails to rupture and discharge sperm cells in the synergid of a *fer* or *lorelei* female gametophyte (Huck *et al.*, 2003; Capron *et al.*, 2008). The receptor-like protein kinase encoded by *FER* is expressed in the synergid cells and localized to the filiform apparatus, consistent with a role in communication with the pollen tube (Escobar-Restrepo *et al.*, 2007). As FER is a receptor-like kinase, this suggests that synergids need to receive a signal from the pollen tube to trigger pollen tube growth arrest and sperm release. Further work is needed to identify the presumptive signals received by the FER receptor from the pollen tube, and that sent by the synergid to arrest the pollen tube, and the relationship of this signaling to the LORELEI GPI-anchored protein, which is also expressed in the synergids (Capron *et al.*, 2008).

FER is a member of a small clade of receptor-like kinase genes and, recently, two closest paralogs, *ANX1* and *ANX2*, are found to be expressed in the pollen, unlike the synergid expression of *FER* (Boisson-Dernier *et al.*, 2009; Miyazaki *et al.*, 2009). Single *anx1* and *anx2* mutants are normal, but the double mutant pollen tube failed to reach the female, and the pollen tube ruptures abnormally, suggesting that the ANX1/2 function prevents premature pollen tube rupture. Another gene acting on the pollen side of the interaction is the *GCS1/HAP2* gene (Mori *et al.*, 2006; von Besser *et al.*, 2006). GCS1 was identified as a membrane protein of generative cells from pollen of *Lilium longiflorum* (Mori *et al.*, 2006). The allelic *gcs1/hap2* (*hapless2*) mutants produce pollen with tubes showing defects in reaching the ovules, and *gcs1/hap2* sperm cells fail to fertilize the female gametes (Mori *et al.*, 2006; von Besser *et al.*, 2006). Therefore, the GCS1/HAP2 membrane protein might also be important for cell interaction prior to fertilization.

The process of double fertilization requires the fusion between a sperm cell and the central cell. Does the central cell play any role in attracting the pollen tube? Evidence for such a role is provided by the Arabidopsis *CCG* (*CENTRAL CELL GUIDANCE*) gene, which is expressed only in the central cell, yet mutation of this gene results in failure of pollen tube attraction to the micropyle. The *CCG* gene encodes a protein containing a zinc beta-ribbon domain found in transcription factors (Chen *et al.*, 2007). Therefore, the function of CCG is probably to regulate the expression of genes that are required for signaling to the pollen tube. Besides CCG and the synergid-expressed genes, it appears that the membrane protein GEX3 expressed in the egg cell can also cause micropylar pollen tube guidance defects in *Arabidopsis* (Alandete-Saez *et al.*, 2008; Chen *et al.*, 2007), which suggests that signaling mechanisms between the egg cell and the synergids might also be important for this process.

In short, molecular genetic investigations have uncovered several components of cell–cell signaling that act in the synergids, in the sperm, and in regulating central cell gene expression. Cell biological studies also have provided evidence for sub-cellular localization of FER and the homologous ANX1 and ANX2 that consistent with their roles in pollen–synergid interactions. In the future, the identification of additional components for the cell signaling pathways that are crucial for successful double fertilization should lead to a mechanistic understanding of this important process which is unique to flowering plants.

5. Concluding Remarks

The realization that the alternation of generations is integral to the plant life cycle was made nearly a generation ago. However, nearly all the subsequent advances in plant biology have focused on the diploid sporophytic generation. As discussed in this review, this situation has changed considerably over the past few years. Several large-scale forward mutant screens, in combination with expression profiling, are uncovering genes that contribute to fundamental processes in pollen and embryo sac development, and double fertilization. Major findings include the identification of genes and mechanisms that play key roles in the differentiation of different cell types, pollen tube attraction, and fertilization. A better understanding of gametophyte development is needed for both fundamental and applied aspects of the plant sciences. In addition to their intrinsic importance in gamete production, gametophytic genes contribute to post-fertilization development of the seed. The female gametophyte plays an important role in both embryo and endosperm development through sequestration of maternal factors required by the zygote as well as by epigenetic modifications in the central cell that regulate endosperm development (Sundaresan and Alandete-Saez, 2010). Surprisingly, the male gametophyte also provides critical factors required for embryo development; specifically, transcripts of the interleukin-1 receptor-associated/Pelle-like kinase gene *SSP* (*SHORT SUSPENSOR*), essential for embryogenesis, are delivered only by sperm cells (chapter 1; Bayer *et al.*, 2009). Finally, studies of gametophytes can lead to useful applications in agriculture, such as generating male-sterility through manipulation of pollen development, or introduction of apomixis (reproduction through seeds without meiosis, resulting in maintenance of hybrid combinations of genes) to crop plants through manipulation of embryo sac development. It is notable that certain meiotic mutants affecting female meiosis in Arabidopsis have recently been shown to evade the requirement for meiosis to initiate the gametophytic generation, and make diploid embryo sacs that retain heterozygosity, although these embryo sacs still require fertilization to generate seeds (d'Erfurth *et al.*, 2009; Ravi *et al.*, 2008). The advances discussed above in understanding and manipulation of the hitherto neglected gametophytic generation carry the potential for new strategies in plant breeding that could translate into significant benefits to agriculture in the future.

REFERENCES

Alandete-Saez, M., Ron, M., and McCormick, S. (2008). GEX3, expressed in the male gametophyte and in the egg cell of *Arabidopsis thaliana*, is essential for micropylar pollen tube guidance and plays a role during early embryogenesis. *Mol. Plant* **1**, 586–598.

Albrecht, C., Russinova, E., Hecht, V., Baaijens, E., and de Vries, S. (2005). The *Arabidopsis thaliana* SOMATIC EMBRYOGENESIS RECEPTOR-LIKE KINASES1 and 2 control male sporogenesis. *Plant Cell* **17**, 3337–3349.

Alves-Ferreira, M., Wellmer, F., Banhara, A., Kumar, V., Riechmann, J. L., and Meyerowitz, E. M. (2007). Global expression profiling applied to the analysis of Arabidopsis stamen development. *Plant Physiol.* **145**, 747–762.

Azumi, Y., Liu, D., Zhao, D., Li, W., Wang, G., Hu, Y., and Ma, H. (2002). Homolog interaction during meiotic prophase I in *Arabidopsis* requires the *SOLO DANCERS* gene encoding a novel cyclin-like protein. *EMBO J.* **21**, 3081–3095.

Bartels, C., Urlaub, H., Luhrmann, R., and Fabrizio, P. (2003). Mutagenesis suggests several roles of Snu114p in pre-mRNA splicing. *J. Biol. Chem.* **278**, 28324–28334.

Bayer, M., Nawy, T., Giglione, C., Galli, M., Meinnel, T., and Lukowitz, W. (2009). Paternal control of embryonic patterning in *Arabidopsis thaliana*. *Science* **323**, 1485–1488.

Bemer, M., Wolters-Arts, M., Grossniklaus, U., and Angenent, G. C. (2008). The MADS domain protein DIANA acts together with AGAMOUS-LIKE80 to specify the central cell in Arabidopsis ovules. *Plant Cell* **20**, 2088–2101.

Boisson-Dernier, A., Frietsch, S., Kim, T. H., Dizon, M. B., and Schroeder, J. I. (2008). The peroxin loss-of-function mutation abstinence by mutual consent disrupts male-female gametophyte recognition. *Curr. Biol.* **18**, 63–68.

Boisson-Dernier, A., Roy, S., Kritsas, K., Grobei, M. A., Jaciubek, M., Schroeder, J. I., and Grossniklaus, U. (2009). Disruption of the pollen-expressed *FERONIA* homologs *ANXUR1* and *ANXUR2* triggers pollen tube discharge. *Development* **136**, 3279–3288.

Bonhomme, S., Horlow, C., Venzon, D., de Laissardiere, S., Guyon, A., Ferault, M., Marchand, M., Bechtold, N., and Pelletier, G. (1998). T-DNA mediated disruption of essential gametophytic genes in Arabidopsis is unexpectedly rare and cannot be inferred from segregation distortion alone. *Mol. Gen. Genet.* **260**, 444–452.

Book, A. J., Smalle, J., Lee, K. H., Yang, P., Walker, J. M., Casper, S., Holmes, J. H., Russo, L. A., Buzzinotti, Z. W., Jenik, P. D., and Vierstra, R. D. (2009). The RPN5 subunit of the 26s proteasome is essential for gametogenesis, sporophyte development, and complex assembly in Arabidopsis. *Plant Cell* **21**, 460–478.

Borges, F., Gomes, G., Gardner, R., Moreno, N., McCormick, S., Feijó, J. A., and Becker, J. D. (2008). Comparative transcriptomics of Arabidopsis sperm cells. *Plant Physiol.* **148**, 1168–1181.

Brink, R. A. (1925). Some reflections on double fertilization, from its discovery to the present. *Genetics* **10**, 359–394.

Brown, R., and Lemmon, B. (1991). Pollen development in orchids. *Protoplasma* **165**, 155–166.

Brown, R., and Lemmon, B. (1992). Cytoplasmic domain: a model for spatial control of cytokinesis in reproductive cells of plants. *EMSA Bull.* **22**, 48.

Brownfield, L., Hafidh, S., Borg, M., Sidorova, A., Mori, T., and Twell, D. (2009a). A plant germline-specific integrator of sperm specification and cell cycle progression. *PLoS Genet.* **5**, e1000430.

Brownfield, L., Hafidh, S., Durbarry, A., Khatab, H., Sidorova, A., Doerner, P., and Twell, D. (2009b). Arabidopsis DUO POLLEN3 is a key regulator of male germline development and embryogenesis. *Plant Cell* **21**, 1940–1956.

Brukhin, V., Curtis, M. D., and Grossniklaus, U. (2005). The angiosperm female gametophyte: no longer the forgotten generation. *Curr. Sci.* **89**, 1844–1852.

Canales, C., Bhatt, A. M., Scott, R., and Dickinson, H. (2002). EXS, a putative LRR receptor kinase, regulates male germline cell number and tapetal identity and promotes seed development in Arabidopsis. *Curr. Biol.* **12**, 1718–1727.

Capron, A., Gourgues, M., Neiva, L. S., Faure, J. E., Berger, F., Pagnussat, G., Krishnan, A., Alvarez-Mejia, C., Vielle-Calzada, J. P., Lee, Y. R., Liu, B., and Sundaresan, V. (2008). Maternal control of male-gamete delivery in Arabidopsis involves a putative GPI-anchored protein encoded by the LORELEI gene. *Plant Cell* **20**, 3038–3049.

Carles, C. C., and Fletcher, J. C. (2003). Shoot apical meristem maintenance: The art of a dynamic balance. *Trends Plant Sci.* **8**, 394–401.

Chambers, C., and Shuai, B. (2009). Profiling microRNA expression in *Arabidopsis* pollen using microRNA array and real-time PCR. *BMC Plant Biol.* **9**, 87.

Chang, L., Ma, H., and Xue, H. (2009). Functional conservation of the meiotic genes *SDS* and *RCK* for male meiosis in the monocot rice. *Cell Res.* **19**, 768–782.

Chen, Z., Hafidh, S., Poha, S. H., Twell, D., and Berger, F. (2009). Proliferation and cell fate establishment during Arabidopsis male gametogenesis depends on the Retinoblastoma protein. *Proc. Natl. Acad. Sci. U.S.A.* **106**, 7257–7262.

Chen, Y. H., Li, H. J., Shi, D. Q., Yuan, L., Liu, J., Sreenivasan, R., Baskar, R., Grossniklaus, U., and Yang, W. C. (2007). The central cell plays a critical role in pollen tube guidance in Arabidopsis. *Plant Cell* **19**, 3563–3577.

Chen, Z., Tan, J.L.H., Ingouff, M., Sundaresan, V., and Berger, F. (2008). Chromatin assembly factor 1 regulates the cell cycle but not cell fate during male gametogenesis in *Arabidopsis thaliana*. *Development* **135**, 65–73.

Cheng, Y., Dai, X., and Zhao, Y. (2006). Auxin biosynthesis by the YUCCA flavin monooxygenases controls the formation of floral organs and vascular tissues in Arabidopsis. *Genes Dev.* **20**, 1790–1799.

Cheng, H., Qin, L., Lee, S., Fu, X., Richards, D. E., Cao, D., Luo, D., Harberd, N. P., and Peng, J. (2004). Gibberellin regulates Arabidopsis floral development via suppression of DELLA protein function. *Development* **131**, 1055–1064.

Christensen, C. A., King, E. J., Jordan, J. R., and Drews, G. N. (1997). Megagametogenesis in Arabidopsis wild type and the Gf mutant. *Sex. Plant Reprod.* **10**, 49–64.

Christensen, C. A., Subramanian, S., and Drews, G. N. (1998). Identification of gametophytic mutations affecting female gametophyte development in *Arabidopsis thaliana. Dev. Biol.* **202**, 136–151.

Clark, S. E., Williams, R. W., and Meyerowitz, E. M. (1997). The *CLAVATA1* gene encodes a putative receptor kinase that controls shoot and floral meristem size in Arabidopsis. *Cell* **89**, 575–585.

Colcombet, J., Boisson-Dernier, A., Ros-Palau, R., Vera, C. E., and Schroeder, J. I. (2005). Arabidopsis SOMATIC EMBRYOGENESIS RECEPTOR KINASES1 and 2 are essential for tapetum development and microspore maturation. *Plant Cell* **17**, 3350–3361.

Coury, D. A., Zhang, C., Ko, A., Skaggs, M. I., Christensen, C. A., Drews, G. N., Feldmann, K. A., and Yadegari, R. (2007). Segregation distortion in Arabidopsis gametophytic factor1 (gfa1) mutants is caused by a deficiency of an essential splicing factor. *Sex. Plant Reprod.* **20**, 87–97.

d'Erfurth, I., Jolivet, S., Froger, N., Catrice, O., Novatchkova, M., and Mercier, R. (2009). Turning meiosis into mitosis. *PLoS Biol.* **7**, e1000124.

DeYoung, B. J., Bickle, K. L., Schrage, K. J., Muskett, P., Patel, K., and Clark, S. E. (2006). The CLAVATA1-related BAM1, BAM2 and BAM3 receptor kinase-like proteins are required for meristem function in *Arabidopsis. Plant J.* **45**, 1–16.

Dharmasiri, N., Dharmasiri, S., Weijers, D., Lechner, E., Yamada, M., Hobbie, L., Ehrismann, J. S., Jurgens, G., and Estelle, M. (2005). Plant development is regulated by a family of auxin receptor F box proteins. *Dev. Cell* **9**, 109–119.

Diboll, A., and Larson, D. (1996). An electron microscopic study of the mature megagametophyte in *Zea mays. Am. J. Bot.* **53**, 391–402.

Doelling, J. H., Phillips, A. R., Soyler-Ogretim, G., Wise, J., Chandler, J., Callis, J., Otegui, M. S., and Vierstra, R. D. (2007). The ubiquitin-specific protease subfamily UBP3/UBP4 is essential for pollen development and transmission in Arabidopsis. *Plant Physiol.* **145**, 801–813.

Ebel, C., Mariconti, L., and Gruissem, W. (2004). Plant retinoblastoma homologues control nuclear proliferation in the female gametophyte. *Nature* **429**, 776–780.

Escobar-Restrepo, J. M., Huck, N., Kessler, S., Gagliardini, V., Gheyselinck, J., Yang, W. C., and Grossniklaus, U. (2007). The FERONIA receptor-like kinase mediates male–female interactions during pollen tube reception. *Science* **317**, 656–660.

Evans, M. M. (2007). The indeterminate gametophyte1 gene of maize encodes a LOB domain protein required for embryo Sac and leaf development. *Plant Cell* **19**, 46–62.

Feldmann, K. A., Coury, D. A., and Christianson, M. L. (1997). Exceptional segregation of a selectable marker (KanR) in Arabidopsis identifies genes important for gametophytic growth and development. *Genetics* **147**, 1411–1422.

Frank, A. C., and Johnson, M. A. (2009). Expressing the diphtheria toxin A subunit from the *HAP2(GCS1)* promoter blocks sperm maturation and produces single sperm-like cells capable of fertilization. *Plant Physiol.* **151**, 1390–1400.

Friedman, W. E., Madrid, E. N., and Williams, J. H. (2008). Origin of the fittest and survival of the fittest: relating female gametophyte development to endosperm genetics. *Int. J. Plant Sci.* **169**, 79–92.

Gallois, J. L., Guyon-Debast, A., Lecureuil, A., Vezon, D., Carpentier, V., Bonhomme, S., and Guerche, P. (2009). The Arabidopsis proteasome RPT5 subunits are essential for gametophyte development and show accession-dependent redundancy. *Plant Cell* **21**, 442–459.

Galweiler, L., Guan, C., Muller, A., Wisman, E., Mendgen, K., Yephremov, A., and Palme, K. (1998). Regulation of polar auxin transport by AtPIN1 in Arabidopsis vascular tissue. *Science* **282**, 2226–2230.

Grant-Downton, R., Hafidh, S., Twell, D., and Dickinson, H. G. (2009b). Small RNA pathways are present and functional in the angiosperm male gametophyte. *Mol. Plant.* **2**, 500–512.

Grant-Downton, R., Le Trionnaire, G., Schmid, R., Rodriguez-Enriquez, J., Hafidh, S., Mehdi, S., Twell, D., and Dickinson, H. (2009a). MicroRNA and tasiRNA diversity in mature pollen of *Arabidopsis thaliana. BMC Genomics* **10**, 643.

Gross-Hardt, R., Kagi, C., Baumann, N., Moore, J. M., Baskar, R., Gagliano, W. B., Jurgens, G., and Grossniklaus, U. (2007). LACHESIS restricts gametic cell fate in the female gametophyte of Arabidopsis. *PLoS Biol.* **5**, e47.

Grossniklaus, U., and Schneitz, K. (1998). The molecular and genetic basis of ovule and megagametophyte development. *Semin. Cell Dev. Biol.* **9**, 227–238.

Guilfoyle, T. J., and Hagen, G. (2007). Auxin response factors. *Curr. Opin. Plant Biol.* **10**, 453–460.

Guo, F., Huang, B. Q., Han, Y., and Zee, S. Y. (2004). Fertilization in maize indeterminate gametophyte1 mutant. *Protoplasma* **223**, 111–120.

Hamant, O., Ma, H., and Cande, Z. W. (2006). Genetics of meiotic prophase I in plants. *Annu. Rev. Plant Biol.* **57**, 267–302.

Harashima, H., Shinmyo, A., and Sekine, M. (2007). Phosphorylation of threonine 161 in plant cyclin-dependent kinase A is required for cell division by activation of its associated kinase. *Plant J.* **52**, 435–448.

Higashiyama, T., Yabe, S., Sasaki, N., Nishimura, Y., Miyagishima, S., Kuroiwa, H., and Kuroiwa, T. (2001). Pollen tube attraction by the synergid cell. *Science* **293**, 1480–1483.

Honys, D., and Twell, D. (2003). Comparative analysis of the Arabidopsis pollen transcriptome. *Plant Physiol.* **132**, 640–652.

Hord, C.L.H., Chen, C., DeYoung, B. J., Clark, S. E., and Ma, H. (2006). The *BAM1/BAM2* receptor-like kinases are important regulators of early *Arabidopsis* anther development. *Plant Cell* **18**, 1667–1680.

Hord, C.L.H., Sun, Y., Pillitteri, L. J., Torii, K. U., Wang, H., Zhang, S., and Ma, H. (2008). Regulation of *Arabidopsis* early anther development by the mitogen-activated protein kinases, MPK3 and MPK6, and the ERECTA and related receptor-like kinases. *Mol. Plant* **1**, 645–658.

Horowitz, D. S., Kobayashi, R., and Krainer, A. R. (1997). A new cyclophilin and the human homologues of yeast Prp3 and Prp4 form a complex associated with U4/U6 snRNPs. *RNA* **3**, 1374–1387.

Howden, R., Park, S. K., Moore, J. M., Orme, J., Grossniklaus, U., and Twell, D. (1998). Selection of T-DNA-tagged male and female gametophytic mutants by segregation distortion in Arabidopsis. *Genetics* **149**, 621–631.

Huang, B. Q., and Russell, S. (1992). Female germ unit: organization, isolation, and function. *Int. Rev. Cytol.* **140**, 233–291.

Huang, B. Q., and Sheridan, W. F. (1994). Female gametophyte development in maize: microtubular organization and embryo sac polarity. *Plant Cell* **6**, 845–861.

Huck, N., Moore, J. M., Federer, M., and Grossniklaus U. (2003). The *Arabidopsis* mutant *feronia* disrupts the female gametophytic control of pollen tube reception. *Development* **130**, 2149–2159.

Ito, T., Nagata, N., Yoshiba, Y., Ohme-Takagi, M., Ma, H., and Shinozaki, K. (2007). Arabidopsis MALE STERILITY1 encodes a PHD-type transcription factor and regulates pollen and tapetum development. *Plant Cell* **19**, 3549–3562.

Ito, T., Wellmer, F., Yu, H., Das, P., Ito, N., Alves-Ferreira, M., Riechmann, J. L., and Meyerowitz, E. M. (2004). The homeotic protein AGAMOUS controls microsporogenesis by regulation of SPOROCYTELESS. *Nature* **430**, 356–360.

Iwakawa, H., Shinmyo, A., and Sekine, M. (2006). Arabidopsis CDKA;1, a cdc2 homologue, controls proliferation of generative cells in male gametogenesis. *Plant J.* **45**, 819–831.

Jia, G., Liu, X., Owen, H. A., and Zhao, D. (2008). Signaling of cell fate determination by the TPD1 small protein and EMS1 receptor kinase. *Proc. Natl. Acad. Sci. U.S.A.* **105**, 2220–2225.
Johnston, A. J., Matveeva, E., Kirioukhova, O., Grossniklaus, U., and Gruissem, W. (2008). A dynamic reciprocal RBR-PRC2 regulatory circuit controls *Arabidopsis* gametophyte development. *Curr. Biol.* **18**, 1680–1686.
Johnston, A. J., Meier, P., Gheyselinck, J., Wuest, S. E., Federer, M., Schlagenhauf, E., Becker, J. D., and Grossniklaus, U. (2007). Genetic subtraction profiling identifies genes essential for Arabidopsis reproduction and reveals interaction between the female gametophyte and the maternal sporophyte. *Genome Biol.* **8**, R204.
Jones-Rhoades, M. W., Borevitz, J. O., and Preuss, D. (2007). Genome-wide expression profiling of the Arabidopsis female gametophyte identifies families of small, secreted proteins. *PLoS Genet.* **3**, 1848–1861.
Kasahara, R. D., Portereiko, M. F., Sandaklie-Nikolova, L., Rabiger, D. S., and Drews, G. N. (2005). MYB98 is required for pollen tube guidance and synergid cell differentiation in Arabidopsis. *Plant Cell* **17**, 2981–2992.
Kim, H. J., Oh, S. A., Brownfield, L., Hong, S. H., Ryu, H., Hwang, I., Twell, D., and Nam, H. G. (2008). Control of plant germline proliferation by SCF(FBL17) degradation of cell cycle inhibitors. *Nature* **455**, 1134–1137.
Klattenhoff, C., and Theurkauf, W. (2008). Biogenesis and germline functions of piRNAs. *Development* **135**, 3–9.
Lalanne, E., Michaelidis, C., Moore, J. M., Gagliano, W., Johnson, A., Patel, R., Howden, R., Vielle-Calzada, J. P., Grossniklaus, U., and Twell, D. (2004). Analysis of transposon insertion mutants highlights the diversity of mechanisms underlying male progamic development in Arabidopsis. *Genetics* **167**, 1975–1986.
Lampard, G. R., Lukowitz, W., Ellis, B. E., and Bergmann, D. C. (2009). Novel and expanded roles for MAPK signaling in Arabidopsis stomatal cell fate revealed by cell type-specific manipulations. *Plant Cell* **21**, 3506–3517.
Latrasse, D., Benhamed, M., Henry, Y., Domenichini, S., Kim, W., Zhou, D. X., and Delarue, M. (2008). The MYST histone acetyltransferases are essential for gametophyte development in Arabidopsis. *BMC Plant Biol.* **8**, 121.
Le, Q., Gutierrez-Marcos, J. F., Costa, L. M., Meyer, S., Dickinson, H. G., Lorz, H., Kranz, E., and Scholten, S. (2005). Construction and screening of subtracted cDNA libraries from limited populations of plant cells: a comparative analysis of gene expression between maize egg cells and central cells. *Plant J.* **44**, 167–178.
Lewis, J. (2008). From signals to patterns: space, time, and mathematics in developmental biology. *Science* **322**, 399–403.
Li, J., Wen, J., Lease, K. A., Doke, J. T., Tax, F. E., and Walker, J. C. (2002). BAK1, an Arabidopsis LRR receptor-like protein kinase, interacts with BRI1 and modulates brassinosteroid signaling. *Cell* **110**, 213–222.
Liu, J., Zhang, Y., Qin, G., Tsuge, T., Sakaguchi, N., Luo, G., Sun, K., Shi, D., Aki, S., Zheng, N., Aoyama, T., Oka, A., *et al.* (2008). Targeted degradation of the cyclin-dependent kinase inhibitor ICK4/KRP6 by RING-type E3 ligases is essential for mitotic cell cycle progression during Arabidopsis gametogenesis. *Plant Cell* **20**, 1538–1554.
Ma, H. (2005). Molecular genetic analysis of microsporogenesis and microgametogenesis. *Annu. Rev. Plant Biol.* **56**, 393–434.
Ma, H. (2006). A molecular portrait of Arabidopsis meiosis. In: Somerville, C. R., Meyerowitz, E. M., Dangl, J., and Stitt, M. (Eds.), *The Arabidopsis Book.* American Society of Plant Biologists, Rockville, MD, doi/10.1199/tab.0009, http://www.aspb.org/publications/arabidopsis/.
Maeda, E., and Miyake, H. (1997). Ultrastructure of antipodal cells of rice (*Oryza sativa*) before anthesis with special reference to concentric configuration of endoplasmic reticula. *J. Crop. Sci.* **66**, 488–496.

Maheshwari, P. (1950). *An Introduction to Embryology of Angiosperms*. McGray-Hill, New York.

Marton, M. L., Cordts, S., Broadhvest, J., and Dresselhaus, T. (2005). Micropylar pollen tube guidance by egg apparatus 1 of maize. *Science* **307**, 573–576.

Miyazaki, S., Murata, T., Sakurai-Ozato, N., Kubo, M., Demura, T., Fukuda, H., and Hasebe, M. (2009) ANXUR1 and 2, sister genes to FERONIA/SIRENE, are male factors for coordinated fertilization. *Curr Biol.* **19**, 1327–1331.

Mol, R., Matthys-Rochon, E., and Dumas, C. (1994). The kinetics of cytological events during double fertilization in *Zea mays* L. *Plant J.* **5**, 197–206.

Moll, C., von Lyncker, L., Zimmermann, S., Kagi, C., Baumann, N., Twell, D., Grossniklaus, U., and Gross-Hardt, R. (2008). CLO/GFA1 and ATO are novel regulators of gametic cell fate in plants. *Plant J.* **56**, 913–921.

Mori, T., Kuroiwa, H., Higashiyama, T., and Kuroiwa, T. (2006). *GENERATIVE CELL SPECIFIC 1* is essential for angiosperm fertilization. *Nat. Cell Biol.* **8**, 64–71.

Muyt, A. D., Mercier, R., Mézard, C., and Grelon, M. (2009). Meiotic recombination and crossovers in plants. *Genome Dyn.* **5**, 14–25.

Nam, K. H., and Li, J. (2002). BRI1/BAK1, a receptor kinase pair mediating brassinosteroid signaling. *Cell* **110**, 203–212.

Nonomura, K., Miyoshi, K., Eiguchi, M., Suzuki, T., Miyao, A., Hirochika, H., and Kurata, N. (2003). The MSP1 gene is necessary to restrict the number of cells entering into male and female sporogenesis and to initiate anther wall formation in rice. *Plant Cell* **15**, 1728–1739.

Nonomura, K., Morohoshi, A., Nakano, M., Eiguchi, M., Miyao, A., Hirochika, H., and Kurata, N. (2007). A germ cell specific gene of the ARGONAUTE family is essential for the progression of premeiotic mitosis and meiosis during sporogenesis in rice. *Plant Cell* **19**, 2583–2594.

Okuda, S., Tsutsui, H., Shiina, K., Sprunck, S., Takeuchi, H., Yui, R., Kasahara, R. D., Hamamura, Y., Mizukami, A., Susaki, D., Kawano, N., Sakakibara, T., *et al.* (2009). Defensin-like polypeptide LUREs are pollen tube attractants secreted from synergid cells. *Nature* **458**, 357–361.

Olmedo-Monfil, V., Duran-Figueroa, N., Arteaga-Vazquez, M., Demesa-Arevalo, E., Autran, D., Grimanelli, D., Slotkin, R. K., Martienssen, R. A., and Vielle-Calzada, J. P. (2010). Control of female gamete formation by a small RNA pathway in Arabidopsis. *Nature* **464**, 628–632.

Pagnussat, G. C., Alandete-Saez, M., Bowman, J. L., and Sundaresan, V. (2009). Auxin-dependent patterning and gamete specification in the Arabidopsis female gametophyte. *Science* **324**, 1684–1689.

Pagnussat, G. C., Yu, H. J., Ngo, Q. A., Rajani, S., Mayalagu, S., Johnson, C. S., Capron, A., Xie, L. F., Ye, D., and Sundaresan, V. (2005). Genetic and molecular identification of genes required for female gametophyte development and function in Arabidopsis. *Development* **132**, 603–614.

Pagnussat, G. C., Yu, H. J., and Sundaresan, V. (2007). Cell-fate switch of synergid to egg cell in Arabidopsis eostre mutant embryo sacs arises from misexpression of the BEL1-like homeodomain gene BLH1. *Plant Cell* **19**, 3578–3592.

Palanivelu, R., Brass, L., Edlund, A. F., and Preuss, D. (2003). Pollen tube growth and guidance is regulated by *POP2*, an *Arabidopsis* gene that controls GABA levels. *Cell* **114**, 47–59.

Portereiko, M. F., Lloyd, A., Steffen, J. G., Punwani, J. A., Otsuga, D., and Drews, G. N. (2006). AGL80 is required for central cell and endosperm development in Arabidopsis. *Plant Cell* **18**, 1862–1872.

Punwani, J. A., Rabiger, D. S., and Drews, G. N. (2007). MYB98 positively regulates a battery of synergid-expressed genes encoding filiform apparatus localized proteins. *Plant Cell* **19**, 2557–2568.

Punwani, J. A., Rabiger, D. S., Lloyd, A., and Drews, G. N. (2008). The MYB98 subcircuit of the synergid gene regulatory network includes genes directly and indirectly regulated by MYB98. *Plant J.* **55**, 406–414.

Ravi, M., Marimuthu, M. P., and Siddiqi, I. (2008). Gamete formation without meiosis in Arabidopsis. *Nature* **451**, 1121–1124.

Reddy, T. V., Kaur, J., Agashe, B., Sundaresan, V., and Siddiqi, I. (2003). The DUET gene is necessary for chromosome organization and progression during male meiosis in Arabidopsis and encodes a PHD finger protein. *Development* **130**, 5975–5987.

Rotman, N., Durbarry, A., Wardle, A., Yang, W. C., Chaboud, A., Faure, J. E., Berger, F., and Twell, D. (2005). A novel class of MYB factors controls sperm-cell formation in plants. *Curr. Biol.* **15**, 244–248.

Rotman, N., Rozier, F., Boavida, L., Dumas, C., Berger, F., and Faure, J. E. (2003). Female control of male gamete delivery during fertilization in *Arabidopsis thaliana*. *Curr. Biol.* **13**, 432–436.

Sandaklie-Nikolova, L., Palanivelu, R., King, E. J., Copenhaver, G. P., and Drews, G. N. (2007). Synergid cell death in Arabidopsis is triggered following direct interaction with the pollen tube. *Plant Physiol.* **144**, 1753–1762.

Schiefthaler, U., Balasubramanian, S., Sieber, P., Chevalier, D., Wisman, E., and Schneitz, K. (1999). Molecular analysis of NOZZLE, a gene involved in pattern formation and early sporogenesis during sex organ development in *Arabidopsis thaliana*. *Proc. Natl. Acad. Sci. U.S.A.* **96**, 11664–11669.

Sheridan, W. F., Avalkina, N. A., Shamrov, I. I., Batygina, T. B., and Golubovskaya, I. N. (1996). The mac1 gene: controlling the commitment to the meiotic pathway in maize. *Genetics* **142**, 1009–1020.

Sheridan, W. F., Golubeva, E. A., Abrhamova, L. I., and Golubovskaya, I. N. (1999). The mac1 mutation alters the developmental fate of the hypodermal cells and their cellular progeny in the maize anther. *Genetics* **153**, 933–941.

Shpak, E. D., McAbee, J. M., Pillitteri, L. J., and Torii, K. U. (2005). Stomatal patterning and differentiation by synergistic interactions of receptor kinases. *Science* **309**, 290–293.

Slotkin, R. K., Vaughn, M., Borges, F., Tanurdzić, M., Becker, J. D., Feijó, J. A., and Martienssen, R. A. (2009). Epigenetic reprogramming and small RNA silencing of transposable elements in pollen. *Cell* **136**, 461–472.

Sorensen, A. M., Kröber, S., Unte, U. S., Huijser, P., Dekker, K., and Saedler, H. (2003). The *Arabidopsis ABORTED MICROSPORES* (*AMS*) gene encodes a MYC class transcription factor. *Plant J.* **33**, 413–423.

Sprunck, S., Baumann, U., Edwards, K., Langridge, P., and Dresselhaus, T. (2005). The transcript composition of egg cells changes significantly following fertilization in wheat (*Triticum aestivum* L.). *Plant J.* **41**, 660–672.

Srilunchang, K. O., Krohn, N. G., and Dresselhaus, T. (2010). DiSUMO-like DSUL is required for nuclei positioning, cell specification and viability during female gametophyte maturation in maize. *Development* **137**, 333–345.

Steffen, J. G., Kang, I. H., Macfarlane, J., and Drews, G. N. (2007). Identification of genes expressed in the Arabidopsis female gametophyte. *Plant J.* **51**, 281–292.

Sundaresan, V., and Alandete-Saez, M. (2010). Pattern formation in miniature: the female gametophyte of flowering plants. *Development* **137**, 179–189.

Suzuki, M., Kamide, Y., Nagata, N., Seki, H., Ohyama, K., Kato, H., Masuda, K., Sato, S., Kato, T., Tabata, S., Yoshida, S., and Muranaka, T. (2004). Loss of function of 3-hydroxy-3-methylglutaryl coenzyme A reductase 1 (HMG1) in Arabidopsis leads to dwarfing, early senescence and male sterility, and reduced sterol levels. *Plant J.* **37**, 750–761.

Takeda, T., Amano, K., Ohto, M.-A., Nakamura, K., Shusei Sato, S., Kato, T., Tabata, S., and Ueguchi, C. (2006). RNA interference of the *Arabidopsis* putative transcription factor

TCP16 gene results in abortion of early pollen development. *Plant Mol. Biol.* **61**, 165–177.

Thorstensen, T., Grini, P. E., Mercy, I. S., Alm, V., Erdal, S., Aasland, R., and Aalen, R. B. (2008). The Arabidopsis SET-domain protein ASHR3 is involved in stamen development and interacts with the bHLH transcription factor ABORTED MICROSPORES (AMS). *Plant Mol. Biol.* **66**, 47–59.

Ulmasov, T., Murfett, J., Hagen, G., and Guilfoyle, T. J. (1997). Aux/IAA proteins repress expression of reporter genes containing natural and highly active synthetic auxin response elements. *Plant Cell* **9**, 1963–1971.

von Besser, K., Frank, A. C., Johnson, M. A., and Preuss, D. (2006). *Arabidopsis HAP2 (GCS1)* is a sperm-specific gene required for pollen tube guidance and fertilization. *Development* **133**, 4761–4769.

von Malek, B., van der Graaff, E., Schneitz, K., and Keller, B. (2002). The Arabidopsis male-sterile mutant *dde2-2* is defective in the *ALLENE OXIDE SYNTHASE* gene encoding one of the key enzymes of the jasmonic acid biosynthesis pathway. *Planta* **216**, 187–192.

Wang, G., Kong, H., Sun, Y., Zhang, X., Zhang, W., Altman, N., dePamphilis, C. W., and Ma, H. (2004). Genome-wide analysis of cyclin family in *Arabidopsis* and comparative phylogenetic analysis of plant cyclin-like proteins. *Plant Physiol.* **135**, 1084–1099.

Wang, H., Ngwenyama, N., Liu, Y., Walker, J. C., and Zhang, S. (2007). Stomatal development and patterning are regulated by environmentally responsive mitogen-activated protein kinases in Arabidopsis. *Plant Cell* **19**, 63–73.

Webb, M. C., and Gunning, E. S. (1994). Embryo sac development in *Arabidopsis thaliana. Sex. Plant Reprod.* **7**, 153–163.

Wijeratne, A. J., Zhang, W., Sun, Y., Liu, W., Albert, R., Zheng, Z., Oppenheimer, D. G., Zhao, D., and Ma, H. (2007). Differential gene expression in Arabidopsis wild-type and mutant anthers: insights into anther cell differentiation and regulatory networks. *Plant J.* **52**, 14–29.

Wilson, Z. A., and Zhang, D. B. (2009). From Arabidopsis to rice: pathways in pollen development. *J. Exp. Bot.* **60**, 1479–1492.

Wuest, S. E., Vijverberg, K., Schmidt, A., Weiss, M., Gheyselinck, J., Lohr, M., Wellmer, F., Rahnenfuhrer, J., von Mering, C., and Grossniklaus, U. (2010). Arabidopsis female gametophyte gene expression map reveals similarities between plant and animal gametes. *Curr. Biol.* **20**, 506–512.

Xu, J., Yang, C., Yuan, Z., Zhang, D., Gondwe, M. Y., Ding, Z., Liang, W., Zhang, D., and Wilson, Z. A. (2010). The ABORTED MICROSPORES regulatory network is required for postmeiotic male reproductive development in *Arabidopsis thaliana. Plant Cell* **22**, 91–107.

Yang, M., Hu, Y., Lodhi, M., McCombie, W. R., and Ma, H. (1999a). The Arabidopsis SKP1-LIKE1 gene is essential for male meiosis and may control homologue separation. *Proc. Natl. Acad. Sci. U.S.A.* **96**, 11416–11421.

Yang, S. L., Jiang, L. X., Puah, C. S., Xie, L. F., Zhang, X. Q., Chen, L. Q., Yang, W. C., and Ye, D. (2005). Overexpression of *TAPETUM DETERMINANT1* alters the cell fates in the Arabidopsis carpel and tapetum via genetic interaction with *EXCESS MICROSPOROCYTES1/EXTRA SPOROGENOUS CELLS. Plant Physiol.* **139**, 186–191.

Yang, H., Kaur, N., Kiriakopolos, S., and McCormick, S. (2006a). EST generation and analyses towards identifying female gametophyte-specific genes in *Zea mays* L. *Planta* **224**, 1004–1014.

Yang, X., Makaroff, C. A., and Ma, H. (2003b). The Arabidopsis MALE MEIOCYTE DEATH1 gene encodes a PHD-finger protein that is required for male meiosis. *Plant Cell* **15**, 1281–1295.

Yang, W. C., Shi, D. Q., and Chen, Y. H. (2010). Female gametophyte development in flowering plants. *Annu. Rev. Plant Biol.* **61**, 89–108.

Yang, X., Timofejeva, L., Ma, H., and Makaroff, C. A. (2006b). The Arabidopsis SKP1 homolog ASK1 controls meiotic chromosome remodeling and release of chromatin from the nuclear membrane and nucleolus. *J. Cell Sci.* **119**, 3754–3763.

Yang, C., Vizcay-Barrena, G., Conner, K., and Wilson, Z. A. (2007). MALE STERILITY1 is required for tapetal development and pollen wall biosynthesis. *Plant Cell* **19**, 3530–3548.

Yang, S. L., Xie, L. F., Mao, H. Z., Puah, C. S., Yang, W. C., Jiang, L., Sundaresan, V., and Ye, D. (2003a). Tapetum determinant1 is required for cell specialization in the Arabidopsis anther. *Plant Cell* **15**, 2792–2804.

Yang, W. C., Ye, D., Xu, J., and Sundaresan, V. (1999b). The SPOROCYTELESS gene of Arabidopsis is required for initiation of sporogenesis and encodes a novel nuclear protein. *Genes Dev.* **13**, 2108–2117.

Yang, W. C., Ye, D., Xu, J., and Sundaresan, V. (1999c). The SPOROCYTELESS gene of Arabidopsis is required for initiation of sporogenesis and encodes a novel nuclear protein. *Genes Dev.* **13**, 2108–2117.

Ye, Q., Zhu, W., Li, L., Zhang, S., Yin, Y., Ma, H., and Wang, X. (2010). Brassinosteroids control male fertility by regulating the expression of key genes involved in Arabidopsis anther and pollen development. *Proc. Natl. Acad. Sci. U.S.A.* **107**, 6100–6105.

Yu, H. J., Hogan, P., and Sundaresan, V. (2005). Analysis of the female gametophyte transcriptome of Arabidopsis by comparative expression profiling. *Plant Physiol.* **139**, 1853–1869.

Zhang, W., Sun, Y., Timofejeva, L., Chen, C., Grossniklaus, U., and Ma, H. (2006). Control of *Arabidopsis* tapetum development by *DYSFUNCTIONAL TAPETUM 1 (DYT1)* encoding a putative bHLH transcription factor. *Development* **133**, 3085–3095.

Zhao, Y., Christensen, S. K., Fankhauser, C., Cashman, J. R., Cohen, J. D., Weigel, D., and Chory, J. (2001). A role for flavin monooxygenase-like enzymes in auxin biosynthesis. *Science* **291**, 306–309.

Zhao, D., Han, T., Risseeuw, E., Crosby, W. L., and Ma, H. (2003). Conservation and divergence of ASK1 and ASK2 gene functions during male meiosis in *Arabidopsis thaliana*. *Plant Mol. Biol.* **53**, 163–173.

Zhao, X., de Palma, J., Oane, R., Gamuyao, R., Luo, M., Chaudhury, A., Herve, P., Xue, Q., and Bennett, J. (2008). OsTDL1A binds to the LRR domain of rice receptor kinase MSP1, and is required to limit sporocyte numbers. *Plant J.* **54**, 375–387.

Zhao, D.-Z., Wang, G.-F., Speal, B., and Ma, H. (2002). The *EXCESS MICROSPOROCYTES1* gene encodes a putative leucine-rich repeat receptor protein kinase that controls somatic and reproductive cell fates in the *Arabidopsis* anther. *Genes Dev.* **16**, 2021–2031.

Zhao, D., Yang, X., Quan, L., Timofejeva, L., Rigel, N. W., Ma, H., and Makaroff, C. A. (2006). ASK1, a SKP1 homolog, is required for nuclear reorganization, presynaptic homolog juxtaposition and the proper distribution of cohesin during meiosis in Arabidopsis. *Plant Mol. Biol.* **62**, 99–110.

Zhu, J., Chen, H., Li, H., Gao, J. F., Jiang, H., Wang, C., Guan, Y. F., and Yang, Z. N. (2008). Defective in *Tapetal Development and Function 1* is essential for anther development and tapetal function for microspore maturation in Arabidopsis. *Plant J.* **55**, 266–277.

Subject Index

Note: Page numbers followed by '*f*' indicates a figure and '*t*' indicates a table

Q

R

Contents of Previous Volumes

Volume 47

Volume 48

Volume 49

Volume 54

Volume 55

Volume 56

Volume 57

Volume 58

Volume 59

Volume 60

Volume 61

Volume 62

Volume 65

Volume 66

Volume 67

Volume 68

Volume 69

Volume 70

Volume 71

Volume 74

Volume 75

Volume 76

Volume 77

Volume 81

Volume 84

Volume 85

Volume 86

Volume 87

Volume 88

Volume 89

Volume 90

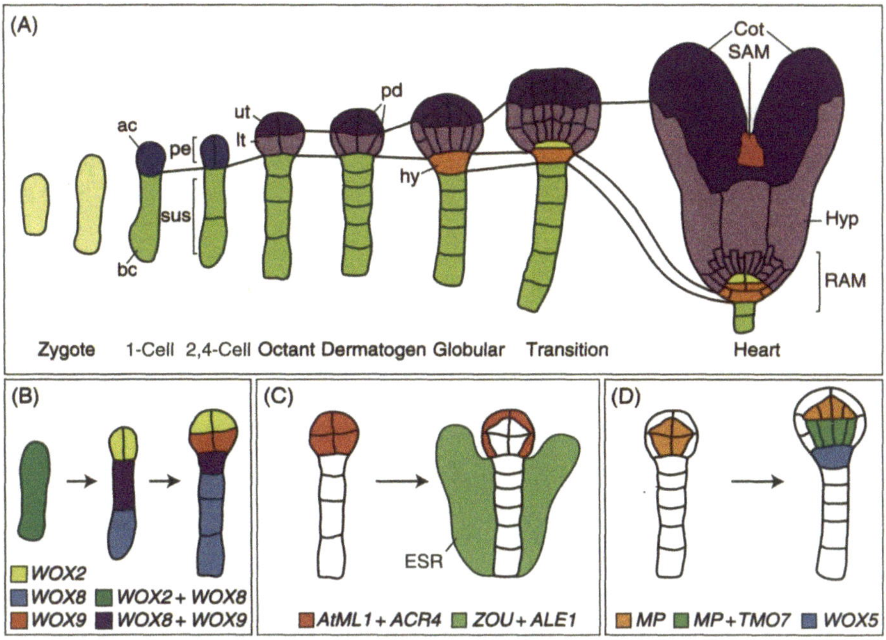

Cristina I. Llavata Peris *et al.*, Figure 1.3 Embryo pattern formation and cell specification. (A) Stages of embryogenesis and cell lineages. The lines between embryos highlight the lineage relationships between different stages of embryogenesis. Stage nomenclature is given below each embryo. Colors represent clonally related regions. After fertilization the zygote elongates and undergoes an asymmetrical division generating a small apical cell (ac) and a larger basal cell (bc). From this point, two regions with different developing programs will be established, the embryonic proembryo (pe) and the extra-embryonic suspensor (sus). All cells in the four-celled proembryo divide to generate an upper tier (ut) and a lower tier (lt). Subsequently, a radial pattern is generated in the eight-cell proembryo when a layer of protoderm (pd) is established. At the globular stage, the uppermost cell of the suspensor is specified to become the hypophysis (hy). Through an asymmetrical division, the hypophysis gives rise to a lens-shaped cell from which the quiescent center will be generated and a basal cell from which stem cells and columella cells of the root tip will derive. Further divisions generate primordia for all seedling structures at the heart stage, including cotyledons (cot), shoot apical meristem (SAM), root apical meristem (RAM), and the hypocotyl (Hyp). (B–D) Molecular markers of cell fate decisions during pattern landmarks. Colors indicate the expression of genes as described in the box legends. (B) While *WOX2* and *WOX8* are coexpressed in the zygote, their mRNAs are found in different cells after two rounds of division. Combined with *WOX9* expression, different combinations of *WOX* transcripts mark each cell. (C) In the octant stage, *ATML1* and *ACR4* are both expressed in the entire proembryo. Upon division, their transcripts are only found in the protoderm. *ZOU* and *ALE1* transcripts are found in the embryo-surrounding endosperm region (ESR) and contribute to protoderm differentiation. (D) At the dermatogen stage, *MP* is expressed in all inner cells and remains so during two division rounds later. At the globular stage, the MP target *TMO7* is expressed only in lower tier inner cells, while the *WOX5* transcript is found in the hypophysis.

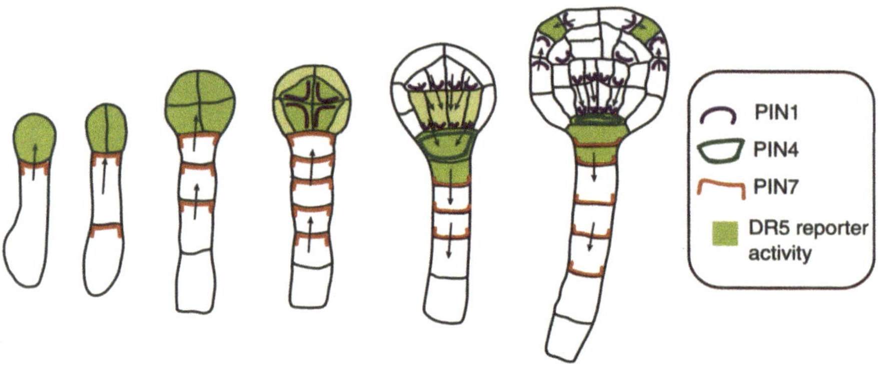

Cristina I. Llavata Peris *et al.*, Figure 1.5 Auxin transport and response in *Arabidopsis* embryogenesis. Activity of the *DR5–GFP* reporter (green) and localization of PIN1 (blue lines), PIN4 (green lines), and PIN7 (red lines) during early embryogenesis. PIN7 is polar localized in the basal cell after zygote division and remains so until the globular stage (for stages, see Fig. 1.3). This is associated with *DR5–GFP* activity in the proembryo. PIN1 becomes polarly localized toward the basal end of the inner cells in the globular stage. This promotes auxin transport to the hypophysis (note *DR5–GFP* activity). Simultaneously, PIN7 polarity is switched to the basal end in suspensor cells and PIN4 is activated in the hypophysis. This constellation of PIN proteins and *DR5–GFP* activity is maintained at later stages. During the globular stage, PIN1 localization becomes polarized toward the flanks of the proembryo apex, eliciting new *DR5–GFP* activity maxima. Arrows indicate auxin flux direction as deduced from PIN polarity.

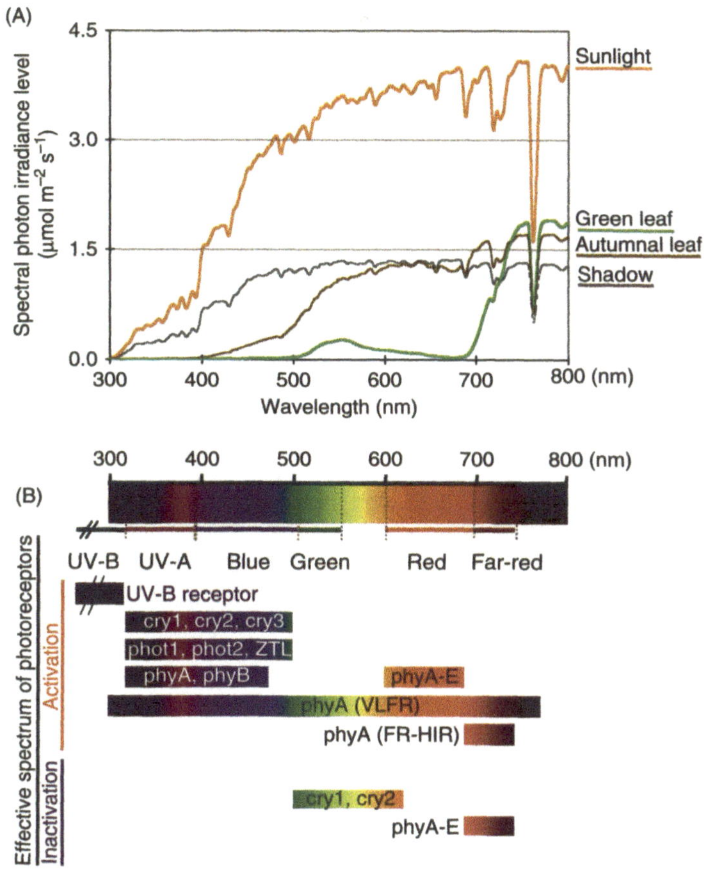

Chitose Kami *et al.*, Figure 2.1 The spectral photon irradiance of natural-light environments and effective spectrum of photoreceptors. (A) All of spectra were measured by spectroradiometer (LI-1800; Li-Cor, Lincoln, NE) in Nara, Japan (May 2003, 15:00, fine weather). Unfiltered sunlight (Sunlight), the shadow of building (Shadow), sunlight transmitted through a green *Pueraria lobata* leaf and light brown *P. lobata* leaf (Autumnal leaf). (B) Effective spectrum of photoreceptors for activation and inactivation.

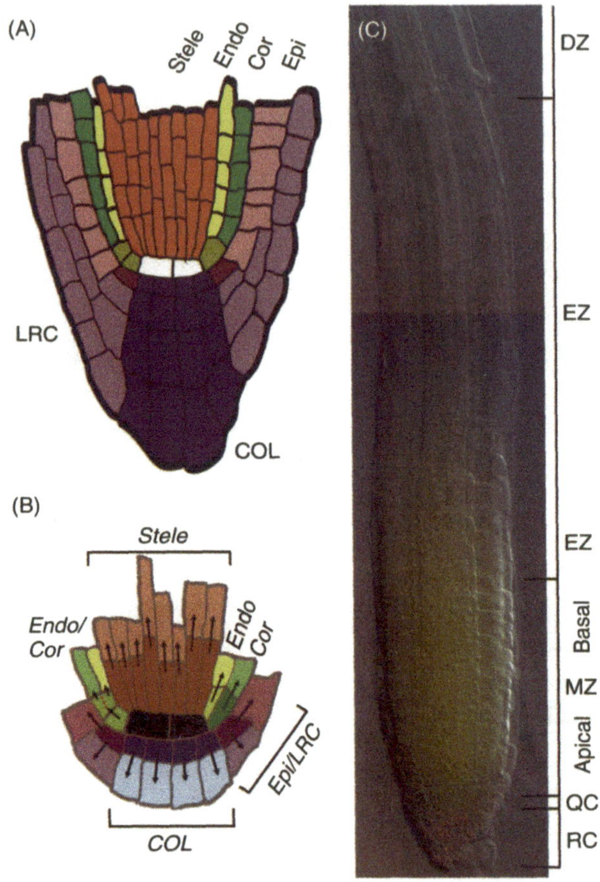

Tom Bennett and Ben Scheres, Figure 3.1 (A) Schematic diagram showing tissues in the *Arabidopsis* root meristem; red, stele (pericycle and vasculature); yellow, endodermis (Endo); green, cortex (Cor); pink, epidermis (Epi); light purple, lateral root cap (LRC); blue, columella (COL); white, quiescent center (QC). (B) Schematic diagram showing stem cell types and division planes in the *Arabidopsis* root meristem. On the left hand side, the diagram shows the organization in young *Arabidopsis* roots, with a common endodermal–cortical stem cell (Endo/Cor); on the right hand side, the organization in older roots is shown, with separate Endo and Cor stem cells. Stem cell daughters are shown in lighter shading; red, stele; mustard, Endo/Cor; yellow, Endo; green, Cor; dark purple, epidermal/lateral root cap stem cells (Epi/LRC); pink, Epi; light purple, LRC; blue, COL; black, QC. (C) Differential contrast interference light micrograph showing longitudinal organization of zones in the *Arabidopsis* root meristem; DZ, differentiation zone; EZ, elongation zone; MZ, meristem zone; QC, quiescent center; RC, root cap.

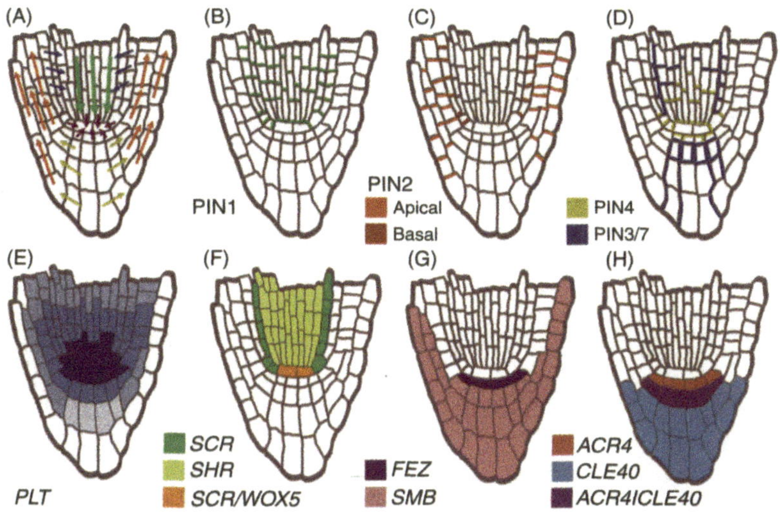

Tom Bennett and Ben Scheres, Figure 3.3 (A) Simplified model of auxin flux within the *Arabidopsis* root meristem. Basal auxin transport (green arrows) from the shoot enters the root meristem and is concentrated in the quiescent center (QC) area (purple arrows). In the columella, auxin is transported laterally (gold arrows) and then transported apically in the epidermis and lateral root cap (red arrows). Finally, apically transported auxin is recycled into the basal auxin stream by inward lateral transport (blue arrows). (B) Schematic diagram showing expression domain and protein localization of PIN1 in the root meristem. (C) Schematic diagram showing expression domain and protein localization of PIN2 in the root meristem; apical localization in epidermal and lateral root cap cells, and basal localization in cortical cells. (D) Simplified schematic diagram showing expression domains and protein localizations of PIN3, PIN4, and PIN7 in the root meristem. (E) Simplified schematic diagram showing graded expression of PLETHORA (PLT) proteins in the root meristem corresponds with cell division activity. (F) Schematic diagram showing expression of *SHORT-ROOT* (*SHR*), *SCARECROW* (*SCR*), and *WOX5* genes in the root meristem. (G) Schematic diagram showing expression of the FEZ protein in stem cells and SMB protein in mature root cap cells. (H) Schematic diagram showing expression of the CLE40 and ACR4 proteins in the root cap.

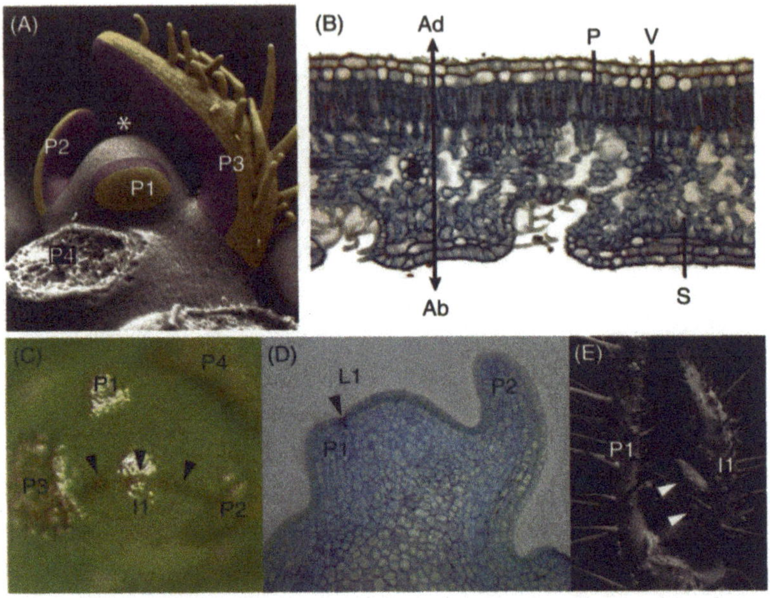

Catherine A. Kidner and Marja C. P. Timmermans, Figure 5.1 Meristem-regulated polarization of the leaf. (A) Scanning electron micrograph (SEM) of a tomato vegetative SAM showing the youngest (P1) and increasingly older (P2 and onward) leaf primordia surrounding the meristem (asterisks). Note that the adaxial/upper side of leaves (purple) develops in closer proximity to the SAM tip than the abaxial/lower side (yellow). (B) Transverse section through a *Nerium* leaf illustrating the differentiation of distinct cell types within the adaxial and abaxial domains. Palisade mesophyll (p) cells form beneath the adaxial epidermis, whereas spongy mesophyll (s) cells differentiate abaxially. Within the vasculature (v), water-bearing xylem cells differentiate adaxial to sugar-bearing phloem cells. (C) A vegetative tomato shoot apex showing ablation of the L1 layer (black arrowheads) separating the incipient leaf (I1) from the meristem. (D) Longitudinal section of a tomato apex showing an L1-restricted incision (black arrowhead) that separates the P1 primordium from the meristem. (E) SEM of a tomato apex a day after the P1 has been surgically separated from the SAM. The P1 develops as a radial, abaxialized primordium (black arrowheads show the scar from the ablation). In contrast, a primordium that developed in contact with the meristem (I1) is dorsiventrally flattened and develops leaflet pairs (white arrowheads). The SEM image in A was kindly provided by C. Kuhlemeier, and images C–E are reprinted with permission from Reinhardt et al. (Reinhardt *et al.*, 2005).

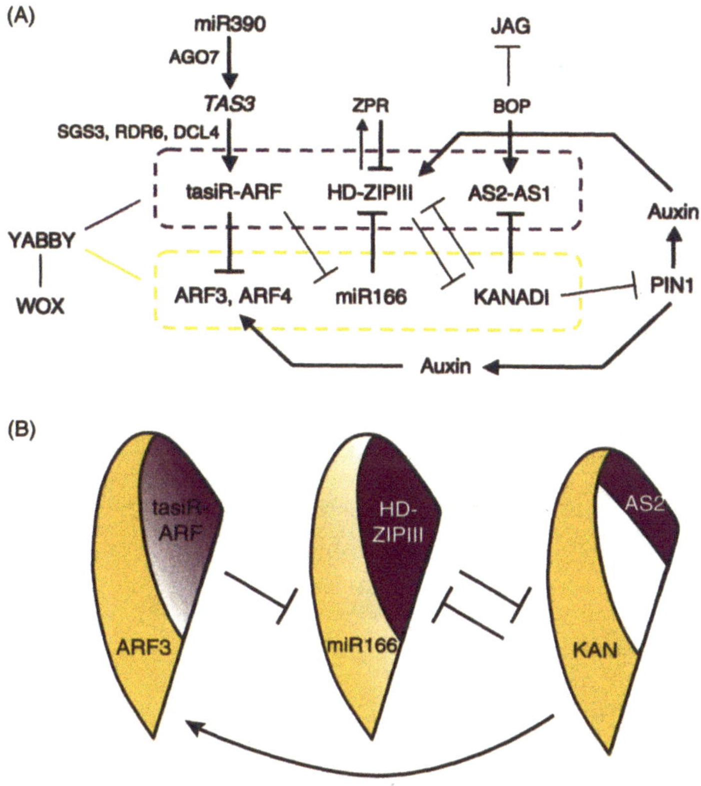

Catherine A. Kidner and Marja C. P. Timmermans, Figure 5.3 Genetic interactions in leaf polarity. (A) Specification of organ polarity involves three conserved interlocking genetic pathways: the *TAS3* ta-siRNA–ARF3/4, miR166–HD-ZIPIII, and AS1-AS2–KAN pathways. These pathways are distinguished by redundancies and by mutual antagonistic interactions between adaxial and abaxial components. Key determinants of adaxial identity are encircled in purple, and abaxial determinants in yellow. Heavy lines indicate direct relationships, light lines indirect. Colored lines show the general effects of each cell fate on YABBY expression, which varies between species. (B) Spatial arrangement of the key components in the three genetic pathways involved in leaf polarity in the P1 leaf. Gradients of tasiR-ARF and miR166 generate sharply defined expression domains of ARF and HD-ZIPIII genes, respectively. Further mutual inhibition between the abaxial and adaxial zones sharpens and defines the boundary.

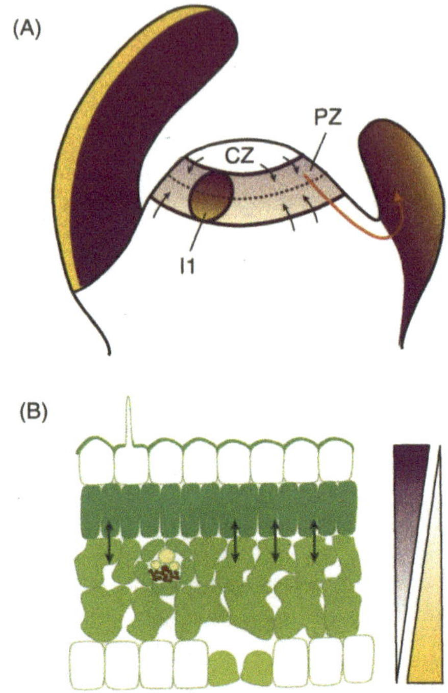

Catherine A. Kidner and Marja C. P. Timmermans, Figure 5.4 Mobile signals in the establishment and maintenance of leaf polarity. (A) The peripheral zone (PZ) is envisioned to be patterned into apical/centric and basal/outer regions based on positional signals derived from the HD-ZIPIII expressing CZ and the internode region below the PZ, which is marked by KAN and miR166 expression. The incipient leaf (I1) forms at an auxin maximum in the PZ and is divided in adaxial and abaxial domains by this prepattern in the PZ. As the leaf develops, the meristem grows away from it, but a signal through the L1 (long red arrow) maintains this initial polarity until the P2 stage of organ development when maintenance mechanisms within the organ are in place. (B) Mutually antagonistic interactions between cell-autonomous polarity determinants contributes to the maintenance of organ polarity, but is unlikely sufficient to define a precise boundary between adaxial and abaxial organ domains. Superimposed intercellular signals derived from the adaxial and abaxial sides can provide positional inputs to place and refine the adaxial–abaxial boundary. Gradients of adaxially derived mobile tasiR-ARFs (purple) and abaxially-derived miR166 (yellow) are candidates for such signals. Additionally, cell–cell communication that allows cells to acquire information about the identity of their neighbors and to coordinate cell division in the different cell layers (small arrows) is required to coordinate blade outgrowth at the adaxial–abaxial boundary.

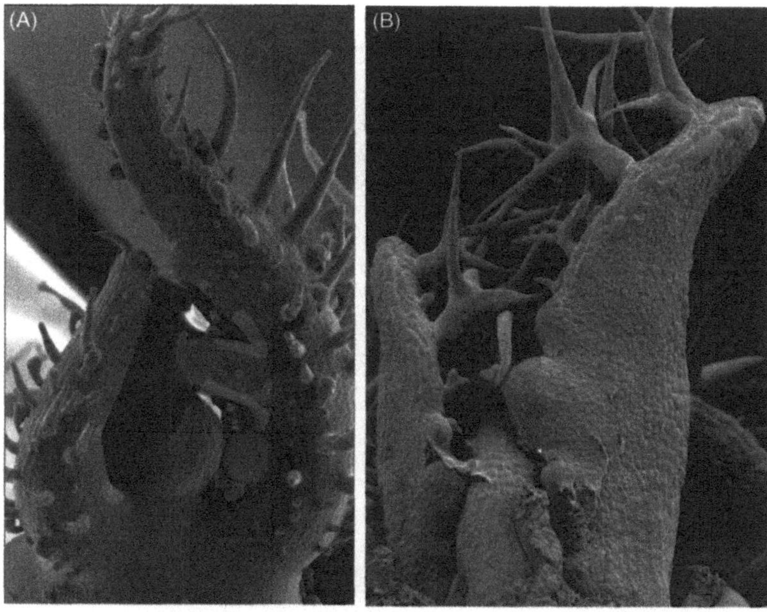

Daniel Koenig and Neelima Sinha, Figure 6.2 Comparison of marginal dissection in compound and simple leaves. (A) Scanning electron micrograph of a tomato shoot apex. (B) Scanning electron micrograph of a developing *A. thaliana* leaf. Purple indicates the activity of growth-promoting factors such as auxin and *LYR*. Orange indicates the activity of repressing factors like *E* and *NAM/CUC* genes. As leaflets elongate, *NAM/CUC* expression remains restricted (dark orange), while *E* is likely to be expanded to all areas with low auxin expression.

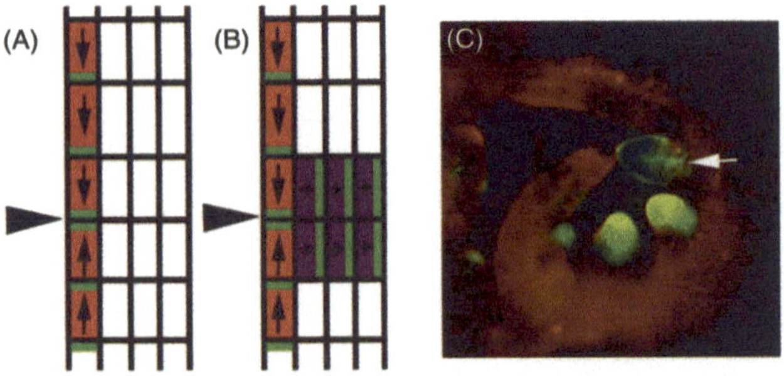

Daniel Koenig and Neelima Sinha, Figure 6.3 Specification of leaflet or lobe initiation by auxin. (A) Formation of auxin peaks (arrowhead) by epidermal PIN1 convergence. The flow of auxin is indicated by the arrows, and the green lines indicate PIN1 localization. Red colored cells are the epidermal cell file. The leaflet or lobe primordium will initiate at this convergence point. (B) Subsequent to epidermal convergence, internal cells below the convergence adopt internal PIN1 localization and specify the cell files that will eventually differentiate into vascular tissue (purple cells). Thus, the two processes are developmental linked. (C) Fluorescence imaging of *PIN1:PIN1::GFP* expression in tomato showing upregulation in leaflets and lobes. The provascular expression of GFP is indicated by the arrow.

Holger Breuninger and Michael Lenhard, Figure 7.2 *Arabidopsis* rosette expressing the *pCYCB1;1::CDB-GUS* reporter that marks mitotic cells. Only a small fraction of the plant is actively dividing. During organ growth, proliferation ceases along an arrest front from distal to proximal, as indicated by the loss of reporter expression at the tip of the outgrowing leaf.

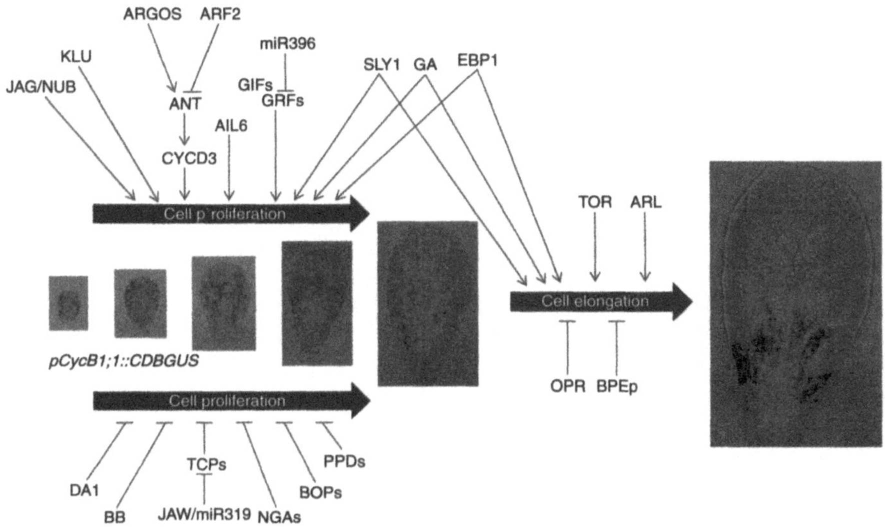

Holger Breuninger and Michael Lenhard, Figure 7.3 Pathways controlling cell proliferation and cell elongation. Central pictures show *pCycB1;1:GUS* expression in progressively older developing petals as a model organ. During development, cell division is restricted to the central part of the growing petal and ceases at later stages, after which the organ grows merely by cell elongation. Genetic pathways promoting cell proliferation or elongation are shown in the upper part; growth-restricting pathways are shown in the lower part of the figure. See also Table 7.1. (*pCycB1;1* expression pictures are taken from Disch *et al.*, 2006, copyright Current Biology).

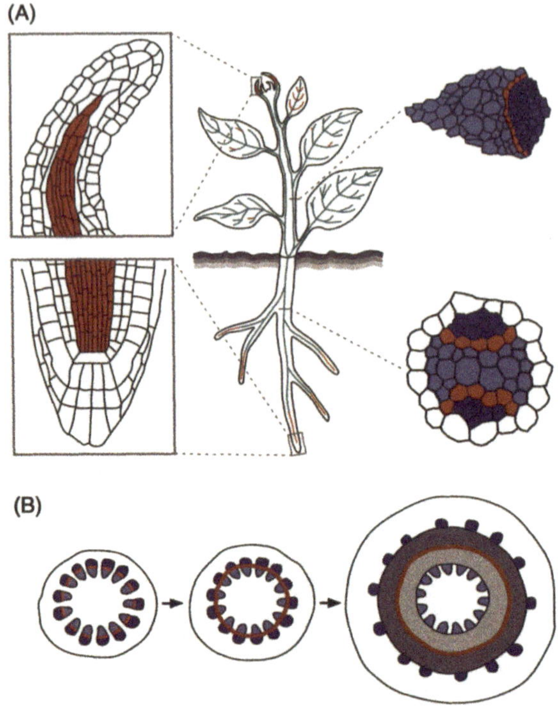

Enrico Scarpella and Ykä Helariutta, Figure 8.1 The vascular system of plants and its origin. (A) The vascular system is a network of continuous vascular strands (green) extending throughout the entire plant body. At maturity, vascular strands represent the organization of two vascular tissues, the phloem (dark blue) and the xylem (light blue). In the stem vascular strands of most plant species, the phloem pole is located at the outer side and the xylem pole at the inner side of the strand. In root vascular strands, the xylem forms a solid central core with projections extending toward the periphery of the vascular strand, and the phloem is organized in separate strands alternating with the xylem ridges. Both vascular tissues differentiate in immature parts of plant organs from a common precursor tissue, the procambium (brown), recognizable as uninterrupted files of terminally aligned, elongated cells. In many species, a layer of undifferentiated procambial cells separates phloem and xylem in mature strands. Redrawn from: Fig. 8, Girolami, G. (1954), Leaf histogenesis in *Linum usitatissimum, Am. J. Bot.* 41, 264–273, with kind permission of the Botanical Society of America; Fig. 2J,K, Scarpella, E., and Meijer, A. H. (2004), Pattern formation in the vascular system of monocot and dicot plant species, *New Phytologist* 164, 209–242, with kind permission of John Wiley and Sons; Figs. 1C, 2A, B; van den Berg, C., Willemsen, V., Hage, W., Weisbeek, P., and Scheres, B. (1995), Cell fate in the Arabidopsis root meristem determined by directional signalling. *Nature* 378, 62–65, with kind permission of Nature Publishing Group and Macmillan Publishers. (B) During stem growth by radial thickening, the residual procambium within vascular strands and parenchyma cells between vascular strands resume division activity and originate a hollow cylinder—a circle in the transverse section shown here—of cambium tissue, from which secondary phloem (dark gray) and xylem (light gray) differentiate. Arrows temporally connect successive stages of secondary vascular tissue formation. A formally similar process occurs during root secondary growth.

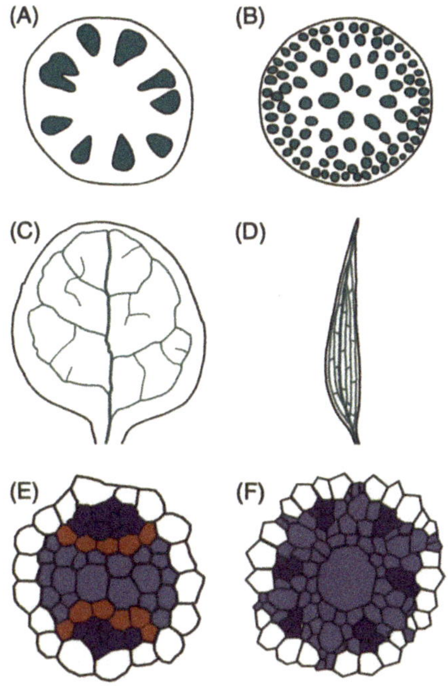

Enrico Scarpella and Ykä Helariutta, Figure 8.2 Levels of vascular organization. Vascular patterns in stems (A,B), leaves (C,D), and roots (E,F) of dicots (A,C,E) and monocots (B,C,F) as examples of organ- and species-specific vascular organizations. Two levels of spatial organization are evident in the structure of the vascular system: the longitudinal pattern, which is the three-dimensional array of vascular strands within an organism; and the radial pattern, which is the arrangement of phloem and xylem within each vascular strand. Both levels of organization are organ specific, as reflected in the different positions of vascular and nonvascular tissues in stems, leaves, and roots of the same species (compare A with C and E, and B with D and F). However, within a given organ, species-specific cues further limit the variety of possible vascular organizations. For example, in most dicot stems (A), vascular strands (green) are arranged in a circle at the periphery of the organ, while, in most monocots (B), they are scattered throughout the organ. Further, in dicot leaves (C), veins (green) are connected in a reticulate pattern, while, in monocots (D), they are aligned in a striate pattern. Finally, in the root, vascular tissues are organized in a central vascular cylinder, in which the xylem (dark blue) forms a solid central core with extensions toward the periphery of the vascular cylinder, while the phloem (light blue) is organized in separate strands alternating with the xylem arms. The number of xylem ridges varies in different species; typically, dicots have up to four and monocots have more than four. Furthermore, procambial cells (brown) separate xylem and phloem in dicot, but not in monocot, roots. Note, however, that while vascular differentiation must be continuous, while the orientations of the cells within each strand must be precisely integrated to ensure conductivity, and while the arrangements and compositions of vascular strands must be constrained to ensure functionality in the local organ context, the precise route and arrangement of vascular strands cannot be predicted. Redrawn from: Fig. 1A, Donner, T. J., and Scarpella, E. (2009), Auxin-transport-dependent leaf vein formation, Botany-Botanique 87, 678–684, with kind permission of National Research Council Canada Research Press; Fig. 2D,H, Scarpella, E., and Meijer, A. H. (2004), Pattern formation in the vascular system of monocot and dicot plant species, *New Phytol.* 164, 209–242, with kind permission of John Wiley and Sons.

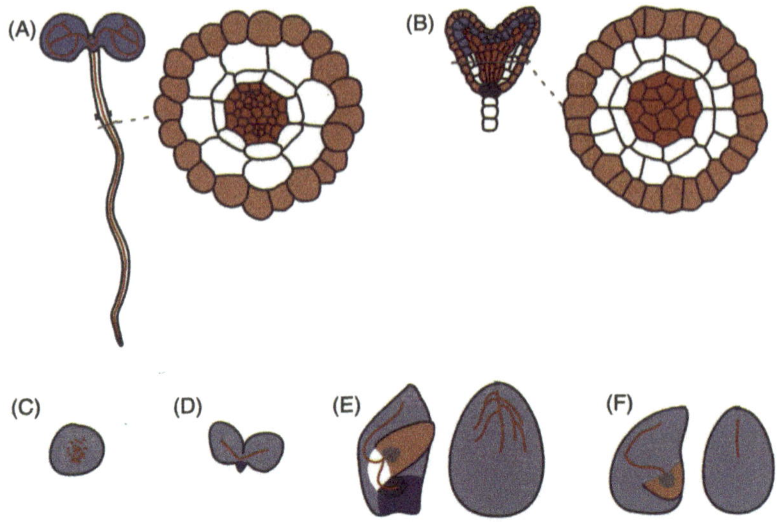

Enrico Scarpella and Ykä Helariutta, Figure 8.4 Vascular strand formation in embryos. (A, B) Embryonic origin of seedling structures. The reproducibility of Arabidopsis embryo development enables tracing the origin of seedling organs and tissues (A) to progenitor cells in the early embryo. Although cells in most tissues will only complete differentiation after germination, the complexity of the tissue pattern in the heart-stage embryo (B) basically equals that of the seedling. Colors identify corresponding regions in seedling and embryo. Longitudinal subdivision: shoot apical meristem (light gray), cotyledons (light blue), embryo axis (white), root apical meristem (dark gray), vascular strands (brown). Radial subdivision: epidermis (light brown), ground tissue (white), vascular strand (brown). (C) Auxin transport-defective Arabidopsis seedling. Pharmacological or genetic interference with auxin transport during embryogenesis results in seedlings completely lacking apical–basal polarity and with randomly oriented vascular cells. (D) Auxin response or biosynthesis Arabidopsis mutants. The vascular system is strongly reduced; hypocotyl and root are missing and a conical structure devoid of vascular tissues (dark blue) forms instead. (E) Mature wild-type rice embryo. Left, median longitudinal section; right, adaxial view. A procambial strand is positioned at the center of the C-shaped embryo axis (white). The shoot and root apical meristem (light and dark gray, respectively) are surrounded by the coleoptile (light brown) and coleorhiza (dark blue), respectively, which are not present in the dicot embryo. The vascular system of the scutellum, which represents the single cotyledon, is composed of a primary procambial strand that branches at the apex into a series of secondary strands. Occasionally, tertiary strands branch from secondary strands. (F) Mature embryo of the auxin-insensitive *ral1* mutant of rice. Left, median longitudinal section; right, adaxial view. The embryo axis with its central procambial strand is absent; the vascular system of the scutellum is limited to a short primary procambial strand and higher-order strands are completely missing. Redrawn from: Fig. 3B,C,D, Berleth, T., and Scarpella, E. (2004), Polar signals in vascular development, In "Polarity in Plants" (K. Lindsey, Ed.), pp. 264–287, Blackwell Publishing Ltd, Oxford with kind permission of Wiley-Blackwell; Fig. 2E, Dolan, L., Janmaat, K., Willemsen, V., Linstead, P., Poethig, S., Roberts, K., and Scheres, B. (1993), Cellular-Organization of the Arabidopsis-Thaliana Root. *Development* 119, 71–84, Fig. 3G, Mayer, U., Buttner, G., and Jürgens, G. (1993), Apical–Basal Pattern-Formation in the Arabidopsis Embryo—Studies on the Role of the Gnom Gene, *Development* 117, 149–162, Fig. 1J, Scheres, B., Dilaurenzio, L., Willemsen, V., Hauser, M. T., Janmaat, K., Weisbeek, P., and Benfey, P. N. (1995), Mutations Affecting the Radial Organization of the Arabidopsis Root Display Specific Defects Throughout the Embryonic Axis, *Development* 121, 53–62, with kind permission of the Company of Biologists.

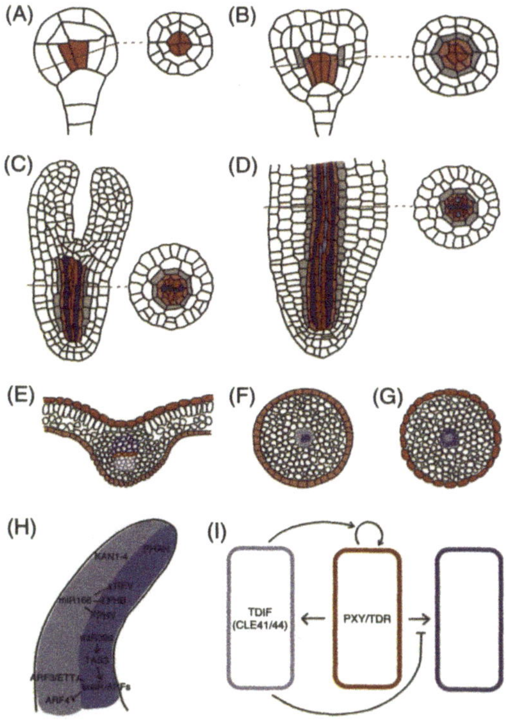

Enrico Scarpella and Ykä Helariutta, Figure 8.6 Radial vascular patterning. (A–D) Schematic longitudinal (left) and transverse (right) representation of the radial vascular pattern of the embryo axis at selected stages of embryogenesis: early globular (A), triangular (B), torpedo (C), and mature embryo (D), respectively. Brown, apparently undifferentiated procambial cells; light brown, pericycle; dark blue, protoxylem; light blue, phloem; dark gray, metaxylem; light gray, endodermis. (E–G) Schematic representation of adaxial–abaxial pattern in wild-type (E), abaxialized (F), and adaxialized (G) leaf, respectively. Brown fill: adaxial epidermis; light brown fill: abaxial epidermis; dark blue line: xylem; light blue line: phloem; brown line: procambium. (H) Model of interactions among leaf adaxial (dark blue) and abaxial (light blue) identity promoting factors. Antagonistic interactions between polarity determinants generate mutually exclusive adaxial and abaxial domains that define organ polarity throughout leaf development. (I) Model of CLE receptor/ligand system maintaining procambial cell activity in a non-cell-autonomous manner. Phloem (light blue) secretes the TDIF/CLE41/44 peptide, which is perceived by the receptor PXY/TDR in procambial cells (brown); this signal promotes proliferation of procambium and inhibits its differentiation into xylem (dark blue). Redrawn from: Figs. 1A,C,E,F,N,O,S,T, Scheres, B., Dilaurenzio, L., Willemsen, V., Hauser, M. T., Janmaat, K., Weisbeek, P., and Benfey, P. N. (1995), Mutations Affecting the Radial Organization of the Arabidopsis Root Display Specific Defects Throughout the Embryonic Axis, *Development* 121, 53–62, Fig. 2K, McConnell, J. R., and Barton, M. K. (1998), Leaf polarity and meristem formation in Arabidopsis, *Development* 125, 2935–2942, Fig. 2A,F, Waites, R., and Hudson, A. (1995), *PHANTASTICA*—a Gene Required for Dorsoventrality of Leaves in *Antirrhinum majus*, *Development* 121, 2143–2154, by kind permission of the Company of Biologists; Fig. 4M, Hirakawa, Y., Shinohara, H., Kondo, Y., Inoue, A., Nakanomyo, I., Ogawa, M., Sawa, S., Ohashi-Ito, K., Matsubayashi, Y., and Fukuda, H. (2008), Non-cell-autonomous control of vascular stem cell fate by a CLE peptide/receptor system, Proc *Natl. Acad. Sci. U.S.A.* 105, 15208–15213, with kind permission of the National Academy of Science, U.S.A.

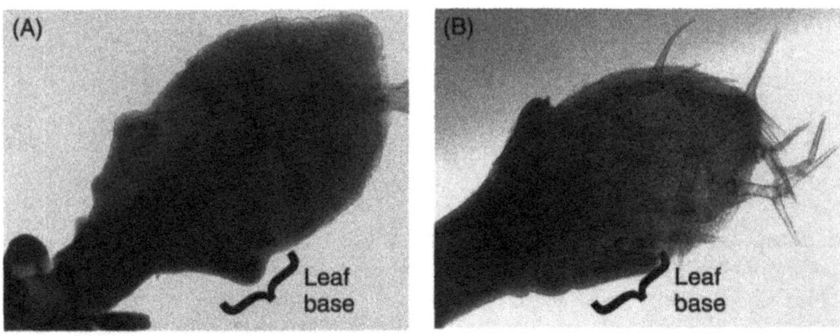

Rachappa Balkunde *et al.*, Figure 10.2 Expression pattern of GL1 and TRY. (A) GL1:GUS leaf showing the wild-type expression pattern of GL1. (B) TRY:GUS leaf showing the wild-type expression pattern of TRY. Trichomes are initiated at the leaf base. Note that both genes exhibit ubiquitous expression in the basal region of the leaf and elevated levels in trichomes.

FSC
www.fsc.org
MIX
Papier | Fördert gute Waldnutzung
FSC® C083411

Kolibri 360 GmbH
Ferdinand-Jühlke-Straße 7
99095 Erfurt, Deutschland
produktsicherheit@kolibri360.de